RELIABILITY AND RISK MODELS

WILEY SERIES IN QUALITY & RELIABILITY ENGINEERING
*and related titles**

SERIES EDITOR: Andre Kleyner

Reliability and Risk Models: Setting Reliability Requirements, Second Edition
Michael Todinov

Applied Reliability Engineering and Risk Analysis: Probabilistic Models and Statistical Inference
Ilia B. Frenkel, Alex Karagrigoriou, Anatoly Lisnianski and Andre V. Kleyner

Design for Reliability
Dev G. Raheja and Louis J. Gullo (editors)

Effective FMEAs: Achieving Safe, Reliable, and Economical Products and Processes Using Failure Mode and Effects Analysis
Carl Carlson

Failure Analysis: A Practical Guide for Manufacturers of Electronic Components and Systems
Marius Bâzu and Titu Băjenescu

Reliability Technology: Principles and Practice of Failure Prevention in Electronic Systems
Norman Pascoe

Improving Product Reliability: Strategies and Implementation
Mark A. Levin and Ted T. Kalal

Test Engineering: A Concise Guide to Cost-effective Design, Development and Manufacture
Patrick D. T. O'Connor

Integrated Circuit Failure Analysis: A Guide to Preparation Techniques
Friedrich Beck

Measurement and Calibration Requirements: For Quality Assurance to ASO 9000
Alan S. Morris

Electronic Component Reliability: Fundamentals, Modelling, Evaluation and Assurance
Finn Jensen

Six Sigma: Advanced Tools for Black Belts and Master Black Belts*
Loon Ching Tang, Thong Ngee Goh, Hong See Yam and Timothy Yoap

Secure Computer and Network Systems: Modeling, Analysis and Design*
Nong Ye

RELIABILITY AND RISK MODELS

SETTING RELIABILITY REQUIREMENTS

SECOND EDITION

Michael Todinov
Oxford Brookes University, UK

This edition first published 2016

First Edition published in 2005

Registered Office
John Wiley & Sons, Ltd, The Atrium, Southern Gate, Chichester, West Sussex, PO19 8SQ, United Kingdom

For details of our global editorial offices, for customer services and for information about how to apply for permission to reuse the copyright material in this book please see our website at www.wiley.com.

Library of Congress Cataloging-in-Publication Data

Todinov, Michael
Reliability and risk models setting reliability requirements / Michael Todinov. – Second edition.
pages cm. – (Wiley series in quality & reliability engineering)
Includes bibliographical references and index.
ISBN 978-1-118-87332-8 (cloth)
1. Reliability (Engineering)–Mathematical models. 2. Risk assessment–Mathematics. I. Title.
TA169.T65 2005
620′.00452015118–dc23

2015016646

A catalogue record for this book is available from the British Library.

Set in 9.5/11.5pt Times by SPi Global, Pondicherry, India
Printed and bound in Singapore by Markono Print Media Pte Ltd

1 2016

To Prolet

Contents

Series Preface

The Wiley Series in Quality & Reliability Engineering aims to provide a solid educational foundation for researchers and practitioners in the field of quality and reliability engineering and to expand the knowledge base by including the latest developments in these disciplines.

The importance of quality and reliability to a system can hardly be disputed. Product failures in the field inevitably lead to losses in the form of repair cost, warranty claims, customer dissatisfaction, product recalls, loss of sale and, in extreme cases, loss of life.

Engineering systems are becoming increasingly complex with added capabilities, options and functions; however, the reliability requirements remain the same or even growing more stringent. This challenge is being faced by design and manufacturing improvements and to no lesser extent by advancements in system reliability modelling. Also, the recent developments of functional safety standards (IEC 61508, ISO 26262, ISO 25119 and others) caused an uptick in interest to system reliability modelling and risk assessment as it applies to product safety.

This book *Reliability and Risk Models* is the second and comprehensively updated edition of the work, which has already gained a wide readership among reliability practitioners and analysts. It presents a foundation and advanced topics in reliability modelling successfully merging statistical-based approach with advanced engineering principles. It offers an excellent mix of theory, practice, applications and common sense engineering, making it a perfect addition to this book series.

The purpose of the Wiley book series is also to capture the latest trends and advancements in quality and reliability engineering and influence future development in these disciplines. As quality and reliability science evolves, it reflects the trends and transformations of the technologies it supports. A device utilising a new technology, whether it be a solar power panel, a stealth aircraft or a state-of-the-art medical device, needs to function properly and without failures throughout its mission life. New technologies bring about new failure mechanisms, new failure sites and new failure modes. Therefore, continuous advancement of the physics of failure combined with a multidisciplinary approach is essential to our ability to address those challenges in the future.

In addition to the transformations associated with changes in technology, the field of quality and reliability engineering has been going through its own evolution developing new techniques and methodologies aimed at process improvement and reduction of the number of design- and manufacturing-related failures.

Risk assessment continues to enhance reliability analysis for an increasing number of applications, addressing not only the probability of failure but also the quantitative consequences of that failure.

Life cycle engineering concepts are expected to find wider applications to reduce life cycle risks and minimise the combined cost of design, manufacturing, quality, warranty and service.

Additionally, continuous globalisation and outsourcing affect most industries and complicate the work of quality and reliability professionals. Having various engineering functions distributed around the globe adds a layer of complexity to design coordination and logistics. Also, moving design and production into regions with little knowledge depth of design and manufacturing processes, with a less robust quality system in place and where low cost is often the primary driver of product development affects company's ability to produce reliable and defect-free products.

Despite its obvious importance, quality and reliability education is paradoxically lacking in today's engineering curriculum. Very few engineering schools offer degree programmes or even a sufficient variety of courses in quality or reliability methods. Therefore, the majority of the quality and reliability practitioners receive their professional training from colleagues, professional seminars, publications and technical books. The lack of formal education opportunities in this field greatly emphasises the importance of technical publications for professional development.

We are confident that this book as well as this entire book series will continue Wiley's tradition of excellence in technical publishing and provide a lasting and positive contribution to the teaching and practice of reliability and quality engineering.

Dr. Andre V. Kleyner,
Editor of the Wiley Series in Quality & Reliability Engineering

Preface

A common tendency in many texts devoted to reliability is to choose either a statistical-based approach to reliability or engineering-based approach. Reliability engineering, however, is neither reliability statistics nor solely engineering principles underlying reliable designs. Rather, it is an amalgam of reliability statistics, theoretical principles and techniques and engineering principles for developing reliable products and reducing technical risk. Furthermore, in the reliability literature, the emphasis is commonly placed on reliability prediction than reliability improvement. Accordingly, the intention of this second edition is to improve the balance between the statistical-based approach and the engineering-based approach.

To demonstrate the necessity of a balanced approach to reliability and engineering risk, a new chapter (Chapter 11) has been devoted exclusively to principles and techniques for improving reliability and reducing engineering risk. The need for unity between the statistical approach and the engineering approach is demonstrated by the formulated principles, some of which are rooted in reliability statistics, while others rely on purely engineering concepts. The diverse risk reduction principles prompt reliability and risk practitioners not to limit themselves to familiar ways of improving reliability and reducing risk (such as introducing redundancy) which might lead to solutions which are far from optimal.

Using appropriate combinations of statistical and physical principles brings a considerably larger effect. The outlined key principles for reducing the risk of failure can be applied with success not only in engineering but in diverse areas of the human activity, for example in environmental sciences, financial engineering, economics, medicine, etc.

Critical failures in many industries (e.g. in the nuclear or deep-water oil and gas industry) can have disastrous environmental and health consequences. Such failures entail loss of production for very long periods of time and extremely high costs of the intervention for repair. Consequently, for industries characterised by a high cost of failure, setting quantitative reliability requirements must be driven by the cost of failure. There is a view held even by some risk experts that there is no need for setting reliability requirements. The examples in Chapter 16 demonstrate the importance of reliability requirements not only for minimising the probability of unsatisfied demand below a maximum acceptable level but also for providing an optimal balance between reliability and cost. Furthermore, many technical failures with disastrous consequences to the environment could have been easily prevented by adopting cost-of-failure-based reliability requirements for critical components.

Common, as well as little known reliability and risk models and their applications are discussed. Thus, a powerful generic equation is introduced for determining the probability of safe/failure states dependent on the relative configuration of random variables, following a homogeneous Poisson process in a finite domain. Seemingly intractable reliability problems can be solved easily using this equation which reduces a complex reliability problem to simpler problems. The equation provides a basis for the new reliability measure introduced in Chapter 16, which consists of a combination of specified minimum separation distances between random variables in a finite interval and the probability with which they must exist. The new reliability measure is at the heart of a technology for setting quantitative reliability requirements based on minimum event-free operating periods or minimum failure-free operating periods (MFFOP). A number of important applications of the new reliability measure are also considered such as limiting the probability of a collision of demands from customers using particular resource for a specified time and the probability of overloading of supply systems from consumers connecting independently and randomly.

It is demonstrated that even for a small number of random demands in a finite time interval, the probability of clustering of two or more random demands within a critical distance is surprisingly high and should always be accounted for in risk assessments.

Substantial space in the book has been allocated for load–strength (demand–capacity) models and their applications. Common problems can easily be formulated and solved using the load–strength interference concept. On the basis of counterexamples, a point has been made that for non-Gaussian distributed load and strength, the popular reliability measures 'reliability index' and 'loading roughness' can be completely misleading. In Chapter 6, the load–strength interference model has been generalised, with the time included as a variable. The derived equation is in effect a powerful model for determining reliability associated with an overstress failure mechanism.

A number of new developments made by the author in the area of reliability and risk models since the publication of the first edition in 2005 have been reflected in the second edition. Such is, for example, the revision of the Weibull distribution as a model of the probability of failure of materials controlled by defects. On the basis of probabilistic reasoning, thought experiments and real experiments, it is demonstrated in Chapter 13 that contrary to the common belief for more than 60 years, the Weibull distribution is a fundamentally flawed model for the probability of failure of materials. The Weibull distribution, with its strictly increasing function, is incapable of approximating a constant probability of failure over a loading region. The present edition also features an alternative of the Weibull model based on an equation which does not use the notions 'flaws' and 'locally initiated failure by flaws'. The new equation is based on the novel concept 'hazard stress density'. A simple and easily reproduced experiment based on artificial flaws provides a strong and convincing experimental proof that the distribution of the minimum breaking strength associated with randomly distributed flaws does not follow a Weibull distribution.

Another important addition in the second edition is the comparative method for improving reliability introduced in Chapter 14. Calculating the absolute reliability built in a product is often an extremely difficult task because in many cases reliability-critical data (failure frequencies, strength distribution of the flaws, fracture mechanism, repair times) are simply unavailable for the system components. Furthermore, calculating the absolute reliability may not be possible because of the complexity of the physical processes and physical mechanisms underlying the failure modes, the complex influence of the environment and the operational loads, the variability associated with reliability-critical design parameters and the non-robustness of the prediction models. Capturing and quantifying these types of uncertainty, necessary for a correct prediction of the reliability of the component, is a formidable task which does not need to be addressed if a comparative reliability method is employed, especially if the focus is on reliability improvement. The comparative methods do not rely on reliability data to improve the reliability of components and are especially suited for developing new designs, with no failure history.

In the second edition, the coverage of physics-of-failure models has been increased by devoting an entire chapter (Chapter 12) to 'fast fracture' and 'fatigue' – probably the two failure modes accounting for most of the mechanical failures.

The conditions for the validity of common physics-of-failure models have also been presented. A good example is the Palmgren–Miner rule. This is a very popular model in fatigue life predictions, yet no comments are made in the reliability literature regarding the cases for which this rule is applicable. Consequently, in Chapter 7, a discussion has been provided about the conditions that must be in place so that the empirical Palmgren–Miner rule can be applied for predicting fatigue life.

A new chapter (Chapter 18) has been included in the second edition which shows that the number of activities in a risky prospect is a key consideration in selecting a risky prospect. In this respect, the maximum expected profit criterion, widely used for making risk decisions, is shown to be fundamentally flawed, because it does not consider the impact of the number of risk–reward activities in the risky prospects.

The second edition also includes a new chapter on optimal allocation of resources to achieve a maximum reduction of technical risk (Chapter 19). This is an important problem facing almost all industrial companies and organisations in their risk reduction efforts, and the author felt that this problem needs to be addressed. Chapter 19 shows that the classical (0–1) knapsack dynamic programming approach for optimal allocation of safety resources could yield highly undesirable solutions, associated with significant waste of resources and very little improvement in the risk reduction. The main reason for this problem is that the standard knapsack dynamic programming approach has been devised to maximise the total value derived from items filling space with no intrinsic value. The risk reduction budget however, does have intrinsic value and its efficient utilisation is just as important as the maximisation of the total removed risk. Accordingly, a new formulation of the optimal resource allocation model has been proposed where the weighted sum of the total removed risk and the remaining budget is maximised.

Traditional approaches invariably require investment of resources to improve the reliability and availability of complex systems. The last chapter however, introduces a method for maximising the system reliability and availability at no extra cost, based solely on permutations of interchangeable components. The concept of well-ordered parallel–series systems has been introduced, and a proof has been provided that a well-ordered parallel–series system possesses the highest possible reliability.

The second edition also includes a detailed introduction into building reliability networks (Chapter 1). It is shown that the conventional reliability block diagrams based on undirected edges cannot adequately represent the logic of operation and failure of some engineering systems. To represent correctly the logic of operation and failure of these engineering systems, it is necessary to include a combination of directed and undirected edges, multiple terminal nodes, edges referring to the same component and negative-state edges.

In Chapter 17, the conventional reliability analysis has been challenged. The conventional reliability analysis is based on the premise that increasing the reliability of a system will always decrease the losses from failures. It is demonstrated that this is valid only if all component failures are associated with similar losses. In the case of component failures associated with very different losses, a system with larger reliability is not necessarily characterised by smaller losses from failures. This counter-intuitive result shows that the cost-of-failure reliability analysis requires a new generation of reliability tools, different from the conventional tools.

Contrary to the classical approach which always starts the reliability improvement with the component with the smallest reliability in the system, the risk-based approach may actually start with the component with the largest reliability in the system if this component is associated with big risk of failure. This defines the principal difference between the classical approach to reliability analysis and setting reliability requirements and the cost-of-failure-based approach.

Accordingly, in Chapter 17, a new methodology and models are proposed for reliability analysis and setting reliability requirements based on the cost of failure. Models and algorithms are introduced for limiting the risk of failure below a maximum acceptable level and for guaranteeing a minimum availability level. Setting reliability requirements at a system level has been reduced to determining the intersection of the hazard rate upper bounds which deliver the separate requirements.

The assessment of the upper bound of the variation from multiple sources has been based upon a result introduced rigorously in Chapter 4 referred to as 'upper bound variance theorem'. The exact upper bound of the variance of properties from multiple sources is attained from sampling not more than two sources. Various applications of the theorem are presented. It is shown how the upper bound variance theorem can be used for developing robust six-sigma products, processes and operations.

Methods related to assessing the consistency of a conjectured model with a data set and estimating the model parameters are also discussed. In this respect, a little known method for producing unbiased and precise estimates of the parameters in the three-parameter power law has been presented in Chapter 5.

All algorithms are presented in pseudocode which can be easily transformed into a programming code in any programming language. A whole chapter has been devoted to Monte Carlo simulation techniques and algorithms which are subsequently used for solving reliability and risk analysis problems.

The second edition includes two new chapters (Chapters 9 and 10) featuring various applications of the Monte Carlo simulation: revealing reliability during shock loading, virtual testing, optimal replacement of components, evaluating the reliability of complex systems and virtual accelerated life testing. Virtual testing is an important application of the Monte Carlo simulation aimed at improving the reliability of common assemblies.

The proposed Monte Carlo simulation approach to evaluating the reliability of complex systems avoids the drawbacks of commonly accepted methods based on cut sets or path sets. A method is also proposed for virtual accelerated testing of complex systems. The method permits extrapolating the life of a complex system from the accelerated lives of its components. This makes the expensive task of building test rigs for life testing of complex engineering systems unnecessary and reduces drastically the amount of time and resources needed for accelerated life testing of complex systems.

The second edition includes also a diverse set of exercises and worked examples illustrating the content of the chapters. The intention was to reveal the full range of applications of the discussed models and make the book useful for test and exam preparation.

By trying to find the balanced mix between theory, physics and application, my desire was to make the book useful to researchers, consultants, students and practising engineers. This text assumes limited familiarity with probability and statistics. Most of the required probabilistic concepts have been summarised in Appendices A and B. Other concepts have been developed in the text, where necessary.

In conclusion, I thank the editing and production staff at John Wiley & Sons, Ltd for their excellent work and particularly the project editor Mr Clive Lawson for his help and cooperation. I also thank the production manager Shiji Sreejish and her team at Spi Global for the excellent copyediting and typesetting. Thanks also go to many colleagues from universities and the industry for their useful suggestions and comments.

Finally, I acknowledge the immense help and support from my wife, Prolet, during the preparation of the second edition.

Michael Todinov
Oxford 2015

1

Failure Modes: Building Reliability Networks

1.1 Failure Modes

According to a commonly accepted definition (IEC, 1991), *reliability* is 'the ability of an entity to perform a required function under given conditions for a given time interval'. A system or component is said to have a failure *if the service it delivers to the user deviates from the specified one*, for example, if the system stops production. System failures or component failures usually require immediate corrective action (e.g. intervention for repair or replacement), in order to return the system or component into operating condition. Each failure is associated with losses due to the cost of intervention, the cost of repair and the cost of lost production.

Failure mode is the way a system or a component fails to function as intended. It is the effect by which failure is observed. The physical processes leading to a particular failure mode will be referred to as *failure mechanism*. It is important to understand that the same failure mode (e.g. fracture of a component) can be associated with different failure mechanisms. Thus, the fracture of a component could be the result of a *brittle fracture* mechanism, *ductile fracture* mechanism or *fatigue failure* mechanism involving nucleation and slow propagation of a fatigue crack. In each particular case, the failure mechanism behind the failure mode 'fracture' is different.

Apart from fracture, other examples of failure modes are 'short circuit', 'open circuit', 'overheating of an electrical or mechanical component', excessive noise and vibration, leakage from a seal, excessive deformation, excessive wear, misalignment which causes a loss of precision, contamination, etc.

Design for reliability is about preventing failure modes from occurring during the specified lifetime of the product. Suppose that the space of all design parameters is denoted by Ω (see Figure 1.1) and the component is characterised by n distinct failure modes. Let $A_1, A_2, \ldots, A_n$ denote the domains of values for the design variables which prevent the first failure mode, the second failure mode and the nth failure mode, respectively.

The intersection $A_1 \cap A_2 \cap \cdots \cap A_n$ of these domains will prevent all failure modes from occurring. An important objective of the design for reliability is to specify the design variables so that they all belong to the intersection domain. This prevents from occurring any of the identified failure modes.

In order to reduce the risk of failure of a product or a process, it is important to recognise their failure modes as early as possible in order to enable execution of design modifications and specific actions

Reliability and Risk Models: Setting Reliability Requirements, Second Edition. Michael Todinov.

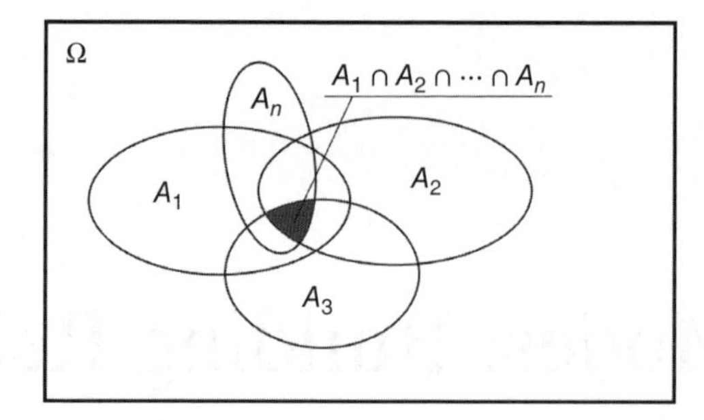

Figure 1.1 Specifying the controllable design variables to be from the intersection domain will prevent all n failure modes

reducing the risk of failure. The benefits from identifying and eliminating failure modes are improved reliability of the product/process, improved safety, reduced warranty claims and other potential losses from failures. It is vital that identifying the failure modes and the required design modifications for their elimination is made during the *early stages* of the design. Design modifications during the early stages of the design are much less costly compared to design modifications executed during the late stages of the design.

Systematic procedures for identifying possible failure modes in a system and evaluating their impact have already been developed. The best known method is the failure mode and effects analysis abbreviated as FMEA, developed in 1963 by NASA (National Aeronautics and Space Administration) for the Apollo project. The method has subsequently been applied in aerospace and aeronautical engineering, nuclear industry, electronics industry, automotive industry and software development. Many literary resources concerning this method are related to the American Military Standard (MIL-STD-1629A, 1977). The fundamental idea behind FMEA is to discover as many as possible potential failure modes, evaluate their impact, identify failure causes and outline controls and actions limiting the risks associated with the identified failure modes. The extension of FMEA which includes *criticality analysis* is known as failure mode and effects criticality analysis (FMECA):

- The inductive approach is an important basic technique for identifying possible failure modes at a system level. It consists of considering sequentially the failure modes of all parts and components building the system and tracking their effect on the system's performance.
- The deductive approach is another important basic technique which helps to identify new failure modes. It consists of considering an already identified failure mode at a system level and investigating what else could cause this failure mode or contribute to it.

Other techniques for identifying potential failure are:

- A systematic analysis of common failure modes by using check lists. An example of a simple check list which helps to identify a number of potential failure modes in mechanical equipment is the following:
  Are components sensitive to variations of load?
  Are components resistant against variations of temperature?
  Are components resistant against vibrations?
  Are components resistant to corrosion?
  Are systems/assemblies robust against variation in their design parameters?
  Are parts sensitive to precise alignment?
  Are parts prone to misassembly?
  Are parts resistant to contamination?
  Are components resistant against stress relaxation?

- Using past failures in similar cases. For many industries, a big weight is given to databases of the type 'lessons learned' which help to avoid failure modes causing problems in the past. Lessons learned from past failures have been useful to prevent failure modes in the oil and gas industry, the aerospace industry and nuclear industry.
- Playing devil's advocate. Probing what could possibly go wrong. Asking lots of 'what if' questions.
- Root cause analysis. Reveals processes and conditions leading to failures. Physics of failure analysis is a very important method for revealing the genesis of failure modes. The root cause analysis often uncovers a number of unsuspected failure modes.
- Assumption analysis. Consists of challenging and testing common assumptions about the followed design procedures, manufacturing, usage of the product, working conditions and environment.
- Analysis of the constraints of the systems. The analysis of the technical constraints of the system, the work conditions and the environment often helps to discover new failure modes.
- Asking not only questions about what could possibly go wrong but also questions how to make the system malfunction. This is a very useful technique for discovering rare and unexpected failure modes.
- Using creativity methods and tools for identifying failure modes in new products and processes (e.g. brainstorming, TRIZ, lateral thinking, etc.)

Before discovering failure modes is attempted, it is vital to understand the basic processes in the system and how the system works. In this respect, building a functional block diagram and specifying the required functions of the system are very important.

The functional diagram shows how the components or process steps are interrelated.

For example, the required system function from the generic lubrication system in Figure 1.2 is *to supply constantly clean oil at a specified pressure, temperature, debit, composition and viscosity to contacting moving parts*. This function is required in order to (i) reduce wear, (ii) remove heat from friction zones and cool the contact surfaces, (iii) clean the contact surfaces from abrasion particles and dirt and (iv) protect from corrosion the lubricated parts. Not fulfilling any of the required components of the system function constitutes a system failure.

The system function is guaranteed by using components with specific functions. The sump is used for the storage of oil. The oil filter and the strainer are used to maintain the oil cleanliness. Maintaining the correct oil pressure is achieved through the pressure relieve valve, and maintaining the correct oil temperature is achieved through the oil cooler. The oil pump is used for maintaining the oil debit, and the oil galleries are used for feeding the oil to the contacting moving parts.

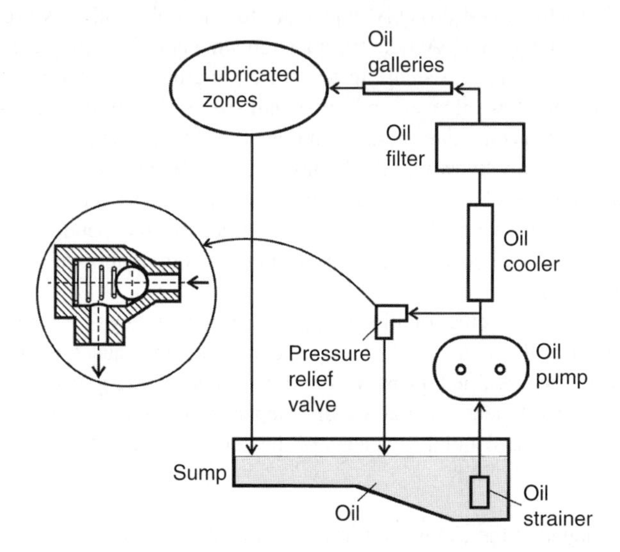

Figure 1.2 Functional block diagram of a lubrication system

The inductive approach for discovering failure modes at a system level starts from the failure modes of the separate components and tracks their impact on the system's performance. Thus, a clogged oil filter leads to a drop of the oil pressure across the oil filter and results in low pressure of the supplied lubricating oil. A low pressure of the supplied lubricating oil constitutes a system failure because supplying oil at the correct pressure is a required system's function.

A mechanical damage of the oil filter prevents the retention of suspended particles in the oil and leads to a loss of the required system function 'supply of clean oil to the lubricated surfaces'.

If the pressure relief valve is stuck in open position, the oil pressure cannot build up and the pressure of the supplied oil will be low, which constitutes a system failure. If the pressure relief valve is stuck in closed position, the oil pressure will steadily build up, and this will lead to excessive pressure of the supplied oil which also constitutes a system failure. With no pressure relief mechanism, the high oil pressure could destroy the oil filter and even blow out the oil plugs.

A cooler lined up with deposited plaques or clogged with debris is characterised by a reduced heat transfer coefficient and leads to decreased cooling capability and a 'high temperature of the supplied oil' which constitutes a system failure. Failure of the cooling circuit will have a similar effect. Clogging the cooler with debris will simultaneously lead to an increased temperature and low pressure of the supplied oil due to the decreased cooling capability and the pressure drop across the cooler.

Excessive wear of the oil pump leads to low oil pressure, while a broken oil pump leads to no oil pressure. Failure of the sump leads to no oil pressure; a blocked oil strainer will lead to a low pressure of the supplied oil.

Blockage of the oil galleries, badly designed oil galleries or manufacturing defects lead to loss of the required system function 'delivering oil at a specified debit to contacting moving parts'.

Oil contamination due to inappropriate storage, oil degradation caused by oxidation or depletion of additives and the selection of inappropriate oil lead to a loss of the required system function 'supplying clean oil with specified composition and viscosity'.

The deductive approach for discovering failure modes at a system level starts with asking questions what else could possibly cause a particular failure mode at a system level or contribute to it and helps to discover contributing failure modes at a component level.

Asking, for example, the question what can possibly contribute to a too low oil pressure helps to discover the important failure mode 'too large clearances between lubricated contact surfaces due to wear out'. It also helps to discover the failure mode 'leaks from seals and gaskets' and 'inappropriate oil with high viscosity being used'.

Asking the question what could possibly contribute to a too high oil pressure leads to the cause 'incorrect design of the oil galleries'. Asking the question what could possibly contribute to a too high oil temperature leads to the cause 'a small amount of circulating oil in the system' which helps to reveal the failure modes 'too low oil level' and 'too small size of the sump'. Undersized sumps lead to a high oil temperature which constitutes a failure mode at the system level.

A common limitation of any known methodology for identifying failure modes is that there is no guarantee that all failure modes have been identified. A severe limitation of some traditional methodologies (e.g. FMEA) is that they treat failure modes of components independently and cannot discover complex failure modes at system level which appear only if a combination of several failure modes at a component level is present.

Another severe limitation of some traditional approaches is that they (e.g. FMEA) cannot discover failure modes dependent on the timing or clustering of conditions and causes. If a number of production units demand independently specified quantity of particular resource (e.g. water steam) for a specified time, the failure mode 'insufficient resource supply' depends exclusively on the clustering of random demands during the time interval and the capacity of the generator centrally supplying the resource.

Exercise

Discover the failure modes of the clevis joint in the figure. The clevis is subjected to a constant axial tensile loading force P (Figure 1.3).

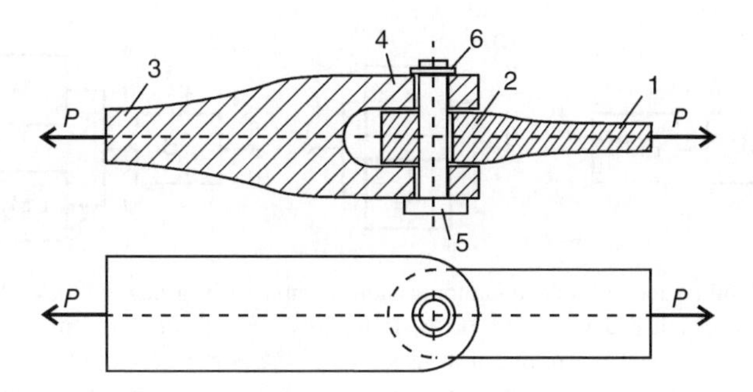

Figure 1.3 A clevis joint

Solution

Shear failure modes:

- Shear failure of the pin 5
- Shear failure of the eye 2
- Shear failure of the clevis 4

Compressive failure modes:

- Compressive failure of the pin 5 due to excessive bearing pressure of the eye 2
- Compressive failure of the pin 5 due to excessive bearing pressure of the clevis 4
- Compressive failure of the clevis 4 due to excessive bearing pressure of the pin 5
- Compressive failure of the eye 2 due to excessive bearing pressure of the pin 5

Tensile failure modes:

- Tensile failure of the blade in zone 1, away from the eye 2
- Tensile failure in zone 3, away from the clevis 4
- Tensile failure of the blade in the area of the eye 2
- Tensile failure in the area of the clevis 4

Other failure modes:

- Bending of the pin 5
- Failure of the clip 6

Thirteen failure modes have been listed for this simple assembly. The analysis in Samuel and Weir (1999), for example, reported only eight failure modes. Preventing all 13 failure modes means specifying the controllable design variables to be from the intersection of the domains which prevent each listed failure mode (Figure 1.1)

1.2 Series and Parallel Arrangement of the Components in a Reliability Network

The operation logic of engineering systems can be modelled by reliability networks, which in turn can be modelled conveniently by graphs. The nodes are notional (perfectly reliable), whereas the edges correspond to the components and are unreliable.

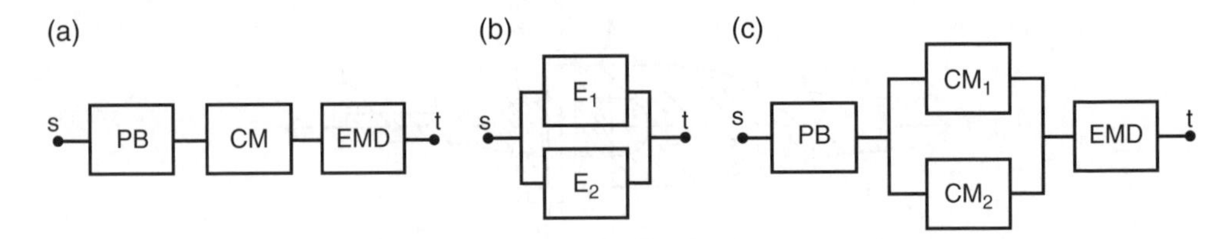

Figure 1.4 (a) Reliability network of a common system composed of a power block (PB), a control module (CM) and an electromechanical device (EMD). (b) Reliability network of a system composed of two power generators E_1 and E_2; the system is working if at least one of the power generators is working. (c) Reliability network of a simple production system composed of power block (PB), two control modules (CM_1 and CM_2) and an electromechanical device (EMD)

The common system in Figure 1.4a consists of a power block (PB), control module (CM) and an electromechanical device (EMD).

Because the system fails whenever any of the components fails, the components are said to be logically arranged in series. The next system in Figure 1.4b is composed of two power generators E_1 and E_2 working simultaneously. Because the system is in working state if at least one of the generators is working, the generators are said to be logically arranged in parallel.

The simple system in Figure 1.4c fails if the power block (PB) fails or if the electromechanical device (EMD) fails or if both control modules CM_1 and CM_2 fail.

However, failure of control module CM_1 only does not cause a system failure. The redundant control module CM_2 will still maintain control over the electromechanical device and the system will be operational.

The system is operational if and only if in its reliability network a path through working components exists from the start node s to the terminal node t; (Figure 1.4).

Reliability networks with a single start node (s) and a single end node (t) can also be interpreted as single-source–single-sink flow networks with edges with integer capacity. The system is in operation if and only if, on demand, a unit flow can be sent from the source s to the sink t (Figure 1.4). In this sense, reliability networks with a single start node and a single end node can be analysed by the algorithms developed for determining the reliability of the throughput flow of flow networks (Todinov, 2013a).

1.3 Building Reliability Networks: Difference between a Physical and Logical Arrangement

Commonly, the reliability networks do not match the functional block diagram of the modelled system. This is why an emphasis will be made on building reliability networks.

The fact that the components in a particular system are logically arranged in series does not necessarily mean that they are logically arranged in series. Although the physical arrangement of the seals in Figure 1.5a is in series, their logical arrangement with respect to the failure mode 'leakage in the environment' is in parallel (Figure 1.5b). Indeed, leakage in the environment is present only if both seals fail.

Conversely, components may be physically arranged in parallel, with a logical arrangement in series. This is illustrated by the seals in Figure 1.6. Although the physical arrangement of the seals is in parallel, their logical arrangement with respect to the failure mode *leakage in the environment* is in series. Leakage in the environment is present if at least one seal stops working (sealing).

Reliability networks are built by using the top-down approach. The system is divided into several large blocks, logically arranged in a particular manner. Next, each block is further detailed into several

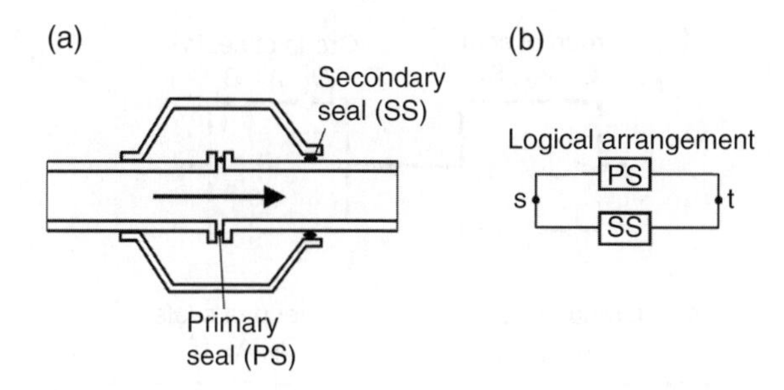

Figure 1.5 Seals that are (a) physically arranged in series but (b) logically arranged in parallel

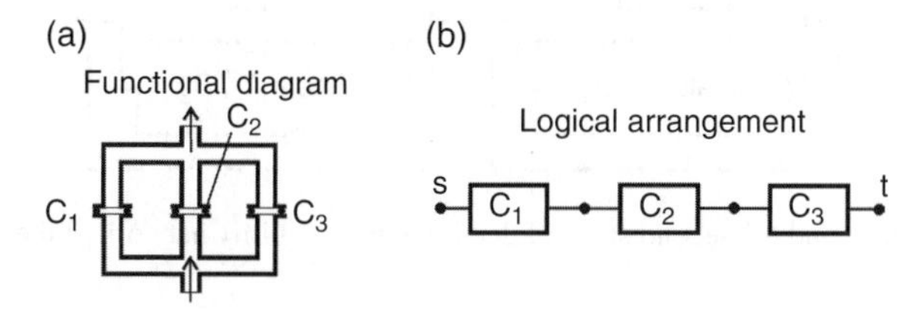

Figure 1.6 The seals are (a) physically arranged in parallel but (b) logically in series

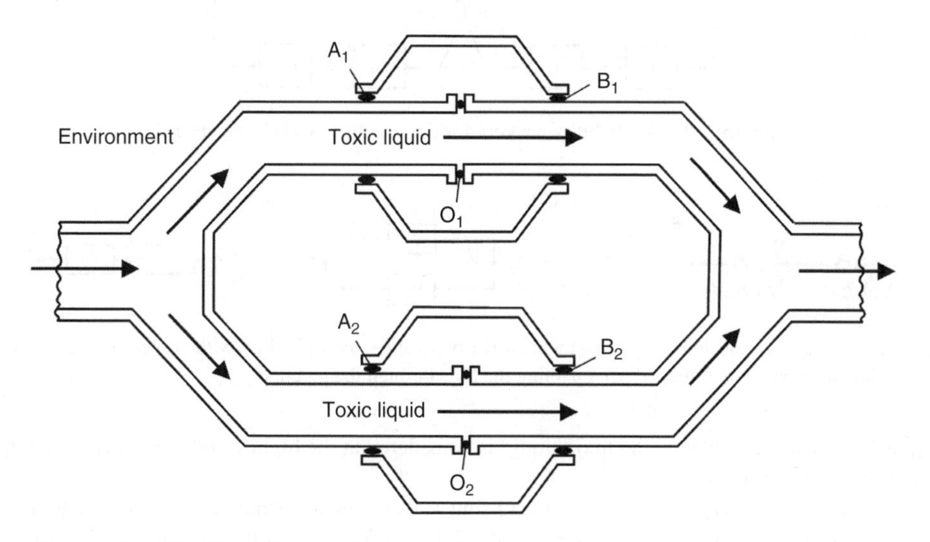

Figure 1.7 A functional diagram of a system of seals isolating toxic liquid from the environment

smaller blocks. These blocks are in turn detailed and so on, until the desired level of indenture is achieved for all blocks.

This approach will be illustrated by the system in Figure 1.7, which represents toxic liquid travelling along two parallel pipe sections. The O-ring seals 'O_1', and 'O_2' are sealing the flanges; the pairs of seals (A_1, B_1) and (A_2, B_2) are sealing the sleeves.

The first step in building the reliability network of the system in Figure 1.7 is to note that despite that physically, the two groups of seals (O_1, A_1, B_1) and (O_2, A_2, B_2) are arranged in parallel, they are arranged logically in series with respect to the function 'preventing a leak to the environment' because both of the two groups of seals must prevent the toxic liquid from escaping in the environment

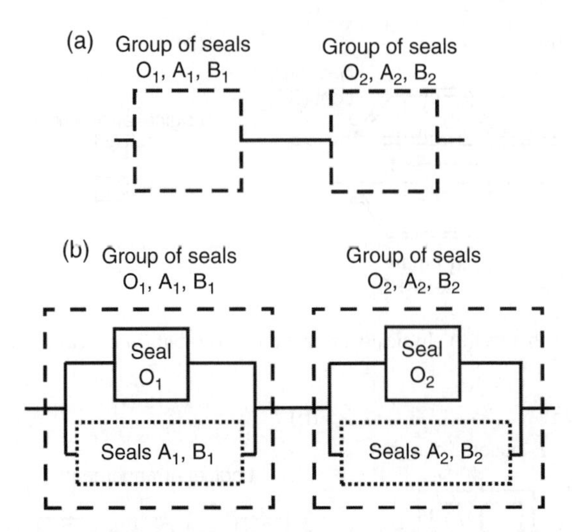

Figure 1.8 (a) First stage and (b) second stage of detailing the reliability network of the system in Figure 1.7

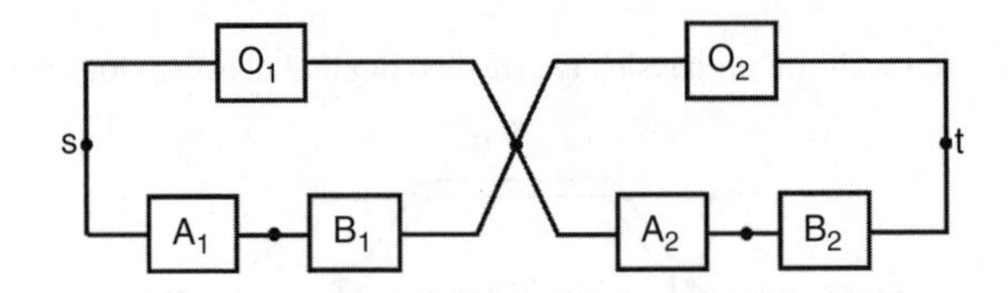

Figure 1.9 A reliability network for the system of seals in Figure 1.7

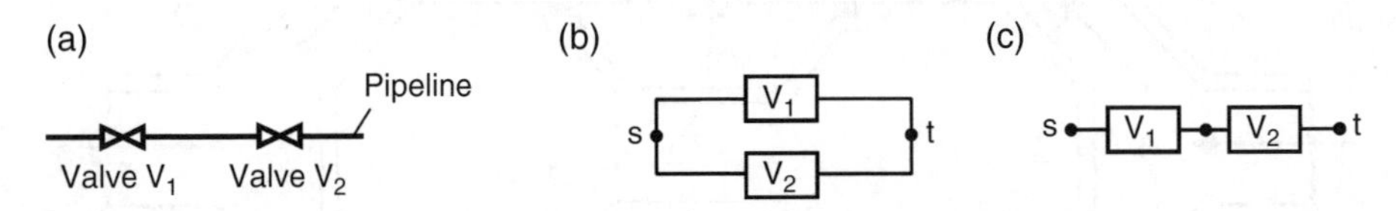

Figure 1.10 Physical and logical arrangement of (a) two valves on a pipeline with respect to the functions. (b) Stopping the production fluid and (c) 'enabling the flow through the pipeline'

(Figure 1.8a). Failure to isolate the toxic liquid is considered at the highest indenture level – the level of the two groups of seals.

Within each of the two groups of seals, the O-ring seal is logically arranged in parallel with the pair of seals (A, B) on the sleeves (Figure 1.8b). Indeed, it is sufficient that the O-ring seal 'O_1' works or the pair of seals (A_1, B_1) works to guarantee that the first group of seals (O_1, A_1, B_1) will prevent a release of toxic liquid in the environment.

Finally, within the pair of seals (A_1, B_1), both seals 'A_1' and 'B_1' must work in order to guarantee that the pair of seals (A_1, B_1) works. The seals A_1 and B_1 are therefore logically arranged in series. This reasoning can be extended for the second group of seals, and the reliability network of the system of seals is as shown in Figure 1.9.

The next example features two valves on a pipeline, physically arranged in series (Figure 1.10). Both valves are initially open. With respect to stopping the production fluid in the pipeline, on demand, the valves are arranged in parallel (Figure 1.10b). Now suppose that both valves are initially closed. With respect to enabling the flow through the pipeline, on demand, the valves are logically arranged in series (Figure 1.10c).

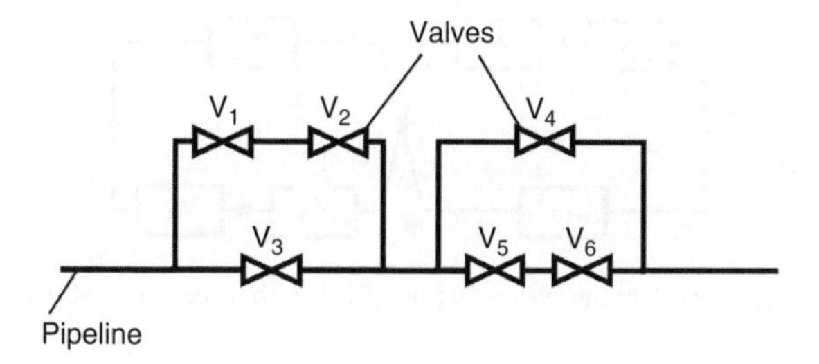

Figure 1.11 A functional diagram of a system of valves

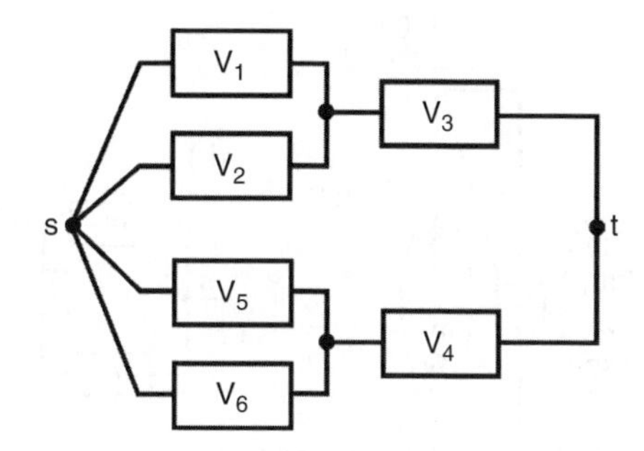

Figure 1.12 The reliability network of the system in Figure 1.9

Indeed, to stop the flow through the pipeline, at least one of the valves must work on demand; therefore, the valves are logically arranged in parallel with respect to the function 'stopping the production fluid'. On the other hand, if both valves are initially closed, to enable the flow through the pipeline, both valves must open on demand; hence, in this case, the logical arrangement of the valves is in series (Figure 1.10c).

Example

Figure 1.11 features the functional diagram of a system of pipes with six valves, working independently from one another, all of which are initially open. Each valve is characterised by a certain probability that if a command for closure is sent, the valve will close and stop the fluid passing through its section. Construct the reliability network of this system with respect to the function 'stopping the flow through the pipeline'.

Solution

The reliability network related to the function stopping the flow in the pipeline is given in Figure 1.11. The blocks of valves (V_1, V_2, V_3) and the block of valves (V_4, V_5, V_6) are logically arranged in parallel because the flow through the pipeline is stopped if either block stops the flow. The block of valves (V_1, V_2, V_3) stops the flow if both groups of valves (V_3) and (V_1, V_2) stop the flow in their corresponding sections. Therefore, the groups (V_1, V_2) and V_3 are logically arranged in series. The group of valves (V_1, V_2) stops the flow if either valve V_1 or V_2 stops the flow in the common section. Therefore, the valves V_1 and V_2 are logically arranged in parallel.

Similar reasoning applies to the block of valves V_4, V_5 and V_6. The reliability network of the system in Figure 1.11 is given in Figure 1.12.

The operational logic of the system has been modelled by a set of perfectly reliable nodes (the filled circles in Figure 1.12) and unreliable edges connected to the nodes.

Interestingly, for the function stopping the fluid in the pipeline, valves or blocks of valves arranged in series in the functional diagram are arranged in parallel in the reliability network. Accordingly, valves or blocks arranged in parallel in the functional diagram are arranged in series in the reliability network.

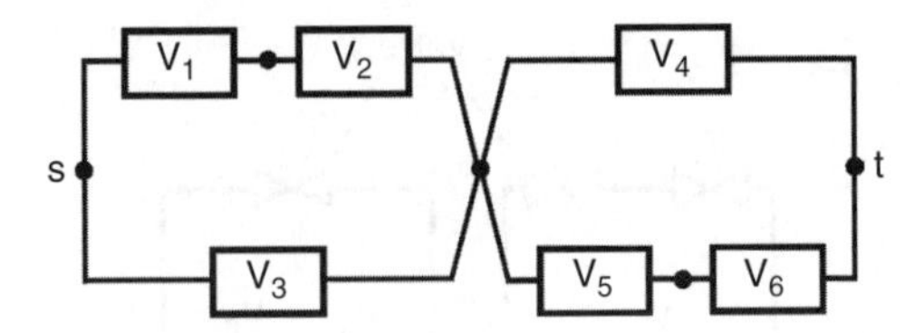

Figure 1.13 The reliability network of the system in Figure 1.9, with respect to the function 'letting flow through the pipeline'

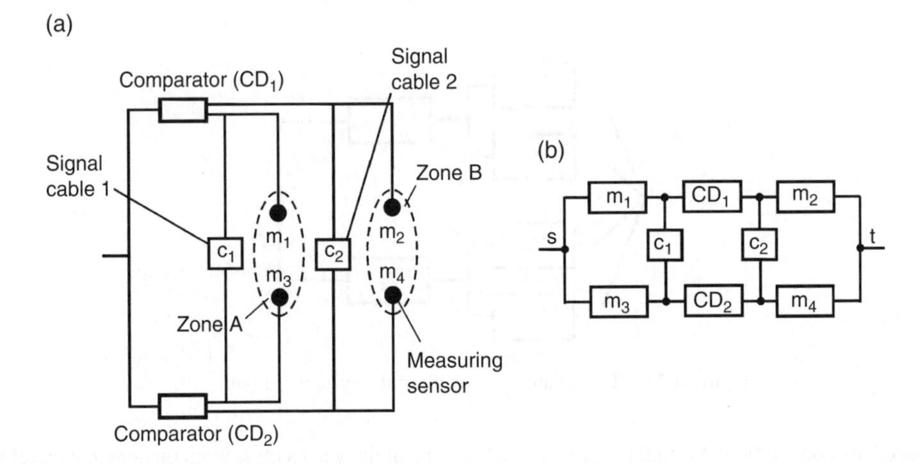

Figure 1.14 (a) A safety-critical system based on comparing measured quantities in two zones and (b) its reliability network

There are also cases where the physical arrangement coincides with the logical arrangement. Consider again the system of valves in Figure 1.11, with all valves initially closed. With respect to the function 'letting flow (any amount of flow) through the pipeline' (the valves are initially closed), the reliability network in Figure 1.13 mirrors the functional diagram in Figure 1.11.

1.4 Complex Reliability Networks Which Cannot Be Presented as a Combination of Series and Parallel Arrangements

Many engineering systems have reliability networks that cannot be described in terms of combinations of series–parallel arrangements. The safety-critical system in Figure 1.14a is such a system. The system compares signals from sensors reading the value of a parameter (pressure, concentration, temperature, water level, etc.) in two different zones. If the difference in the parameter levels characterising the two zones exceeds a particular critical value, a signal is issued by a special device (comparator).

Such generic comparators have a number of applications. If, for example, the measurements indicate a critical concentration gradient between the two zones, the signal may operate a device which eliminates the gradient. In the case of a critical differential pressure, for example, the signal may be needed to open a valve which will equalise the pressure. In the case of a critical temperature gradient measured by thermocouples in two zones of the same component, the signal may be needed to interrupt heating/cooling in order to limit the magnitude of the thermal stresses induced by the thermal gradient. In the case of a critical potential difference measured in two zones of a circuit, the signal may activate a switch protecting the circuit.

The complex safety-critical system in Figure 1.14a compares the temperature (pressure) in two different zones (A and B) measured by the sensors (m_1, m_2, m_3 and m_4). If the temperature (pressure) difference is greater than a critical value, the difference is detected by one of the comparators (control devices) CD_1 or CD_2, and a signal is sent which activates an alarm. The two comparators and the two pairs of sensors have been included to increase the robustness of the safety-critical system. For the same purpose, the signal cables c_1 and c_2 have been included, whose purpose is to increase the connectivity between the sensors and the comparators. If, for example, sensors m_1, m_2 and comparator CD_2 have failed, the system will still be operational. Because of the existence of signal cables, the measured parameter levels by the remaining operational sensors m_3 and m_4 will be fed to comparator CD_1 through the signal cables c_1 and c_2 (Figure 1.14a). If excessive difference in the parameter levels characterising the two zones exists, the comparator CD_1 will activate the alarm. If sensors m_1 and m_4 fail, comparator CD_1 fails and signal cable c_1 fails, the system is still operational because the excessive difference in the measured levels will be detected by sensors m_3 and m_2 and through the working signal cable c_2 will be fed to comparator CD_2.

The system will be operational whenever an s–t path through working components exists in the reliability network in Figure 1.14b. The reliability network in Figure 1.14b cannot be reduced to combinations of series, parallel or series–parallel arrangements. Telecommunication systems and electronic control systems may have very complex reliability networks which cannot be represented with series–parallel arrangements.

1.5 Drawbacks of the Traditional Representation of the Reliability Block Diagrams

1.5.1 Reliability Networks Which Require More Than a Single Terminal Node

Traditionally, reliability networks have been presented as networks with a single start node s and a single terminal node t (Andrews and Moss, 2002; Billinton and Allan, 1992; Blischke and Murthy, 2000; Ebeling, 1997; Hoyland and Rausand, 1994; Ramakumar, 1993). This traditional representation, however, is insufficient to model the failure logic of many engineering systems. There are systems whose logic of failure description requires more than a single terminal node. Consider, for example, the safety-critical system in Figure 1.15 that consists of a power supply (PS), power cable (PC), block of four switches (S_1, S_2, S_3 and S_4) and four electric motors (M_1, M_2, M_3 and M_4).

In the safety-critical system, all electric motors must be operational on demand. Typical examples are electric motors driving fans or pumps cooling critical devices, pumps dispensing water in case of fire, life support systems, automatic shutdown systems, control systems, etc. The reliability on demand of the system in Figure 1.15a can be improved significantly by making the inexpensive low-reliability components redundant (the power cable and the switches) (Figure 1.15b). For the system in Figure 1.15b, the electric motor M_1, for example, will still operate if the power cable PC or the switch S_1 fails because power supply will be maintained through the alternative power cable PC′ and the switch S_1'. The same applies for the rest of the electric motors. The power supply to an electric motor will fail only if both power supply channels fail. The reliability network of the system in Figure 1.15b is given in Figure 1.16. It has one start node s and four terminal nodes t_1, t_2, t_3 and t_4. The system is in working state if a path through working components exists between the start node s and each of the terminal nodes t_1, t_2, t_3 and t_4.

The reliability network in Figure 1.16 is also an example of a system which cannot be presented as a series–parallel system. It is a system with complex topology.

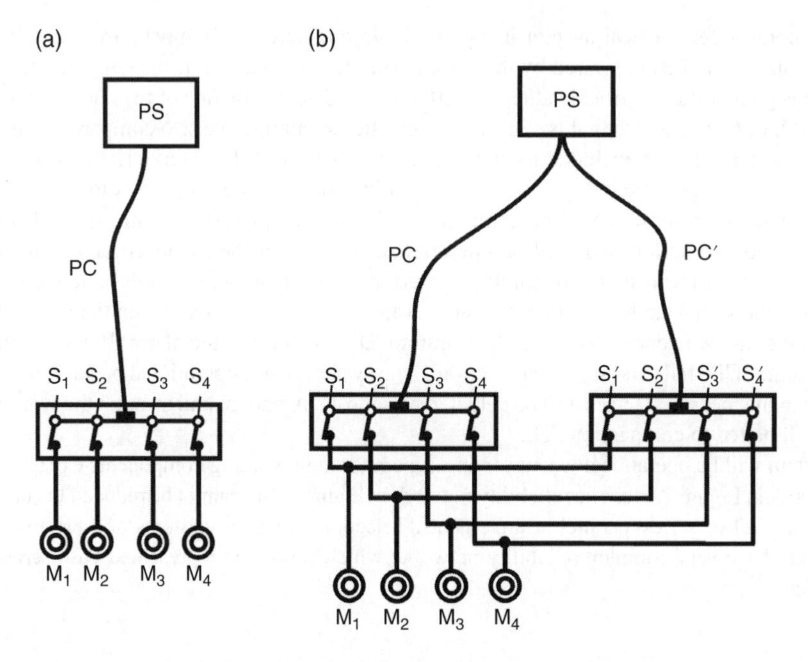

Figure 1.15 A functional diagram of a power supply to four electric motors (a) without redundancy and (b) with redundancy

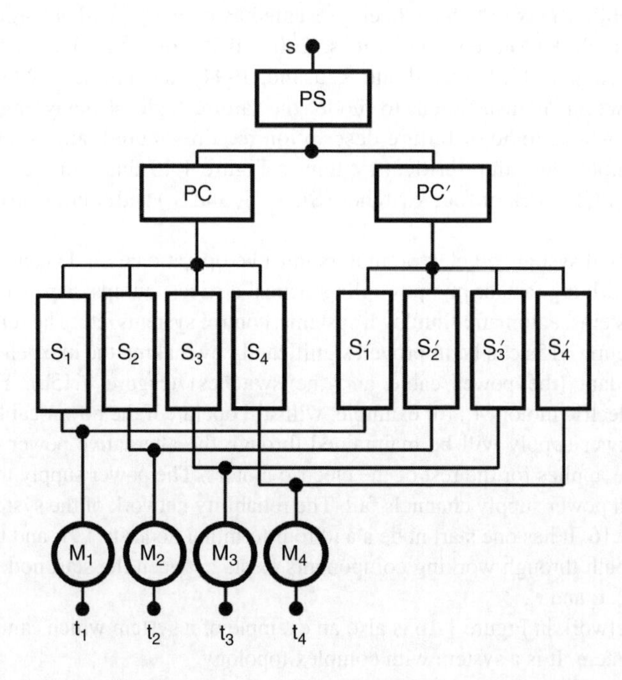

Figure 1.16 A reliability network of the safety-critical system from Figure 1.15b

1.5.2 *Reliability Networks Which Require the Use of Undirected Edges Only, Directed Edges Only or a Mixture of Undirected and Directed Edges*

Commonly, in traditional reliability networks, only undirected edges are used (Andrews and Moss, 2002; Billinton and Allan, 1992; Blischke and Murthy, 2000; Ebeling, 1997; Hoyland and Rausand, 1994; Ramakumar, 1993). This traditional representation is often insufficient to model correctly the logic of system's operation and failure. Often, introducing directed edges is necessary to emphasise that the edge can be traversed in one direction but not in the opposite direction. Consider, for example, the electronic control system in Figure 1.17a, which consists of a control module CM, electronic control switches K_1–K_4 and four controlled devices S_1–S_4.

Assume for the sake of simplicity that the connecting cables are perfectly reliable. As a result, the reliability of the system in Figure 1.17 is determined by the reliability of the control module, the electronic control switches and the controlled devices. Suppose that a signal sent by the control module must reach all four controlled devices S_1–S_4. The reliability of the system is defined as 'the probability that a control signal from the control module CM will reach every single controlled device and all controlled devices will be in working state'.

Similar to the power supply system from Figure 1.15, the reliability of the control system in Figure 1.17a can be improved significantly by making some of the components redundant (e.g. the control module and the electronic control switches) and by providing dual control channels to each controlled device. As a result, from the system in Figure 1.17a, the system in Figure 1.17b is obtained. For the system in Figure 1.17b, for example, the controlled device S_1 will still receive the controlling signal if the control module CM_1 or the switch K_1 fails. The control signal will be received through the alternative control module CM_2 and the switch K_5. The same applies to the rest of the controlled devices. The control signal will not be received only if both control channels fail.

Despite the seeming similarity between the reliability network in Figure 1.18 of the control system and the reliability network in Figure 1.16 of the power supply system, there are essential differences. The power supply system in Figure 1.15b, for example, will be fully operational after the failure of power cable PC′ and switches S_2, S_3 and S_4 (see Figure 1.19a). In contrast, after the failure of control module CM_2 and switches K_2, K_3 and K_4, only device S_1 will receive the control signal. This is because unlike the current in the power supply system, the control signal transmitted to device S_1 cannot reach the other controlled devices by travelling backwards, through the electronic control switch K_5. This backward path has been forbidden *by introducing directed edges in the reliability network.*

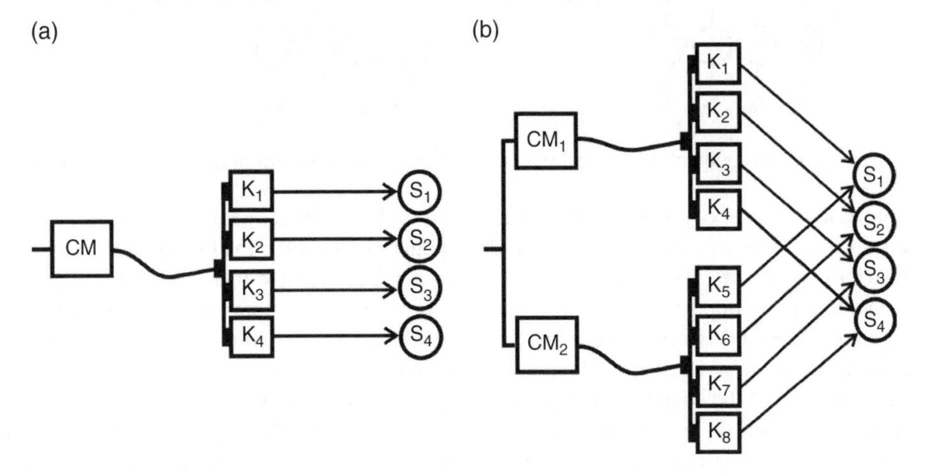

Figure 1.17 An example of a control system including control modules, switches and controlled devices: (a) a single-control system and (b) a dual-control system

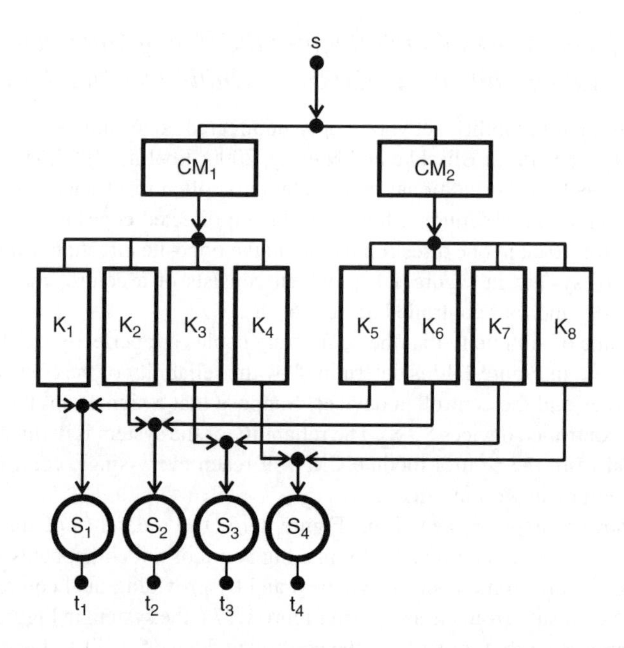

Figure 1.18 A reliability network of the control system from Figure 1.17b

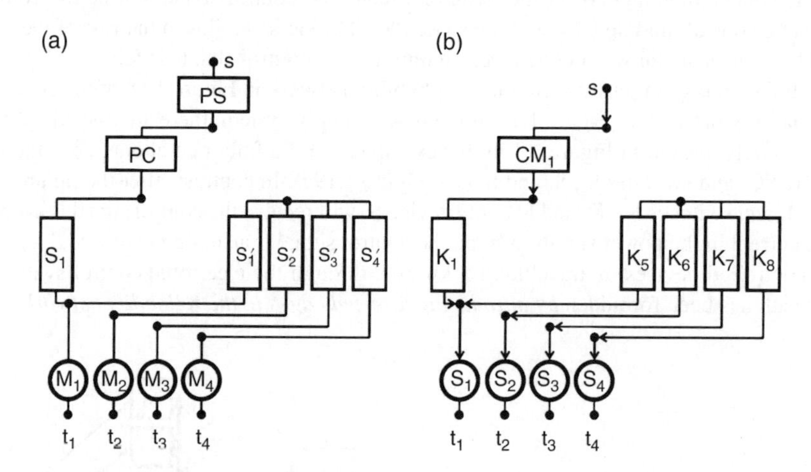

Figure 1.19 An illustration of the difference between the reliability networks in (a) Figures 1.16 and (b) 1.18

A unique sequence of edges between the start node s of the reliability network and any of the terminal nodes will be referred to as *a path*. Edges which point into the direction of traversing the path will be referred to as *forward edges*, edges without direction will be referred to as *undirected edges*, while edges pointing in the opposite direction of the path traversal will be referred to as *backward edges*. A valid path in a reliability network connecting the start node with any of the terminal nodes can have forward edges or undirected edges or both, but it cannot have backward edges.

Thus, in Figure 1.20a, edge (i, j) is a forward edge, while edge (j, k) is a backward edge, and no transition can be made from node j to node k. The sequence of edges between the start node s and the terminal node t of Figure 1.20a is not a valid connecting path. The sequence of edges in Figure 1.20b, however, is a valid s–t path because it consists of forward and undirected edges only.

The next example features a system where both directed and undirected edges are necessary for describing correctly the logic of system operation. The safety-critical system in Figure 1.21 features two power

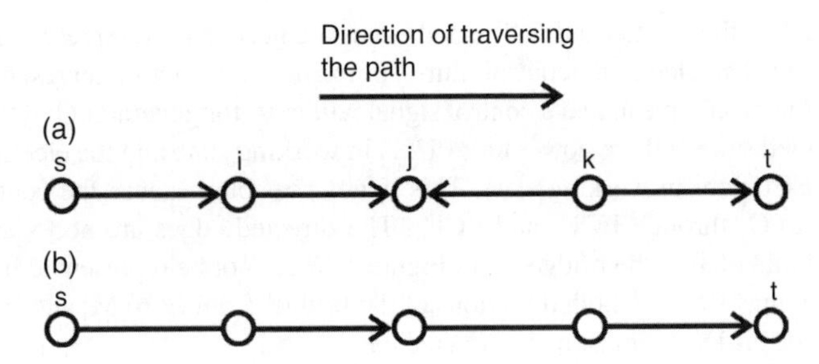

Figure 1.20 The sequence of edges in (a) does not constitute a valid connecting path because of the backward edge (j, k). The sequence of edges in (b) constitutes a valid connecting path

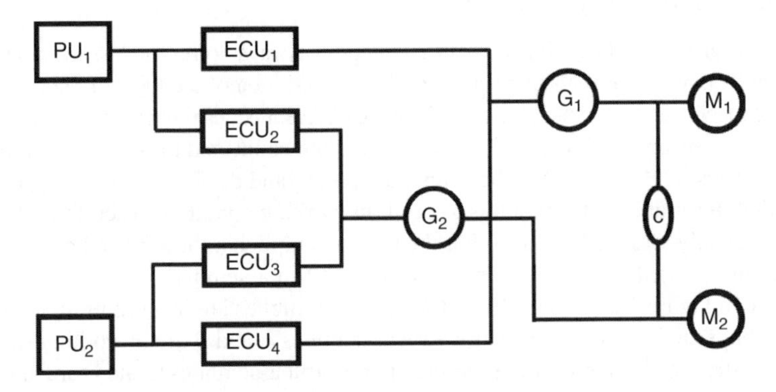

Figure 1.21 Two power generators G_1 and G_2 powering two electric motors M_1 and M_2. The power generators are controlled by four electronic control units ECU_1–ECU_4, powered by the units PU_1 and PU_2

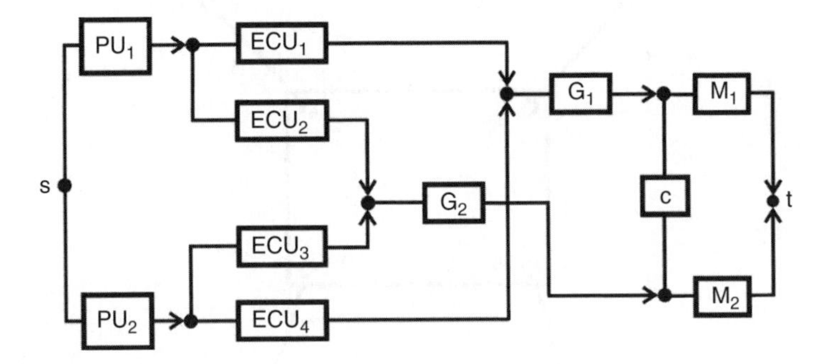

Figure 1.22 Reliability network of the system in Figure 1.21. Both directed and undirected edges are necessary to correctly represent the logic of system's operation

generators G_1 and G_2 delivering current to two electric motors M_1 and M_2. The system is in operation when at least a single electric motor is in operation. The identical, independently working power generators G_1 and G_2 are controlled by four identical electronic control units ECU_1, ECU_2, ECU_3 and ECU_4 powered by two power units PU_1 and PU_2 (Figure 1.21). The redundant electronic control units guarantee that the control over the generators will be maintained even if some of the control units have failed.

To further reduce the risk of system failure, a bridge (power cable) c has also been included. The bridging power cable 'c' guarantees the system's operation in the case where both the electric motor M_1 and the power generator G_2 are in failed state at the end of a specified time interval or in the case where both the electric motor M_2 and the power generator M_1 are in failed state.

The reliability network of the system from Figure 1.21 is given in Figure 1.22.

As can be verified, both directed and undirected edges are necessary to represent correctly the logic of system's operation. The electronic control units, for example, cannot be represented by undirected edges. Otherwise, this would mean that a control signal will exist for generator G_1 if the power unit PU_1 and ECU_4 are in failed state and the power unit PU_2 is in working state and the electronic control units ECU_1, ECU_2 and ECU_3 are in working state. This is not possible because the control signal cannot travel from ECU_3 to G_1 through ECU_2 and ECU_1. The directed edges are necessary *to forbid* such redirection. On the other hand, the bridge 'c' in Figure 1.22 cannot be represented by a directed edge, because the current must travel in both directions of the bridge, from G_1 to M_2 and from G_2 to M_1. The edge representing the bridge 'c' must be undirected edge.

1.5.3 Reliability Networks Which Require Different Edges Referring to the Same Component

In the traditional reliability block diagrams, different edges always correspond to different components. The next example, however, reveals that sometimes, the description of the logic of operation and failure, even for simple mechanical systems, cannot avoid using different edges referring to the same component.

The mechanical system in Figure 1.23 consists of a plate connected through the pin joints a_2, b_2, c_2 and d_2 and the struts A, B, C and D to the supports a_1, b_1, c_1 and d_1. For a strut to support the plate, it is necessary that the strut and its pin joints to be all in working condition. Therefore, the strut and its pin joints are logically arranged in series. For the sake of simplicity, the strut and both of its pin joints are aggregated and treated as a single component called 'strut assembly'.

The structure in Figure 1.23 is stable if all four strut assemblies are in working state, if any three of the strut assemblies are in working state or if strut assemblies A and B are in working state. In the rest of the cases, the structure collapses. For example, if only strut assemblies C and D are in working state,

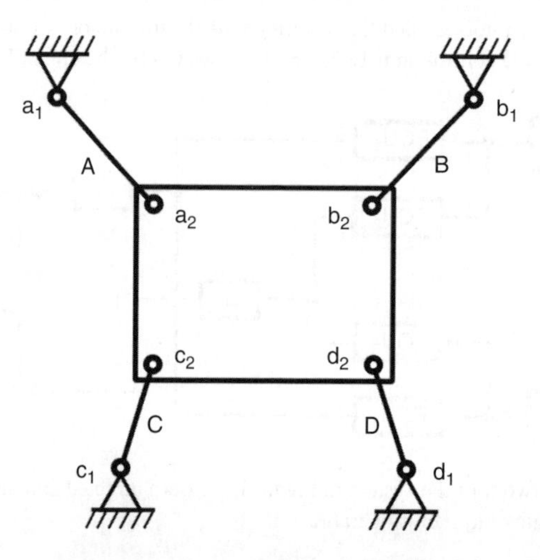

Figure 1.23 A simple mechanical structure

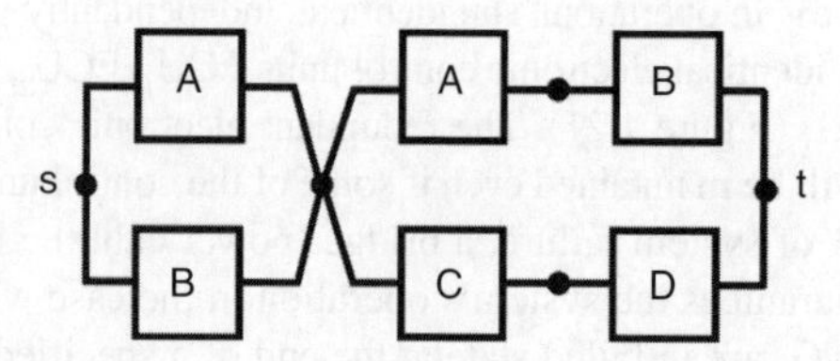

Figure 1.24 The reliability network of the structure in Figure 1.23

the structure collapses. The structure also collapses if only strut assemblies C and B are in working state or if only strut assemblies D and A are operational.

The reliability block diagram of the mechanical structure is shown in Figure 1.24. As can be seen, even for this simple mechanical system, to represent correctly the logic of reliable operation, it is necessary that different edges refer to the same components A and B in the reliability network.

It must be pointed out that in the reliability network from Figure 1.24, the edges marked by A and B *cannot be treated as statistically independent components* because whenever an edge labelled A is in a failed/working state, the other edge also labelled A is in a failed/working state. The same applies to the edges labelled B. Consequently, the reliability of this system cannot be determined through the well-known analytical relationships working for systems with parallel–series arrangement. The reliability of such systems however can be determined easily by using the Monte Carlo simulation technique described in Chapter 10.

1.5.4 Reliability Networks Which Require Negative-State Components

Traditional reliability block diagrams do not deal with negative-state components – components which provide connection between their nodes in the reliability network only if they are in a failed state. An example of a reliability network which requires a negative-state component can be given with the system for transportation of toxic gas in Figure 1.25 through parallel pipes with flanges. The system includes a pump (P) control module (CM) toxic gas sensors (TS_1 and TS_2) and seals (O_1, O_2). To protect personnel in the case of toxic gas release from the seals O_1 and O_2 of the flanges, an enclosure sleeve ES has been added, sealed by the seals K_1, K_2 and K_3. If a toxic gas escapes in the enclosure sleeve ES from the flange seals O_1 or O_2, it is expected that sensor TS_1 or sensor TS_2 will detect the toxic gas release and through the power cut off control module CM will cut the power to the pump (P) and the supply of toxic gas will stop. Stopping the toxic gas supply by cutting the power to the pump prevents the formation of dangerous concentration of toxic gas in the environment. Only one working sensor is needed for the activation of the control module. If the active protection system based on sensors fails to operate, the only remaining barrier to the formation of a dangerous concentration of toxic gas and the environment are the seals K_1, K_2 and K_3.

It is assumed that the enclosure sleeve ES is a perfectly reliable component.

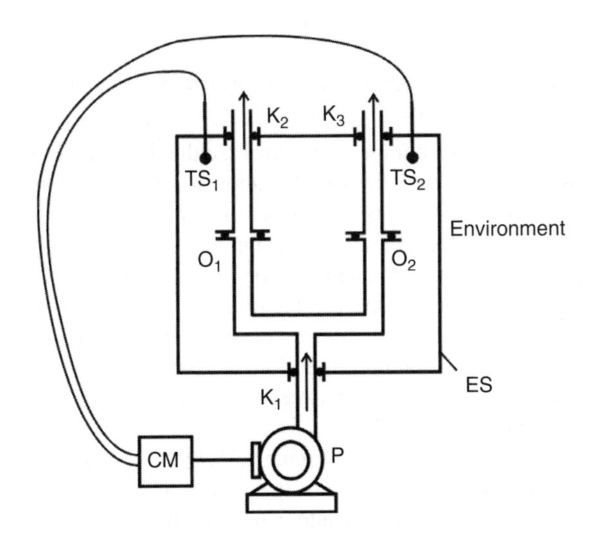

Figure 1.25 A system supplying toxic fluid

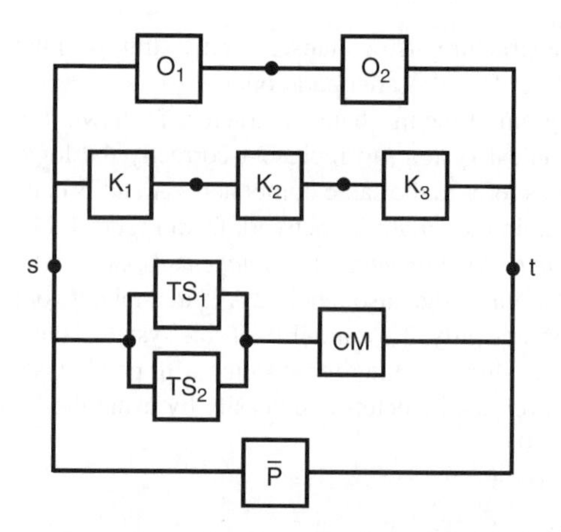

Figure 1.26 Reliability network of the system from Figure 1.25

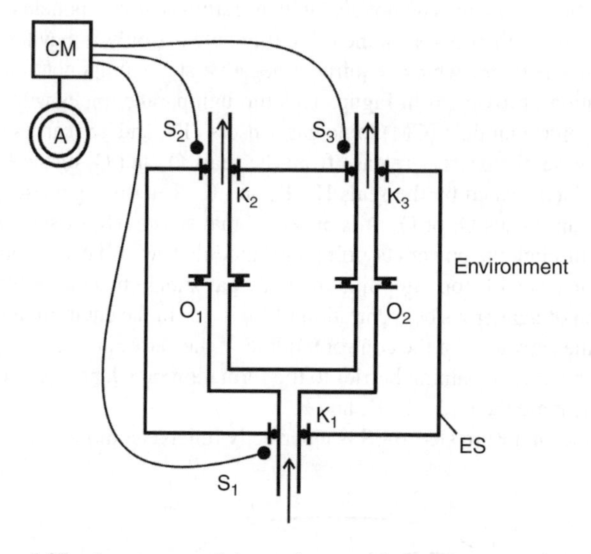

Figure 1.27 A system supplying toxic gas with three sensors and an alarm

In order to isolate the toxic fluid from the environment, either both seals O_1 and O_2 work (seal) or the toxic fluid release is sensed and the power to the pump is cut off or all three seals K_1, K_2 and K_3 work. The power to the pump is cut off if at least one of the sensors TS_1 or TS_2 detects the toxic fluid release and the control module works. The state of the pump does not affect the reliability network of the switching off branch, and this is why the pump is not present there. The state of the pump however does affect the reliability network with respect to the function "prevention of a toxic fluid release in the environment". If the pump is in a failed state, the environment is automatically protected because toxic fluid is no longer supplied. The pump is therefore logically arranged in parallel, as a negative-state component which provides connection between the start node s and the terminal node t only when the pump is not working (Figure 1.26).

Consider now a modification of the system in Figure 1.25. In the case of a leak of toxic gas from the flanges and from the seals K_1, K_2 or K_3, the role of the sensors S_1, S_2 and S_3 is to detect the toxic gas release and to trigger the control module CM into activating the alarm A. The sensors can detect a toxic gas release only locally, in the immediate vicinity of the seal they are attached to (Figure 1.27).

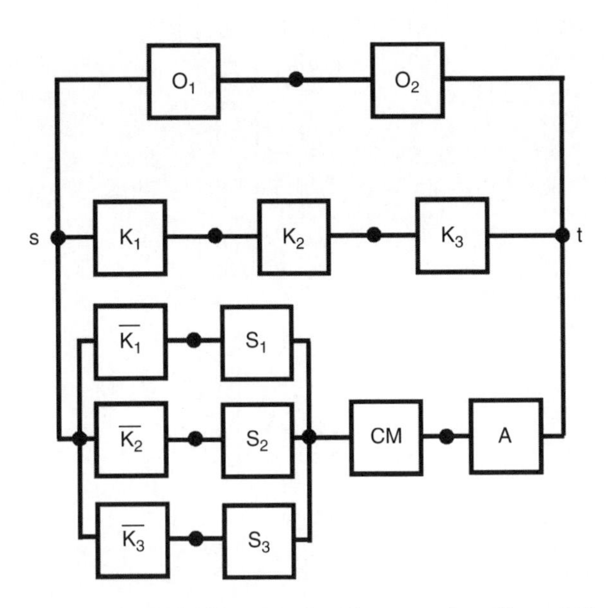

Figure 1.28 Reliability network of the system from Figure 1.27

In order to isolate the toxic gas from the environment, either both seals O_1 and O_2 work (seal) or all three seals K_1, K_2 and K_3 work. Therefore, the block of O-seals and the block of K-seals are logically arranged in parallel. In the case of failure of any of the K-seals, the alarm can be activated if the control module CM, the alarm and the corresponding sensor are in working state. The correct logical arrangement of the components is given in Figure 1.28. The components $\overline{K}_1$, $\overline{K}_2$ and $\overline{K}_3$ are negative-state components. They provide connection between their corresponding nodes only when component K_1, K_2 or K_3 is in failed state, respectively. When component K_1, K_2 or K_3 is in working state, the negative-state component provides no connection between its nodes. Because of the statistical dependence of a component and its negative-state counterpart, Monte Carlo simulation methods are needed for analyzing reliability networks where components and their negative-state counterparts are both present.

2

Basic Concepts

2.1 Reliability (Survival) Function, Cumulative Distribution and Probability Density Function of the Times to Failure

In the mathematical sense, reliability is measured by the *probability* that a system or a component will work without failure during a specified time interval (0, t) under specified operating conditions and environment (Figure 2.1).

The probability $P(T > t)$ that the time to failure T will be greater than a specified time t is given by the *reliability function* $R(t) = P(T > t)$, also referred to as *survival function.* The reliability function is a monotonic non-increasing function, always unity at the start of life ($R(0) = 1$, $R(\infty) = 0$). It is linked with the cumulative distribution function $F(t)$ of the time to failure by $R(t) = 1 - F(t)$: reliability = 1 – probability of failure. If T is the time to failure, $F(t)$ gives the probability $P(T \le t)$ that the time to failure T will be smaller than the specified time t, or in other words, the probability that the system or component will fail before time t.

The probability density function of the time to failure is denoted by $f(t)$. It describes how the failure probability is spread over time. In the infinitesimal interval $t, t + dt$, the probability of failure is $f(t)\,dt$. The probability of failure in any specified time interval $t_1 \le T \le t_2$ is

$$P(t_1 \le T \le t_2) = \int_{t_1}^{t_2} f(t)dt \tag{2.1}$$

Basic properties of the probability density of the time to failure are as follows: (i) $f(t)$ is always non-negative and (ii) the total area beneath $f(t)$ is always equal to one: $\int_0^\infty f(t)dt = 1$. This is because $f(t)$ is a probability distribution, that is, the probabilities of all possible outcomes for the time to failure must add up to unity (the item will certainly fail).

The cumulative distribution function of the time to failure is related to the failure density function by

$$f(t) = \frac{dF(t)}{dt} \tag{2.2}$$

Reliability and Risk Models: Setting Reliability Requirements, Second Edition. Michael Todinov.

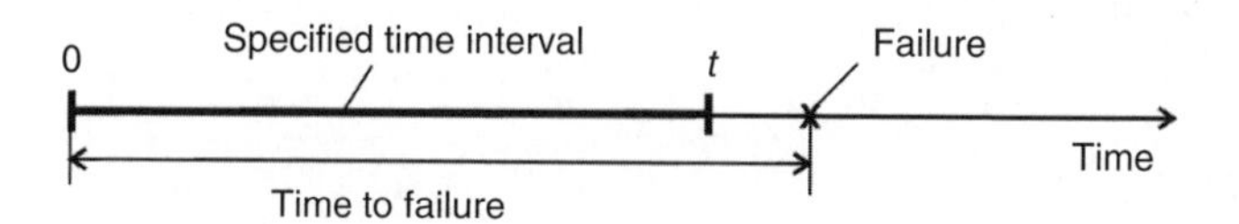

Figure 2.1 Reliability is measured by the probability that the time to failure will be greater than a specified time t

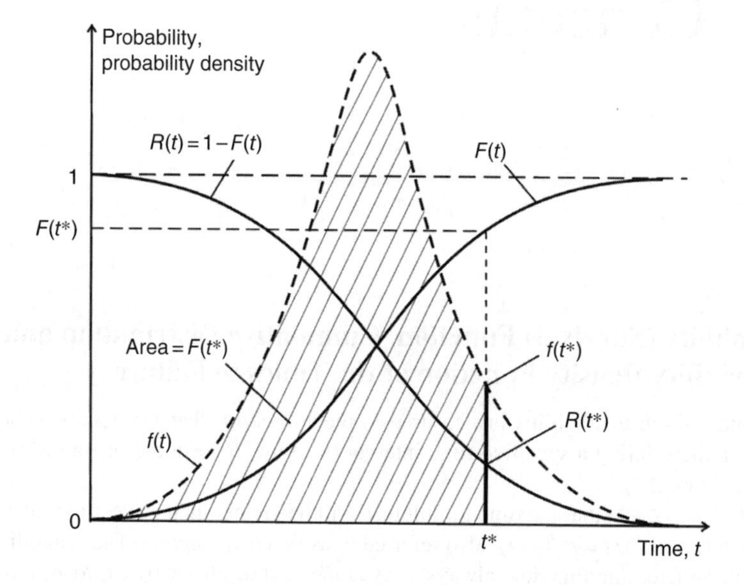

Figure 2.2 Reliability function, cumulative distribution function of the time to failure and failure density function

From expression (2.2), the probability that the time to failure will be smaller than a specified value t is

$$F(t) = P(T \leq t) = F(t) = \int_0^t f(v)dv \tag{2.3}$$

where v is a dummy integration variable: $F(\infty) = \int_0^\infty f(v)dv = 1$, $F(0) = 0$. Because $f(t)$ is non-negative, its integral $F(t)$ is a monotonic non-decreasing function of t (Figure 2.2). The value $F(t^*) = \int_0^{t^*} f(v)dv$ of the cumulative distribution function at time t^* gives the area beneath the probability density function $f(t)$ until time t^* (Figure 2.2).

The link between the reliability function $R(t)$, cumulative distribution function $F(t)$ and probability density function $f(t)$ is illustrated in Figure 2.2.

$P(t_1 < T \leq t_2)$ is the probability of failure between times t_1 and t_2:

$$P(t_1 < T \leq t_2) = \int_{t_1}^{t_2} f(v)dv = F(t_2) - F(t_1) \tag{2.4}$$

The hatched area in Figure 2.3 is equal to the difference $F(t_2) - F(t_1)$ and gives the probability that the time to failure T will be between t_1 and t_2.

A comprehensive discussion related to the basic reliability functions has been provided by Grosh (1989).

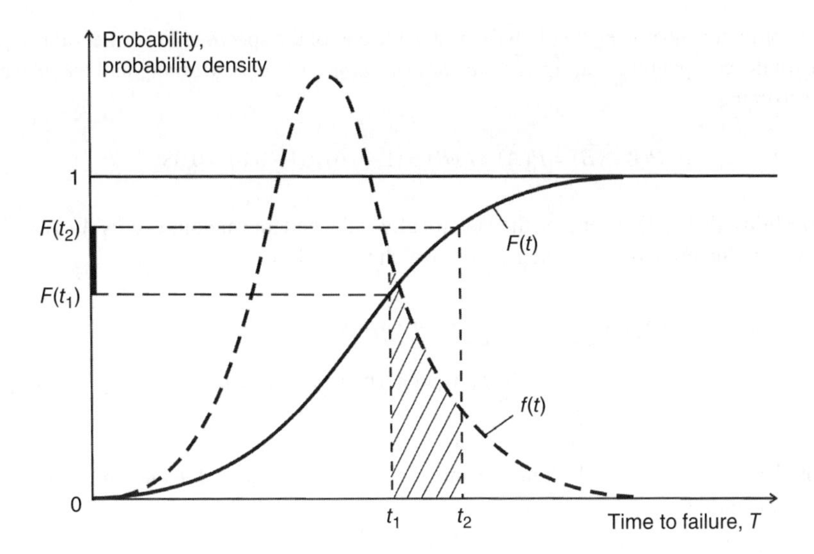

Figure 2.3 Cumulative distribution and probability density function of the time to failure

2.2 Random Events in Reliability and Risk Modelling

Two very common approaches in reliability and risk modelling are based on random events and random variables.

2.2.1 Reliability and Risk Modelling Using Intersection of Statistically Independent Random Events

Statistically independent events are present *when the outcome of any of the events has no effect on the outcomes of other events*. In other words, the events are not related in any way to one another.

The probability that n statistically independent random events $A_1, A_2, \ldots, A_n$ will occur simultaneously is given by the probability of their intersection, which is equal to the product of the individual probabilities of the events:

$$P(A_1 \cap A_2 \cap \cdots \cap A_n) = P(A_1) \times P(A_2) \times \cdots \times P(A_n) \quad (2.5)$$

Example

Suppose that two power generators (an old one and a new one) have been installed and work independently from each other as part of a power supply system. The reliability of the old generator associated with a specified time interval is 0.6. The reliability of the new generator associated with the same time interval is 0.8.

Suppose that it is required to determine the probability that there will exist *a full power supply* at the end of the specified time interval. This common problem can be solved using random events. Let A denote the event *the old power generator will be working* at the end of the time interval and B denote the event *the new power generator will be working* at the end of the time interval.

The probability that there will be a full power supply at the end of the time interval is equal to the probability of the event $A \cap B$ that both generators will be working. Because events A and B are statistically independent, according to Equation 2.5, the probability of their intersection $A \cap B$ is

$$P(A \cap B) = P(A) \times P(B) = 0.6 \times 0.8 = 0.48 \quad (2.6)$$

The probability that no power supply will exist at the end of the specified time interval is equal to the probability of the compound event $\bar{A} \cap \bar{B}$: *the old generator will not be working* and *the new generator will not be working*:

$$P(\bar{A} \cap \bar{B}) = P(\bar{A}) \times P(\bar{B}) = (1-0.6) \times (1-0.8) = 0.08 \tag{2.7}$$

The probability that at least one of the statistically independent random events $A_1, A_2, \ldots, A_n$ will occur is given by the probability of their union, which is equal to

$$\begin{aligned} P(A_1 \cup A_2 \cup \cdots \cup A_n) &= \sum_{i=1}^{n} P(A_i) - \sum_{i<j}\sum P(A_i)P(A_j) \\ &+ \sum_{i<j<k}\sum\sum P(A_i)P(A_j)P(A_k) - \cdots + (-1)^{n+1} P(A_1)P(A_2)\cdots P(A_n) \end{aligned} \tag{2.8}$$

Equation 2.8 is easily derived from the more general expression 2.9 known as *the inclusion–exclusion expansion*:

$$\begin{aligned} P(A_1 \cup A_2 \cup \cdots \cup A_n) &= \sum_{i=1}^{n} P(A_i) - \sum_{i<j}\sum P(A_i \cap A_j) \\ &+ \sum_{i<j<k}\sum\sum P(A_i \cap A_j \cap A_k) - \cdots + (-1)^{n+1} P(A_1 \cap A_2 \cap \cdots \cap A_n) \end{aligned} \tag{2.9}$$

The rule for the expansion can be summarised by the following steps:

1. Add the probabilities of all single events. This means that the probability of the intersections of any pair of events is added twice and should be subtracted.
2. Subtract the probabilities of all double intersections from the previous result. Since the contribution of the intersection of any three events has been added three times through the single events and subsequently has been subtracted three times from the twofold intersections, the probabilities of all threefold intersections must be added.
3. For higher-order intersections, the terms with odd number of intersecting events are added, while the terms with even number of intersecting events are subtracted from the sum.

For three statistically independent events, Equation 2.8 reduces to

$$\begin{aligned} P(A_1 \cup A_2 \cup A_3) &= P(A_1) + P(A_2) + P(A_3) - P(A_1)P(A_2) - P(A_2)P(A_3) \\ &- P(A_1)P(A_3) + P(A_1)P(A_2)P(A_3) \end{aligned} \tag{2.10}$$

For two statistically independent events A_1 and A_2, the probability that at least one event will occur is given by

$$P(A_1 \cup A_2) = P(A_1) + P(A_2) - P(A_1)P(A_2) \tag{2.11}$$

Suppose now that in the previous example, it is required to determine the probability that there will be a power supply at the end of the specified time interval. The probability of a power supply is equal to the probability of the event $A \cup B$ that *at least one generator will be working*. According to Equation 2.11, the probability of the union $A \cup B$ of the statistically independent events A and B is

$$P(A \cup B) = P(A) + P(B) - P(A)P(B) = 0.6 + 0.8 - 0.6 \times 0.8 = 0.92$$

2.2.2 *Reliability and Risk Modelling Using a Union of Mutually Exclusive Random Events*

The mutually exclusive random events cannot occur simultaneously, and the probability of their simultaneous occurrence is zero. If one of them occurs, the other does not occur, and vice versa. Consequently, mutually exclusive events are not statistically independent. The events *A* 'the device will work at the end of the time interval (0, *t*)' and *B* 'the device will not work at the end of the time interval *t*' are mutually exclusive events. They cannot occur simultaneously.

If *n* devices are working independently during the time interval (0, *t*), the events A_0 (no device works at the end of the time interval), A_1 (exactly one device works at the end of the time interval), A_2 (exactly two devices work at the end of the time interval), …, A_n (all *n* devices work at the end of the time interval) are statistically independent events. Exactly one of these events can occur at the end of the time interval, and if it occurs, the rest of the events do not.

The probability that at least one of several mutually exclusive events will occur is equal to the sum of the probabilities of the separate events:

$$P(A_1 \cup A_2 \cup \cdots \cup A_n) = P(A_1) + P(A_2) + \cdots + P(A_n) \tag{2.12}$$

A very important special case of mutually exclusive events are the complementary events *A* and $\bar{A}$. The complementary events are mutually exclusive, and in addition, they are exhaustive which means that one of them will certainly occur. A common example are the events *A* (the device will work at the end of the time interval (0, *t*)) and the event $\bar{A}$ (the device will be in a failed state at the end of the time interval (0, *t*)). Because the complementary events are exhaustive (one of them will certainly occur), their probabilities add up to unity:

$$P(A \cup \bar{A}) = P(A) + P(\bar{A}) = 1 \tag{2.13}$$

The probability of an event can therefore be determined by subtracting the probability of its complementary event from unity:

$$P(A) = 1 - P(\bar{A}) \tag{2.14}$$

Equation 2.14 has a very important application. The probability of the event *A* that at least one of *n* independent generators will be working can be determined from the probability of its complementary event $\bar{A}$ – none of the *n* generators will be working. If the probabilities of the separate generators working are denoted by $P(A_i)$, $i = 1, \ldots, n$, the probability of the complementary event $\bar{A}$ is given by

$$P(\bar{A}) = [1 - P(A_1)] \times \cdots \times [1 - P(A_n)]$$

The probability of the event that at least one of the *n* generators will be working is given by

$$P(A) = 1 - [1 - P(A_1)] \times \cdots \times [1 - P(A_n)] \tag{2.15}$$

Note that the same probability could have been estimated by using Equation 2.8. However, for even a moderately large number of events *n*, the application of the inclusion–exclusion expansion formula becomes intractable because it leads to expressions with a very large number of terms. Consequently, for more than two events, the probability of a union of independent events should be calculated by using Equation 2.15.

Suppose now that in the previous example, it is required to determine the probability of *insufficient power supply*. This is equivalent to determining the probability of the event *exactly one generator will be working*. This event can be presented as a union $(A \cap \bar{B}) \cup (\bar{A} \cap B)$ of the following events: $A \cap \bar{B}$ – *the old generator will be working* and *the new generator will not be working* and $\bar{A} \cap B$ – *the new generator will be working* and *the old generator will not be working*. If one of these events occurs, the other does not; therefore, they are mutually exclusive events.

According to the Equation 2.12 related to a probability of a union of mutually exclusive events,

$$\begin{aligned} P(A \cap \bar{B} \cup \bar{A} \cap B) &= P(A \cap \bar{B}) + P(\bar{A} \cap B) = P(A)P(\bar{B}) + P(\bar{A})P(B) \\ &= 0.6 \times (1-0.8) + (1-0.6) \times 0.8 = 0.44 \end{aligned} \tag{2.16}$$

which is the probability of insufficient power supply.

The next example involves more than two mutually exclusive events.

Example

Four bolts are holding a flange (Figure 2.4). Each bolt has a probability $p=0.35$ of containing a particular fault which leads to a premature failure of the bolt.

If only a single bolt fails, the flange connection is operational. If any pair of diametrically opposite bolts (e.g. bolts 1,3 or 2,4) fail but the other two diametrically opposite bolts are operational, the flange connection is also operational. The flange connection fails prematurely if all bolts contain the fault, if exactly three bolts contain the fault or if exactly two bolts contain the fault, but they are not diametrically opposite bolts (e.g. bolts 1,2 or 3,4 or 2,3 or 1,4). Determine the probability that the flange will not fail prematurely due to the presence of the fault in some of the bolts.

Solution

If each bolt contains the fault with probability p, it does not contain the fault with probability $1-p$. The flange connection will not fail prematurely due to the fault being present in some of the bolts if any of the following mutually exclusive events is present:

(a) No bolt contains the fault. $P(A) = (1-p)^4$.

(b) Exactly one bolt contains the fault. The probability of this event is $P(B) = 4p(1-p)^3$, which is a sum of probabilities of four mutually exclusive events:

  (i) Bolt 1 contains the fault and the rest of the bolts do not (probability $p(1-p)(1-p)(1-p) = p(1-p)^3$).

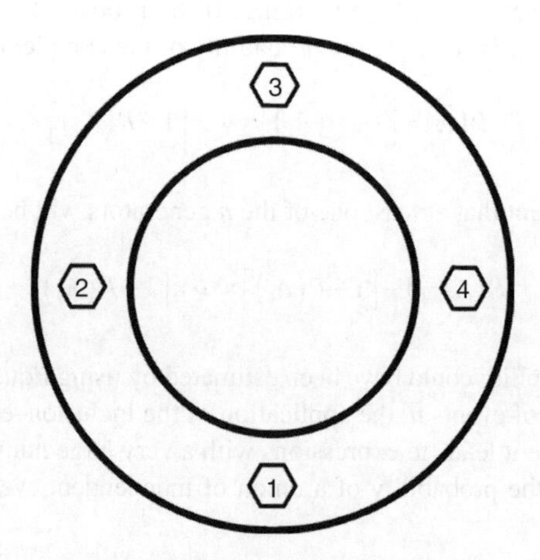

Figure 2.4 A flange with four bolts, each containing a particular fault with certain probability

(ii) Bolt 2 contains the fault and the rest of the bolts do not (probability $(1-p)p(1-p)(1-p) = p(1-p)^3$).
(iii) Bolt 3 contains the fault and the rest of the bolts do not (probability $(1-p)(1-p)p(1-p) = p(1-p)^3$).
(iv) Bolt 4 contains the fault and the rest of the bolts do not (probability $(1-p)(1-p)(1-p)p = p(1-p)^3$).

(c) Exactly two diametrically opposite bolts contain the fault, but the other two bolts do not. This event is characterised by a probability $P(C) = 2p^2(1-p)^2$, which is a sum of the probabilities of two mutually exclusive events:
(i) Bolts 1 and 3 contain the fault and bolts 2 and 4 do not, the probability of which is $p^2(1-p)^2$.
(ii) Bolts 2 and 4 contain the fault and bolts 1 and 3 do not, the probability of which is $(1-p)^2p^2$.

Because events A, B and C are mutually exclusive events, the probability of their union is equal to the sum of their probabilities:

$$P(A \cup B \cup C) = P(A) + P(B) + P(C) = (1-p)^4 + 4p(1-p)^3 + 2p^2(1-p)^2$$

Substituting $p = 0.35$ gives

$$P(A \cup B \cup C) = (1-p)^2[(1-p)^2 + 4p(1-p) + 2p^2] = 0.67$$

for the probability that the flange connection will not fail prematurely because of the fault being present in some of the bolts.

2.2.3 *Reliability of a System with Components Logically Arranged in Series*

Another important application of the intersection of statistically independent random events is the practically important case of a system composed of statistically independent components, arranged logically in series (Figure 2.5).

Let S denote the event *the system will be working* and C_k denote the event *the kth component will be working*. For the series arrangement in Figure 2.5, event S is an intersection of all events C_k, $k = 1, 2, \ldots, n$, because the system will be working only if *all* of the components work. Consequently,

$$S = C_1 \cap C_2 \cap \cdots \cap C_n \tag{2.17}$$

According to the equation related to a probability of an intersection of statistically independent events (Appendix A), the probability that the system will be working is

$$P(S) = P(C_1) \times P(C_2) \times P(C_3) \times \cdots \times P(C_n) \tag{2.18}$$

equal to the product of the probabilities that the separate components will be working. If $R = P(S)$ is the reliability of the system and $R_k = P(C_k)$ is the reliability of the kth component, the reliability of a system with components logically arranged in series is Bazovsky (1961)

$$R = R_1 \times R_2 \times \cdots \times R_n \tag{2.19}$$

Two important conclusions can be made from this expression. The larger the number of components is, the lower is the reliability of the arrangement.

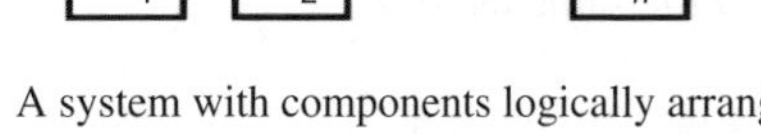

C_1 — C_2 — ••• — C_n

Figure 2.5 A system with components logically arranged in series

Figure 2.6 An extra component added to a series arrangement

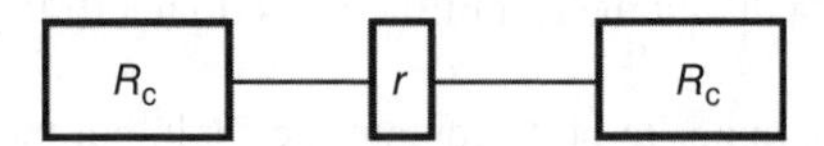

Figure 2.7 Two very reliable components with reliabilities $R_c \approx 1$, connected through an unreliable interface with reliability r

Indeed, if an extra component C_{n+1} with reliability R_{n+1} is added as shown in Figure 2.6, the reliability of the arrangement becomes $R_1 \times R_2 \times \cdots \times R_n \times R_{n+1}$, and since $R_{n+1} < 1$,

$$R_1 \times R_2 \times \cdots \times R_n > (R_1 \times R_2 \times \cdots \times R_n) \times R_{n+1} \tag{2.20}$$

Another important observation is that the reliability of a series arrangement is smaller than the reliability R_k of the least reliable component k:

$$R_1 \times R_2 \times \cdots \times R_k \times \cdots \times R_n < R_k \tag{2.21}$$

This fact has important practical implications. It means that the reliability of a series arrangement cannot be improved beyond the reliability of the lest reliable component, unless the reliability of the least reliable component is improved. If a reliability improvement on a system level is to be made, the reliability improvement efforts should be focused on improving the reliability of the least reliable component first, not on improving the reliability of components with already high reliability.

Consider a common practical example related to two very reliable components with high reliabilities $R_c \approx 1$, connected through an interface of relatively low reliability $r << R_c$ (Figure 2.7).

According to Equation 2.19, the reliability of the arrangement

$$R = R_c \times r \times R_c \approx r \tag{2.22}$$

is approximately equal to the reliability of its weakest link, the interface r. In order to improve substantially the reliability of the arrangement, the reliability of the interface must be increased. One of the reasons why so many failures occur at interfaces is the circumstance that often interfaces are not manufactured to match the reliability of the corresponding components.

An alternative expression for the reliability of n identical, statistically independent components arranged logically in series is

$$R = (1-p)^n \tag{2.23}$$

where p is the probability of failure of a single component. This equation provides an insight into the link between the complexity of a system and its reliability.

Indeed, if Equation 2.23 is presented as $R = \exp[n\ln(1-p)]$. For small probabilities of failure $p << 1$, $\ln(1-p) \approx -p$ and the reliability of the arrangement becomes

$$R = \exp(-pn) \tag{2.24}$$

As can be verified from Equation 2.24, if the number n of components in the system is increased by a factor of k, in order to maintain the reliability of the system, the probability of failure p of a single component must be decreased by the same factor k.

It must be pointed out that, for a relatively large number of components logically arranged in series, the error associated with reliability predictions based on Equation 2.19 can be significant. Indeed, suppose that a large number n of identical components have been logically arranged in series and each component has a reliability r. Since the reliability of the system is $R = r^n$, a small relative error $\Delta r/r$ in estimating the reliability r of the individual components gives rise to a large relative error $\Delta R/R \approx n\Delta r/r$ in the predicted reliability of the system. This point is important and will be illustrated by a simple numerical example.

Assume that the reliability estimate related to an individual component varies in the relatively small range $0.96 \le r \le 0.98$. If $r = 0.96$ is taken as a basis for predicting the reliability of a system including 35 components logically arranged in series, the calculated reliability is $R = 0.96^{35} \approx 0.24$. If $r = 0.98$ is taken as a basis for the reliability prediction, the calculated reliability is $R = 0.98^{35} \approx 0.49$, more than twice compared to the previous estimate!

This example can also be interpreted in the following way. If 35 identical components with reliabilities 0.96 are logically arranged in series, a relatively small increase in the reliability of a component results in a large system reliability increase. The application of the formula related to a series arrangement is not restricted to hardware components only. Assume that the successful accomplishment of a project depends on the successful accomplishment of 35 identical and independent tasks. If a person accomplishes successfully a separate task with probability 0.96, an investment in additional training which increases this probability to only 0.98 would make a big impact on the probability of accomplishing the project.

In another example, suppose that a single component can fail due to n statistically independent failure modes. The event S: *the component will survive time t* can be presented as an intersection of the events S_i: *the component will survive the ith failure mode*, $S = S_1 \cap S_2 \cap \cdots \cap S_n$.

The probability $P(S)$ of the event S is the reliability $R(t)$ of the component associated with the time interval $(0, t)$, and is given by the product

$$R(t) = P(S_1) \cap P(S_2) \cap \cdots \cap P(S_n) = R_1(t)R_2(t)\cdots R_n(t) \tag{2.25}$$

where $R_i(t)$, $i = 1, 2, \ldots, n$, is the probability that the component will survive the ith failure mode. Since the probabilities of failure before time t, associated with the separate failure modes, are given by $F_i(t) = 1 - R_i(t)$, the probability of failure of the component before time t is

$$F(t) = 1 - R(t) = 1 - \prod_{k=1}^{n} \left[1 - F_k(t)\right] \tag{2.26}$$

2.2.4 Reliability of a System with Components Logically Arranged in Parallel

For independently working components logically arranged in parallel (Figure 2.8), the event S (the system will be working) is a union of the events C_k: *the kth component will be working*, $k = 1, 2, \ldots, n$, because the system will be working if at least one component works.

Consequently, event S can be presented as the union of events C_k:

$$S = C_1 \cup C_2 \cup \cdots \cup C_n \tag{2.27}$$

Simpler expressions are obtained if the reasoning is in terms of system failure ($\bar{S}$) rather than system success (S). For a parallel logical arrangement, the event $\bar{S}$ (system failure) is an intersection of events $\bar{C}_k$, $k = 1, 2, \ldots, n$, denoting non-working states for the components, because the system will fail only if all of the components fail:

$$\bar{S} = \bar{C}_1 \cap \bar{C}_2 \cap \cdots \cap \bar{C}_n \tag{2.28}$$

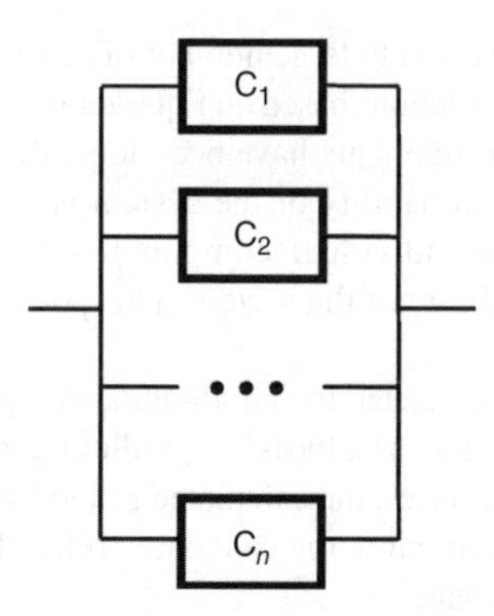

Figure 2.8 Components logically arranged in parallel

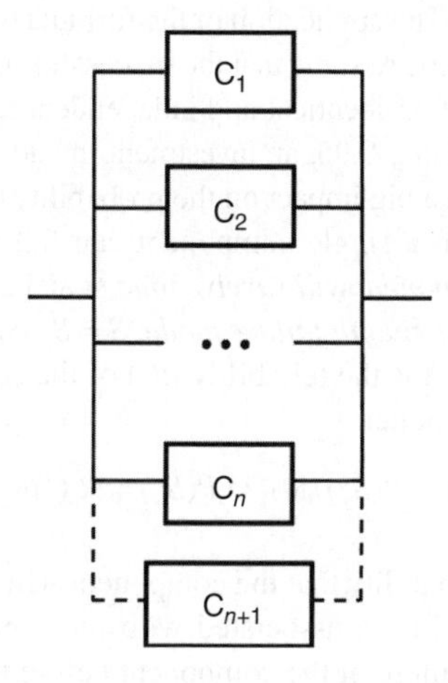

Figure 2.9 An extra component added to a parallel arrangement

The probability of system failure is $P(\bar{S}) = P(\bar{C}_1) \times P(\bar{C}_2) \times \cdots \times P(\bar{C}_n)$. Notice that while the reliability of a series arrangement is a product of the reliabilities of the components, the probability of failure of a parallel arrangement is a product of the probabilities of failure of the components.

Since the reliability of the system is $R = 1 - P(\bar{S})$ and the reliabilities of the components are R_i, $i = 1, 2, \ldots, n$, the reliability of the parallel arrangement (Bazovsky, 1961) becomes

$$R = 1 - (1 - R_1) \times (1 - R_2) \times \cdots \times (1 - R_n) \tag{2.29}$$

Two important conclusions can be made from this expression. *The larger the number of components logically arranged in parallel, the larger is the reliability of the system.*

Indeed, if an extra component C_{n+1} with reliability R_{n+1} is added as shown in Figure 2.9, the reliability of the arrangement becomes $R = 1 - (1 - R_1) \times \cdots \times (1 - R_n) \times (1 - R_{n+1})$, and since $R_{n+1} < 1$,

$$1 - (1 - R_1) \times \cdots \times (1 - R_n) < 1 - (1 - R_1) \times \cdots \times (1 - R_n) \times (1 - R_{n+1}) \tag{2.30}$$

The second conclusion is that *the reliability of the parallel arrangement is larger than the reliability of its most reliable component*. In other words, for a parallel arrangement, the relationship

$$1 - (1 - R_1) \times \cdots \times (1 - R_i) \times \cdots \times (1 - R_n) > R_i \tag{2.31}$$

holds. Indeed, because the inequality

$$(1-R_i)\left[1-(1-R_1)\times\cdots\times(1-R_{i-1})\times(1-R_{i+1})\times\cdots\times(1-R_n)\right]>0$$

is always true, the inequality (2.31) follows immediately from it after multiplying by $1–R_i$ and rearranging.

2.2.5 *Reliability of a System with Components Logically Arranged in Series and Parallel*

A system with components logically arranged in series and parallel can be reduced in complexity in stages, as shown in Figure 2.10. In the first stage, the components in parallel with reliabilities R_1, R_2 and R_3 are reduced to an equivalent component with reliability $R_{123}=1-(1-R_1)(1-R_2)(1-R_3)$ (Figure 2.10a). The components in parallel with reliabilities R_4 and R_5 are reduced to an equivalent component with reliability $R_{45}=1-(1-R_4)(1-R_5)$, and the components in series with reliabilities R_6, R_7 and R_8 are reduced to an equivalent component with reliability $R_{678}=R_6\times R_7\times R_8$ (Figure 2.10b). As a result, in the second stage, the equivalent reliability network *B* is obtained (Figure 2.10b). Next, the reliability network *B* is further simplified by reducing equivalent components with reliabilities R_{123} and R_{45} to a single equivalent component with reliability $R_{12345}=R_{123}\times R_{45}$. The final result is the trivial reliability network *C*, whose reliability is $R=1-(1-R_{12345})(1-R_{678})$.

For the system of two components logically arranged in series in Figure 2.11a, two different ways of increasing the reliability can be considered: (i) by including redundancy at a system level (Figure 2.11b) and (ii) by including redundancies at a component level (Figure 2.11c). The cost of two extra components is the same in each of the considered variants.

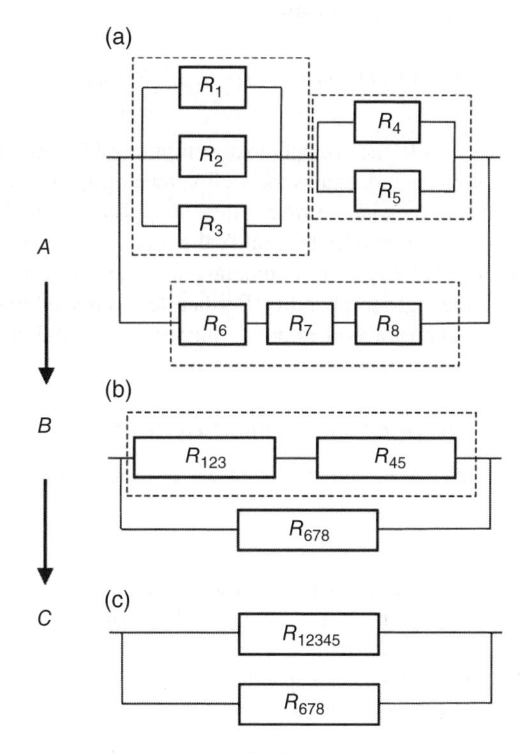

Figure 2.10 Reducing the complexity of a reliability network *A* by transforming it into intermediate networks *B* and *C*

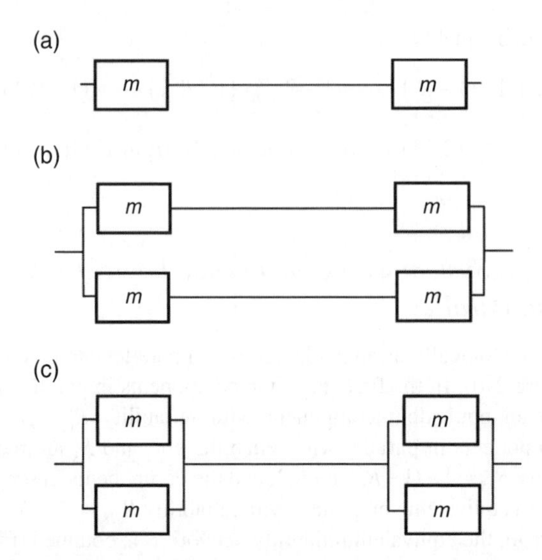

Figure 2.11 (a) A simple series arrangement and two ways (b and c) of increasing its reliability

The system reliabilities characterising the two variants can be compared. Arrangement (b) is characterised by reliability

$$R_1 = 1 - \left(1 - m^2\right)\left(1 - m^2\right) = m^2\left(2 - m^2\right) \tag{2.32}$$

while arrangement (c) is characterised by reliability

$$R_2 = \left[1 - (1 - m)(1 - m)\right]^2 = m^2(2 - m)^2 \tag{2.33}$$

Because $R_2 - R_1 = 2m^2(m-1)^2 > 0$, the arrangement in Figure 2.11b has a smaller reliability than that in Figure 2.11c. This example demonstrates the well known principle that *redundancy at a component level is more effective than redundancy at a system level* (Barlow and Proschan, 1965).

Note that no reliability data were necessary to identify the superior alternative. Comparing the algebraic expressions related to the reliability of the competing arrangements can be used for selecting the arrangement characterised by the highest reliability. Further details related to this approach are given in Chapter 14 which discusses comparative methods for improving reliability.

2.2.6 *Using Finite Sets to Infer Component Reliability*

Often, reliability of components can be inferred from basic operations over sets. This technique is illustrated by the next example.

Example

For a large batch of capacitors, it is known that 15% of the capacitors are out of tolerances, 7% have a fault (a short circuit or an open circuit fault) and 3% are both out of tolerances and with a fault. What is the probability that a selected capacitor will be within the tolerances and with no fault present?

Denoting with event A (out of tolerances capacitor) and with event B (faulty capacitor) for the probability of the union of these events, we have

$$P(A \cup B) = P(A) + P(B) - P(A \cap B)$$

where $P(A) = 0.15$, $P(B) = 0.07$ and $P(A \cap B) = 0.03$. The probability of the event $P(\bar{A} \cap \bar{B})$, a capacitor both within the tolerance limits and with no fault present, is given by

$$P(\bar{A} \cap \bar{B}) = 1 - P(A \cup B)$$

Substituting the numbers gives

$$P(\bar{A} \cap \bar{B}) = 1 - (0.15 + 0.07 - 0.03) = 0.81$$

2.3 Statistically Dependent Events and Conditional Probability in Reliability and Risk Modelling

Statistically dependent events are present if the outcome of one of the events affects the outcome of other events. Statistically dependent events in reliability and risk modelling can be illustrated by the next simple example.

The probability of the intersection of two statistically dependent events A and B is given by the product of the probability $P(A)$ of one of the events and the probability $P(B|A)$ of the second event given that the first has occurred:

$$P(A \cap B) = P(A)P(B \mid A) \tag{2.34}$$

Alternatively, $P(A \cap B) = P(B)P(A \mid B)$.

If the outcome of event A does not affect the outcome of event B, and vice versa, the events are statistically independent. In this case, $P(B \mid A) = P(B)$ and $P(A \mid B) = P(A)$ hold, and the probability of the intersection of the two events is given by the formula $P(A \cap B) = P(A)P(B)$ which is the probability of intersection of statistically independent events.

Example

In a batch of 10 capacitors, four are defective. Two capacitors are selected randomly from the batch and installed in an electronic device. Both capacitors must be non-defective for the electronic device to work properly. Determine the probability that two non-defective capacitors will be installed in the electronic device.

The electronic device will work properly only if both capacitors are non-defective. Let A denote the event *the first capacitor is non-defective* and B denote the event *the second capacitor is non-defective.* Event $\bar{A}$ denotes that the fist selected capacitor is defective.

The probability that both capacitors will be non-defective is given by Equation 2.34, where the probability $P(B \mid A)$ is the conditional probability of occurrence of event B, given that event A has occurred. In the example, $P(B \mid A)$ is the probability of selecting a non-defective second capacitor, given that the first selected capacitor is non-defective. Clearly, the probability $P(B \mid A)$ of selecting a non-defective second capacitor depends on the outcome of the first selection. If a non-defective first capacitor has been selected, the number of remaining non-defective capacitors is 5 and $P(B \mid A) = 5/9$. If a defective first capacitor has been selected, the number of remaining non-defective capacitors is 6, and the probability of selecting a second non-defective capacitor given that the first selected capacitor is faulty is $P(B \mid \bar{A}) = 6/9 = 2/3$. The dependency of the outcome of the second selection on the outcome of the first selection is denoted by $P(B \mid A)$ or $P(B \mid \bar{A})$.

The probability that two non-defective capacitors will be selected is given by the conditional probability formula:

$$P(A \cap B) = P(A)P(B \mid A) \tag{2.35}$$

Since $P(A) = 6/10 = 3/5$ and $P(B \mid A) = 5/9$,

$$P(A \cap B) = P(A)P(B \mid A) = (3/5) \times (5/9) = 1/3.$$

The probability that three non-defective capacitors will be selected is given by the probability of an intersection of three statistically dependent events:

$$P(A \cap B \cap C) = P(A)P(B \mid A)P(C \mid AB) \tag{2.36}$$

where $P(A) = 3/5$ is the probability that the first selected capacitor will be non-defective; $P(B \mid A) = 5/9$ is the probability that the second selected capacitor will be non-defective given that the first selected capacitor is non-defective and $P(C \mid AB) = 4/8 = 1/2$ is the probability that the third selected capacitor will be non-defective given that the first two selected capacitors are non-defective.

As a result, the probability that the three selected capacitors will be non-defective becomes

$$P(A \cap B \cap C) = P(A) \times P(B \mid A) \times P(C \mid AB) = \frac{3}{5} \times \frac{5}{9} \times \frac{1}{2} = \frac{1}{6}.$$

Equation 2.36 is easily generalised by induction for n statistically dependent events.

For a component exhibiting n failure modes which are not statistically independent, the reliability is equal to the product

$$R(t) = P(S_1) \times P(S_2 \mid S_1) \times P(S_3 \mid S_1 S_2) \times \cdots \times P(S_n \mid S_1 S_2 \cdots S_{n-1}) \tag{2.37}$$

where $P(S_k \mid S_1 S_2 \cdots S_{k-1})$ is the probability of surviving the kth failure mode given that no failure has been initiated by the first (S_1), the second (S_2), …, and the $(k-1)$th failure mode (S_{k-1}). An application of this formula to determining the reliability for two statistically dependent failure modes, 'failure initiated by individual flaws' and 'failure caused by clustering of flaws within a small critical distance', can be found in Todinov (2005).

The conditional probability $P(A \mid B)$ is determined by taking the ratio of the probability of simultaneous occurrence of events A and B to the probability of occurrence of the conditioning event B:

$$P(A \mid B) = \frac{P(A \cap B)}{P(B)} \tag{2.38}$$

Equation 2.38 gives the probability $P(A \mid B)$ of event A given that event B has occurred.

This relationship follows from a probabilistic reasoning based on a large number of sequential trials or observations. To calculate the conditional probability of event A, the number $N_{A \cap B}$ of simultaneous appearance of events A and B must be counted and divided to the number N_B of appearances of the conditioning event only:

$$P(A \mid B) = \frac{N_{A \cap B}}{N_B} \tag{2.39}$$

Equation 2.39 can also be presented as $P(A \mid B) = (N_{A \cap B}/N) : (N_B/N)$, where N is the total number of trials or observations. This representation yields Equation 2.38.

If event A is the conditioning event, Equation 2.38 becomes

$$P(B \mid A) = \frac{P(A \cap B)}{P(A)} \tag{2.40}$$

Equation 2.40 gives the probability $P(B \mid A)$ of event B given that event A has occurred.

There is an essential difference between conditional probability and probability not conditioned on the outcome of a particular event.

Let event A denote the event that a random demand of a particular resource, at any time, will be satisfied and $P(A)$ be the probability of satisfying the demand. Let event B denote the event that a random demand will occur during daytime, in the interval (6.00 am to 18.00 pm) ($P(B) = 1/2$). Event $\bar{B}$ denotes the event that a random demand will occur during night-time, in the interval (the period 18.00 pm to 6.00 am) ($P(\bar{B}) = 1/2$).

Suppose that during daytime the demand for the resource is high and the probability of obtaining the resource $P(A \mid B)$ is small. Conversely, the demand of the same resource during night-time (the period 18.00 pm to 6.00 am) ($P(B) = 1/2$) is very small, and the probability that a random demand will be satisfied is very high. Clearly, in this case, $P(A) > P(A \mid B)$, that is, the probability that the demand will be satisfied irrespective of the time of the day is greater than the probability that the demand will be satisfied given that it occurs during daytime.

In a number of cases, the conditional probability formula can be applied to extract important information such as in the next example.

Example

A detection method discovers a fault with probability 0.6, given that the fault is present in the component. If the component contains no fault, the method cannot classify it as faulty.

For a particular batch of components, the inspection method indicates that 20% of the components are faulty.

What is the actual percentage of faulty components in the batch?

Solution

Let A be the event 'the inspection method detects a faulty component' and B be the event 'the component is faulty'.

In 20% of the inspections, a fault has been present and the inspection method detects a faulty component. Rearranging the conditional probability formula (2.38) gives

$$P(B) = \frac{P(A \cap B)}{P(A \mid B)}$$

where $P(B)$ is the probability of a faulty component, $P(A \cap B) = 0.2$ is the probability that the component is faulty and the inspection method detects it as 'faulty' and $P(A|B) = 0.6$ is the probability that the detection method will indicate a faulty component given that the component is faulty. Substituting the number yields that approximately 33% of the components in the batch are faulty:

$$P(B) = \frac{0.2}{0.6} = 0.33$$

Conditional probabilities can be used in making correct risk estimates in critical situations, which can be illustrated by the following example.

Example

Each of the two independently working sensors survives one week of operation in high-temperature conditions with probability 0.7. The correct operation of at least one of the sensors is needed for controlling a particular chemical process.

At the end of the week, a test circuit indicates that at least one of the sensors has failed. What is the probability that the other sensor has also failed?

If A denotes the event 'the first sensor works' and B denotes the event that the second sensor works, the probability that at least one sensor has failed is

$$P(\bar{A} \cup \bar{B}) = P(\bar{A}) + P(\bar{B}) - P(\bar{A}) \times P(\bar{B}) = 0.3 + 0.3 - 0.3 \times 0.3 = 0.51$$

The probability that both sensors have failed, given that at least one sensor has failed is

$$\frac{P(\bar{A} \cap \bar{B})}{P(\bar{A} \cup \bar{B})} = \frac{0.3 \times 0.3}{0.51} = 0.176$$

As can be seen, the conditional probability that both sensors will be in a failed state given that at least one sensor has failed is almost twice the absolute probability $P(\bar{A} \cap \bar{B} = 0.3 \times 0.3 = 0.09$ that both sensors will be in a failed state at the end of the week. The knowledge that at least one of the sensors has failed increased significantly the likelihood that the other sensor has also failed. A decision based on this more pessimistic probability estimate, in the light of the indication given by the test circuit, is of superior quality compared to a decision based on the optimistic probability of 9%.

2.4 Total Probability Theorem in Reliability and Risk Modelling. Reliability of Systems with Complex Reliability Networks

The total probability theorem has a fundamental importance to reliability and risk modelling. Consider the following common engineering problem.

Electronic components are delivered by n suppliers $A_1, A_2, \ldots,$ and A_n. The market shares of the suppliers are $p_1, p_2, \ldots, p_n$, respectively, $\sum_{i=1}^{n} p_i = 1$. The probabilities characterising the separate suppliers, that the life of their electronic components will be greater than a required expected life of Y years, are $y_1, y_2, \ldots, y_n$, respectively. What is the probability that a purchased component will have expected life larger than Y years?

This problem can be solved easily using the total probability theorem. Let B denote the event *the purchased component will have expected life greater than* Y years and $A_1, A_2, \ldots, A_n$ denote the events *the component comes from the first, the second*, …, or *the nth supplier*, respectively. Event B occurs whenever any of the mutually exclusive and exhaustive events A_i occurs. Events A_i form a partition of the sample space (Figure 2.12): $A_i \cap A_j = \varnothing$ if $i \neq j$, $A_1 \cup A_2 \cup \cdots \cup A_n = \Omega$, and $P(A_1) + P(A_2) + \ldots + P(A_n) = 1$.

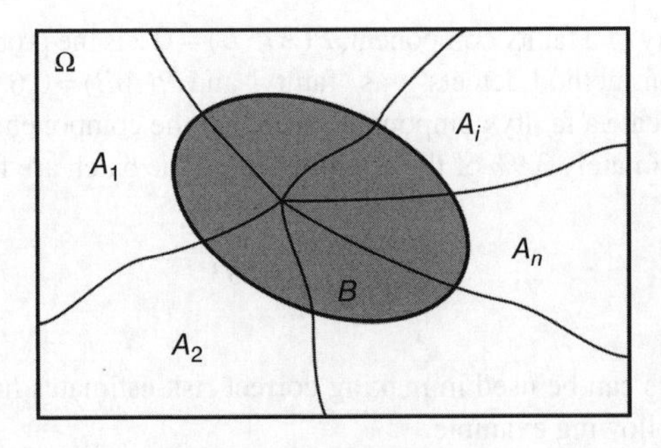

Figure 2.12 Venn diagram representing events related to the expected life of electronic components from n suppliers

The events $A_1 \cap B, A_2 \cap B, \ldots, A_n \cap B$ are also mutually exclusive. Their union constitutes event B:

$$B = (A_1 \cap B) \cup (A_2 \cap B) \cup \cdots \cup (A_n \cap B).$$

Because B is a union of mutually exclusive events, its probability is given by the sum of the probabilities of the mutually exclusive events:

$$P(B) = P(A_1)P(B \mid A_1) + P(A_2)P(B \mid A_2) + \cdots + P(A_n)P(B \mid A_n) \tag{2.41}$$

Equation 2.41 is known as the *total probability formula/theorem.*

In the considered specific example, the probabilities $P(A_i)$ are equal to the market shares p_i of the individual suppliers: $p_i = P(A_i)$, $i = 1, 2, \ldots, n$. Because the conditional probabilities $P(B \mid A_1) = y_1$, $P(B \mid A_2) = y_2, \ldots$, and $P(B \mid A_n) = y_n$ are known, according to the total probability formula (2.41), the probability that a purchased component will have expected life greater than Y years is

$$P(B) = p_1 y_1 + p_2 y_2 + \cdots + p_n y_n \tag{2.42}$$

From the total probability theorem, an important special case can be derived if the probability space is partitioned into two complementary events only. Let events A and $\bar{A}$ be complementary: $A \cap \bar{A} = 0$, $A \cup \bar{A} = \Omega$, $P(A) + P(\bar{A}) = 1$. Because these events partition the probability space, then either A occurs and B occurs or A does not occur and B occurs. The probability of event B according to the total probability theorem is

$$P(B) = P(B \mid A)P(A) + P(B \mid \bar{A})P(\bar{A}) \tag{2.43}$$

Equation 2.43 forms the basis of the *decomposition method* for solving reliability problems and will be illustrated by a generic engineering application.

Example

A comparator similar to the one considered in Chapter 1 includes four identical measuring devices (thermocouples, manometers, voltmeters, etc.) which measure a particular quantity (temperature, pressure and voltage) in two separate zones (A and B) of a component (Figure 2.13).

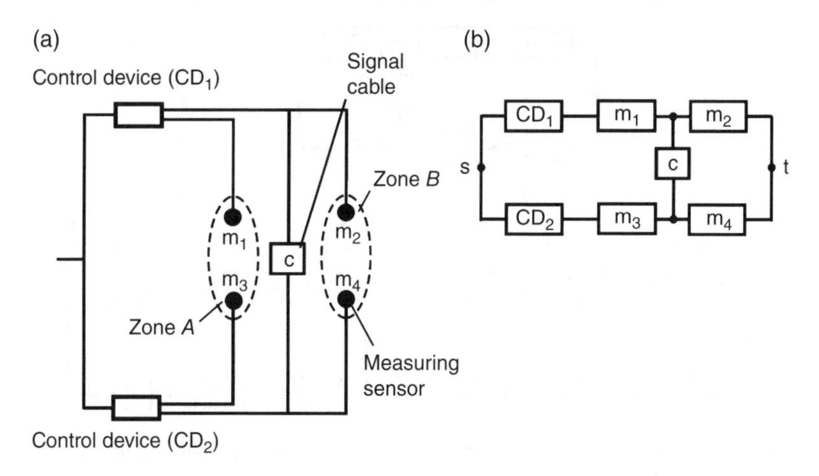

Figure 2.13 (a) A functional diagram and (b) reliability network of the generic comparator

There are also two identical control devices which compare the readings from the measuring devices and send a signal when a critical difference in the measured quantity is registered between the two zones.

Only one of the control devices is required to register a critical difference in the measured quantity from the two zones. Once a critical difference is registered, an alarm is triggered. A signal cable is used to transfer data between the control devices. The reliability on demand of each measuring device is m, the reliability on demand of the signal cable is c and the reliability of each control device is d. The question of interest is the probability that a signal for triggering the alarm will be sent in case of a critical difference in the measured quantity from zones A and B.

The logical arrangement of the components in the generic comparator can be represented by the reliability network in Figure 2.13b (see Chapter 1 for details regarding the functionality of the comparator and building its reliability network).

Clearly, the probability that there will be a signal if a critical difference in the measured quantity exists is equal to the probability of existence of a path through working components between the start node s and the terminal node t (Figure 2.13b). Let event S denote the event *the comparator is working on demand*, C be the event *the signal cable is working on demand* and $\bar{C}$ be the event *the signal cable is not working on demand.* Depending on whether the signal cable is working, the initial reliability network in Figure 2.13 decomposes into two reliability networks (Figure 2.14).

According to the decomposition method, the probability $P(S)$ that the comparator will be working on demand is equal to the sum of the probabilities that the comparator will be working, given that the signal cable is working and the probability that the comparator will be working given that the signal cable is not working (Figure 2.15):

$$P(S) = P(S|C)P(C) + P(S|\bar{C})P(\bar{C}) \tag{2.44}$$

Figure 2.14a and b give the reliability networks corresponding to events C, *the signal cable is working*, and $\bar{C}$, *the signal cable is not working*.

The reliability of the network in Figure 2.14a is given by

$$P(S|C) = \left[1-(1-dm)^2\right] \times \left[1-(1-m)^2\right] = m^2 d(2-md)(2-m),$$

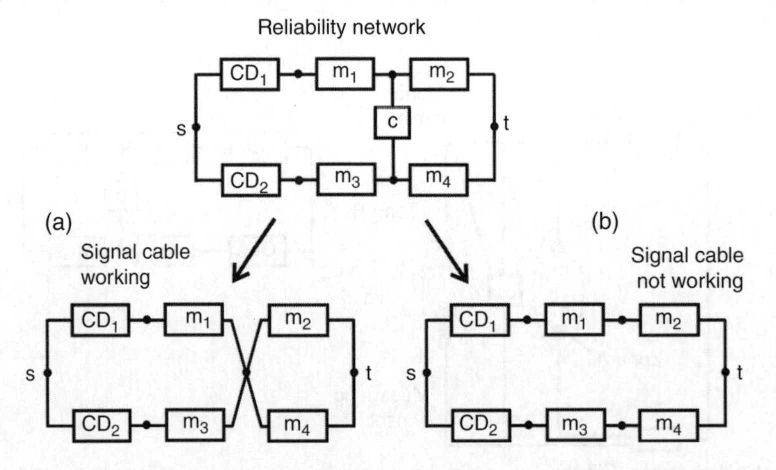

Figure 2.14 Depending on whether the signal cable is working, the initial reliability network of the comparator decomposes into two reliability networks (a) and (b)

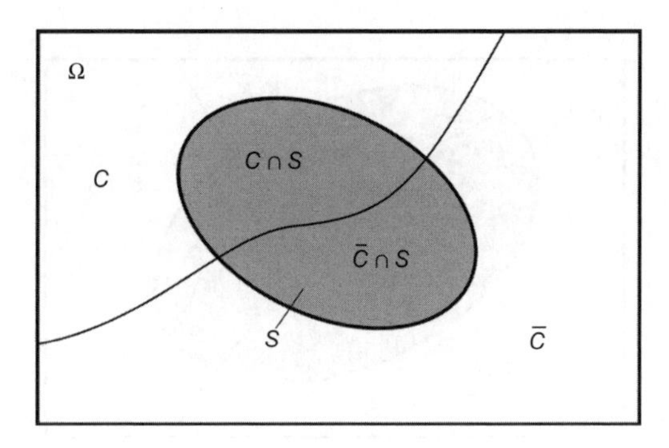

Figure 2.15 The event S (the comparator works on demand) is the union of two mutually exclusive events: $C \cap S$ and $\bar{C} \cap S$

while the reliability of the network in Figure 2.14b is given by

$$P(S \mid \bar{C}) = 1 - (1 - dm^2)^2 = dm^2(2 - dm^2)$$

Because $P(C) = c$ and $P(\bar{C}) = 1 - c$, the probability that the comparator will be working on demand becomes

$$P(S) = c \times [m^2 d(2 - md)(2 - m)] + (1 - c) \times [dm^2(2 - dm^2)] \tag{2.45}$$

If any of the product networks is a non-trivial network, by selecting another key component, the decomposition method could be applied to decompose the network into two simpler networks and so on, until trivial networks are reached whose reliability can be evaluated easily.

Alternatively, two key components K_1 and K_2 instead of one could be selected for the decomposition. The probability that the system will be in working state can be presented as a sum of the probabilities of four mutually exclusive events, which correspond to the four distinct combinations of states, related to the key components (Figure 2.16):

$$\begin{aligned} P(S) = {} & P(S \mid K_1 K_2)P(K_1)P(K_2) + P(S \mid K_1 \bar{K}_2)P(K_1)P(\bar{K}_2) \\ & + P(S \mid \bar{K}_1 K_2)P(\bar{K}_1)P(K_2) + P(S \mid \bar{K}_1 \bar{K}_2)P(\bar{K}_1)P(\bar{K}_2) \end{aligned} \tag{2.46}$$

where K_1 and $\bar{K}_1$ denote the working and failed state of the first key component and K_2 and $\bar{K}_2$ denote the working and failed state of the second key component.

Example

Evaluating reliability by a decomposition on the state of two selected key components will be illustrated with the system in Figure 1.17b whose reliability network is shown in Figure 2.17.

The system is in operation whenever on demand, a directed path exists from node s to each of the four terminal nodes t_1, t_2, t_3 and t_4 (Figure 2.17).

Despite that the system is rather complex, its reliability can be revealed easily by conditioning the reliability network in Figure 2.17, on the state of the control modules CM_1 and CM_2. There are four different, mutually exclusive outcomes: (i) control module 1 working, control module 2 working; (ii) control module 1 working, control module 2 not working; (iii) control module 1 not working, control module 2 working and (iv) control module 1 not working, control module 2 not working. Depending on these outcomes, the resultant networks are shown in Figure 2.18. The unreliable components have been replaced with unreliable edges whose reliability is equal to the reliability of the

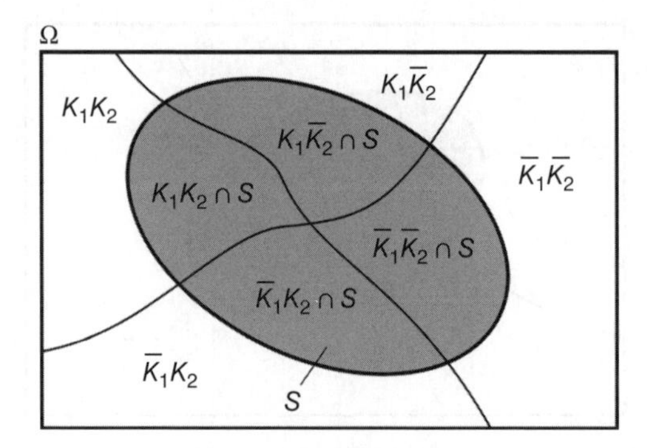

Figure 2.16 The event S (system is working on demand) can be presented as a union of four mutually exclusive events

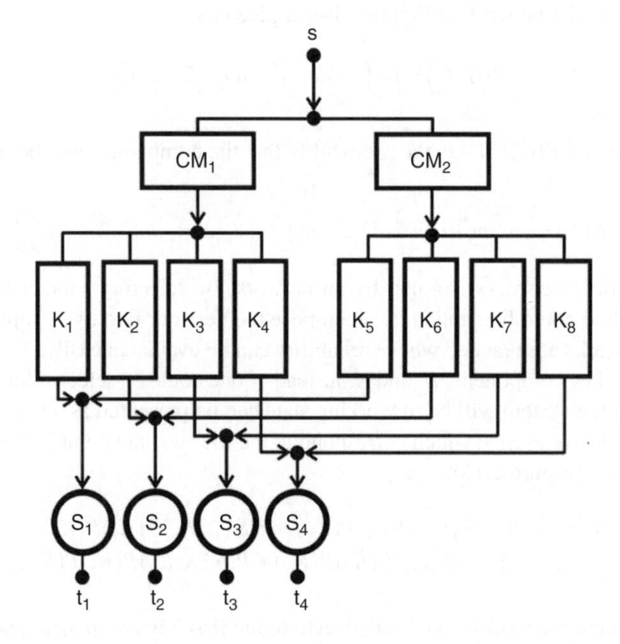

Figure 2.17 A dual control system including control modules CM_1 and CM_2, switches K_1–K_8 and operating devices S_1–S_4. The event S (system is working on demand) can be presented as a union of four mutually exclusive events

corresponding components. The reliability of the identical switches has been denoted by k while the reliability of the controlled devices by d. The reliability of each of the control modules is c.

The nodes are perfectly reliable and cannot fail.

The reliability network in Figure 2.18a corresponds to an outcome where both control modules CM_1 and CM_2 are working. As a series–parallel network, the probability of existence of directed paths to each of the terminal nodes is $d^4\left[1-(1-k)^2\right]^4$. The network in Figure 2.18b corresponds to each of the two outcomes where exactly one of the control modules is working. The network in Figure 2.18b is also a series–parallel network, for which the probability of existence of a path to each terminal node is k^4d^4. The network corresponding to the state where both control modules CM_1 and CM_2 are in a

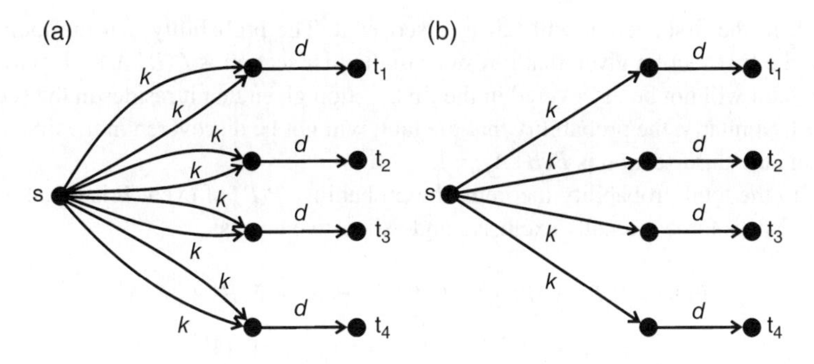

Figure 2.18 Basic resultant reliability networks as a result of the conditioning of the initial reliability network on the states of the control modules CM_1 and CM_2: (a) both control modules working and (b) exactly one of the control modules working

failed state has not been drawn. For this network, the start node s is isolated, with no path through working components to any of the terminal nodes.

Since the probability that a control module is working is c and the probability that a control module is not working is $1-c$, conditioning on the four distinct outcomes, formed by the state of the two control modules, yields

$$P(S)=c^2\times d^4\left[1-(1-k)^2\right]^4+c(1-c)\times k^4d^4+(1-c)c\times k^4d^4+(1-c)(1-c)\times 0,$$

which can be written as

$$P(S)=c^2d^4\left[1-(1-k)^2\right]^4+2c(1-c)\times k^4d^4$$

for the probability that, on demand, there will be directed paths from the start node to each of the terminal nodes.

In many cases, the total probability theorem is a useful device for evaluating the probabilities related to an event which cannot occur simultaneously in more than one place (the places of occurrence are mutually exclusive). This application will be illustrated with the following example.

Example

A system contains three identical sections. The probability of discovering a critical fault in any particular section if the fault is in the section is 0.8. The system has stopped operation due to the occurrence of a critical fault in one of the three sections. If the fault is equally likely to be in any of the three sections, calculate the probability that after inspecting the first section of the system, the fault will not be discovered there.

Solution

Let A_i $(i=1, 2, 3)$ denote the event the fault is in the ith section. Because the fault is equally likely to be in any of the three sections, $P(A_1)=P(A_2)=P(A_3)=1/3$.

Let B be the event 'after inspecting the first section, the fault will not be discovered there'.

This event is a union of the following three mutually exclusive and exhaustive events: (i) the fault is in the first section, and the search in the first section will fail to discover it; (ii) the fault is in the second section, and the search in the first section will fail to discover it; and (iii) the fault is in the third section, and the search in the first section will fail to discover it. The probability that the fault will not be discovered in the first section given that it resides in the first section is $P(B \mid A_1) = 1 - 0.8$. The probability that the fault will not be discovered in the first section given that it resides in the second section is $P(B \mid A_2) = 1$. Similarly, the probability that the fault will not be discovered in the first section given that it resides in the third section is $P(B \mid A_3) = 1$.

According to the total probability theorem, the probability $P(B)$ of event B is equal to the sum of probabilities of these three mutually exclusive and exhaustive events:

$$P(B) = P(B \mid A_1)P(A_1) + P(B \mid A_2)P(A_2) + P(B \mid A_3)P(A_3)$$
$$= (1 - 0.8) \times \left(\frac{1}{3}\right) + 1 \times \left(\frac{1}{3}\right) + 1 \times \left(\frac{1}{3}\right) = 0.73$$

In some important cases where the total probability is known, the total probability theorem can be used to estimate the unconditional probability of one of the events. Here is an example illustrating this application.

Example

A detection method indicates a fault with probability 0.7, given that the fault is present in the component and gives a false indication of a fault with probability 0.1, given that the fault is not present in the component. For a particular batch of components, the inspection method indicates that 35% of the components are faulty. What is the actual percentage of faulty components in the batch?

Solution

Let B denote the event *the method indicates a faulty component* and A denote the event *the component is faulty*.

Because events A and $\bar{A}$ are complementary, they partition the probability space, $P(A) + P(\bar{A}) = 1$ and $P(A \cap \bar{A}) = 0$. According to the total probability theorem,

$$P(B) = P(B \mid A)P(A) + P(B \mid \bar{A})P(\bar{A})$$

Because $P(\bar{A}) = 1 - P(A)$, this equation can also be presented as

$$P(B) = P(B \mid A)P(A) + P(B \mid \bar{A})[1 - P(A)]$$

The probabilities $P(B) = 0.35$, $P(B \mid A) = 0.7$ and $P(B \mid \bar{A}) = 0.1$ are known. The unknown probability is $P(A)$. From the last equation, the following expression is obtained for the unknown probability $P(A)$:

$$P(A) = \frac{P(B) - P(B \mid \bar{A})}{P(B \mid A) - P(B \mid \bar{A})}$$

Substituting the known values in this expression gives

$$P(A) = \frac{0.35 - 0.1}{0.7 - 0.1} = 0.417$$

There are approximately 42% faulty components in the batch.

2.5 Reliability and Risk Modelling Using Bayesian Transform and Bayesian Updating

2.5.1 Bayesian Transform

Suppose now that $A_i, i = 1,\ldots,n$ are n mutually exclusive and exhaustive events, where $P(A_i)$ are the prior probabilities of A_i before testing. B is an observation characterised by a probability $P(B)$. Let $P(B \mid A_i)$ denote the probability of the observation B, given that event A_i has occurred. From the definition of conditional probability (see also Appendix A),

$$P(A_i \mid B) = \frac{P(A_i \cap B)}{P(B)} = \frac{P(A_i)P(B \mid A_i)}{P(B)} \tag{2.47}$$

Since $P(B) = \sum_{i=1}^{n} P(A_i)P(B \mid A_i)$, the *Bayes' formula* (*Bayes' transform*) is obtained (DeGroot, 1989):

$$P(A_i \mid B) = \frac{P(A_i)P(B \mid A_i)}{\sum_{i=1}^{n} P(A_i)P(B \mid A_i)} \tag{2.48}$$

The Bayesian formula 2.48 is useful, because $P(B \mid A_i)$ is easier to calculate than $P(A_i \mid B)$. Its application in reliability and risk modelling will be illustrated by two basic examples. The first example is similar to an example related to diagnostic tests discussed by Parzen (1960).

Example

The probability of a critical fault in an electronic safety equipment is 0.01. It is known that in 90% of the cases, if a fault is present, an alarm will indicate the presence of the fault. It is also known that in 10% of the cases, where no fault is present, there will also be an alarm (a false alarm) indicating the presence of a fault.

What is the probability that a fault will be present given that there has been a fault alarm?

Solution

Let B denote the event *there is a fault alarm* and A the event *a fault is present*.

Because events A and $\bar{A}$ are complementary, they partition the probability space, $P(A) + P(\bar{A}) = 1$ and $P(A \cap \bar{A}) = 0$. According to the total probability theorem,

$$P(B) = P(B \mid A)P(A) + P(B \mid \bar{A})P(\bar{A})$$

From the Bayes' formula,

$$P(A \mid B) = \frac{P(B \mid A)P(A)}{P(B)} = \frac{P(B \mid A)P(A)}{P(B \mid A)P(A) + P(B \mid \bar{A})P(\bar{A})} \tag{2.49}$$

After substituting the numerical values, the probability

$$P(A \mid B) = \frac{0.90 \times 0.01}{0.90 \times 0.01 + 0.10 \times 0.99} = 0.083$$

is obtained. Note that in only about 8% of the cases where there has been a fault alarm a fault is actually present.

Exercise

In a large batch of seals, 30% of the seals are faulty. Given that a seal is fault-free, it passes a pressure test with probability 0.9. If the seal is faulty, this probability is only 0.4. Calculate the probability that from two seals, each of which has passed the pressure test, at least one of them will be faulty.

2.5.2 Bayesian Updating

Very often, prior belief or statistics exists about the distribution of a particular random parameter Θ. The possible values of the parameter Θ are $\theta_1, \theta_2, \ldots, \theta_n$ and the probabilities with which these values are accepted are $P(\theta_1), P(\theta_2), \ldots, P(\theta_n)$. This will be referred to as 'prior distribution' of the random parameter Θ.

The likelihoods $P(\eta \mid \Theta = \theta_i)$ related to a particular experimental outcome η given that the parameter Θ accepts a particular value θ_i $(i = 1,2,\ldots,n)$ are also known. If an outcome η of the experiment is available, the prior distribution of the parameter Θ can be updated by using *Bayesian updating*.

According to the total probability theorem, the probability $P(\eta)$ of the outcome η is given by

$$P(\eta) = \sum_{i=1}^{n} P(\eta \mid \theta_i) P(\theta_i) \tag{2.50}$$

Applying the Bayes' transform gives updated probability distribution of the parameter Θ:

$$P(\theta_i \mid \eta) = \frac{P(\eta \mid \theta_i) P(\theta_i)}{P(\eta)} \tag{2.51}$$

Reliability and risk modelling using Bayesian updating will be illustrated by an example.

Example

A supplier produces a particular electrical power device whose strength against lightning and switching surges is tested by a high-impulse voltage test. Because of uncontrolled variation in the manufacturing process, the supplier produces high-reliability batches which survive the high-impulse voltage test with probability 0.9 and low-reliability batches which survive the high-impulse voltage test with probability 0.6.

A batch of electrical power devices has been delivered to a customer, but it is not known whether the devices are from a high-reliability or a low-reliability batch. What is the probability that a particular batch will be a high-reliability batch, given that the device taken from the batch has survived the test?

Solution

The random discrete parameter Θ standing for the reliability of the batch can accept values θ_H or θ_L which stand for (i) high-reliability and (ii) low-reliability batch, correspondingly.

Because, before the test, it is unknown whether the batch is a high- or low-reliability batch, the probabilities of the events θ_H and θ_L that the device is from a high- or low-reliability batch can be assumed to be $P(\theta_H) = 0.5$ and $P(\theta_L) = 0.5$, respectively. Let η_1 denote the event 'the device has survived the high-voltage test'. After the test, the probabilities $P(\theta_i)$ can be modified (updated) formally through the Bayes' theorem, in the light of the outcome η_1 from the high-voltage test. Since the total probability of surviving the test is

$$P(\eta_1) = P(\eta_1 \mid \theta_H) P(\theta_H) + P(\eta_1 \mid \theta_L) P(\theta_L) = 0.9 \times 0.5 + 0.6 \times 0.5 = 0.75.$$

The probability $P(\theta_H | \eta_1)$ that the component comes from a high-reliability batch given that it has survived the test is

$$P(\theta_H | \eta_1) = \frac{P(\eta_1 | \theta_H)P(\theta_H)}{P(\eta_1)} = \frac{0.9 \times 0.5}{0.75} \approx 0.6$$

Thus, as result of the test, our confidence that the batch is a high-reliability batch increased from $P(\Theta = \theta_H) = 0.5$ to $P(\Theta = \theta_H | \eta_1) = 0.6$. The new belief about the probability distribution related to the reliability of the batch Θ is now $P(\Theta = \theta_H | \eta_1) = 0.6$ and $P(\Theta = \theta_L | \eta_1) = 0.4$. This is the posterior distribution of the parameter Θ, obtained after the outcome of the experiment η_1.

Suppose that a second experiment η_2 is made and the new tested device from the same batch also survives the high-impulse voltage test. The posterior distribution obtained after the first test ($P(\Theta = \theta_H | \eta_1) = 0.6$; $P(\Theta = \theta_L | \eta_1) = 0.4$) can now be used as a new prior distribution for the reliability of the batch Θ ($P(\Theta = \theta_H) = 0.6$; $P(\Theta = \theta_L) = 0.4$).

The total probability that a component from the same batch will survive the second test is

$$P(\eta_2) = P(\eta_2 | \theta_H)P(\theta_H) + P(\eta_2 | \theta_L)P(\theta_L) = 0.9 \times 0.6 + 0.6 \times 0.4 = 0.78.$$

The probability $P(\theta_H | \eta_1)$ that the batch of components is a high-reliability batch given that a second component from the batch has survived the test is

$$P(\Theta = \theta_H | \eta_2) = \frac{P(\eta_2 | \theta_H)P(\theta_H)}{P(\eta_2)} = \frac{0.9 \times 0.6}{0.78} \approx 0.69.$$

Thus, as result of the second test, our confidence that the batch of devices is a high-reliability batch increased from $P(\theta_H) = 0.6$ to $P(\theta_H | \eta_1) = 0.69$. The new belief about the probability distribution related to the reliability of the batch Θ after the second test is now $P(\Theta = \theta_H | \eta_2) = 0.69$ and $P(\Theta = \theta_L | \eta_2) = 0.31$. This is the posterior distribution of the reliability of the batch Θ obtained after the outcome from the second test η_2. This posterior distribution can now be assumed to be the new prior distribution, and further tests can be conducted.

In this way, the Bayesian updating technique provides an important mechanism for revising prior knowledge (belief) about the distribution of a particular parameter with the outcomes of experiments. The result is new, more precise knowledge about the distribution of the parameter and more precise assessment of reliability and risk.

3

Common Reliability and Risk Models and Their Applications

3.1 General Framework for Reliability and Risk Analysis Based on Controlling Random Variables

The factors controlling reliability are material strength, operating loads, dimensional design parameters, voltage, current, distributions of defects, residual stresses, service conditions (e.g. extremes in temperature) and environmental effects (e.g. corrosion). These factors are commonly associated with a great deal of uncertainty and they are essentially random factors.

Each random factor controlling reliability can be modelled by a discrete or a continuous random variable which will be referred to as *controlling random variable*. The controlling random variables can in turn be functions of other random variables. Strength, for example, is a controlling random variable which is a function of material properties, design configuration and dimensions:

$$Strength = F(material\ properties, design\ configuration, dimensions)$$

Modelling based on random variables is a powerful technique in reliability. Some of the properties of random variables and operations with random variables are discussed in Appendix B. The general algorithmic framework for reliability and risk analysis, based on random variables, can be summarised in the following steps:

- Identify all basic random factors controlling reliability and risk.
- Define controlling random variables corresponding to the basic factors.
- Select appropriate statistical models for the controlling random variables.
- Update the model parameters in the light of new observations.
- Build a reliability and risk model incorporating the statistical models of the controlling random variables.
- Test the quality of the model (e.g. by conducting sensitivity analysis).
- Solve the model using analytical or numerical techniques.
- Generate uncertainty bounds of the results predicted from the model, for example, by a Monte Carlo simulation or probability calculus.

Reliability and Risk Models: Setting Reliability Requirements, Second Edition. Michael Todinov.

3.2 Binomial Model

Consider the following common engineering example.

Example

A system supplying cooling fluid for a chemical reactor consists of five identical pumps connected in parallel as shown in Figure 3.1. The pumps work and fail independently from one another. The capacity of each pump is 10 litres per second, and the probability that a pump will be in working state on demand is 0.95. To control the chemical reaction, the pumps must supply at least 30 litres per second. What is the probability that on demand there will be at least 30 litres per second fluid supply?

Another similar example is the following.

Example

Suppose that a system for detecting a particular harmful chemical substance is composed of $n=5$ identical sensors, detecting a chemical release with probability $p=0.8$, independently of one another. In order to avoid a false alarm, at least $m=2$ sensors must detect the chemical release in order to activate a shut-down system (Figure 3.2). What is the probability that the shut-down system will be activated in case of a release of the harmful substance?

A common feature of these problems is:

- A fixed number of identical trials.
- Each trial results either in success (e.g. the component is working) or failure (e.g. the component is not working).
- All trials are statistically independent, that is, the probability that a component will be working does not depend on the state (e.g. working or failed) of other components.
- The probability of success in each trial (the probability that a component will be working) is the same.

These common features define the so-called *binomial experiment*. The number of successful outcomes from a binomial experiment is given by the *binomial distribution* – a probability distribution of fundamental importance.

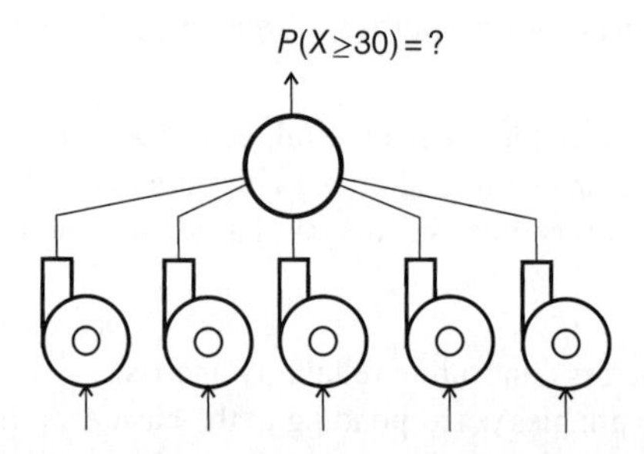

Figure 3.1 Functional diagram of the system supplying cooling fluid

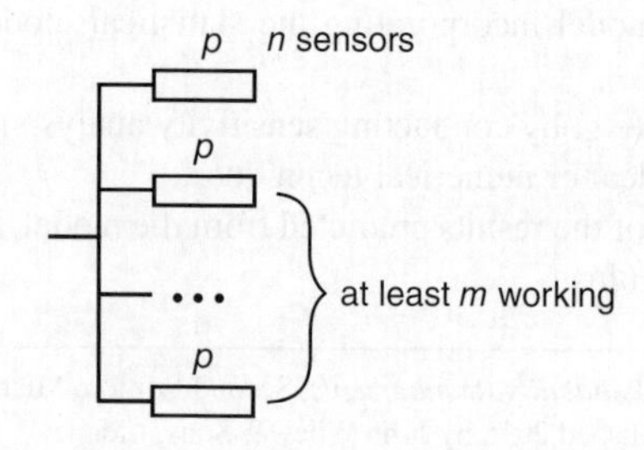

Figure 3.2 A system which works if at least m components work

In mathematical terms, if X is a discrete random variable denoting the number of successes during n trials, its distribution is given by

$$f(X=x)=\frac{n!}{x!(n-x)!}p^x(1-p)^{n-x} \tag{3.1}$$

$(x=0,1,2,\ldots,n)$, where $f(X=x)$ stands for the probability of exactly x successes out of n trials.

Suppose that there are $n=4$ statistically independent trials and the probability of success in each trial is p. Consider all sequences which lead to $X=2$ number of successes. There are six mutually exclusive sequences which lead to two successes out of four trials. If '1' denotes 'success' and '0' denotes 'failure', the mutually exclusive distinct sequences are 0011, 0101, 0110, 1001, 1010 and 1100. The number of these sequences is equal to the number of combinations of two out of four, which is equal to $4!/(2!\times 2!)=6$.

Because each success occurs with probability p, the probability of each distinct sequence is the same, equal to $p^2(1-p)^2$, because two successes are combined with two failures. At the end of any particular set of n trials, *only one sequence can occur*. Therefore, the separate sequences are mutually exclusive events. The probability of two successes out of n trials is equal to the probability of the union of all events leading to two successes out of n trials. However, the separate events (sequences) leading to two successes out of four trials are mutually exclusive, and the probability of their union is equal to the sum of their probabilities. Consequently, the probability of two successes out of four trials is equal to $6\times p^2(1-p)^2$.

This result is easily generalised for x successes out of n trials. Indeed, the probability of any particular sequence of x successes and $n-x$ failures is given by the product $p^x(1-p)^{n-x}$ of probabilities characterising n statistically independent events: x successes each characterised by a probability p and $n-x$ failures, each characterised by a probability $1-p$. There are $n!/[x!(n-x)!]$ distinct sequences yielding x successes and $n-x$ failures. The realisations of the particular sequences are mutually exclusive, that is, only one particular sequence can occur at the end of the n trials. The sum of the probabilities characterising all mutually exclusive sequences yielding x successes and $n-x$ failures is given by $[n!/(x!(n-x)!)]\times p^x(1-p)^{n-x}$ which is Equation 3.1.

In Figure 3.3, the binomial distribution has been illustrated by three binomial experiments, each of which involves 10 sequential, statistically independent trials. The binomial experiments are characterised by a constant probability of success in each trial, equal to $p=0.1$ for the first binomial experiment, $p=0.5$ for the second binomial experiment and $p=0.8$ for the third binomial experiment.

The probability of obtaining a number of successes greater than or equal to a particular number m is

$$P(X\geq m)=\sum_{x=m}^{n}\frac{n!}{x!(n-x)!}p^x(1-p)^{n-x} \tag{3.2}$$

Equation 3.2 gives in fact the sum of the probabilities of the following mutually exclusive events: 'exactly m successes at the end of the n trials', whose probability is

$$\frac{n!}{m!(n-m)!}p^m(1-p)^{n-m}$$

'exactly $m+1$ successes at the end of the n trials', whose probability is

$$\frac{n!}{(m+1)![n-(m+1)]!}p^{m+1}(1-p)^{n-(m+1)}$$

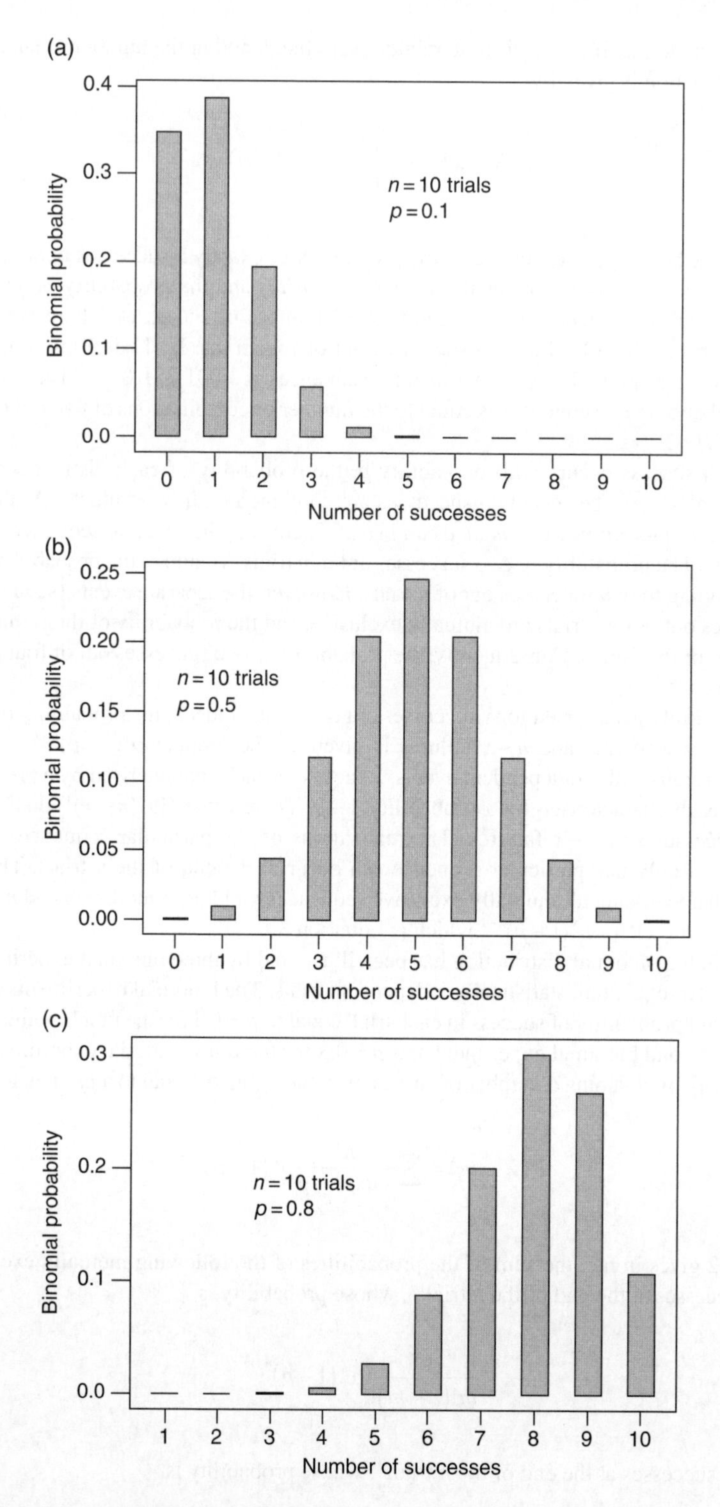

Figure 3.3 Binomial probability density distributions associated with three binomial experiments with the same number of trials $n=10$ and different probability of success p in a single trial: (a) $p=0.1$, (b) $p=0.5$ and (c) $p=0.8$

… and 'exactly n successes at the end of the n trials', whose probability is

$$\frac{n!}{n!(n-n)!}p^n(1-p)^{n-n} = p^n$$

(note that $0! = 1$).

The mean of a random variable X which follows a binomial distribution with a number of trials n and probability of success in each trial p is $E(X) = np$, and the variance is $V(X) = np(1-p)$ (Miller and Miller, 1999).

Going back to the 'chemical reactor example' stated that at the beginning of this chapter, the probability that the total amount of supplied cooling fluid on demand will be at least 30 l/s is equal to the probability that at least three pumps will be working at the time of demand. Substituting the numerical values $n=5$, $m=3$ and $p=0.95$ in Equation 3.2 results in

$$P(X \geq 3) = \sum_{x=3}^{5} \frac{5!}{x!(5-x)!} 0.95^x (1-0.95)^{5-x} \approx 0.9988$$

For a binomial experiment involving n trials, the probability that the number of successes will be smaller than or equal to a specified number $0 \leq r \leq n$ is given by the *binomial cumulative distribution function*

$$P(X \leq r) = \sum_{x=0}^{r} \frac{n!}{x!(n-x)!} p^x (1-p)^{n-x} \tag{3.3}$$

Equation 3.3 is in fact a sum of probabilities of the following mutually exclusive events at the end of the n trials: 'zero successes', characterised by a probability

$$\frac{n!}{0!(n-0)!}p^0(1-p)^{n-0} = (1-p)^n$$

'exactly one success', characterised by a probability

$$\frac{n!}{1!(n-1)!}p^1(1-p)^{n-1}$$

… and 'exactly r successes', characterised by a probability

$$\frac{n!}{r!(n-r)!}p^r(1-p)^{n-r}$$

Going back to the 'chemical sensors example', the number of sensors X detecting the toxic gas release can be modelled as an outcome from a binomial experiment with parameters $n = 5$ and $p = 0.80$. The probability of at least $m=2$ sensors working is $P(X \geq 2) = 1 - P(X \leq 1)$.

Since

$$P(X \leq 1) \equiv F(1) = \sum_{x=0}^{1} \frac{5!}{x!(5-x)!} 0.8^x (1-0.8)^{5-x} = 0.0067$$

the probability of detecting the gas leak is

$$P(X \geq 2) = 1 - P(X \leq 1) = 0.9933$$

3.2.1 Application: A Voting System

An important application of the binomial distribution is in constructing and evaluating *voting systems*.

Suppose that a component A receiving a particular input produces an error with probability p (Figure 3.4). The probability of an error can be reduced by creating a voting system. Voting is based on replicating the initial component A to n identical components, each of which receives the same input as the original component (Figure 3.4).

Each component operates independently from the others, and with probability p, the component produces an error in its output. All outputs from the separate components are collected by a voter device V (Figure 3.4). Suppose that the output of the voter device is determined by the majority vote of the components' outputs. In other words, in order for the voter to produce an error output, more than half of the components must produce an error output. For the special case of $n = 2k + 1$ identical components, at least $k+1$ outputs must be error outputs. The distribution of the number of error outputs X is given by the binomial distribution 3.1. The probability that the number or error outputs will be greater than or equal to $k+1$ is given by

$$P(X \geq k+1) = 1 - P(X \leq k) = 1 - \sum_{x=0}^{k} \frac{n!}{x!(n-x)!} p^x (1-p)^{n-x} \tag{3.4}$$

For $n = 11$ and $p = 0.1$, for example, the probability of erroneous output is

$$P(X \geq 6) = 1 - \sum_{x=0}^{5} \frac{n!}{x!(n-x)!} 0.1^x (1-0.1)^{n-x} = 0.0003 \tag{3.5}$$

As a result, the relatively high probability of an error output of $p = 0.1$ characterising a single component has been decreased 333 times by using a voting system.

Exercise

A particular manufacturer produces capacitors, 12% of which are defective. Five capacitors are purchased from this manufacturer.

What is the probability that among the purchased capacitors, there will be at least four non-defective capacitors?

What is the minimum number of capacitors that need to be purchased so that the probability that there will be at least four non-defective capacitors is at least 99%?

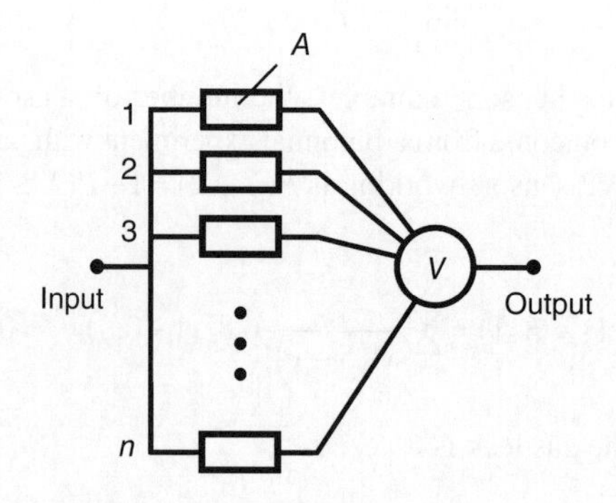

Figure 3.4 A voting system reduces significantly the probability of an error output

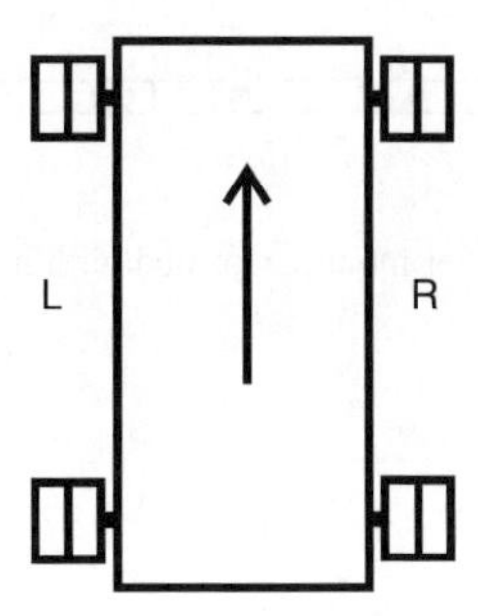

Figure 3.5 An eight-wheel vehicle which is operational if at least three tyres on the left side (L) and at least three tyres on the right side (R) are operational

Exercise

The eight-wheel vehicle on Figure 3.5 is on a mission. With a reasonable accuracy, the failures of the separate tyres can be considered to be statistically independent events. Each tyre has a chance of 20% to suffer failure during the mission. To be capable of travelling without stopping, at least three tyres on the left side and at least three tyres on the right side of the vehicle must be operational. Calculate the probability that the vehicle will fulfil its mission without stopping.

3.3 Homogeneous Poisson Process and Poisson Distribution

A binomial experiment is considered, where the number of trials n tends to infinity and the probability of success p in each trial tends to zero in such a way that the mean np of the binomial distribution remains finitely large. Assume that the trials in the binomial experiment are performed within n infinitesimally small time intervals with length Δ. A single experiment in each interval is performed, which results in either success or failure (Figure 3.6). The empty cells in Figure 3.6 correspond to 'failure', while the filled cells correspond to 'success'.

If the probability of success is equal to p and the number of cells is n, the expected number of successes is np. If λ is the number density of the successes, alternatively, the expected number of successes is also equal to λt, where λ is the number density of the successes per unit time interval. Consequently, the relationship

$$\lambda t = np$$

holds on the finite time interval $(0, t)$. The probability density function of the binomial distribution can be presented as

$$\begin{aligned} f(X=x) &= \frac{n!}{x!(n-x)!}\left(\frac{\lambda t}{n}\right)^x (1-p)^{n-x} \\ &= \frac{n(n-1)\cdots(n-x+1)}{n^x} \times \frac{(\lambda t)^x}{x!} \times \frac{(1-p)^n}{(1-p)^x} \end{aligned} \tag{3.6}$$

Since n tends to infinity and p tends to zero, $(1-p)^n = \exp[n\ln(1-p)] \approx \exp(-np) = \exp(-\lambda t)$. The number of successes x is finitely large; therefore, $(1-p)^x \approx \exp(-xp) \approx \exp(-0) = 1$.

Since

$$\lim_{n\to\infty}\left[\frac{n(n-1)\cdots(n-x+1)}{n^x}\right] = \lim_{n\to\infty}\left[\left(1-\frac{1}{n}\right)\left(1-\frac{2}{n}\right)\cdots\left(1-\frac{x-1}{n}\right)\right] = 1,$$

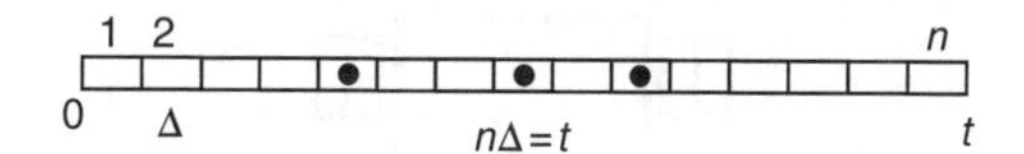

Figure 3.6 Trials of a binomial experiment performed within small time intervals with lengths Δ

finally,

$$f(X = x) = \frac{(\lambda t)^x \exp(-\lambda t)}{x!}, \quad x = 0,1,2,\ldots \tag{3.7}$$

is obtained for the probability of exactly x successes in n trials. This is the *probability density function of the Poisson distribution* describing the distribution of the number of occurrences from a *homogeneous Poisson process*. The homogeneous Poisson process is a limiting case of a binomial experiment with parameters n and $p = \lambda t / n$ when $n \to \infty$. In other words, a binomial experiment with a large number of trials n and a small probability of success p in each trial can be approximated reasonably well by a homogeneous Poisson process with intensity $\lambda = np / t$.

The homogeneous Poisson process is an important model for random events. It exists whenever the following conditions are fulfilled:

- The numbers of occurrences in non-overlapping intervals/regions are statistically independent.
- The probability of an occurrence in intervals/regions of the same size is the same and depends only on the size of the interval/region but does not depend on its location.
- The probability of more than one occurrence in a vanishingly small interval/region is negligible.

For a homogeneous Poisson process, the intensity is constant ($\lambda = \text{const}$) and so is the mean number λt of occurrences in the interval 0, t. Since the Poisson distribution is a limiting case of the binomial distribution, its mean $E(X) = np = \lambda t$ and its variance $V(X) = \lambda t$ are the same. Indeed, because if $n \to \infty$, $p \to 0$ and $np = \lambda t$, then $V(X) = np(1-p) \to \lambda t$.

If a homogeneous Poisson process with intensity λ is present, the distribution of the number of successes (occurrences) in the time interval (0, t) is given by the Poisson distribution (3.7). The probability of r or fewer occurrences in the finite time interval (0, t) is given by the *cumulative Poisson distribution function*:

$$P(X \le r) \equiv F(r) = \sum_{x=0}^{r} \frac{(\lambda t)^x}{x!} \exp(-\lambda t) \tag{3.8}$$

The Poisson process and the Poisson distribution are used frequently as statistical models (Thompson, 1988). The homogeneous Poisson process, for example, can be used as a model of randomly distributed defects in a spatial domain. In this case, the random occurrences are locations of defects. Various other applications of the homogeneous Poisson process are shown in Figure 3.7.

Example

Failures of a repairable system are described well by a homogeneous Poisson process with intensity five average number of failures per year (365 days).

What is the probability that the system will not fail in the next 24 hours?

Solution

The failure density is $\lambda = 5/365 = 0.0137$ day^{-1}. For the expected number of failures during 1 day of operation (24 hours), $\lambda t = 0.0137 \times 1 = 0.0137$ is obtained.

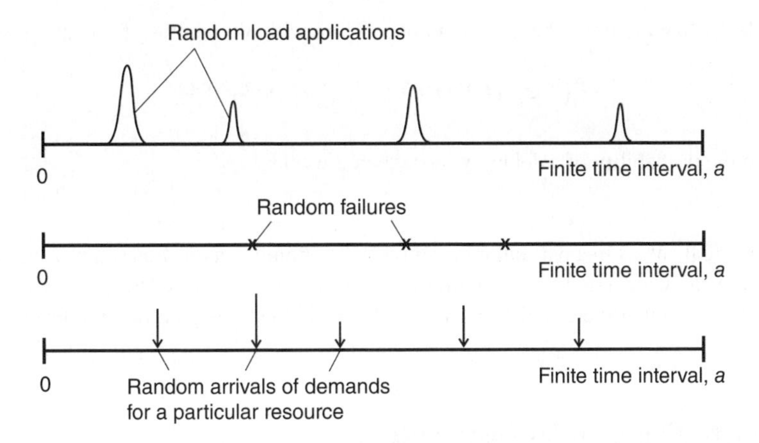

Figure 3.7 Various applications of the homogeneous Poisson process

According to Equation 3.7, the probability that the system will not fail in the next 24 hours is

$$f(0) = \frac{0.0137^0}{0!}\exp(-0.0137) = 0.986$$

Exercise

A system incorporates a number of identical electric motors which suffer random failures following a homogeneous Poisson process, with density three failures per year, characterising all electrical motors in operation. If an electric motor fails and a spare electric motor is available, the downtime for replacement of the electric motor is only 1 hour. If no spare electric motor is available, the ordering of a new electric motor and its shipping take more than 2 weeks during which time the system is not operating. To reduce the risk of delay in the system's operation, one electric motor is kept as a spare part. What is the probability that within 6 months of operation of the system, there will be a delay by more than 2 weeks to replace a failed electric motor?

Solution

The probability of a downtime of more than 2 weeks is equal to the probability of two or more than two failures during 6 months of operation. The probability $P(X \geq 2)$ that for a period of 6 months of operation, there will be two or more than two failures is equal to 1 – the probability $P(X \leq 1)$ that there will be no more than one failure: $P(X \geq 2) = 1 - P(X \leq 1)$.

For 6 months of operation, $\lambda t = (3/12)\times 6 = 1.5$.

The probability $P(X \leq 1)$ can be calculated from the cumulative Poisson distribution:

$$\begin{aligned}
P(X \leq 1) &= f(0) + f(1) \\
&= \frac{(\lambda t)^0 \exp(-\lambda t)}{0!} + \frac{(\lambda t)^1 \exp(-\lambda t)}{1!} \\
&= \exp(-\lambda t)\times\left(\frac{(\lambda t)^0}{0!} + \frac{(\lambda t)^1}{1!}\right) \\
&= \exp(-1.5)\times\left(1 + \frac{(1.5)^1}{1!}\right) \\
&= \exp(-1.5)\times(1+1.5) \approx 0.558
\end{aligned}$$

The probability that there will be more than one failure within 6 months of operation is

$$P(X \geq 2) = 1 - P(X \leq 1) = 1 - 0.558 = 0.44$$

This is also the probability of a delay by more than 2 weeks.

Exercise

Failures of identical valves in a particular system follow a homogeneous Poisson process with density 0.8 failures per year. Calculate the minimum number of spare valves needed to guarantee with probability of at least 95% that there will be a spare valve available after each failure, during 2.5 years of operation.

3.4 Negative Exponential Distribution

Suppose that the occurrences of the random events in any specified time interval with length t is a homogeneous Poisson process with density λ. The probability $f(x)$ of x occurrences in the interval 0, t is then given by the Poisson distribution:

$$f(X = x) = \frac{(\lambda t)^x}{x!} \exp(-\lambda t)$$

The probability that there will be no occurrences in the time interval (0, t) is obtained by substituting $x = 0$ (zero number of occurrences) in the above equation:

$$f(X = 0) = \frac{(\lambda t)^0}{0!} \exp(-\lambda t) = \exp(-\lambda t)$$

Consequently, the probability of at least one occurrence in the time interval (0, t) is given by $1 - \exp(-\lambda t)$. This is also the probability that the time T to the first occurrence will be smaller than t:

$$P(T \leq t) \equiv F(t) = 1 - \exp(-\lambda t) \tag{3.9}$$

As result, the distribution of the time to the first occurrence is given by the *negative exponential distribution* (3.9). If the times between the occurrences follow the negative exponential distribution, the number of occurrences follows a homogeneous Poisson process and vice versa.

Suppose that a component/system experiences random shocks which follow a homogeneous Poisson process with density λ. The time intervals between the shocks $t_1, t_2 - t_1, t_3 - t_2, \ldots$ in the finite time interval with length a (Figure 3.8) then follow the negative exponential distribution $F(t) = 1 - \exp(-\lambda t)$.

Suppose that each shock exceeds the strength of the component/system and causes failure. The reliability R associated with a finite time interval with length t is then equal to the probability $R = \exp(-\lambda t)$ that there will be no shocks within the specified time t. Consequently, the time to failure of the component will be given by the negative exponential distribution (3.9).

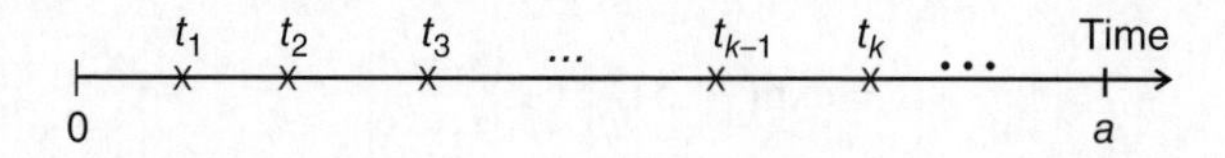

Figure 3.8 Times of successive failures in a finite time interval with length a

This is an important application of the negative exponential distribution for modelling the times to failure of components and systems which fail whenever a random load exceeds the strength of the component/system.

Another reason for the importance of the negative exponential distribution is that it is an approximate limit failure law for complex systems containing a large number of components which fail independently and whose failures lead to a system failure (Drenick, 1960).

The probability density function of the time to failure is obtained by differentiating the cumulative distribution function (3.9) with respect to time t $(f(t) = dF(t)/dt)$ (Figure 3.9):

$$f(t) = \lambda \exp(-\lambda t) \tag{3.10}$$

The negative exponential distribution applies whenever the probability of failure in a small time interval practically does not depend on the age of the component. It describes the distribution of the time to failure of a component/system characterized by a failure density λ.

3.4.1 *Memoryless Property of the Negative Exponential Distribution*

If the time to failure of a component follows the negative exponential distribution, the probability that the component will fail within a specified time interval is the same, irrespective of whether the component has been used for some time or has just been put in use. In other words, the probability that the life of the component will be greater than $t + \Delta t$, given that the component has survived time t, does not depend on the age t of the component. The component is as good as new.

Indeed, let A denote the event *the component will survive time* $t + \Delta t$. Let B denote the event *the component has survived time* t. The probability $P(A \mid B)$ that the component will survive time $t + \Delta t$ given that it has survived time t can be determined from the conditional probability formula $P(A \mid B) = P(A \cap B) / P(B)$:

$$P(T \geq t + \Delta t \mid T > t) = \frac{R(t + \Delta t)}{R(t)} = \frac{\exp[-\lambda(t + \Delta t)]}{\exp(-\lambda t)} = \exp(-\lambda \Delta t) \tag{3.11}$$

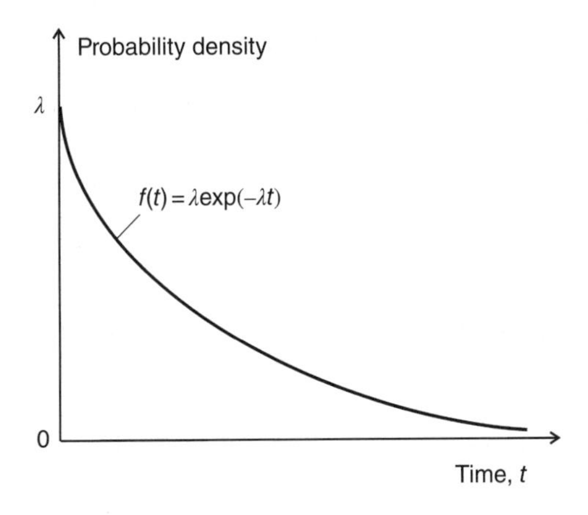

Figure 3.9 Probability density function of the negative exponential distribution

where $R(t) = \exp(-\lambda t)$ is the probability that the component will survive time t. From Equation 3.11, it follows that the probability that the component will survive a time interval $t + \Delta t$ given that it has survived time t is always equal to the probability that the component will survive the time interval Δt, starting from the start of operation ($t = 0$). This is the so-called *memoryless property* of the negative exponential distribution.

3.5 Hazard Rate

Suppose that the probability of failure in the elementary time interval $t, t + \Delta t$, given that the component has survived time t, depends on the age of the component and is given by $h(t)\Delta t$, where $h(t)$ will be referred to as *hazard rate*. Again, let the time interval t be divided into a large number n of small intervals with lengths $\Delta t = t / n$, such as shown in Figure 3.10.

Consider the nth time interval. By definition, the probability of failure in this interval, given that all previous time intervals have been survived, is $h_n \Delta t$, where h_n approximates the hazard rate in the nth time interval. Consequently, the probability of surviving the nth interval given that all previous time intervals have been survived is $1 - h_n \Delta t$. Let A denote the event 'surviving all $n-1$ elementary time intervals and B denote the event surviving the last (nth) elementary time interval. According to the formula $P(A \cap B) = P(A) \times P(B \mid A)$ expressing the probability of intersection of two dependent events A and B, the absolute probability of surviving the last time interval R_n is a product of the probability $P(A) = R_{n-1}$ of surviving $n-1$ elementary time intervals and the conditional probability $P(B \mid A) = 1 - h_n \Delta t$ of surviving the last elementary time interval, given that all previous $n-1$ elementary time intervals have been survived:

$$R_n = R_{n-1} \times (1 - h_n \Delta t) \tag{3.12}$$

The same reasoning can be applied to define R_{n-1} – the probability of surviving the $(n-1)$st elementary time interval

$$R_{n-1} = R_{n-2} \times (1 - h_{n-1} \Delta t) \tag{3.13}$$

Finally,

$$R_2 = R_1 \times (1 - h_2 \Delta t) \tag{3.14}$$

and

$$R_1 = (1 - h_1 \Delta t) \tag{3.15}$$

are obtained, where $h_{n-1}, h_{n-2}, \ldots, h_1$ approximate the hazard rate function $h(t)$ in the corresponding small time intervals with lengths Δt.

Substituting back Equation 3.15 into Equation 3.14 and so on, until a substitution into Equation 3.12 is finally made, gives

$$R(t) \equiv R_n = (1 - h_1 \Delta t)(1 - h_2 \Delta t) \times \cdots \times (1 - h_n \Delta t) \tag{3.16}$$

for the probability R_n that the component will survive the time interval $(0, t)$.

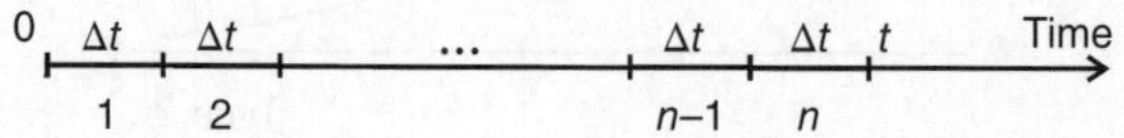

Figure 3.10 A time interval $(0, t)$ divided into n small intervals with lengths Δt

Equation 3.16 can also be presented as

$$R(t) = \exp\{\ln[(1-h_1\Delta t)(1-h_2\Delta t)\times\cdots\times(1-h_n\Delta t)]\} = \exp\left[\sum_{i=1}^{n}\ln(1-h_i\Delta t)\right]$$

For $\Delta t \to 0$, $\ln[1-h_i\Delta t] \approx -h_i\Delta t$ and considering that for $\Delta t \to 0$, $\sum_{i=1}^{n} h_i\Delta t \to \int_0^t h(v)dv$, Equation 3.16 becomes

$$R(t) = \exp\left(-\int_0^t h(v)dv\right) \tag{3.17}$$

where ν is a dummy integration variable. The integral $H(t) = \int_0^t h(v)dv$ in Equation 3.17 is also referred to as *cumulative hazard rate*. Using the cumulative hazard rate, reliability can be presented as (Barlow and Proschan, 1975)

$$R(t) = \exp(-H(t)) \tag{3.18}$$

Reliability $R(t)$ can be increased by decreasing the hazard rate $h(t)$, which decreases the value of the cumulative hazard rate $H(t)$. Correspondingly, the cumulative distribution of the time to failure becomes

$$F(t) = 1 - \exp(-H(t)) \tag{3.19}$$

If the hazard rate $h(t)$ increases with age, the cumulative distribution of the time to failure is known as an *increasing failure rate* (IFR) distribution. Alternatively, if the hazard rate $h(t)$ is a decreasing function of t, the cumulative distribution of the time to failure is known as a *decreasing failure rate* (DFR) distribution.

If the hazard rate is constant, the negative exponential distribution is obtained from Equation 3.17. The negative exponential distribution is a *constant failure rate* (CFR) distribution. Indeed, for $h(t) = \lambda = \text{const}$, the cumulative hazard rate becomes

$$H(t) = \int_0^t \lambda dv = \lambda t \tag{3.20}$$

and the cumulative distribution function of the time to failure is given by the negative exponential distribution (3.9). Consequently, the time to failure of a component is given by the negative exponential distribution if the probability $\lambda\Delta t$ that a component/system will fail within the elementary time interval $t, t+\Delta t$, given that the component has survived time t, does not depend on the age t of the component.

This is a realistic assumption which holds for many electrical and mechanical components which have passed an initial period of work but have not yet entered a stage of wearout and degradation. Failures of such components are caused by random factors (e.g. random overstress) whose frequency does not depend on the age of the component.

The hazard rate can also be presented as a function of the probability density $f(t)$ of the time to failure. Indeed, the probability of failure in the time interval $t, t+\Delta t$ is given by $f(t)\Delta t$, which is equal to the probability $R(t)\times h(t)\Delta t$ of the compound event that the component will survive time t and after that will fail in the small time interval $t, t+\Delta t$ (Figure 3.11).

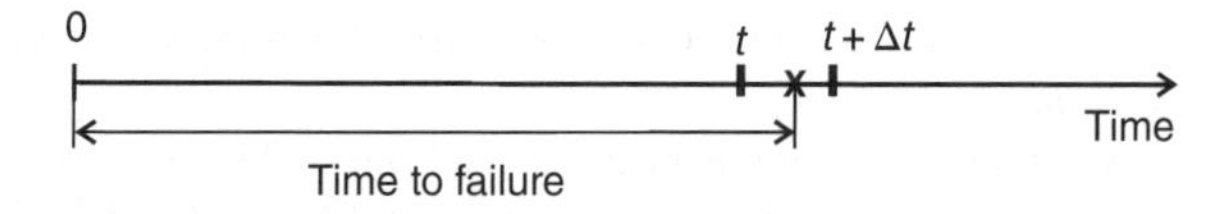

Figure 3.11 Time to failure in the small time interval $t, t+\Delta t$

Equating the two probabilities results in

$$f(t)=R(t)h(t) \tag{3.21}$$

from which

$$h(t)=\frac{f(t)}{R(t)} \qquad 3.22$$

Since $f(t)=-R'(t)$, where $R(t)$ is the reliability function, Equation 3.22 can also be presented as

$$\frac{R'(t)}{R(t)}=-h(t) \tag{3.23}$$

Integrating both sides of Equation 3.23 gives

$$\ln R(t)=-\int_0^t h(v)dv+C$$

Applying the initial condition $R(0)=1$ yields $C=0$ for the integration constant and expressing $R(t)$ from the above equation yields Equation 3.17, which *provides an important link between the reliability associated with the time interval (0, t) and the hazard rate function.*

Example
What is the expected number of failures among 100 electrical components during 1000 hours, if each component has a linear hazard rate function $h(t)=10^{-6}t$?

Solution
According to Equation 3.17,

$$R(t)=\exp\left(-\int_0^t h(v)dv\right)=\exp\left(-10^{-6}\int_0^t v\,dv\right)=\exp\left(-10^{-6}\frac{t^2}{2}\right)$$

For $t=1000$ hours, the reliability is $R(1000)=\exp(-1/2)\approx 0.61$. Since the probability of failure is $F(1000)=1-R(1000)=0.39$, the expected number of failures among the 100 components is $100\times 0.39=39$.

3.5.1 Difference between Failure Density and Hazard Rate

There exists a fundamental difference between the failure density $f(t)$ and the hazard rate $h(t)$. Consider an initial population of N_0 items. The proportion $\Delta n/N_0$ of items from the initial number N_0 that will fail within the time interval $t,t+\Delta t$ is given by $\Delta n/N_0=f(t)\Delta t$. *This is also the absolute probability of failure in the time interval $t,t+\Delta t$ of a randomly selected item from the initial population of N_0 items.*

Suppose that $N(t)$ gives the number of items in service at time t. The proportion $\Delta n/N(t)$ of items in service that will fail in the time interval $t,t+\Delta t$ is given by $\Delta n/N(t)=h(t)\Delta t$. *This is the probability of failure in the time interval $t,t+\Delta t$ of a randomly selected item from the population of $N(t)$ items* *which have survived time t*. In this sense, the probability of failure $h(t)\Delta t$ in the time interval $t, t+\Delta t$ is a conditional probability (conditional on surviving the time t). The probability of failure of an item is related to the items in service, not to the initial population of items and this constitutes the fundamental difference between the hazard rate and the failure density.

If age has no effect on the probability of failure, the hazard function $h(t)$ will be constant ($h(t)=\lambda=\text{const}$), and the same proportion $dn(t)/n(t)$ of items in service is likely to fail within the interval $t, t+dt$. Because this proportion is also equal to $-\lambda dt$, the relationship

$$\frac{dn(t)}{n(t)}=-\lambda dt$$

holds, where $dn(t)$ and dt are infinitesimal quantities. After integrating within limits $(0, t)$, the equation

$$n(t)=n_0 \exp(-\lambda t) \tag{3.24}$$

is obtained, where n_0 is the initial number of items. The probability of survival of time t is therefore given by $n(t)/n_0=R(t)=\exp(-\lambda t)$. Consequently, the probability of failure is obtained from $F(t)=1-n(t)/n_0=1-\exp(-\lambda t)$ which is the negative exponential distribution.

3.5.2 Reliability of a Series Arrangement Including Components with Constant Hazard Rates

The reliability of a system with components logically arranged in series is $R=R_1\times R_2\times\cdots\times R_n$ where $R_1=\exp(-\lambda_1 t)$, …, $R_n=\exp(-\lambda_n t)$ are the reliabilities of n components with constant hazard rates λ_1, λ_2, …, λ_n. The failures of the components are statistically independent. Substituting the component reliabilities in the system reliability formula results in

$$R=\exp(-\lambda_1 t)\times\exp(-\lambda_2 t)\times\cdots\times\exp(-\lambda_n t)=\exp(-\lambda t) \tag{3.25}$$

where $\lambda=\sum_{i=1}^{n}\lambda_i$. As a result, the hazard rate of the system is a sum of the hazard rates of the separate components. The times to failure of such a system follow a homogeneous Poisson process with intensity $\lambda=\sum_{i=1}^{n}\lambda_i$. This additive property is the theoretical basis for the widely used *parts count method* for predicting system reliability (Bazovsky, 1961; MIL-HDBK-217F, 1991). The method is suitable for systems including independently working components, logically arranged in series, where failure of any component causes a system failure. If the components are not logically arranged in series, the system hazard rate $\lambda=\sum_{i=1}^{n}\lambda_i$ calculated on the basis of the parts count method is an upper bound of the real system hazard rate. One downside of this approach is that the reliability predictions are too conservative.

3.6 Mean Time to Failure

An important reliability measure is the *mean time to failure* (MTTF), which is the average time to the first failure. Because the absolute probability of failure in an elementary time interval $t, t+dt$ is given by $f(t)dt$, where $f(t)\geq 0$ is the failure density function, the MTTF can be obtained from the integral

$$\text{MTTF}=\int_0^\infty t f(t)dt \tag{3.26}$$

The integral from Equation 3.26 yields the following inequality

$$\int_0^\infty t f(t)dt=\int_0^t v f(v)dv+\int_t^\infty v f(v)dv\geq\int_0^t v f(v)dv+t\int_t^\infty f(v)dv \tag{3.27}$$

where ν is a dummy integration variable. Considering that $R(t) = \int_t^{\infty} f(v)dv$ holds, expression (3.27) can also be written as

$$\text{MTTF} - \int_0^t v f(v) dv \geq tR(t) \tag{3.28}$$

For $t \to \infty$, $\lim\left(\int_0^t vf(v)dv\right) = \text{MTTF}$. In addition, $tR(t) \geq 0$ for $t \to \infty$. Consequently, it can be inferred that for $t \to \infty$, $\lim[tR(t)] = 0$.

The integral in Equation 3.26 can now be presented as $\text{MTTF} = -\int_0^{\infty} t\, dR(t)$ which, after integrating by parts, gives

$$\text{MTTF} = -\left[tR(t)\right]_0^{\infty} + \int_0^{\infty} R(t)dt$$

Considering that, for $t = 0$, $tR(t) = 0$ and, for $t \to \infty$, $\lim[tR(t)] = 0$, the MTTF becomes

$$\text{MTTF} = \int_0^{\infty} R(t)dt \tag{3.29}$$

For a constant hazard rate $\lambda = \text{const}$,

$$\text{MTTF} = \theta = \int_0^{\infty} \exp(-\lambda t)dt = \frac{1}{\lambda}$$

that is, for a negative exponential time to failure distribution, the MTTF is the reciprocal of the hazard rate.

The MTTF from the last equation is valid only for failures characterised by a constant hazard rate. In this case, the probability that a failure will occur earlier than the MTTF is approximately 63%. Indeed,

$$P(T \leq \text{MTTF}) = 1 - \exp(-\lambda\, \text{MTTF}) = 1 - \exp(-1) \approx 0.63$$

Example

The MTTF of an electronic component characterised by a constant hazard rate is MTTF = 50 000 hours. Calculate the probabilities of the following events:

(i) The electronic component will survive continuous service for 1 year.
(ii) The electronic component will fail between the fifth and the sixth year.
(iii) The electronic component will fail within a year given that it has survived the end of the fifth year. Compare this probability with the probability that the component will fail within a year given that it has survived the end of the 10th year.

Solution

(i) Since MTTF = 50000 hours = 5.7 years, the hazard rate of the component is $\lambda = 1/5.7$ years^{-1}. Reliability is determined from $R(t) = \exp(-\lambda t)$, and the probability of surviving 1 year is

$$R(1) = P(T > 1) = e^{-1/5.7} \approx 0.84$$

(ii) The probability that the component will fail between the end of the fifth and the end of the sixth year can be obtained from the cumulative distribution function of the negative exponential distribution:

$$P(5 < T \leq 6) = F(6) - F(5) = \exp(-5/5.7) - \exp(-6/5.7) \approx 0.07$$

(iii) Because of the memoryless property of the negative exponential distribution, the probability that the electronic component will fail within a year, given that it has survived the end of the fifth year, is equal to the probability that the component will fail within a year after having been put in use:

$$P(5 < T \leq 6 \mid T > 5) = P(0 < T \leq 1) = 1 - \exp(-1/5.7) \approx 0.16$$

Similarly, the probability that the component will fail within a year given that it has survived the end of the 10th year is obtained from

$$P(10 < T \leq 11 \mid T > 10) = P(0 < T \leq 1) = 1 - \exp(-1/5.7) \approx 0.16$$

This probability is equal to the probability from the previous example (as it should be) because of the memoryless property of the negative exponential distribution.

3.7 Gamma Distribution

Consider k components whose failures follow a negative exponential time to failure distribution with constant hazard rate λ and a system built with these components. Such is the *k-fold standby system* in Figure 3.12 which consists of k components with identical negative exponential time to failure distributions. Component c_i is switched in immediately after the failure of the working component c_{i-1}. The system fails when all components fail. Because the time to failure of the components follows a negative exponential distribution with hazard rate λ, the number of component failures in the interval $(0,t)$ follows a homogeneous Poisson process with intensity λ.

The distribution of the time to failure of this system coincides with the distribution of the time to k component failures and can be derived using the following probabilistic argument. The probability $F(t)$ that there will be a system failure before time t is equal to $1 - R(t)$, where $R(t)$ is the probability that there will be fewer than k component failures before time t. The compound event *fewer than k failures before time t* is composed of the following mutually exclusive events: *no failures before time t, exactly one failure before time t, …, exactly k − 1 failures before time t*. The probability $R(t)$ is then a sum of the probabilities of these mutually exclusive events:

$$R(t) = \exp(-\lambda t) + \frac{(\lambda t)^1}{1!}\exp(-\lambda t) + \frac{(\lambda t)^2}{2!}\exp(-\lambda t) + \cdots + \frac{(\lambda t)^{k-1}}{(k-1)!}\exp(-\lambda t) \qquad (3.30)$$

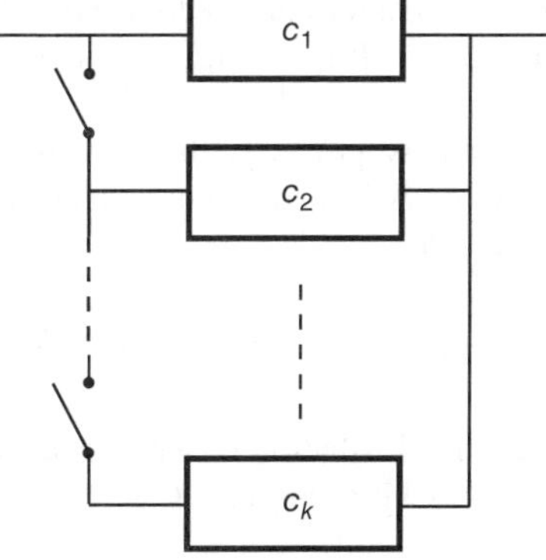

Figure 3.12 A k-fold standby system

Equation 3.30 gives the reliability of a k-fold standby system with perfect switching.

The probability of at least k failures before time t is given by $F(t) = 1 - R(t)$. For unlimited number of standby components,

$$F(t) = 1 - R(t) = \frac{(\lambda t)^k}{k!} \exp(-\lambda t) + \frac{(\lambda t)^{k+1}}{(k+1)!} \exp(-\lambda t) + \cdots \tag{3.31}$$

Differentiating $F(t)$ in Equation 3.31 with respect to t gives the *probability density function of the time to the* kth *failure*:

$$\frac{dF}{dt} = f(t) = \frac{\lambda(\lambda t)^{k-1} \exp(-\lambda t)}{(k-1)!} \tag{3.32}$$

Denoting the MTTF by $\theta = 1/\lambda$, Equation 3.32 becomes

$$f(t) = \frac{t^{k-1} \exp(-t/\theta)}{\theta^k (k-1)!}$$

which is the *gamma density function* $G(\alpha, \beta)$,

$$G(\alpha, \beta) = \frac{t^{\alpha-1} \exp(-t/\beta)}{\beta^\alpha \Gamma(\alpha)} \tag{3.33}$$

with parameters $\alpha = k$ and $\beta = \theta$ ($\Gamma(k) = (k-1)!$; Abramowitz and Stegun, 1972).

As a result, the sum of k statistically independent random variables following the negative exponential distribution with parameter λ follows a gamma distribution $G(k, 1/\lambda)$ with parameters k and $1/\lambda$. The mean $E(X)$ and the variance $V(X)$ of a random variable X following the gamma distribution $G(k, 1/\lambda)$ are $E(X) = k/\lambda$ and $V(X) = k/\lambda^2$, respectively.

If the time between failures of a repairable device follows the negative exponential distribution with MTTF $\theta = 1/\lambda$, the probability density of the time to the kth failure is given by the gamma distribution $G(k, \theta)$ (Eq. 3.33). Here is an alternative formulation: if k components with identically distributed lifetimes following a negative exponential distribution are characterised by a MTTF θ, the sum of the times to failure of the components $T = t_1 + t_2 + \cdots + t_k$ follows a gamma distribution $G(k, \theta)$ with parameters k and θ.

The negative exponential distribution is a special case of a gamma distribution in which $k = 1$. Another important special case of a gamma distribution with parameters k and $\theta = 2$ is the χ^2-distribution $G(k, 2)$ with $2k$ degrees of freedom.

Values of the χ^2-statistics for different degrees of freedom can be found in the χ^2-distribution table in Appendix D. In the table, the area α cut off to the right of the abscissa (Figure 3.13) is given with the relevant degrees of freedom n.

Gamma distributions have an additivity property: the sum of two random variables following gamma distributions $G(k_1, \theta)$ and $G(k_2, \theta)$ is a random variable following a gamma distribution $G(k_1 + k_2, \theta)$ with parameters $k_1 + k_2$ and θ:

$$G(k_1, \theta) + G(k_2, \theta) = G(k_1 + k_2, \theta) \tag{3.34}$$

An important property of the gamma distribution is that if a random variable X follows a gamma distribution $G(k, \theta)$, the product $C \times X$ with a constant C follows the gamma distribution $G(k, C\theta)$ (Grosh, 1989).

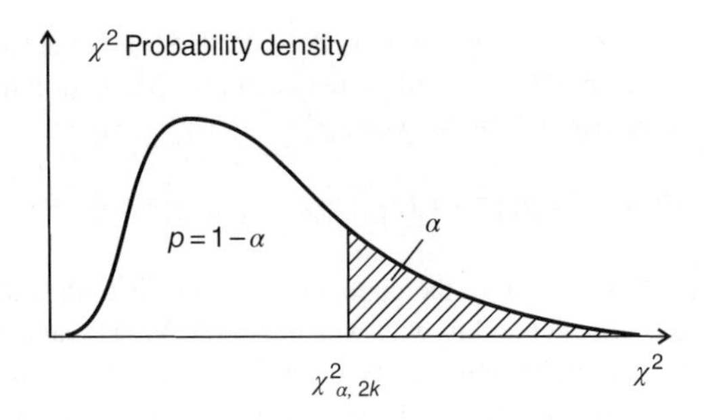

Figure 3.13 Probability density function of the χ^2-distribution

According to this property, since the sum of k times to failure $T = t_1 + t_2 + \cdots + t_k$ follows a gamma distribution $G(k, \theta)$, the distribution of the estimated MTTF $\hat{\theta} = T / k = (t_1 + t_2 + \cdots + t_k) / k$ follows the gamma distribution $G(k, \theta/k)$. Therefore, the quantity $2k\hat{\theta} / \theta$ will follow a χ^2-distribution $G(k, 2)$ with $n = 2k$ degrees of freedom.

An interesting application of the *gamma distribution* is for modelling the distribution of the time to failure of a component, subjected to shocks whose arrivals follow a homogeneous Poisson process with intensity λ. If the component is subjected to partial damage or degradation by each shock and fails completely at the kth shock, the distribution of the time to failure of the component is given by the gamma distribution $G(k, 1/\lambda)$.

Another application of the Gamma distribution will be discussed in the next section related to determining the uncertainty associated with the MTTF determined from a limited number of failure times.

3.8 Uncertainty Associated with the MTTF

Suppose that components characterised by a constant hazard rate have been tested for failures. After failure, the components are not replaced, and the test is truncated on the occurrence of the kth failure, at which point T component hours will have been accumulated. In other words, the total accumulated operational time T includes the sum of the times to failure of all k components. The MTTF can be estimated by dividing the total accumulated operational time T (the sum of all operational times) to the number of failures k:

$$\hat{\theta} = \frac{T}{k} \tag{3.35}$$

where $\hat{\theta}$ is the estimator of the unknown MTTF.

If θ denotes the true value of the MTTF, according to the previous section, the expression $2k\hat{\theta} / \theta$ follows a χ^2-distribution with $2k$ degrees of freedom, where the end of the observation time is at the kth failure. Therefore, the expression $2T/\theta$ follows a χ^2*-distribution*. This property can be used to determine a lower limit for the MTTF which is guaranteed with a specified probability (confidence level). Assume that a bound is required for which a statement is to be made that the true MTTF θ is greater than the bound with probability p. The required bound is obtained by finding the value $\chi^2_{\alpha,2k}$ of the statistics χ^2 which cuts off an area α from the right tail of the χ^2-distribution (Figure 3.13). The required bound is obtained from $\theta^* = 2T / \chi^2_{\alpha,2k}$ where $\alpha = 1 - p$ and $\chi^2_{\alpha,2k}$ is the value of the χ^2-statistics for the selected confidence level $p = 1 - \alpha$ and degrees of freedom $n = 2k$. The probability that the MTTF θ will be greater than the lower bound θ^* is equal to the probability that the χ^2-statistics will be smaller than $\chi^2_{\alpha,2k}$:

$$P(\theta \geq \theta^*) = P(\chi^2 \leq \chi^2_{\alpha,2k}) = 1 - \alpha = p \tag{3.36}$$

Suppose now that $\chi^2_{\alpha 1,2k}$ and $\chi^2_{\alpha 2,2k}$ correspond to two specified bounds θ_1 and θ_2 ($\theta_1 < \theta_2$). The probability $P(\theta_1 \le \theta \le \theta_2)$ that the true MTTF θ will be between the specified bounds θ_1 and θ_2 is equal to the probability that the χ^2-statistics will be between $\chi^2_{\alpha 2,2k}$ and $\chi^2_{\alpha 1,2k}$ (Figure 3.14):

$$P(\theta_1 \le \theta \le \theta_2) = P\left(\chi^2_{\alpha 2,2k} \le \chi^2 \le \chi^2_{\alpha 1,2k}\right) = \alpha 2 - \alpha 1 \tag{3.37}$$

Another question is related to estimating the confidence level which applies to the estimate given by Equation 3.35. In other words, the probability with which the true MTTF value θ will be greater than the estimate $\hat{\theta} = T / k$ is required. This probability can be determined if a value of the χ^2-statistics is calculated first from $\chi^2 = 2T / \hat{\theta}$. Next, for $n=2k$ degrees of freedom from the table in Appendix D, it is determined what value of α gives $\chi^2_{\alpha,2k}$ equal to the calculated $\chi^2 = 2T / \hat{\theta}$. The value $1-\alpha$ is the confidence level with which it can be stated that the true MTTF is greater than the estimate $\hat{\theta}$ (Smith, 1972).

In cases where two bounds are required, between which the MTTF lies with a specified probability p, two values of the χ^2-statistics are needed.

The lower limit $\chi^2_{1-\alpha/2,2k}$ of χ^2 (Figure 3.15) is used to determine the upper confidence limit $\theta_U = 2T / \chi^2_{1-\alpha/2,2k}$ of the confidence interval for the MTTF, while the upper limit $\chi^2_{\alpha/2,2k}$ (Figure 3.15) is used to determine the lower bound $\theta_L = 2T / \chi^2_{\alpha/2,2k}$. The probability that the true MTTF θ will be between θ_L and θ_U is equal to the probability that χ^2 will be between $\chi^2_{1-\alpha/2,2k}$ and $\chi^2_{\alpha/2,2k}$:

$$P(\theta_L \le \theta \le \theta_U) = P\left(\chi^2_{1-\alpha/2,2k} \le \chi^2 \le \chi^2_{\alpha/2,2k}\right)$$

This probability is equal to the specified confidence level

$$p = 1 - \left(\frac{\alpha}{2} + \frac{\alpha}{2}\right) = 1 - \alpha$$

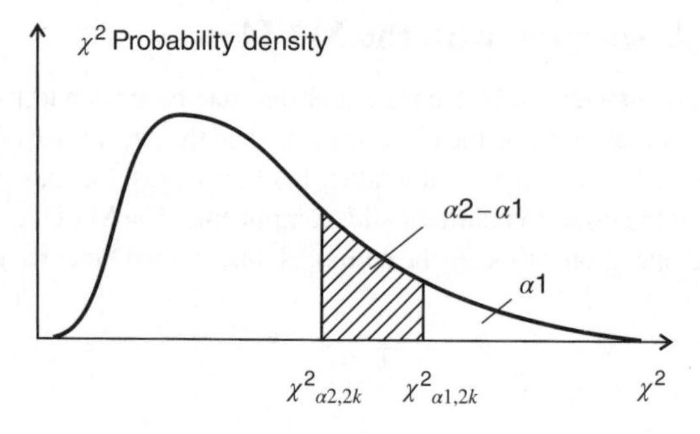

Figure 3.14 The hatched area $\alpha 2 - \alpha 1$ gives the probability that the χ^2-statistics will be between $\chi^2_{\alpha 2,2k}$ and $\chi^2_{\alpha 1,2k}$

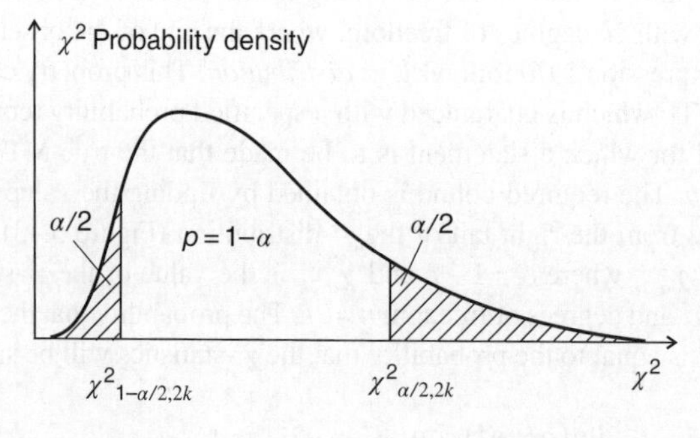

Figure 3.15 Two limits of the χ^2-statistics necessary to determine a confidence interval for the MTTF

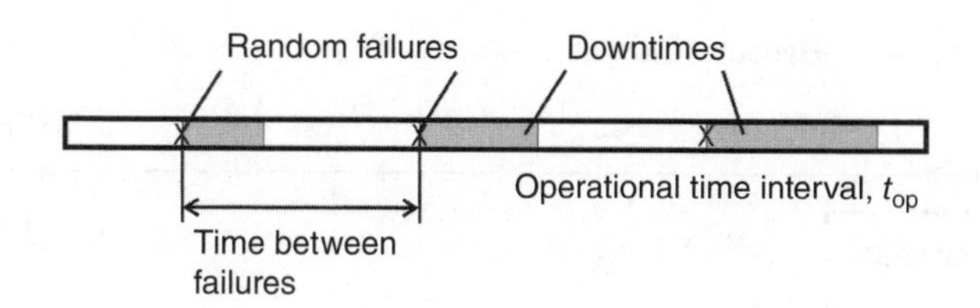

Figure 3.16 Mean time between failures

3.9 Mean Time between Failures

The *mean time between failure* (MTBF) reliability measure is defined for repairable systems. Assume that the failed component is restored to as good as new condition. Let U_i be the duration of the ith operational period (uptime) and D_i be the repair time (downtime) needed to restore the system after failure to as good as new condition (Figure 3.16).

It is assumed that all uptimes U_i come from a common parent distribution and all downtimes D_i come from another common parent distribution. It is also assumed that the sums $X_i = U_i + D_i$ are statistically independent. The MTBF is the expected (mean) value of the sum $X_i = U_i + D_i$ (Trivedi, 2002):

$$\text{MTBF} = E(U_i + D_i) = E(U_i) + E(D_i) \tag{3.38}$$

where $E(U_i)$ and $E(D_i)$ are the expected values of the uptimes and the downtimes correspondingly. Since MTTF = $E(U_i)$ and MTTR = $E(D_i)$, where MTTR is the mean time to repair, the MTBF becomes

$$\text{MTBF} = \text{MTTF} + \text{MTTR} \tag{3.39}$$

Equation 3.39 illustrates the difference between the MTTF and MTBF.

The MTBF can also be approximated by the ratio of the length of the operational time interval t_{op} and the number of failures N_{f} during the operational time:

$$\text{MTBF} \approx \frac{t_{\text{op}}}{N_{\text{f}}} \tag{3.40}$$

3.10 Problems with the MTTF and MTBF Reliability Measures

It must be pointed out that for failures characterised by a non-constant hazard rate, the MTTF reliability measure can be very misleading because a large MTTF can be obtained from failure data where increased frequency of failure (low reliability) at the start of life is followed by very large periods of failure-free operation (high reliability). This will be illustrated by the following numerical example.

Example

Consider components of the same type but from two different manufacturers.

Manufacturer 1 (times to failure, days)
5, 21, 52, 4131, 8032, 12 170 and 16 209

$$\text{MTTF}_1 = \frac{5 + 21 + 52 + 4131 + 8032 + 12170 + 16209}{7} \approx 5803 \text{ days}$$

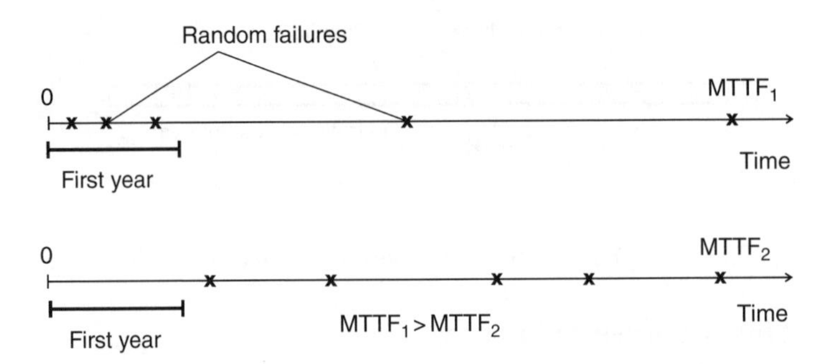

Figure 3.17 Two components of the same type with different MTTFs

Manufacturer 2 (times to failure, days)
412, 608, 823, 1105, 1291 and 1477

$$\mathrm{MTTF}_2 = \frac{412 + 608 + 823 + 1105 + 1291 + 1477}{6} \approx 953 \text{ days}$$

Suppose that reliable work is required during the first year (Figure 3.17). If the component is selected solely on the basis of its MTTF, a component from the first manufacturer, with $\mathrm{MTTF}_1 = 5803$ days will be selected which has a smaller reliability in the first year (Figure 3.17).

This argument is also valid for the MTBF reliability measure. The MTBF reliability measure is misleading for repairable systems characterised by non-constant failure rates.

3.11 BX% Life

An alternative reliability measure which mitigates to some extent the problem with the MTTF measure is the *BX% life. This is the time at which X% of the items put in operation will have failed* (Figure 3.18). This reliability measure is widespread in the ball bearing and roller bearing production where the B10 life is used as a reliability metric. B10 life is the time at which 10% of the population put in operation will have failed.

As can be seen, a high failure frequency (low reliability) at the start of operation of a product will be indicated correctly by a shorter BX% life.

The BX% reliability measure incorporates failures from both the infant mortality region and the useful life region of the bathtub curve. Unlike the MTTF, for the same specified level $X\%$ of the probability of premature failure, a larger BX% life means always a larger reliability.

Finally, the BX% life reliability measure is intuitively clear and more comprehensible to practitioners not trained in probabilistic reasoning.

The problem with the BX% life is its restricted definition based on *a population of items* put in operation. This reliability measure works well for mass-produced mechanical and electronic components. For unique complex systems (e.g. production systems) which are not replicated, the concept percentage of failed population has no real meaning. In addition, the concept percentage of failed components has no meaning for complex systems with redundancy. For a system with 10 components logically arranged in parallel, 90% of the components can fail and the system will still be operational. The BX% reliability measure *is not connected with the concept 'critical failure' discussed in the next section*, and for this reason, it is unsuitable as a system reliability measure.

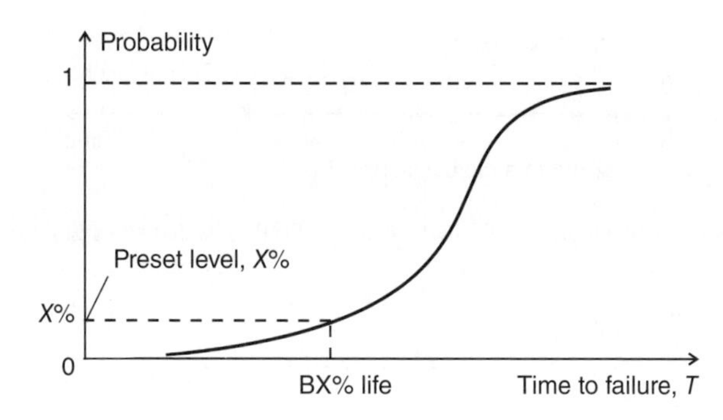

Figure 3.18 BX% life is defined as the time at which $X\%$ of the population of products put in operation will have failed

In the next section, a reliability metric is introduced which is suitable for both mass-produced components and systems and unique systems which are not replicated.

3.12 Minimum Failure-Free Operation Period

Reliability can be interpreted as the probability of surviving a specified *minimum failure-free operation period* (*MFFOP*) without a 'critical failure'. A *critical failure* is a component failure which causes the product/system to stop functioning. Failure of the power supply unit of electromechanical equipment is a typical example of a critical failure.

A *non-critical failure* is a component failure which does not cause the system to stop functioning. A typical example of non-critical failure is the failure of a redundant component or the failure of a minor non-redundant component. In both cases, the component failure does not cause a system failure.

Within the specified MFFOP, there may be only non-critical failures of redundant components which do not cause a system failure (Figure 3.19). The idea behind the MFFOP is to guarantee with a high probability absence of critical failures (causing system to stop functioning). Guaranteeing a specified MFFOP with high probability is equivalent to guaranteeing a large reliability associated with the specified MFFOP time interval.

Section 3.11 demonstrated that for a distribution of the time to failure different from the negative exponential distribution, the MTTF and MTBF reliability measures *can be very misleading*. The component/system with the larger MTTF/MTBF is not necessarily the component/system associated with the larger probability of surviving the specified time interval. This is because a large MTTF/MTBF can result from aggregating times to failure reflecting increased failure frequency at the start of life and very low failure frequency later in life.

Instead of determining the probability of surviving a specified minimum failure-free operation period, a preset level α can be specified ($0 < \alpha < 1$), and the minimum failure-free operation period MFFOP_α corresponding to this level can be determined.

MFFOP_α *is an alternative reliability measure. It is the time within which the probability of a critical failure does not exceed a preset level* α. In other words, $MFFOP_\alpha$ is guaranteed with probability $1 - \alpha$ (Figure 3.20).

The MTTF/MTBF reliability measures are non-misleading only for constant hazard rates. In other words, the correct use of the MTTF/MTBF measures is limited to the useful region of the bathtub curve (see the discussion related to bathtub curve) where the underlying time to failure distribution is the negative exponential distribution.

Unlike the MTTF and MTBF, the MFFOP corresponding to a preset level α is a powerful reliability measure which *does not depend on the underlying time to failure distribution*.

The MFFOP reliability measure has an advantage to the BX life measure because it is valid for both mass-produced components and unique systems which are not replicated.

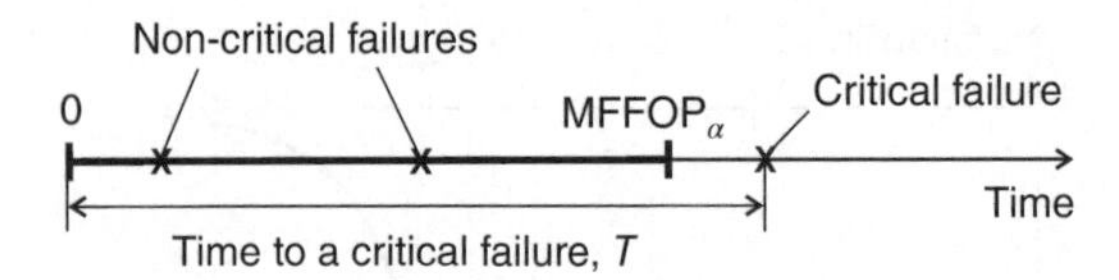

Figure 3.19 Guaranteeing with a high probability $1-\alpha$ an MFFOP$_\alpha$ of specified length, free from critical failures

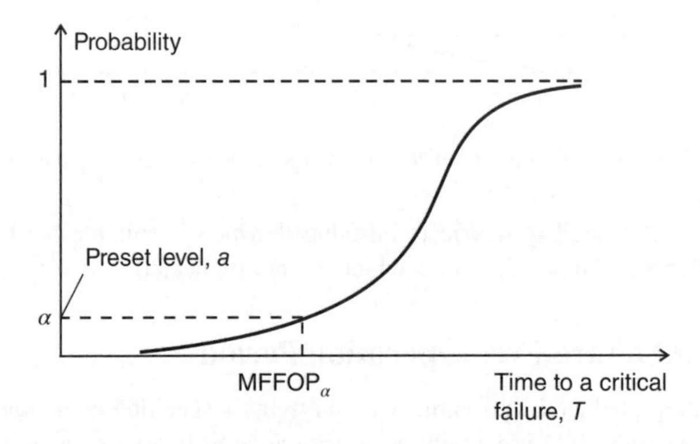

Figure 3.20 The MFFOP$_\alpha$ is the time within which the probability of a critical failure does not exceed a preset level α

An MFFOP of 1.5 years at a preset level of $\alpha = 0.05$ essentially states that with probability 95%, the system/component will survive 1.5 years of continuous operation without a critical failure. The larger the reliability of the system, the larger the MFFOP at a preset level. If a common preset level α has been specified, the minimum failure-free operation periods MFFOP$_{\alpha,i}$ characterising different systems can be compared, and the system with the largest MFFOP$_\alpha$ selected.

Example

A numerical example using the MFFOP measure can be given with a component for which the time to failure follows the negative exponential distribution

$$F(t) = 1 - \exp\left(-\frac{t}{30}\right)$$

where time t is measured in years.

For a preset level $\alpha = 0.05$, solving the equation with respect to the time t yields an MFFOP$_{0.05}$ of 1.54 years (MFFOP$_{0.05} = 1.54$). This is the time period within which the probability of failure does not exceed 5%.

3.13 Availability

3.13.1 Availability on Demand

The probability that the demand at a specified point in time will sample a working state for the component with index i is given by

$$\alpha_i = \frac{\text{MTTF}_i}{(\text{MTTF}_i + \text{MTTR}_i)} \tag{3.41}$$

where α_i denotes *the availability on demand* of the *i*th component, MTTF_i is the mean time to failure of the component and MTTR_i is its mean time to repair. *The availability on demand for a component stands for the expected fraction of time the component is in working state.*

A system composed of components in series is in working state if all components are in working state.

The probability α_i that, on demand, the *i*th component will be in working state is given by Equation 3.41. Consequently, the *availability on demand* α_{ser} of a system whose components are logically arranged in series is

$$\alpha_{\text{ser}} = \prod_{i=1}^{n} \alpha_i \tag{3.42}$$

The availability on demand of a system with components logically arranged in series is smaller than each of the availabilities α_i, characterising the separate components. The availability of the system is smaller than the worst component availability.

From Equation 3.41, it follows that there are two principal ways of increasing the availability on demand α_i of a component: (i) by improving its reliability which increases the MTTF_i or (ii) by improving its maintainability which reduces the mean time to repair MTTR_i.

The *unavailability* $\beta_i = 1 - \alpha_i$ *of a component on demand* stands for *the expected fraction of time during which the component is not working*. Suppose that the availabilities α_i of the components on demand are values very close to unity. The unavailabilities $\beta_i = 1 - \alpha_i$ of the components are then very small values $\beta_i << 1$.

For the unavailability $\beta_{\text{ser}} = 1 - \alpha_{\text{ser}}$ of a system on demand, with components logically arranged in series, the following expression is obtained from Equation 3.42:

$$\beta_{\text{ser}} = 1 - \prod_{i=1}^{n} (1 - \beta_i) \tag{3.43}$$

After expanding the right-hand side of Equation 3.43 and ignoring the second- and higher-order terms of the expansion ($\beta_i << 1$), the unavailability of the system on demand can be approximated by

$$\beta_{\text{ser}} \approx \sum_{i=1}^{n} \beta_i \tag{3.44}$$

A system with components in parallel is available on demand if at least a single component in the system is in working state. The complementary event is 'none of the components are in working state'. Hence, the availability on demand α_{par} of a system with *n* components logically arranged in parallel is

$$\alpha_{\text{par}} = 1 - \prod_{i=1}^{n} (1 - \alpha_i) \tag{3.45}$$

3.13.2 Production Availability

An important performance measure of production systems generating production flow (gas, oil, water, electricity, data, manufactured items, etc.) is the *production availability*. This is *the ratio* ψ *of the expected total throughput flow* $\bar{Q}_r$ *produced during a specified time interval in the presence of component failures*

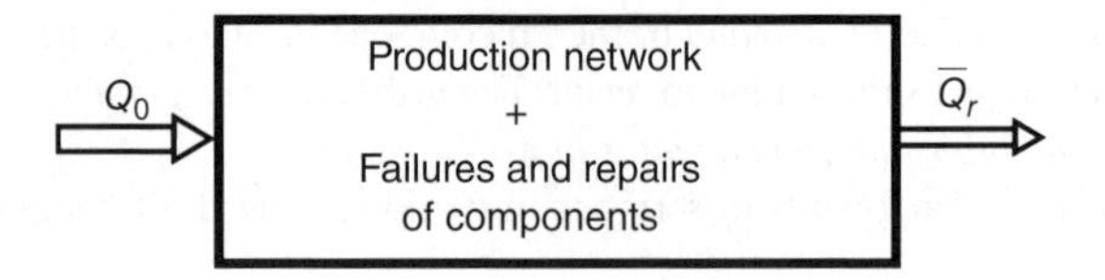

Figure 3.21 For a production network, the expected total throughput flow $\bar{Q}_r$ in the presence of failures is always smaller than the total expected throughput flow Q_0 in the absence of failures

to the total throughput flow Q_0 that could be produced during this time interval, in the absence of failures (Figure 3.21):

$$\psi = \frac{\bar{Q}_r}{Q_0} \tag{3.46}$$

Another related performance characteristic is the cumulative distribution of the production flow Q_r during a specified time interval:

$$\Phi(x) \equiv P(Q_r \le x) \tag{3.47}$$

The *cumulative distribution* $\Phi(x)$ *of the total production flow* reflects the variation of the production flow caused by component failures; it *gives the probability* $P(Q_r \le x)$ *that the total production flow* Q_r *in the presence of component failures, during a specified time interval, will not exceed a specified level x.*

The variation of the total production flow Q_r is a function of the times to failure and times to repair distributions of the components. To reveal the variation of the total throughput flow, a large number of failure–repair histories during the period of operation of the production system must be simulated. The expected production flow in the presence of failures for complex systems can be determined by using special fast simulation algorithms discussed in detail in Todinov (2013a).

3.14 Uniform Distribution Model

Often, the only information available about a parameter X is that it varies between certain limits a and b. In this case, the *uniform distribution* is useful for modelling the uncertainty associated with the parameter X. A random variable X following uniform distribution in the interval $[a, b]$ is characterised by a probability density function:

$$f(x) = \frac{1}{(b-a)} \tag{3.48}$$

for $a \le x \le b$ and $f(x) = 0$, elsewhere (Figure 3.22).

The cumulative distribution function of the uniform distribution is

$$F(x) = \frac{x-a}{b-a} \tag{3.49}$$

for $a < x \le b$, $F(x) = 0$ for $x \le a$ and $F(x) = 1$ for $x > b$ (Figure 3.22). The mean $E(X)$ of a uniformly distributed random variable X is

$$E(X) = \frac{(a+b)}{2} \tag{3.50}$$

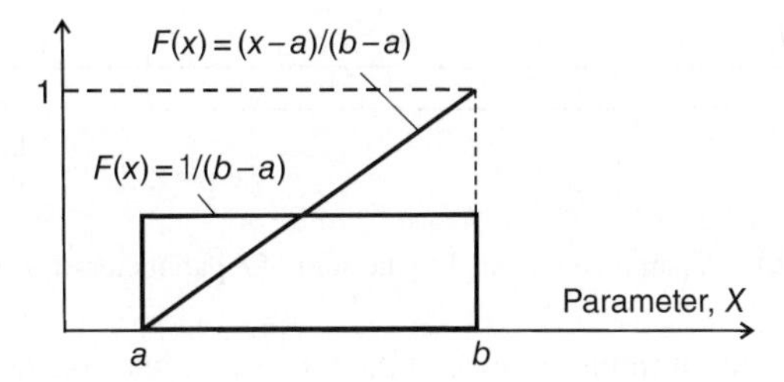

Figure 3.22 Probability density and cumulative distribution function of the uniform distribution

and the variance $V(X)$ is

$$V(X) = \frac{(b-a)^2}{12} \tag{3.51}$$

A uniform distribution for the coordinates of a point is often used to guarantee unbiased sampling during Monte Carlo simulations or to select a random location. For example, a random position of a rectangle on a plane can be simulated if the coordinates of its centre and the rotation angle are uniformly distributed. Sampling from the uniform distribution can also be used to generate times to repair.

The homogeneous Poisson process and the uniform distribution are closely related. An important property of the homogeneous Poisson process, well documented in books on probabilistic modelling (e.g. Ross, 2000), states: Given that n random variables following a homogeneous Poisson process are present in the finite interval $(0, a)$, the coordinates of the random variables are distributed uniformly in the interval $(0, a)$. As a result, in cases where the number of failures following a homogeneous Poisson process in a finite time interval 0, a is known, the failures times are uniformly distributed along the length of the interval 0, a.

3.15 Normal (Gaussian) Distribution Model

Often, the random variable of interest is a sum of a large number of random variables none of which dominates the distribution of the sum. For example, the distribution of a geometrical design parameter (e.g. length) incorporates the additive effects of a large number of factors: temperature variation, cutting tool wear, variations related to the parameters of the control system, etc.

If the number of separate contributions (additive terms) is relatively large and if none of the separate contributions dominate the distribution of their sum, the distribution of the sum (Figure 3.23) can be approximated by a *normal distribution* with mean equal to the sum of the means and variance equal to the sum of the variances of the separate contributions. The variation of a quality parameter in manufacturing often complies well with the normal distribution because its variation is usually a result of the additive effects of multiple small causes, none of which is dominant.

Formulated regarding a sum of statistically independent random variables, the *central limit theorem* states: *The distribution of the sum* $X = \sum_{i=1}^{n} X_i$ *of a large number n of statistically independent random variables* $X_1, X_2, \ldots, X_n$, *none of which dominates the distribution of the sum, approaches a normal (Gaussian) distribution with increasing the number of random variables* (Gnedenko, 1962; DeGroot, 1989).

The sum X has a mean $\mu = \sum_{i=1}^{n} \mu_i$ and variance $V = \sum_{i=1}^{n} V_i$, where μ_i and V_i are the means and the variances of the separate random variables. Even the sum of a relatively small number of statistically independent random variables can be approximated well by the normal distribution. Suppose that a system of independent collinear forces is present, with random magnitudes. According to the central limit theorem,

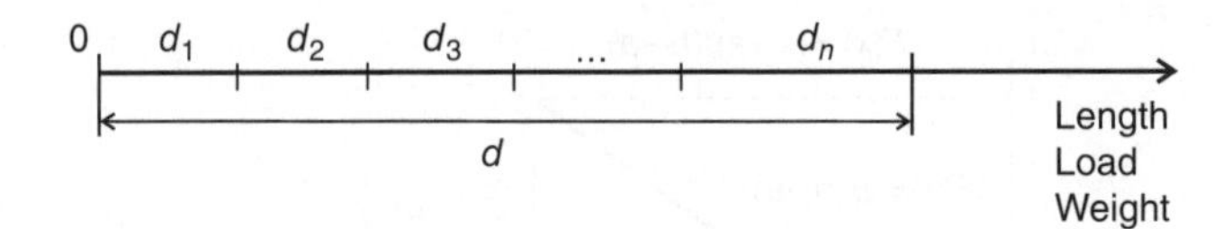

Figure 3.23 A parameter d equal to the sum of n parameters $d_i, i = 1,2,\ldots,n$

for a sufficiently large number of forces, if none of the forces dominate the distribution of the sum, the magnitude of the total force is approximately normally distributed, irrespective of the individual distributions of the individual forces. A similar conclusion can be drawn about the random consumption of a particular resource (water, electricity, gas, etc.) from a large number of independent consumers during a day. If none of the independent consumptions of the individual consumers dominates the total consumption, the sum of all individual consumptions follows approximately the normal distribution, irrespective of the individual distributions characterising the separate consumers.

Formulated with respect to sampling from any particular distribution, the central limit theorem states: *If the independent random variables are identically distributed and have finite variances, the sum of the random variables approaches normal distribution with increasing their number.*

This version of the central limit theorem can be illustrated by an example. Consider n (X_1, X_2, …, X_n) statistically independent uniformly distributed random variables in the interval (0, 1). According to Equations 3.50 and 3.51, their means are $\mu_i = 0.5$ and their variances are $V_i = 1/12$. The sum of the random variables $S = \sum_1^n X_i$ then approaches a normal distribution with mean $\mu = \sum_{i=1}^n 0.5 = n \times 0.5$ and variance $V = \sum_{i=1}^n 1/12 = n/12$. Even $n = 12$ random variables are sufficient to provide a good approximation to normal distribution, and this is used in developing generators of random numbers following a normal distribution.

In another example, the average weight of n items selected randomly from a batch is recorded. The population of all items is characterised by a mean weight μ and standard deviation σ. For a large number n of selected items, the sample mean (the average weight of the selected items) follows approximately a normal distribution with mean μ and standard deviation $\sigma / \sqrt{n}$.

A random variable X, characterised by a probability density function,

$$f(x) = \frac{1}{\sigma\sqrt{2\pi}} \exp\left[-\frac{(x-\mu)^2}{2\sigma^2}\right] \tag{3.52}$$

where $-\infty < x < \infty$ is said to be normally (Gaussian) distributed. The normal distribution of a random variable X is characterised by two parameters, the mean $E(X) = \mu$ and the variance $V(X) = \sigma^2$. After changing the variables by $z = (x - \mu)/\sigma$, the normal distribution transforms into

$$f(z) = \frac{1}{\sqrt{2\pi}} \exp\left(-\frac{z^2}{2}\right) \tag{3.53}$$

which is the probability density function of the standard normal distribution. The new variable $z = (x - \mu)/\sigma$ is normally distributed, with mean

$$E(z) = \frac{1}{\sigma}\left[E(x) - \mu\right] = 0$$

and variance

$$V(z) = \frac{1}{\sigma^2}\left[V(x) - 0\right] = \frac{\sigma^2}{\sigma^2} = 1$$

Equation 3.54 gives the cumulative distribution function of the *standard normal distribution*:

$$\Phi(z) \equiv P(Z \le z) = \int_{-\infty}^{z} \frac{1}{\sqrt{2\pi}} \exp\left(-\frac{u^2}{2}\right) du \tag{3.54}$$

where u is a dummy integration variable. If the probability that X will be smaller than b is required, it can be determined from $P(X \le b) = \Phi(z_2)$ where $z_2 = (b-\mu)/\sigma$. The probability that X will take on values from the interval $[a, b]$ can be determined from $P(a \le X \le b) = \Phi(z_2) - \Phi(z_1)$, where $z_1 = (a-\mu)/\sigma$ and $z_2 = (b-\mu)/\sigma$.

The probability density function and the cumulative distribution function of the standard normal distribution are given in Figure 3.24, where the probability $P(Z \le z = 1) \approx 0.8413$ has been determined (the hatched area beneath the probability density function $f(z)$) using the statistical table in Appendix C. Although the table lists $\Phi(z)$ for nonnegative values $z \ge 0$, it can be used for determining probabilities $P(Z \le -|z|)$ associated with negative values $-|z|$. Considering the symmetry of the standard normal curve, $P(Z \le -|z|) = 1 - P(|z|)$.

Example

The electrical resistance X of an element built in an electrical device is normally distributed, with mean $\mu = 75\Omega$ and a standard deviation $\sigma = 10\,\Omega$:

(i) Calculate the probability that the resistance X will be smaller than $65\,\Omega$.
(ii) Calculate the probability $P(55 \le X \le 65)$ that the resistance X will be between 55 and $65\,\Omega$.

Solution

(i)

$$P(X \le 65) = \Phi\left(\frac{65-75}{10}\right) = \Phi(-1) = 1 - \Phi(1) = 0.158$$

(ii)

$$z_1 = \Phi\left(\frac{55-75}{10}\right) = \Phi(-2); \quad z_2 = \Phi\left(\frac{65-75}{10}\right) = \Phi(-1);$$

$$P(55 \le X \le 65) = \Phi(z_2) - \Phi(z_1) = \Phi(-1) - \Phi(-2) = [1-\Phi(1)] - [1-\Phi(2)]$$
$$= \Phi(2) - \Phi(1) = 0.9772 - 0.8413 = 0.1359$$

An important property holds for a sum of statistically independent, normally distributed random variables: *The distribution of the sum $X = \sum_{i=1}^{n} X_i$ of n statistically independent, normally distributed random variables $X_1, X_2, \ldots, X_n$ is a normally distributed random variable.* The sum X has

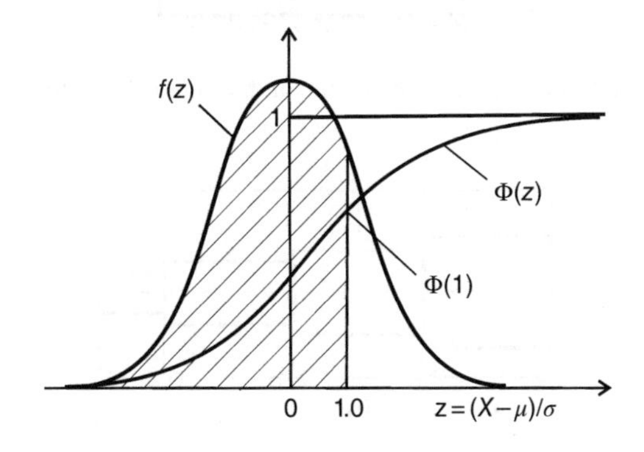

Figure 3.24 Probability density function of the standard normal distribution

mean $\mu = \sum_{i=1}^{n} \mu_i$ equal to the sum of the means μ_i of the random variables and variance $\sigma^2 = \sum_{i=1}^{n} \sigma_i^2$ equal to the sum of the variances σ_i of the separate random variables.

The difference $X = X_1 - X_2$ of two normally distributed random variables with means μ_1 and μ_2 and variances σ_1^2 and σ_2^2 is a normally distributed random variable with mean $\mu = \mu_1 - \mu_2$ and variance $\sigma^2 = \sigma_1^2 + \sigma_2^2$. Indeed, from the properties of expectations and variances of random variables (Appendix B), for the expected value of the difference X, the expression

$$E(X) = E(X_1) + E(-1 \times X_2) = E(X_1) - E(X_2) = \mu_1 - \mu_2$$

holds. For the variance, the expression

$$V(X) = V(X_1) + V(-1 \times X_2) = V(X_1) + (-1)^2 \times V(X_2) = \sigma_1^2 + \sigma_2^2$$

holds.

For normally distributed and statistically independent collinear loads $\pm F_i$ (Figure 3.25) with different directions along the y-axis, the resultant load R is always normally distributed with mean $\mu_R = \sum_{i=1}^{n} \pm F_i$ and variance $\sigma_R = \sum_{i=1}^{n} \sigma_i^2$ for any number of loads and any load directions.

Exercise

The shaft in Figure 3.26 has been produced by cutting the length L from a cylindrical rod and subsequently machining the steps 1, 2, 3 and 4 with lengths L_1, L_2, L_3 and L_4 on a lathe. The length L of the cylindrical rod has a mean of 650 mm and standard deviation 1.5 mm.

Because of imprecision, associated with controlling the lengths of steps 1, 2 and 3, the lengths L_1, L_2 and L_3 vary by following normal distributions with means $\mu_1 = 150\,\text{mm}$, $\mu_2 = 200\,\text{mm}$ and $\mu_3 = 250\,\text{mm}$ and standard deviations $\sigma_1 = \sigma_2 = \sigma_3 = 0.5\,\text{mm}$. The shaft is considered faulty, if the length of step L_4 does not exceed 48 mm. Calculate the percentage of faulty shafts.

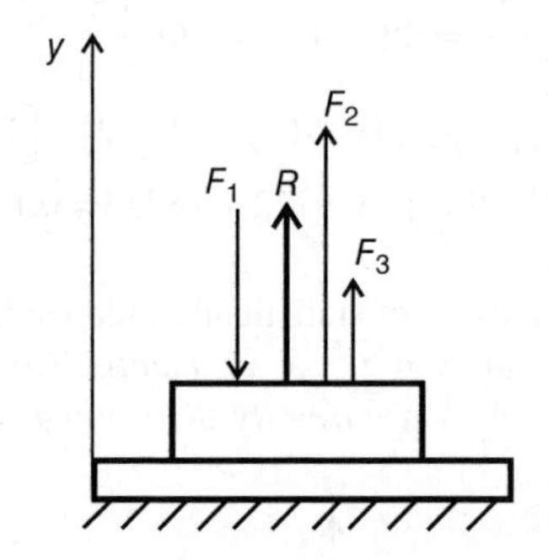

Figure 3.25 A resultant load R equal to the sum of three normally distributed collinear loads

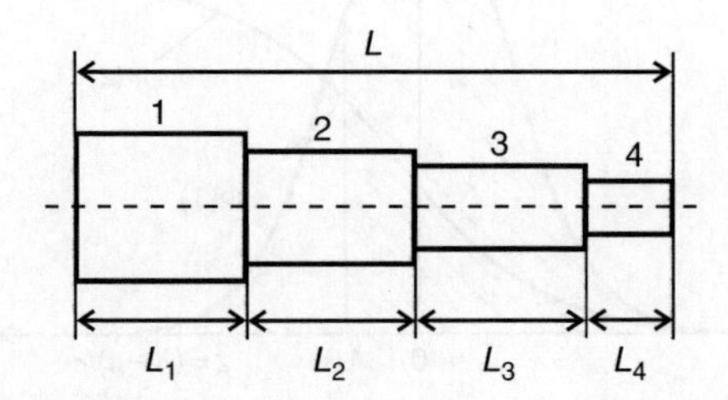

Figure 3.26 A shaft with steps machined on a lathe

Solution

Because $L_4 = L - L_1 - L_2 - L_3$, the expected value (the mean) of the step with length L_4 is $\mu_{L_4} = \mu_L - \mu_1 - \mu_2 - \mu_3 = 650 - 150 - 200 - 250 = 50$ mm. The lengths of the steps are statistically independent; therefore, the variance $V(L_4)$ of the length L_4 is $V(L_4) = V(L) + V(L_1) + V(L_2) + V(L_3) = 1.5^2 + 0.5^2 + 0.5^2 + 0.5^2 = 3$, and the standard deviation is $\sigma_{L_4} = \sqrt{3} = 1.732$ mm. The probability that the length L of step L_4 will not exceed 48 mm can then be determined from

$$P(L \leq 48) = \Phi\left(\frac{48 - 50}{1.732}\right) \approx \Phi(-1.155) = 1 - \Phi(1.155) \approx 0.124$$

Therefore, the percentage of faulty shafts is approximately 12.4%.

Exercise

An experiment has been made, where a very large number of sets of samples have been sequentially generated from the exponential distribution $F(t) = 1 - \exp(-2t)$. Each set k includes 100 sampled values $t_{k1}, t_{k2}, \ldots, t_{k100}$. What is the distribution followed by the means $\mu_k = (1/100)\sum_{i=1}^{100} t_{ki}$ of the sets of sampled values? Determine the mean and standard deviation of this distribution.

3.16 Log-Normal Distribution Model

A random variable X is log-normally distributed if its logarithm $\ln X$ is normally distributed. Suppose that the quantity X is a product

$$X = \prod_{i=1}^{n} Y_i \tag{3.55}$$

of a large number of statistically independent positive quantities $Y_i > 0$, none of which dominates the distribution of the product. The quantities Y_i are characterised by distinct individual distributions. The question of interest is the distribution of X.

Taking logarithm from both sides of Equation 3.55 yields

$$\ln X = \ln Y_1 + \ln Y_2 + \cdots + \ln Y_n \tag{3.56}$$

The logarithms in the right-hand side of Equation 3.56 can be regarded as random variables. According to the central limit theorem, the sum of a large number n of random variables follows a normal distribution. As a result, the sum of the logarithms in the right-hand side of Equation 3.56 follows a normal distribution. This sum, however, is equal to $\ln X$. Consequently, $\ln X$ follows a normal distribution. According to the definition of the log-normal distribution, if $\ln X$ is normally distributed, the random variable X is log-normally distributed.

This proves the multiplicative version of the central limit theorem:

The distribution of a product of statistically independent random variables, none of which dominates the product, approaches a log-normal distribution with increasing the number of random variables.

A basic application of the *log-normal distribution model* is the case where (i) a multiplicative effect of factors controlling reliability is present; (ii) the controlling factors are statistically independent and (iii) the magnitudes of their effects are comparable.

The log-normal distribution model has been used with success for modelling the time to failure controlled by degradation caused by corrosion, wear, erosion, crack growth, chemical reactions and diffusion.

Suppose that failure occurs when the total accumulated damage D from n time intervals of degradation reaches a critical level D_c. Suppose that the amount of new damage at any stage is proportional to the level of existing damage. The increase of degradation damage ΔD_1 in the first time interval is proportional to the quantity of initial damage D_0: $\Delta D_1 = r_1 D_0$ where r_1 is a small random quantity characterising the first time interval of degradation. As a result, the total damage D_1 at the end of the first time interval becomes $D_1 = D_0(1+r_1)$. The increase of degradation (damage) ΔD_2 in the second time interval of degradation is proportional to the quantity of damage D_1 at the start of the second time interval $\Delta D_2 = r_2 D_1$, where r_2 is also a small random quantity characterising the second time interval of degradation. As a result, the total damage D_2 at the end of the second time interval becomes

$$D_2 = D_1(1+r_2) = D_0(1+r_1)(1+r_2)$$

Continuing this reasoning, the total accumulated damage D from n stages of degradation is given by

$$D = D_0 \times (1+r_1) \times (1+r_2) \times \cdots \times (1+r_n) \tag{3.57}$$

The multistage degradation factor $1+r$, which represents the compound effect from many independent degradation stages characterised by random degradation factors $1+r_i$, can be presented as

$$1+r = (1+r_1) \times (1+r_2) \times \cdots \times (1+r_n)$$

and follows a log-normal distribution as a product of statistically independent random variables none of which dominates the product.

The log-normal distribution model is characterised by a probability density function (Figure 3.27)

$$f(x) = \frac{1}{x\sigma\sqrt{2\pi}} \exp\left[\frac{-(\ln x - \mu)^2}{2\sigma^2}\right], \quad x > 0 \tag{3.58}$$

where μ and σ are the mean and the standard deviation of the $\ln x$ data.

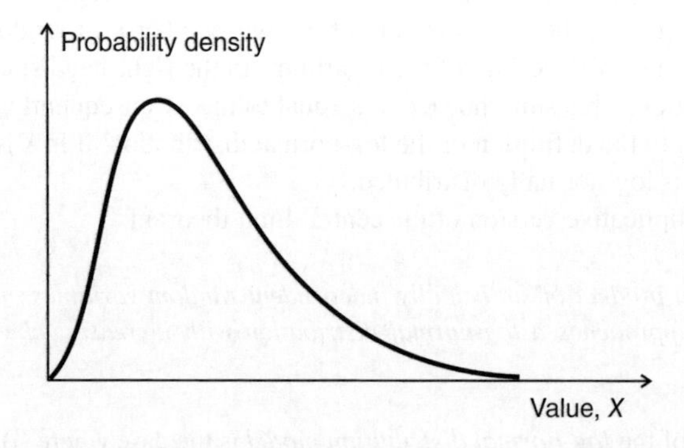

Figure 3.27 Probability density function of the log-normal distribution

An important property of the log-normal model is its reproductive property. A product of any number of log-normal random variables is a log-normal random variable. Indeed, from $X = \prod_{i=1}^{n} Y_i$, the logarithm of X, $\ln X = \sum_{i=1}^{n} \ln Y_i$ is normally distributed because it is a sum of normally distributed random variables. Because $\ln X$ is normally distributed, according to the definition of the log-normal distribution, X is log-normally distributed.

Often, the log-normal distribution is appropriate for modelling material strength. The log-normal distribution can often be used for describing the length of time to repair (Barlow and Proschan, 1965). The repair time distribution is usually skewed, with a long upper tail, which is explained by some problem repairs taking a long time.

3.17 Weibull Distribution Model of the Time to Failure

A popular model for the distribution of the times to failure is the *Weibull distribution model* (Weibull, 1951):

$$F(t) = 1 - \exp\left[-\left(\frac{t - t_0}{\eta}\right)^m\right] \tag{3.59}$$

In Equation 3.59, $F(t)$ is the probability that failure will occur before time t, t_0 is a location parameter or minimum life, η is the characteristic lifetime and m is a shape parameter.

In many cases, the minimum life t_0 is assumed to be zero and the three-parameter Weibull distribution transforms into a *two-parameter Weibull distribution*:

$$F(t) = 1 - \exp\left[-\left(\frac{t}{\eta}\right)^m\right] \tag{3.60}$$

Setting $t = \eta$ in Equation 3.60 gives $F(t) = 1 - e^{-1} \approx 0.632$. In other words, the probability of failure before time $t = \eta$, referred to as *characteristic life*, is 63.2%. Alternatively, the characteristic life corresponds to a time at which 63.2% of the initial population of items has failed.

The Weibull model is also popular for the strength distribution and reliability of brittle materials, but as it will be revealed in Chapter 13, as a model of the breaking strength of materials controlled by random defects, the Weibull distribution is a fundamentally flawed model.

If $m = 1$, the Weibull distribution transforms into the negative exponential distribution $F(t) = 1 - \exp(-\lambda t)$ with parameter $\lambda = 1/\eta$. Differentiating Equation 3.60 with respect to t gives the probability density function of the Weibull distribution:

$$f(t) = \frac{m}{\eta}\left(\frac{t}{\eta}\right)^{m-1} \exp\left[-\left(\frac{t}{\eta}\right)^m\right] \tag{3.61}$$

which has been plotted in Figure 3.28 for different values of the shape parameter m:

As can be verified, the Weibull distribution is very flexible. By selecting different values of the shape parameter m and by varying the scale parameter η, a variety of shapes can be obtained to fit experimental data. Since the hazard rate is defined by $h(t) = f(t)/R(t)$, where $f(t)$ is given by equation (3.61)

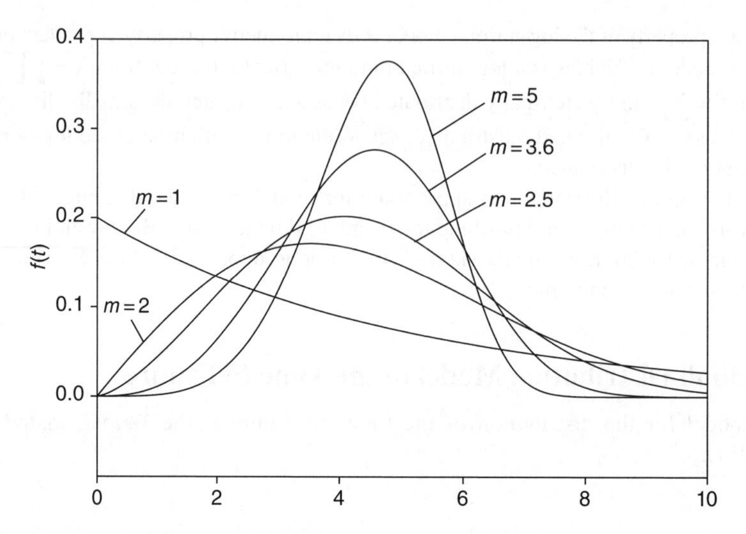

Figure 3.28 Two-parameter Weibull probability density function for different values of the shape parameter m

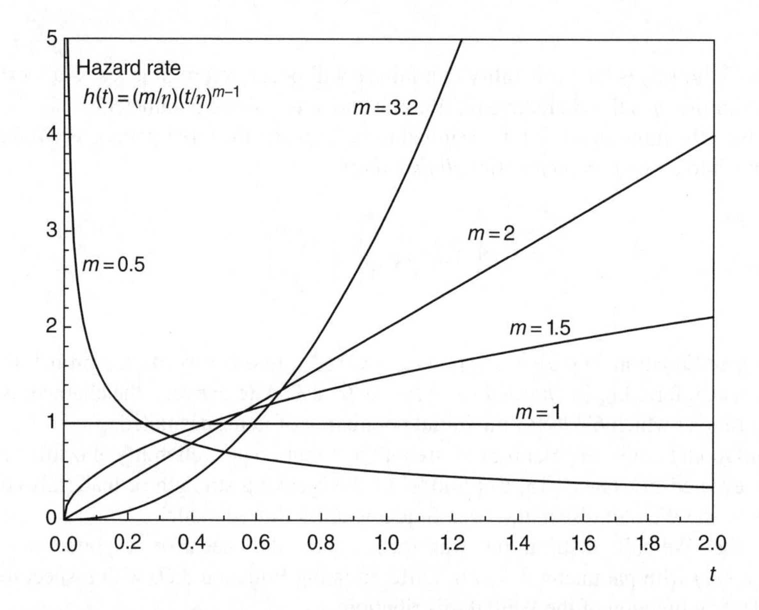

Figure 3.29 Weibull hazard rates for different values of the shape parameter m

and the reliability function is $R(t)=\exp\left[-(t/\eta)^m\right]$, the Weibull hazard rate function becomes

$$h(t)=\left(\frac{m}{\eta}\right)\left(\frac{t}{\eta}\right)^{m-1} \tag{3.62}$$

As can be verified from Equation 3.62, for $m<1$, the hazard rate is decreasing, and for $m>1$, the hazard rate is increasing and $m=1$ corresponds to a constant hazard rate.

Weibull hazard rate functions for different values of the Weibull exponent m and for $\eta=1$ have been plotted in Figure 3.29.

3.18 Extreme Value Distribution Model

Suppose that $X_1, \ldots, X_n$ are n independent observations of a random variable (e.g. load, strength, etc.). Let $X = \max\{X_1, X_2, \ldots, X_n\}$ denote the maximum value among these observations. Provided that the right tail of the random variable distribution decreases at least as fast as that of the negative exponential distribution, the asymptotic distribution of X for large values of n is the type I distribution of extreme values (Gumbel, 1958):

$$F(x) = \exp\left[-\exp\left(-\frac{x-\xi}{\theta}\right)\right] \tag{3.63}$$

This condition is satisfied by most of the reliability distributions: the normal, the log-normal and the negative exponential distribution. The maximum value from sampling an extreme value distribution also follows an extreme value distribution.

The *maximum extreme value distribution model* (Figure 3.30) is often appropriate in the usual case where the maximum load controls the component failure.

The type I extreme value model has two parameters: a *scale parameter* θ and mode ξ. The mean of the distribution is $\mu \approx \xi + 0.57722\,\theta$ and the standard deviation is $\sigma = 1.28255\,\theta$ (Metcalfe, 1994).

Suppose that n statistically independent loads $X_1, X_2, \ldots, X_n$ following the maximum extreme value distribution $F(x)$ have been applied consecutively to a component with strength x. The probability that the component will survive all n loads is given again by the maximum extreme value distribution.

Proof

The probability that a component with strength x will survive all n independent load applications is

$$P(X_1 \le x) \times P(X_2 \le x) \times \cdots \times P(X_n \le x) = [F(x)]^n = \exp\left\{-n\exp\left[-\frac{(x-\xi)}{\theta}\right]\right\}$$

The expression $n\exp[-(x-\xi)/\theta]$ can be presented as

$$n\exp\left(-\frac{x-\xi}{\theta}\right) = \exp\left\{\ln\left[n\exp\left(-\frac{x-\xi}{\theta}\right)\right]\right\} = \exp\left[-\frac{x-(\xi+\theta\ln n)}{\theta}\right]$$

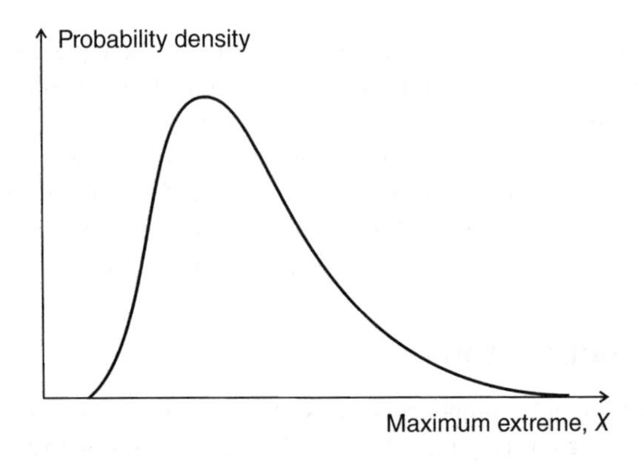

Figure 3.30 Probability density function of the maximum extreme value model

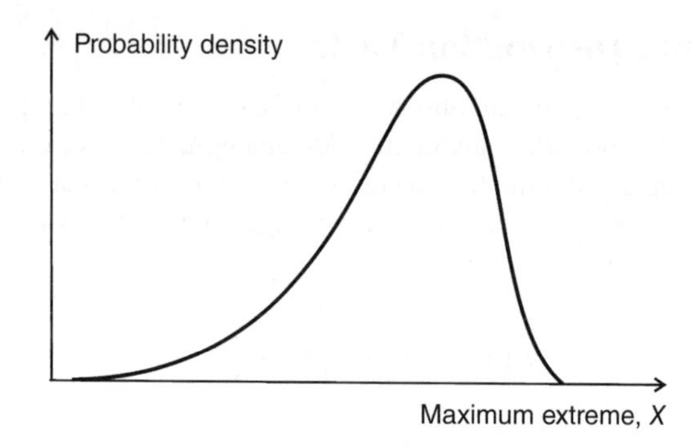

Figure 3.31 Probability density function of the minimum extreme value distribution

As a result,

$$\exp\left[-n\exp\left(-\frac{x-\xi}{\theta}\right)\right] = \exp\left\{-\exp\left[-\frac{\left(x-(\xi+\theta\ln n)\right)}{\theta}\right]\right\}$$

and the distribution of the maximum load becomes

$$F(x) = \exp\left[-\exp\left(-\frac{x-\xi'}{\theta}\right)\right] \tag{3.64}$$

which is a maximum extreme value distribution with the same scale parameter θ and a new displacement parameter (mode) $\xi' = \xi + \theta \ln n$. ■

Let X_1, ..., X_n be n independent observations of a random variable (e.g. load, strength). Let $X = \min\{X_1, X_2, \ldots, X_n\}$ denote the minimum value among these observations. With increasing the number of observations n, the asymptotical distribution of the minimum value follows the *minimum extreme value distribution* (Gumbel, 1958):

$$F(x) = 1 - \exp\left[-\exp\left(\frac{x-\xi}{\theta}\right)\right] \tag{3.65}$$

The minimum extreme value distribution (Figure 3.31) can, for example, be used for describing the distribution of the lowest temperature and the smallest strength. The Weibull distribution is related to the minimum extreme value distribution as the log-normal distribution is related to the normal distribution. If the logarithm of a variable follows the minimum extreme value distribution, the variable follows the Weibull distribution. Indeed, it can be verified that if the transformation $x = \ln z$ is made in Equation 3.65, the new variable z follows the Weibull distribution. References and discussions related to other statistical models used in reliability and risk analysis can be found in Bury (1975) and Trivedi (2002).

3.19 Reliability Bathtub Curve

The hazard rate of non-repairable components and systems follows a curve with bathtub shape, characterised by three distinct regions (Figure 3.32). The first region referred to as *infant mortality region* comprises the start of life and is characterised by an initially high hazard rate which decreases with

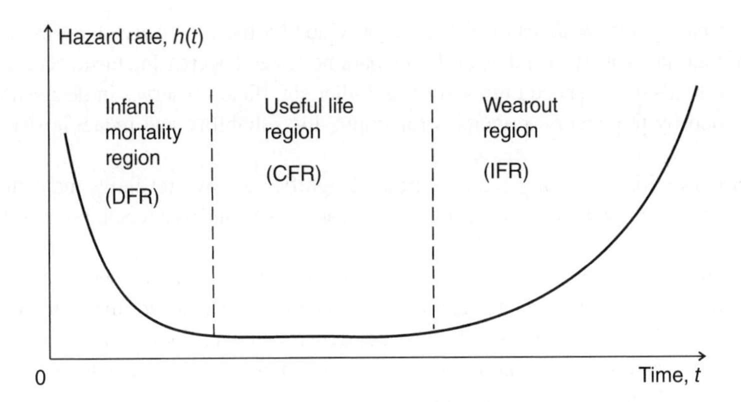

Figure 3.32 Reliability bathtub curve

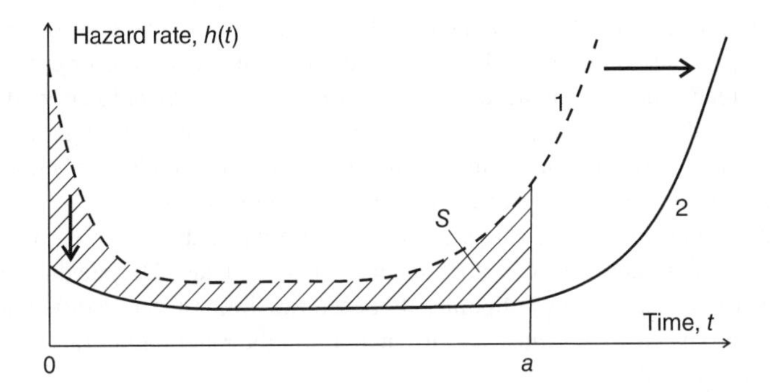

Figure 3.33 Decreasing the hazard rate translates into reducing the expected number of failures in the interval (0, a)

time. Most of the failures in the infant mortality region are quality related and result from inherent defects due to poor design, manufacturing and assembly. A substantial proportion of failures can also be attributed to a human error during installation and operation. Since most substandard components fail during the infant mortality period and, with time, the experience of the personnel operating the equipment increases, the initially high hazard rate gradually decreases. The cumulative distribution of the time to failure is a decreasing failure rate (DFR) distribution.

The second region of the bathtub curve, referred to as *useful life region*, is characterised by approximately constant hazard rate. This is why the negative exponential distribution, which is a constant failure rate (CFR) distribution, is the model of the times to failure in this region. Failures in this region are not due to age, wearout or degradation; they are due to random causes. This is why preventive maintenance in this region has no effect on the hazard rate.

The third region referred to as *wearout region* is characterised by an increasing with age hazard rate due to accumulated wear and degradation of properties (e.g. wear, erosion, corrosion, fatigue and creep). The corresponding cumulative distribution of the times to failure is an increasing failure rate (IFR) distribution.

For a batch of components experiencing a non-constant hazard rate $\lambda(t)$, the integral $\overline{N} = \int_0^a \lambda(t)dt$ gives the expected number of failures in the finite time interval 0, a.

Reliability in the interval (0, a) is given by $R(a) = \exp(-\overline{N})$. Consequently, reducing the hazard rate from curve '1' to curve '2' (Figure 3.33) results in a reduction of the expected number of failures equal to the hatched area S.

In the infant mortality region, the hazard rate can be decreased (curve 2 in Figure 3.33) and reliability increased, by better design, materials, manufacturing, inspection and assembly. Significant reserves in decreasing the hazard rate at the start of life are provided by the root–cause analysis, decreasing the variability of material properties and other design parameters and decreasing the uncertainty associated with the actual loads experienced during service. Other significant reserves in decreasing the hazard rate are provided by the generic principles for improving reliability and reducing risk discussed in Chapter 11.

In the wearout region, reliability can be increased significantly by preventive maintenance consisting of replacing old components. This delays the wearout phase, and as a result, reliability is increased (Figure 3.33).

For a shape parameter $m = 1$, the Weibull distribution transforms into the negative exponential distribution and describes the useful life region of the bathtub curve, where the probability of failure within a specified time interval practically does not depend on age. For components, for which early life failures have been eliminated and a preventive maintenance has been conducted to replace worn parts before they fail, the hazard rate tends to remain constant.

A value of the shape parameter smaller than one ($m < 1$) corresponds to a decreasing hazard rate and indicates infant mortality failures. A value of the shape parameter greater than one ($m > 1$) corresponds to increasing hazard rate and indicates wearout failures. Values in the interval ($1.0 < m \leq 4$) indicate *early wearout failures* caused, for example, by a low cycle fatigue, corrosion or erosion. Values of the shape parameter greater than four indicate *old age wearout* ($m > 4$). Most steep Weibull distributions have a safe period, within which the probability of failure is negligible. The larger the parameter m is, the smaller is the variation of the time to failure. An almost vertical Weibull distribution, with very large m, implies perfect design, quality, control and production (Abernethy, 1994).

Unlike the bathtub curve characterising hardware, the software bathtub curve is usually decreasing, with no wearout region because software does not deteriorate with age. Despite that the rate of appearance of software errors is decreasing with time, after each rewriting (new release) of the software, new faults (bugs) are introduced which cause a sharp increase in the rate of appearance of software errors (Figure 3.34) (Beasley, 1991).

Reducing the number of software faults is a prerequisite for a reduced number of software failures. Unlike hardware where no two pieces of equipment are absolutely identical and therefore there exists a substantial variation in the failure pattern, all copies of a piece of software are identical, and there is no variation in the failure pattern. If a software fault exists, it is present in all copies of the software programme and always causes failure if particular conditions or a combination of input data is present. Software is particularly prone to *common cause faults* if the routines are designed by the same programmer/team. A fault in one of the software modules/paths is also likely to be present in the back-up module.

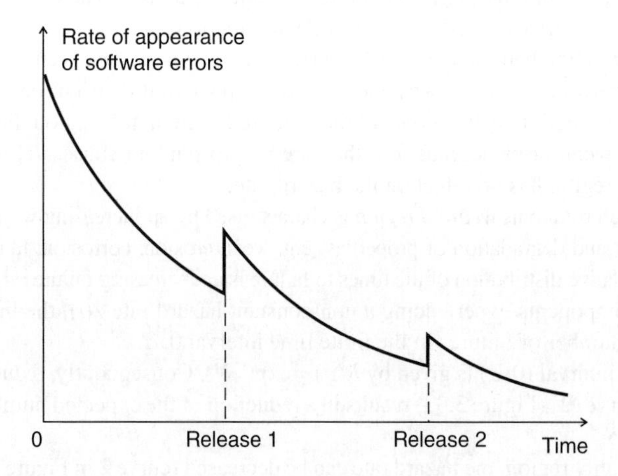

Figure 3.34 Rate of appearance of software errors as a function of time

Human errors are also a significant contributing factor for an increased hazard rate. They are an inevitable part of each stage of the product development and operation: design, manufacturing, installation and operation. Following Dhillon and Singh (1981), human errors can be categorised as (i) errors in design, (ii) operator errors (failure to follow the correct procedures), (iii) errors during manufacturing, (iv) errors during maintenance, (v) errors during inspection and (vi) errors during assembly and handling.

Complex systems, particularly electronic systems, require components with small hazard rates. The more complex the system, the higher the reliability required from the separate components and the lower their hazard rates should be. Indeed, for the sake of simplicity, suppose that a complex electronic block is composed of N identical capacitors, arranged logically in series. This is the logical arrangement of capacitors physically arranged in parallel in a block, against the failure mode 'dielectric breakdown'. The dielectric in any capacitor from the block is subjected to the full potential to which the block of capacitors is charged. The dielectric breakdown is promoted by a high-temperature, ageing and deterioration processes of the dielectric.

If the required reliability of the capacitor block is R_s, the reliability of a single capacitor should be $R_0 = (R_s)^{1/N}$. Clearly, with increasing the number of capacitors N in the block, the reliability R_0 required from the separate capacitors approaches unity. In other words, in order to guarantee the required reliability R_s of the block, the individual capacitors must be highly reliable. In this respect, *the six-sigma quality philosophy* (Harry and Lawson, 1992) is an important approach, based on a production with very small number of defective items (zero defect levels). Adopting a six-sigma process guarantees no more than two defective components out of a billion manufactured, and this is an efficient approach to reducing hazard rates.

4

Reliability and Risk Models Based on Distribution Mixtures

4.1 Distribution of a Property from Multiple Sources

Suppose that items arrive from M different sources in proportions $p_1, p_2, \ldots, p_M$, $\sum_{k=1}^{M} p_i = 1$. A particular property of the items from each source k is characterised by a mean μ_k and variance V_k. Often, of significant practical interest is the variance V of the property characterising the items collected from all sources.

In another example, small samples are taken randomly from a three-component inhomogeneous structure (components A, B and C, Figure 4.1). The probabilities p_1, p_2 and p_3 of sampling the structural constituents A, B and C are equal to their volume fractions ξ_A, ξ_B and ξ_C ($p_1 = \xi_A$; $p_2 = \xi_B$; $p_3 = \xi_C$). Suppose that the three structural constituents A, B and C have volume fractions $\xi_A = 0.55$; $\xi_B = 0.35$ and $\xi_C = 0.1$. The mean yield strengths of the constituents are $\mu_A = 800$ MPa, $\mu_B = 600$ MPa and $\mu_C = 900$ MPa, and the standard deviations are $\sigma_A = 20$ MPa, $\sigma_B = 25$ MPa and $\sigma_C = 10$ MPa, correspondingly.

The question of interest is the variance of the strength from random sampling of the inhomogeneous structure.

It can be demonstrated that in these cases, the property of interest from the different sources can be modelled by a *distribution mixture*.

Suppose that M sources ($i = 1, M$) are sampled with probabilities $p_1, p_2, \ldots, p_M$, $\sum_{i=1}^{M} p_i = 1$. The distributions of the property characterising the individual sources are $F_i(x)$, $i = 1, 2, \ldots, M$, correspondingly. Thus, the probability $F(x) \equiv P(X \le x)$ of the event B (Figure 4.2) *that the randomly sampled property X will not be greater than a specified value x* can be presented as a union of the following mutually exclusive and exhaustive events.

$A_1 \cap B$: *the first source is sampled (event A_1) and the property X is not greater than x* (the probability of this compound event is $p_1 F_1(x)$); $A_2 \cap B$: *the second source is sampled (event A_2) and the property X is not greater than x* (the probability of this compound event is $p_2 F_2(x)$); $A_M \cap B$: *the Mth source is sampled (event A_M) and the property X is not greater than x* (the probability of this compound event is $p_M F_M(x)$). According to the total probability theorem, the probability that the randomly sampled property X will not be greater than a specified value x is

$$F(x) \equiv P(X \le x) = \sum_{k=1}^{M} p_k F_k(x) \tag{4.1}$$

Reliability and Risk Models: Setting Reliability Requirements, Second Edition. Michael Todinov.

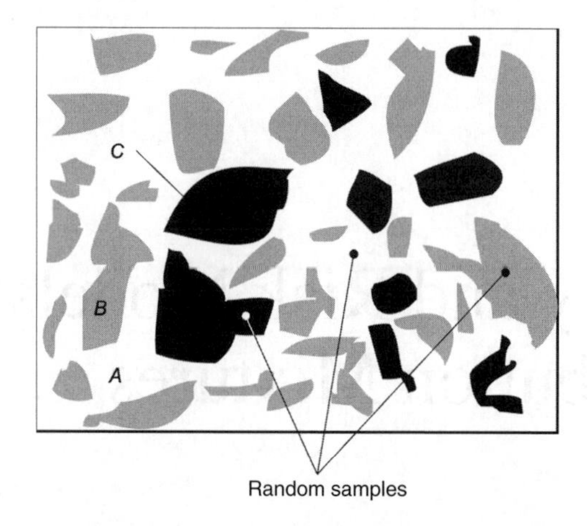

Figure 4.1 Sampling of an inhomogeneous microstructure composed of three structural constituents A, B and C

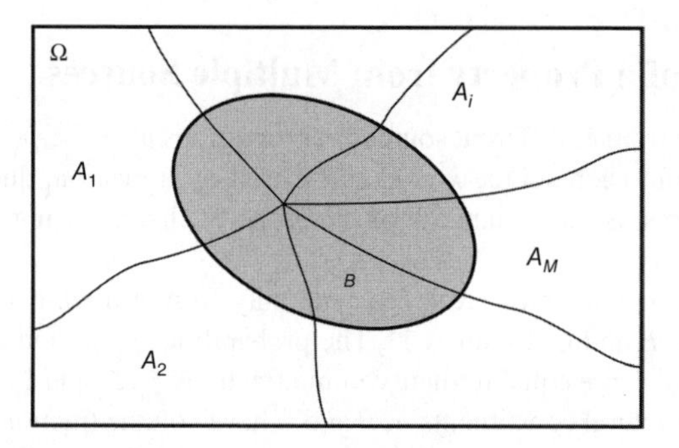

Figure 4.2 The probability of event B is a sum of the probabilities of the mutually exclusive events $A_i \cap B$

which is the cumulative distribution of the property from all sources. $F(x)$ is a mixture of the probability distribution functions $F_k(x)$ characterising the individual sources, scaled by the probabilities p_k, $k = 1, M$ with which they are sampled. After differentiating Equation 4.1, a relationship between the probability densities is obtained:

$$f(x) = \sum_{k=1}^{M} p_k f_k(x) \tag{4.2}$$

Multiplying both sides of Equation 4.2 by x and integrating

$$\int_{-\infty}^{+\infty} x f(x) dx = \sum_{k=1}^{M} p_k \int_{-\infty}^{+\infty} x f_k(x) dx$$

gives

$$\mu = \sum_{k=1}^{M} p_k \mu_k \tag{4.3}$$

for the mean value μ of a property from M different sources characterised by means μ_k (Everitt and Hand, 1981).

4.2 Variance of a Property from Multiple Sources

The variance V of the mixture distribution (4.1) for continuous probability density functions $f_k(x)$ characterising the existing microstructural constituents can be derived as follows:

$$V = \int (x-\mu)^2 f(x)dx = \int (x-\mu_k+\mu_k-\mu)^2 \sum_{k=1}^{M} p_k f_k(x)dx$$

$$= \sum_{k=1}^{M} p_k \left[\int (x-\mu_k)^2 f_k(x)dx + \int 2(x-\mu_k)(\mu_k-\mu) f_k(x)dx + \int (\mu_k-\mu)^2 f_k(x)dx \right]$$

Because the middle integral in the expansion is zero ($\int 2(x-\mu_k)(\mu_k-\mu)f_k(x)dx = 0$), the expression for the variance becomes (Todinov, 2002)

$$V = \sum_{k=1}^{M} p_k \left[V_k + (\mu_k-\mu)^2 \right] \tag{4.4}$$

where V_k, $k = 1, M$ are the variances characterising the M individual distributions. Although Equation 4.4 has a simple form, the grand mean μ of the distribution mixture given by Equation 4.3 is a function of the means μ_k of the individual distributions. An expression for the variance can also be derived as a function only of the pairwise distances between the means μ_k of the individual distributions.

Indeed, substituting expression (4.3) for the mean of a distribution mixture into the term $\sum_{k=1}^{M} p_k(\mu_k-\mu)^2$ of Equation 4.4 gives

$$\sum_{k=1}^{M} p_k(\mu_k-\mu)^2 = \sum_{k=1}^{M} p_k\mu_k^2 - \left(\sum_{k=1}^{M} p_k\mu_k\right)^2 \tag{4.5}$$

The variance of the distribution mixture can now be expressed only in terms of the pairwise differences between the means of the individual distributions. Expanding the right-hand part of Equation 4.5 results in

$$\begin{aligned}
\sum_{k=1}^{M} p_k\mu_k^2 - \left(\sum_{k=1}^{M} p_k\mu_k\right)^2 &= p_1\mu_1^2 + p_2\mu_2^2 + \cdots + p_M\mu_M^2 - p_1^2\mu_1^2 - p_2^2\mu_2^2 - \cdots - p_M^2\mu_M^2 \\
&\quad -2\sum_{i=2}^{M}\sum_{j=1}^{i-1} p_i p_j \mu_i\mu_j \\
&= p_1(p_2+p_3+\cdots+p_M)\mu_1^2 + p_2(p_1+p_3+\cdots+p_M)\mu_2^2 + \cdots \\
&\quad + p_M(p_1+p_2+\cdots+p_{M-1})\mu_M^2 - 2\sum_{i<j} p_i p_j \mu_i\mu_j \\
&= \sum_{i<j} p_i p_j (\mu_i-\mu_j)^2
\end{aligned}$$

because $1-p_1 = p_2+p_3+\cdots+p_M$, $1-p_2 = p_1+p_3+\cdots+p_M$, etc.

Finally, the variance of the distribution mixture (4.4) becomes

$$V = \sum_{k=1}^{M} p_k \left[V_k + (\mu_k-\mu)^2 \right] = \sum_{k=1}^{M} p_k V_k + \sum_{i<j} p_i p_j (\mu_i-\mu_j)^2 \tag{4.6}$$

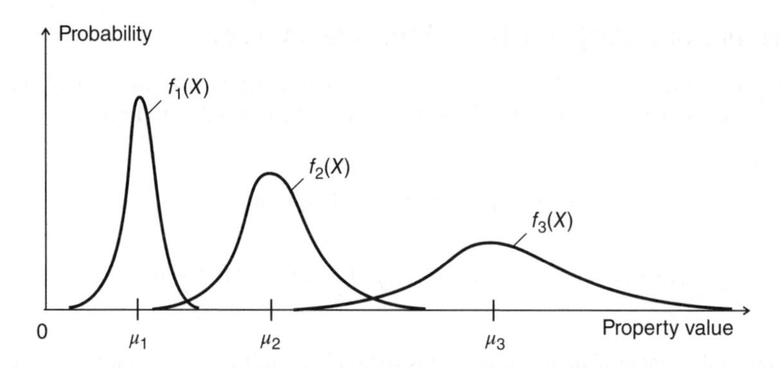

Figure 4.3 Distributions of properties from three different sources

The expansion of $\sum_{i<j} p_i p_j (\mu_i - \mu_j)^2$ has $M(M-1)/2$ number of terms, equal to the number of different pairs (combinations) of indices among M indices. For $M = 2$ individual distributions (sources), Equation 4.6 becomes

$$V = pV_1 + (1-p)V_2 + p(1-p)(\mu_1 - \mu_2)^2 \tag{4.7}$$

For three sources ($M = 3$, Figure 4.3), Equation 4.6 becomes

$$V = p_1 V_1 + p_2 V_2 + p_3 V_3 + p_1 p_2 (\mu_1 - \mu_2)^2 + p_2 p_3 (\mu_2 - \mu_3)^2 + p_1 p_3 (\mu_1 - \mu_3)^2 \tag{4.8}$$

Going back to the problem at the beginning of this chapter, according to Equation 4.3, the mean yield strength of the samples from all microstructural zones is

$$\mu = p_A \mu_A + p_B \mu_B + p_C \mu_C = 0.55 \times 800 + 0.35 \times 600 + 0.10 \times 900 = 740 \text{ MPa}$$

Considering that the variances V_k in Equation 4.8 are the squares of the standard deviations σ_A, σ_B and σ_C characterising the yield strength of the separate microstructural zones, the standard deviation of the yield strength from sampling all microstructural zones becomes

$$\sigma = [p_A \sigma_A^2 + p_B \sigma_B^2 + p_C \sigma_C^2 + p_A p_B (\mu_A - \mu_B)^2 + p_B p_C (\mu_B - \mu_C)^2 + p_A p_C (\mu_A - \mu_C)^2]^{1/2} \tag{4.9}$$

After substituting the numerical values in Equation 4.9, the value $\sigma \approx 108.8$ MPa is obtained for the standard deviation characterising sampling from all microstructural constituents. As can be verified, the value $\sigma \approx 108.8$ MPa is significantly larger than the standard deviations $\sigma_A = 20$ MPa, $\sigma_B = 25$ MPa and $\sigma_C = 10$ MPa characterising sampling from the individual structural constituents.

If Equation 4.6 is examined closely, the reason for the large variance becomes clear. In Equation 4.6, the variance of the distribution mixture has been decomposed into two major components. The first component $\sum_{k=1}^{M} p_k V_k$ including the terms $p_k V_k$ characterises only variation of properties within the separate sources (individual distributions). The second component is the sum $\sum_{i<j} p_i p_j (\mu_i - \mu_j)^2$ and characterises the variation of properties between the separate sources (individual distributions). Assuming that all individual distributions have the same mean μ ($\mu_i = \mu_j = \mu$), the terms $p_i p_j (\mu_i - \mu_j)^2$ in Equation 4.6 become zero and the total variance becomes $V = \sum_{k=1}^{M} p_k V_k$. In other words, the total variation of the property is entirely a within-sources variation (Figure 4.4).

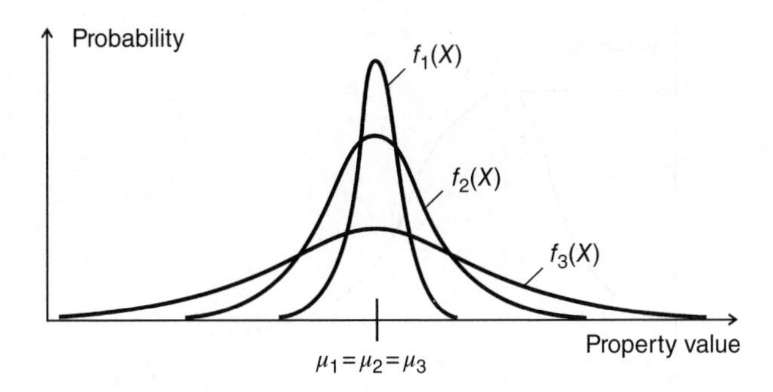

Figure 4.4 Distribution of properties from three different sources with the same mean

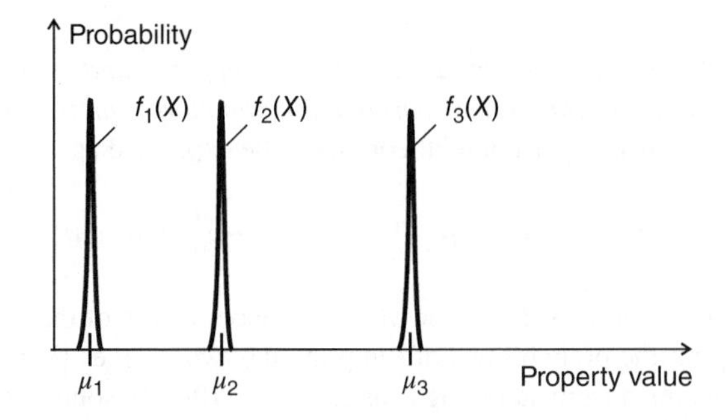

Figure 4.5 Distribution of properties from three different sources with very small variances

From Equation 4.8, for three sources with the same mean, the variance becomes

$$V = p_1V_1 + p_2V_2 + p_3V_3 \tag{4.10}$$

Now, assume that all sources (individual distributions) are characterised by very small variances $V_k \approx 0$ (Figure 4.5). In this case, the within-sources variance can be neglected: $\sum_{k=1}^{M} p_k V_k \approx 0$ and the total variance becomes $V = \sum_{i<j} p_i p_j \left(\mu_i - \mu_j\right)^2$. In other words, the total variation of the property is a 'between-sources variation'.

From Equation 4.8, for the three sources in Figure 4.5 characterised by negligible variances ($V_1 \approx 0$, $V_2 \approx 0$ and $V_3 \approx 0$), the total variance from sampling all sources becomes

$$V \approx p_1p_2\left(\mu_1 - \mu_2\right)^2 + p_2p_3\left(\mu_2 - \mu_3\right)^2 + p_3p_1\left(\mu_3 - \mu_1\right)^2 \tag{4.11}$$

4.3 Variance Upper Bound Theorem

If the mixing proportions p_k are unknown, the variance V in Equation 4.4 cannot be calculated. Depending on the actual mixing proportions p_k, the variance V may vary from the smallest variance V_k characterising one of the sources up to the largest possible variance obtained from sampling a particular combination of sources with appropriate probabilities p_i. A central question is to establish an exact upper bound for the variance of properties from multiple sources, irrespective of the mixing proportions p_k with which the sources are sampled. This upper bound can be obtained using the simple numerical algorithm in Appendix 4.2 which is based on an important result derived rigorously in Appendix 4.1.

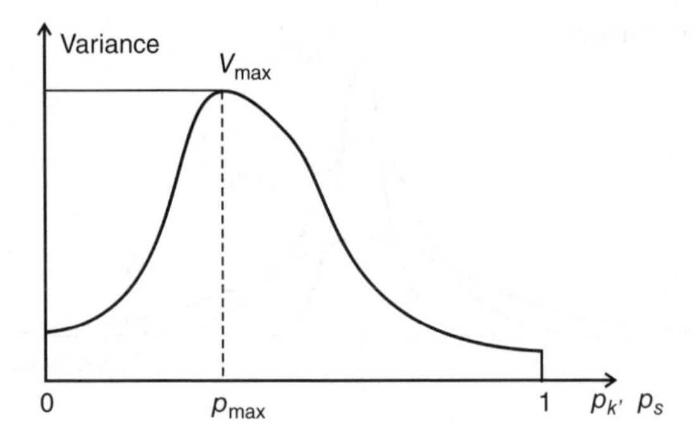

Figure 4.6 Upper bound of the variance from sampling two sources with indices k and s

Variance upper bound theorem: (Todinov, 2003) *The exact upper bound of the variance of properties from sampling multiple sources is obtained from sampling not more than two sources.*

Mathematically, the variance upper bound theorem can be expressed as

$$V_{max} = p_{max} V_k + (1 - p_{max}) V_s + p_{max} (1 - p_{max}) (\mu_k - \mu_s)^2 \tag{4.12}$$

where k and s are the indices of the sources for which the upper bound of the variance is obtained and $0 \le p_{max} \le 1$ and $1 - p_{max}$ are the probabilities of sampling the two sources (Figure 4.6). If $p_{max} = 1$, the maximum variance is obtained from sampling a single source (the kth source) only.

For a large M, determining the upper bound variance by finding directly the global maximum of expression (4.6) regarding the probabilities p_k is a difficult task which can be made easy by using the variance upper bound theorem. The algorithm for finding the maximum variance of the properties in Appendix 4.2 is based on the variance upper bound theorem and consists of checking the variances of all individual sources and the variances from sampling all possible pairs of sources. As a result, finding the upper bound variance of the properties from M sources involves only $M(M+1)/2$ checks, which can be done easily by a computer.

4.3.1 Determining the Source Whose Removal Results in the Largest Decrease of the Variance Upper Bound

The variance upper bound theorem can be applied to identify the source whose removal yields the smallest value of the variance upper bound. The algorithm consists of finding the source (pair of sources) yielding the largest variance. Removing one of these sources will result in the largest decrease of the variance upper bound. If sampling from a single source yields the largest variance, removing this source yields the largest decrease in the variance upper bound. The algorithm will be illustrated by the following numerical example.

Suppose that properties of items from five sources are characterised by individual distributions with variances $V_1 = 208$, $V_2 = 240$, $V_3 = 108$, $V_4 = 102$ and $V_5 = 90$ and means $\mu_1 = 39$, $\mu_2 = 43$, $\mu_3 = 45$, $\mu_4 = 56$ and $\mu_5 = 65$, correspondingly.

The question of interest is removal of which source yields the largest decrease in the variance upper bound.

The global maximum of the variance of properties from these sources is $V_{max\,1} \approx 323.1$, attained from sampling the fifth source with probability $p_{max} \approx 0.41$ and the first source with probability $1 - p_{max} \approx 0.59$. Removing the fifth source yields the largest reduction of the variance upper bound. Indeed, the calculations show that after removing the fifth source, the variance upper bound $V_{max} \approx 241.4$

is attained from sampling the fourth source with probability $p_{max} \approx 0.09$ and the second source with probability $1 - p_{max} \approx 0.91$. The removal of the fifth source yields a value of the variance upper bound which cannot be improved (decreased further) by the removal of any other source instead. This result is particularly useful in cases where the mixing proportions from the sources are unknown and a tight upper bound for the variance is necessary for a conservative estimate of the uncertainty associated with the properties from the sources.

4.4 Applications of the Variance Upper Bound Theorem

4.4.1 Using the Variance Upper Bound Theorem for Increasing the Robustness of Products and Processes

The variance upper bound theorem (VUBT) can be used as a basis for a new worst-case design method aiming at improving the robustness of processes, operations and products originating from multiple sources, *for which the worst performance is closely related to the worst variation of a property* (Todinov, 2009b).

For these processes or products, the common mean is not critical and can be easily adjusted to a specified target value, but deviations from the common mean lead to undesirable performance.

It is a well-known fact that while the mean value of the output of a process can be easily adjusted to a target value, the variance cannot be adjusted so easily. Reducing the variance of a process usually requires fundamental technological changes which need a substantial investment.

Robustness will be defined as *capability of a process or a product to cope with variability with minimal loss of functionality*. This is in line with the Taguchi *on-target engineering* philosophy (Fowlkes and Creveling, 1995) and with the fundamental components of quality defined by Juran and Gryna (1988): (i) the product features and (ii) the product's conformance to those features. Product or process conformance to a required target means that quality is improved when the maximum variation of the output is minimised.

This powerful approach for delivering robust designs is illustrated in Figure 4.7. Product (process) *B* is more robust and performs better than product (process) *A*, because the key output parameter characterising product *B* is more often close to the target (optimum) value than product *A*. Conformance to a customer-defined target also means that quality improves when variation in performance is minimised (Fowlkes and Creveling, 1995).

The selected approach is also in line with the worst-case design philosophy (Pierre, 1986) and the worst-case philosophy of the classical decision theory (Wald, 1950).

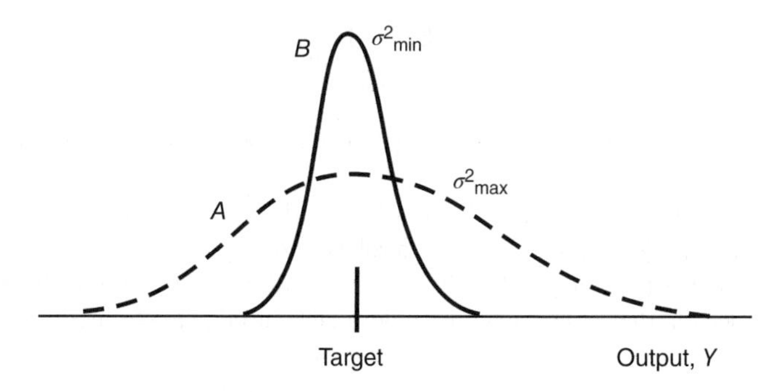

Figure 4.7 Delivering robust designs by decreasing the maximum variance σ^2_{max} to σ^2_{min} through removing sources of variation

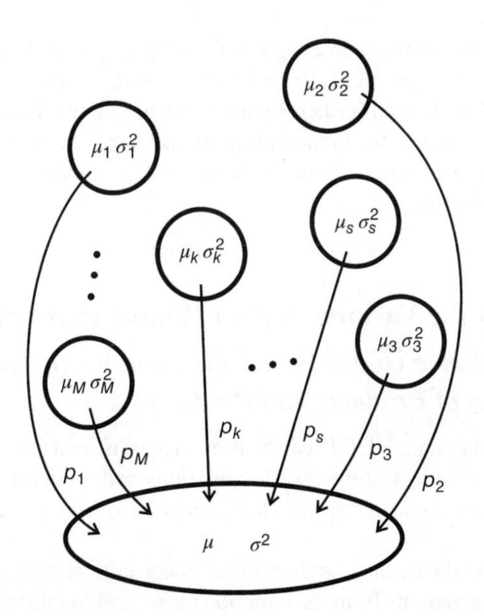

Figure 4.8 Variation of properties from multiple sources

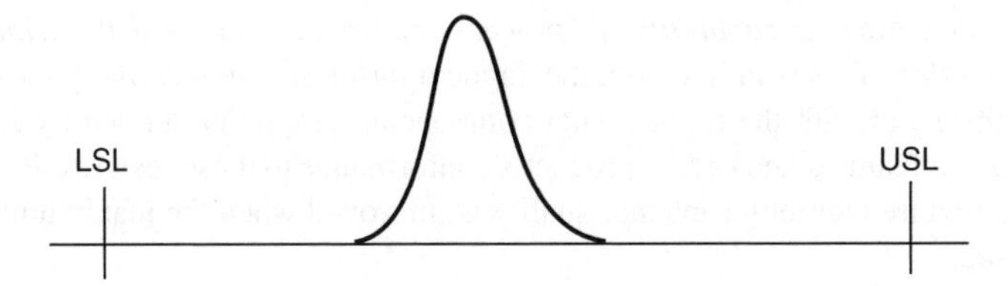

Figure 4.9 The mean of a robust process can shift off-centre and the percentage of faulty items will still remain very low

The variance upper bound theorem (VUBT) can be used for obtaining a conservative non-parametric estimate of the process capability index when the mixing proportions from the separate sources are unknown (Figure 4.8), which is usually the case.

The process capability index is defined as (Montgomery *et al.*, 2001)

$$C_p = \frac{\text{USL} - \text{LSL}}{6\sigma} \tag{4.13}$$

where USL and LSL are the upper and the lower specification limits (Figure 4.9) and σ is the standard deviation of the process.

A large process capability index means that fewer defective or non-conforming units will be produced. A process with a large capability index is a robust process. This means that the process mean can shift off-centre and the percentage of faulty items can still remain very low. If the process is not centred around the mean μ, the actual capability index can be used (Montgomery *et al.*, 2001):

$$C_{pk} = \min\left[\frac{\text{USL} - \mu}{3\sigma}, \frac{\mu - \text{LSL}}{3\sigma}\right] \tag{4.14}$$

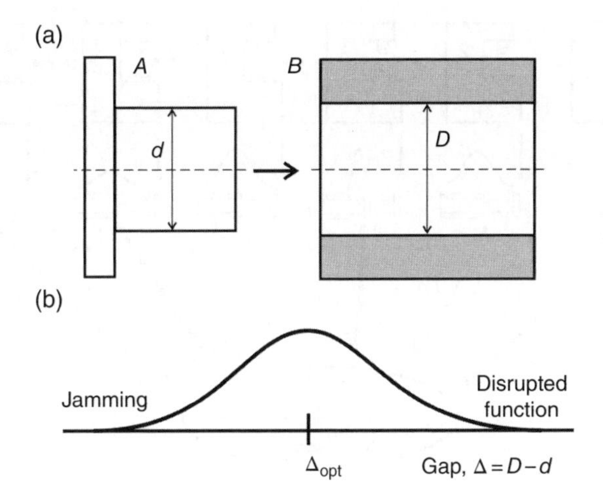

Figure 4.10 Assembly (a) which requires (b) an optimum value Δ_{opt} for the clearance $\Delta = D - d$

A conservative estimate of the process capability index for properties from multiple sources can be obtained by using an upper bound variance estimate σ^2_{max} produced by the algorithm described in Appendix 4.2:

$$C_p^* = \frac{\text{USL} - \text{LSL}}{6\sigma_{max}} \tag{4.15}$$

For a process which has not been centred, a conservative estimate of the actual capability index can be used:

$$C_{pk}^* = \min\left(\frac{USL - \mu}{3\sigma_{max}}, \frac{\mu - LSL}{3\sigma_{max}}\right) \tag{4.16}$$

For the conservative estimates of the capability index, the relationships

$$C_p^* \leq C_p; \; C_{pk}^* \leq C_{pk} \tag{4.17}$$

are valid.

Determining a non-parametric and conservative estimate of the process capability index helps to stabilise the variation of the process within the control limits and reduce the number of faults in the end product. The non-parametric capability index can serve as a basis for comparing, ranking and selecting competing manufacturing processes.

In the assembly from Figure 4.10a, component A with a mean diameter d must fit into component B with mean diameter D. In order to guarantee precision, the clearance $\Delta = D - d$ should not deviate significantly towards values greater than its optimum value Δ_{opt} (Figure 4.10b). On the other hand, in order to avoid a fit failure (inability to fit A into B) or jamming because of insufficient clearance, the difference $\Delta = D - d$ cannot deviate significantly towards values smaller than the optimum value Δ_{opt} (Figure 4.10b). Each machine centre is associated with a specific precision with which the diameters d and D are manufactured. As a result, the reliability-critical clearance $\Delta = D - d$ varies from centre to centre.

Failures are caused by unfavourable tolerance stacks due to excessive variation. Suppose that the ranges within which the diameters d and D vary are $d_{min} \leq d \leq d_{max}$ and $D_{min} \leq D \leq D_{max}$. Jamming, for example, occurs if, in an assembly, the diameter d of part A has deviated towards its upper limit d_{max} and, simultaneously, the inside diameter D of part B has deviated towards its lower limit D_{min}. Conversely,

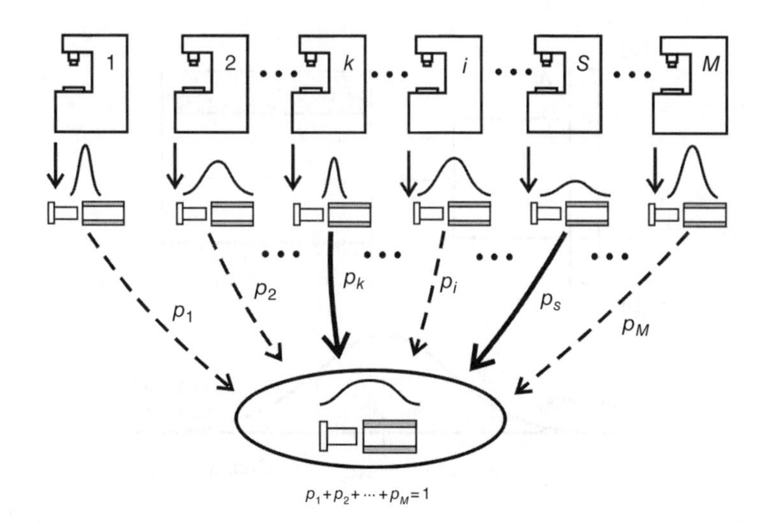

Figure 4.11 The maximum variation of properties of an assembly manufactured from multiple machine centres is obtained from sampling not more than two machine centres

imprecise assembly occurs when the diameter d of part A has deviated towards its lower limit d_{min} and the inside diameter D has deviated towards its upper limit D_{max}. Decreasing the variation of the clearance $\Delta = D - d$ improves the reliability of the assembly. Usually, the variances of the clearance $\Delta = D - d$ characterising each manufacturing centre (Figure 4.11) producing the assemblies are known.

Then, the pair of manufacturing centres yielding the maximum variance of the clearance $\Delta = D - d$ can be determined.

If no source of variation can be removed, the pair of distributions yielding the worst possible variation of the clearance can be identified. Next, the distance between the mean of the worst-case distribution and the optimal target value can be determined. The clearance from all manufacturing centres is then simultaneously increased/decreased by this value in order to adjust to the target value. This can, for example, be achieved by increasing/decreasing the inner diameter D of component B or decreasing/increasing the outer diameter d of component A. A check is finally performed whether the design can accommodate the worst-case variation.

The final step involves a verification whether the design can perform with this worst-case variation. If the design cannot perform with the worst possible variation of a particular parameter, steps are taken to reduce the worst-possible variation of the parameter. If reducing the worst possible variation is not possible or if it is too expensive, a robust design is sought for which the worst-case variation of the parameter causes an acceptable variation of the output characteristic. In short, *the process of creating a more robust product based on the variance upper bound theorem is a process of making the product resistant against the worst-case variation of the design parameters. Without the variance upper bound theorem, the worst variation of the input parameters cannot be estimated. Benign variations will be assumed instead, which result in optimistic predictions regarding the product's performance.*

If sources of variation can be removed, by removing the source resulting in the most significant decrease of the variance upper bound, the variability of the clearance $\Delta = D - d$ will be reduced and the capability index of the manufacturing process increased.

This is a way of achieving a robust manufacturing process, whose variation is under control, irrespective of the actual mixing proportions from the manufacturing centres.

Increasing the Robustness of Electronic Devices

Components building electronic circuits are characterised by properties like resistance, capacitance, inductance, etc. Because of imprecision during manufacturing, the actual magnitudes of these properties deviate from the stated nominal values.

Suppose that these components are part of safety-critical systems containing sensors measuring temperature, pressure, concentration, etc. in two different zones. A difference exceeding a particular threshold triggers an alarm or a shutdown system. Large deviations in the properties of the components building the circuit are undesirable, because they lead to a deteriorated performance of the safety-critical devices.

The components are manufactured by different centres/suppliers. Each centre/supplier is characterised by its individual distribution of the corresponding property. Usually, the variation of the property (resistivity, capacitance, inductance, etc.) associated with the common pool of manufactured components, for given mixing proportions from the suppliers, is not the maximum possible variation that can occur. There exists a particular combination of sources and mixing proportions that yields the largest (worst-case) variation. The variance upper bound theorem makes it possible to determine this worst-case variation, and this will be illustrated by a simple example.

Suppose that resistors are delivered from four suppliers. The mean resistances [Ω] characterising the resistors from the individual suppliers are

$$\mu_R = \{500, 504, 510, 516\}$$

The variances of the resistors characterising the individual suppliers are

$$V_R = \{102, 141, 166, 85\}$$

Suppose that the market shares of the suppliers are $p_R = \{0.15,\ 0.65,\ 0.15, 0.05\}$, where $\sum_{i=1}^{4} p_{Ri=1}$.

For the variance of the resistance of the supplied resistors, Equation 4.4 yields $V \approx 150$. A calculation of the maximum variance however by using the algorithm in Appendix 4.2 reveals $V_{max} \approx 169$, attained from sampling two suppliers only: the first supplier with a mixing proportion (market share) $p = 0.18$ and the third supplier with a mixing proportion (market share) $q = 1 - p = 0.82$.

The designer must make sure that the electronic circuit will operate satisfactorily under the worst possible combination of suppliers, yielding the maximum possible variation of the resistance. If the circuit design cannot perform with the worst possible variation of the resistance, steps are taken to reduce the worst-possible variation of the resistance. If this is not possible or if it is too expensive, a robust circuit design is sought for which the worst-case variation of the resistance does not cause unacceptable performance of the circuit.

Here, we need to point out that the distribution of properties from several suppliers is a distribution mixture, different from the individual distributions characterising the separate suppliers or any particular distribution. For the example discussed earlier, the resultant mixture distribution is different from Gaussian distribution even if the resistance of the components from each individual supplier follows a Gaussian distribution.

4.4.2 Using the Variance Upper Bound Theorem for Developing Six-Sigma Products and Processes

The conservative estimate of the variation of properties can be used for developing robust designs and processes, where the mean output can be easily adjusted. Indeed, *if the design is capable to perform with the worst variation of the design parameters, it will be capable to perform for any other variation of the design parameters, produced by any particular combination of mixing proportions from the sources*. In other words, the design can be made resistant to the worst possible variation of the design parameters.

The algorithm for creating a six-sigma process by using the variance upper bound theorem can be presented with the following steps:

Algorithm 4.1

1. The maximum possible variance of the process is determined, by calculating the variance upper bound σ^2_{max} and the mean μ_{max} corresponding to the largest variance given by Equation 4.6.

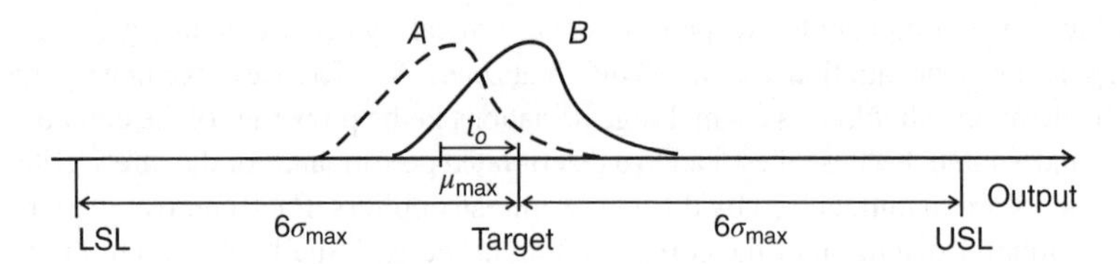

Figure 4.12 Steps to create a six-sigma process/design by using the variance upper bound theorem

2. The mean μ_{max} of the process (parameter) is adjusted on the target value by adding or subtracting a common value (the difference Δ between the mean μ_{max} of the worst-case distribution mixture and the target value). Adding or subtracting a common value t_0 to a distribution alters only its mean and does not alter the variance (Figure 4.12). Adding or subtracting a common value to all individual distributions does not alter their relative position; it alters only the global mean of the distribution mixture defined on these individual distributions.
   In fact, the first two steps are about creating the worst possible variation of the parameter around the target value.
3. The new specification limits are obtained by adding and subtracting $6\sigma_{max}$ using the relationships

$$\text{USL} = \text{Target} + 6\sigma_{max}$$

$$\text{LSL} = \text{Target} - 6\sigma_{max}$$

4. The process is tested whether it is acceptable with variation within the obtained specification limits.

```
5. If (the process output is acceptable)
    then {A six-sigma process has been obtained; Stop.}
    else{
          if (the variation of the process can be reduced at low
cost) then
                {reduce the variation of the process;
                 repeat steps 1-5;
                }
           else {A six-sigma process cannot be produced at low
cost; Stop.}
          }
```

This is a powerful method for delivering six-sigma operations and processes.

Suppose that several operators (e.g. from different shifts) are setting the value of a critical parameter (e.g. length). The parameter, for example, could be a critical distance that defines the position of a part in a manufacturing cell (e.g. the position of the glass in a glazing cell of a car manufacturing plant).

Each operator sets the critical distance with a particular mean and a standard error. In order to achieve the necessary consistency of production control, the production process must accommodate the worst variation of the set position from its target value. The worst variation of the set position is given by the upper bound of the variance σ^2_{max} by applying the variance upper bound theorem. After defining the pair of operators associated with the worst-case variance σ^2_{max} of the set position, the distance t_0 between the mean of the worst-case mixture and the target value is determined (Figure 4.12). Creating a six-sigma operation proceeds along the steps described in Algorithm 4.1.

If $6\sigma_{max}$ operation is produced, the operation is now resistant against this worst-case variation.

Appendix 4.1: Derivation of the Variance Upper Bound Theorem

The upper bound V_{max} of the variance V of a distribution mixture is obtained by maximising the general Equation 4.6 with respect to p_k, $k = 1, M$:

$$V_{max} = \max_{p_1 p_2 \ldots p_M} \left(\sum_{k=1}^{M} p_k V_k + \sum_{i<j} p_i p_j \left(\mu_i - \mu_j\right)^2 \right) \quad (4.A.1)$$

The local extrema of expression (4.A.1) can be determined using Lagrange multipliers, because of the equality constraint $g\left(p_1, \ldots, p_M\right) \equiv \left(\sum_{k=1}^{M} p_k\right) - 1 = 0$. The necessary condition for an extremum of the variance given by expression (4.6) is

$$\frac{\partial V}{\partial p_k} + \frac{\lambda \partial g}{\partial p_k} = 0, \quad k = 1, 2, \ldots, M \quad (4.A.2)$$

where λ is the Lagrange multiplier. These M equations, together with the constraint $\sum_{k=1}^{M} p_k - 1 = 0$, form a system of $M+1$ linear equations in the $M+1$ unknowns $p_1, \ldots, p_M$ and λ:

$$p_1 + p_2 + \cdots + p_M = 1$$

$$\lambda + p_1\left(\mu_k - \mu_1\right)^2 + p_2\left(\mu_k - \mu_2\right)^2 + \cdots + p_{k-1}\left(\mu_k - \mu_{k-1}\right)^2 + p_{k+1}\left(\mu_k - \mu_{k+1}\right)^2 + \cdots + p_M\left(\mu_k - \mu_M\right)^2 = -V_k$$

where $k = 1, \ldots, M$. This linear system can also be presented in a vector form:

$$\mathbf{Ap} = \mathbf{V} \quad (4.A.3)$$

where

$$\mathbf{A} = \begin{pmatrix} 0 & 1 & 1 & 1 & 1 \\ 1 & 0 & \left(\mu_1 - \mu_2\right)^2 & \ldots & \left(\mu_1 - \mu_M\right)^2 \\ 1 & \left(\mu_2 - \mu_1\right)^2 & 0 & \ldots & \left(\mu_2 - \mu_M\right)^2 \\ \ldots & \ldots & \ldots & \ldots & \ldots \\ 1 & \left(\mu_M - \mu_1\right)^2 & \left(\mu_M - \mu_2\right)^2 & \ldots & 0 \end{pmatrix} \quad (4.A.4)$$

$$\mathbf{p} = \begin{pmatrix} \lambda \\ p_1 \\ p_2 \\ \ldots \\ p_M \end{pmatrix} \text{ and } \mathbf{V} = \begin{pmatrix} 1 \\ -V_1 \\ -V_2 \\ \ldots \\ -V_M \end{pmatrix}$$

Matrix **A** is a *Cayley–Menger* matrix (Glitzmann and Klee, 1994). The determinant of this matrix is at the basis of a method for calculating the volume V of an N-simplex (in the N-dimensional space).

Suppose an N-simplex has vertices $v_1, v_2, \ldots, v_{N+1}$ and d_{km}^2 is the squared distance between vertices v_k and v_m. Let the matrix $\mathbf{D}$ be defined as

$$\mathbf{D} = \begin{pmatrix} 0 & 1 & \ldots & 1 \\ 1 & d_{11}^2 & \ldots & d_{1,N+1}^2 \\ \ldots & \ldots & \ldots & \ldots \\ 1 & d_{N+1,1}^2 & \ldots & d_{N+1,N+1}^2 \end{pmatrix}$$

The equation $V^2 = [(-1)^{N+1} / (2^N (N!)^2)]\det(\mathbf{D})$ then gives the volume V of the N-simplex (Glitzmann and Klee, 1994; Sommerville, 1958). From this equation, it is clear that if the Cayley–Menger determinant det(**D**) is zero, the volume of the simplex is also zero. The converse is also true that if the volume of the simplex is zero, the Cayley–Menger determinant is necessarily zero.

As can be verified from the matrix **A** given by (4.A.4), the means μ_i can be considered as first-axis coordinates of the vertices of a simplex in an M-1-dimensional space with other coordinates set to zero. The 'volume' of this simplex is clearly zero except in the one-dimensional case ($M-1=1$) where the 'volume' of the simplex is simply the distance $|\mu_1 - \mu_2|$ between the two means of the individual distributions (sources). As a consequence, the determinant of matrix **A** is always zero for $M > 2$, and the linear system (4.A.3) has no solution.

We shall now prove that no local maximum for the variance exists if exactly $k > 2$ individual distributions (sources) are sampled from a mixture distribution composed of M components (individual distributions). If $k>2$ individual distributions are sampled, the sampling probabilities of these distributions must all be different from zero. Without loss of generality, let us assume that only the first k individual distributions are sampled, with probabilities $p_1 \neq 0,\ p_2 \neq 0,\ \ldots,\ p_k \neq 0,\ \sum_{i=1}^{k} p_i = 1$ $(p_{k+1} = 0, \ldots, p_M = 0)$. Since the k individual distributions also form a mixture distribution and $k > 2$, the linear system (4.A.3) has no solution; therefore, no local maximum exists. This means that the global maximum is attained somewhere on the boundary of the domain $0 \le p_1 \le 1, \ldots, 0 \le p_k \le 1$, either for some $p_s = 0$ or for some $p_t = 1$. The case $p_t = 1$, however, means that the rest of the sampling probabilities must be zero ($p_i = 0$ for $i \neq t$). In both cases, at least one of the sampling probabilities $p_1, p_2, \ldots, p_k$ must be zero. Without loss of generality, suppose that $p_k = 0$. The same reasoning can be applied to the remaining $k-1$ sources. If $k-1 > 2$, some of the $p_1, p_2, \ldots, p_{k-1}$ must be zero, etc.

If $k = 2$ (one-dimensional simplex), the matrix Equation 4.A.3 becomes

$$\begin{pmatrix} 0 & 1 & 1 \\ 1 & 0 & (\mu_1 - \mu_2)^2 \\ 1 & (\mu_1 - \mu_2)^2 & 0 \end{pmatrix} \times \begin{pmatrix} \lambda \\ p_1 \\ p_2 \end{pmatrix} = \begin{pmatrix} 1 \\ -V_1 \\ -V_2 \end{pmatrix} \tag{4.A.5}$$

Because in this case, the Cayley–Menger determinant is equal to $2(\mu_1 - \mu_2)^2$, a solution of the linear system (4.A.5) now exists and a local maximum may be present. The solution of the linear system (4.A.5) is

$$p_1 = 0.5 + \frac{V_1 - V_2}{2(\mu_1 - \mu_2)^2} \tag{4.A.6}$$

$$p_2 = 0.5 - \frac{V_1 - V_2}{2(\mu_1 - \mu_2)^2} \tag{4.A.7}$$

These values of p_1 and p_2 correspond to a local maximum in the domain $0 \le p_1 \le 1, 0 \le p_2 \le 1$, only if $|V_1 - V_2| < (\mu_1 - \mu_2)^2$. If $|V_1 - V_2| \ge (\mu_1 - \mu_2)^2$, the maximum is attained at the boundary of the domain, either for $p_1 = 1, p_2 = 0$ ($V_{max} = V_1$) or for $p_1 = 0, p_2 = 1$ ($V_{max} = V_2$), whichever is greater ($V_{max} = \max\{V_1, V_2\}$). If $|V_1 - V_2| < (\mu_1 - \mu_2)^2$, a local maximum of the variance exists for p_1 and p_2 given by the relationships (4.A.6) and (4.A.7). The maximum value of the variance corresponding to these values is

$$V_{max} = \frac{V_1 + V_2}{2} + \frac{(V_1 - V_2)^2}{4(\mu_1 - \mu_2)^2} + \left(\frac{\mu_1 - \mu_2}{2}\right)^2 \tag{4.A.8}$$

In short, the global maximum of the right-hand side of Equation 4.6 is attained either from sampling a single source/individual distribution, in which case one of the sampling probabilities p_i is unity and the rest are zero ($p_i = 1; p_{j \ne i} = 0$), or from sampling only two individual distributions k and m among all individual distributions composing the mixture distribution. In this case, $p_k \ne 0$ and $p_m \ne 0$ and the rest of the p_i are zero ($p_i = 0$) for $i \ne k$ and $i \ne m$. If $V_{max,k,m}$ denotes the local maximum of the variance from sampling sources (individual distributions) k and m ($k \ne m$), the global maximum V_{max} of the right-hand side of Equation 4.6 can be found from $V_{max} = \max\{V_1, V_2, \ldots, V_M, V_{max,k,m}\}$ where $k = 2, \ldots, M$ and $m = 1, \ldots, k-1$. Since there exist $M \times (M-1)/2$ number of terms $V_{max,k,m}$, the global maximum is determined after $M + M \times (M-1)/2 = M(M+1)/2$ checks. As can be verified from the algorithm presented in Appendix 4.2, the maximum of the variance is determined by two nested loops. The control variable i of the external loop takes on values from 2 to M (the number of sources), while the control variable j of the internal loop takes on values from 1 to $i-1$.

Appendix 4.2: An Algorithm for Determining the Upper Bound of the Variance of Properties from Sampling Multiple Sources

The algorithm is in pseudocode, where the statements in braces {*op1*; *op2*; *op3*; …} are executed as a single block. Detailed description of the pseudocode notation is given in Chapter 8. The variable `max` contains the largest variance at the end of the calculations, and the constant `M` is the number of components (sources) composing the mixture distribution. Variables `k_max` and `m_max` contain the indices of the sources sampling from which yields the largest variance. If the maximum variance is attained from sampling a single source, `k_max` and `m_max` will both contain the index of this source.

Algorithm 4.2

```
max=V[1]; k_max=1; m_max=1;
  for i from 2 to M do
 {
    if (max<V[i]) then do {
                                  max=V[i]; k_max=i; m_max=i;
                                  pk_max=1; pm_max=1;
                                }
  for j=1 to i do
   {
    if | V[i] - V[j] |< (μ[i] - μ[j] )^2 then do {
         candidate_max = V[i] / 2 + V[j] / 2 + (V[i] - V[j] )^2 / (4(μ[i] - μ[j] )^2) + ((μ[i] - μ[j]) / 2)^2;
         if (max<candidate_max) then do
```

```
{
max=candidate_max; k_max=i;
m_max=j;
pk_max = 0.5 + (V[i] - V[j]) / (2 (μ[i] - μ[j])^2);
pm_max=1-pk_max;
}
}
}
}
```

Variables pk_max and pm_max contain the probabilities of sampling the two sources which yield the largest variance. As can be verified, the statements in the internal loop are executed only if the condition $|V[i]-V[j]|<(\mu[i]-\mu[j])^2$ is fulfilled, which indicates a local maximum.

5

Building Reliability and Risk Models

Failure rate and failure mode data are of vital importance for the reliability analysis. For the purposes of the reliability predictions, failure rates should be quoted for specific failure modes.

A reliability data record usually includes the following basic components (fields):

- *Inventory field*: Gives information on the particular component or subsystem
- *Failure mode field*: Describes the particular failure event including its effect on the operational state of the system
- *Failure time field*: Reflects the time to failure
- *Operational and environment conditions*

Industry-specific data are related to a particular industry. An example of an industry-specific data source is the Offshore REliability DAta (OREDA) database (OREDA, 1992) which contains field data. Failure rates are listed for a wide range of offshore hardware components – electromechanical, mechanical and hydraulic components – and for various environmental applications.

5.1 General Rules for Reliability Data Analysis

A basic initial step in reliability data analysis is to verify whether the data have been collected correctly since the quality of field data varies between misleading data and useful data. Without good quality data, there is no quality prediction, despite the amount of sophisticated analysis involved. A proper data collection should include data quality assurance. This involves auditing the source of data and specifying the procedures to be followed during data collection, recording and processing.

After the data collection, the exploratory data analysis is the next step. It involves summarising and examining the data in an exploratory way using plots, charts and histograms. The results from the preliminary data analysis are helpful for the next step, where an appropriate reliability and risk model is formulated. An examination of the data in Figure 5.1 for example, regarding life times, clearly indicates bimodality. There is a well-defined distribution of the lives of the weak sub-population and another

Reliability and Risk Models: Setting Reliability Requirements, Second Edition. Michael Todinov.

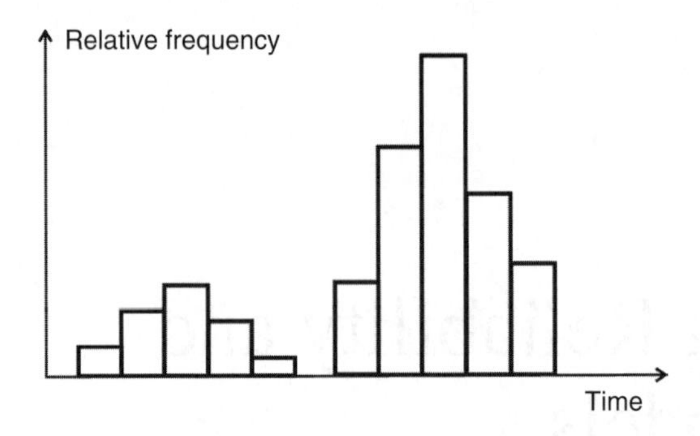

Figure 5.1 Bimodality in life data

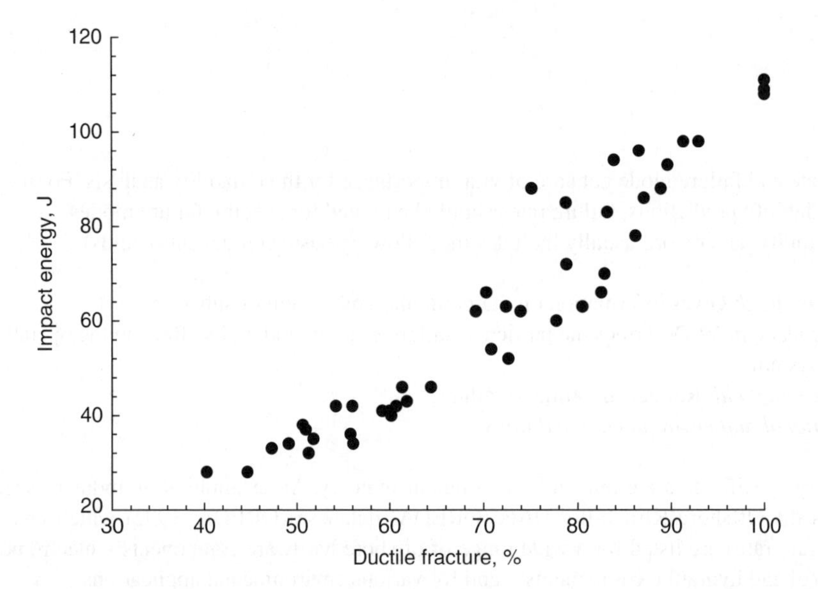

Figure 5.2 Scatter plot revealing a strong positive correlation between the Charpy impact energy and the percentage of ductile fracture (Todinov *et al.*, 2000)

distribution, characterising the lives of the normal population. As a result, none of the classical unimodal distributions will be appropriate for this particular data set.

Another example is presented in Figure 5.2, where the dependence of the Charpy impact energy on the percentage of ductile fracture over the fracture surface of a Charpy specimen was studied using a scatter plot. A strong positive correlation exists between the absorbed impact energy and the percentage of ductile fracture (Todinov *et al.*, 2000). Indeed, ductile fracture is associated with a large amount of absorbed impact energy as opposed to brittle fracture, which is associated with a small amount of absorbed impact energy.

Thorough understanding of the failure mechanisms and factors which control failure is required by the next step – formulation of a reliability (risk) model. It is particularly important to verify whether the selected random variables indeed control reliability (risk). If a well-established model already exists, the model formulation may be reduced to assessing whether the data conform to the model. In some cases, the reliability and risk models are obvious: for example, the homogeneous Poisson process for random arrivals, the Gaussian model for the distribution of the diameter of a machined component, the binomial distribution for statistically independent tests characterised by a constant probability of success in each test, etc. If a suitable model is not obvious, data can be presented in various plots which often suggest a suitable model. Another approach is to use a general model that is likely to include a suitable one as a special case. Such is, for example, the Weibull distribution which includes the negative exponential distribution as a special case.

Once the model has been formulated, it usually involves a number of parameters which need to be estimated from data. Figure 5.3 gives basic types of probability density distributions for the estimates $\hat{a}$ of a particular parameter a. The quality of the estimates varies widely from *unbiased* and *efficient*, characterised by a small error, to *biased* and *inefficient*, characterised by a large error (Figure 5.3).

The characteristics of a good reliability model match closely the characteristics of a good statistical model (Chatfield, 1998). The model should be:

- *Physically based* – providing *insight into the underlying physical mechanism* of the modelled random factor.
- *Parsimonious* – involving the smallest number of parameters necessary to describe the data set.
- *Robust* – small variations in the input data and in the estimates of the model parameters should lead to small variations in the model predictions.
- Capable of correct predictions outside the data range and providing a good fit within the data range.

Model building is not about getting the best fit to the observed data. It is about constructing a model which is *consistent with the underlying physical mechanism* of failure. The ability of a model to give a very good fit to a single data set may indicate little because a good fit can be achieved simply by including more parameters in the model. As a rule, over-parameterised models have poor predictive capability.

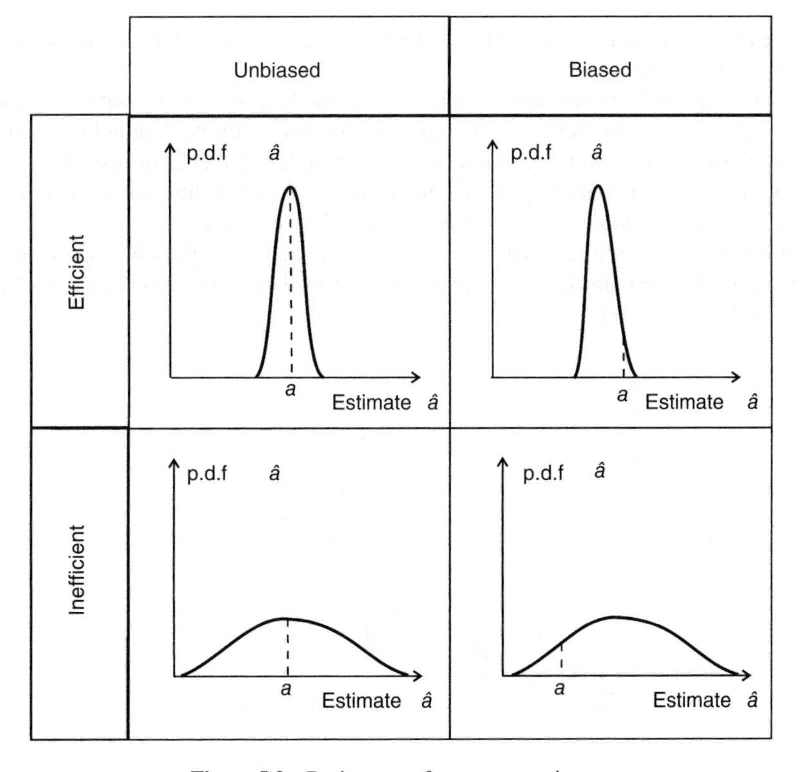

Figure 5.3 Basic types of parameter estimates

This point will be illustrated with Figure 5.4. For a set of $n+1$ points (x_0, y_0), …, (x_n, y_n) with distinct x-coordinates, it is always possible to construct a polynomial $f(x)$ of nth degree which passes through every single point.

For example, the polynomial (known as Lagrange's interpolation formula)

$$f(x) = y_0 L_0(x) + y_1 L_1(x) + \cdots + y_n L_n(x) \tag{5.1}$$

where

$$L_k(x) = \frac{(x-x_0)(x-x_1)\cdots(x-x_{k-1})(x-x_{k+1})\cdots(x-x_n)}{(x_k-x_0)(x_k-x_1)\cdots(x_k-x_{k-1})(x_k-x_{k+1})\cdots(x_k-x_n)}, \quad k = 0,1,\ldots,n$$

is a polynomial of degree n, which can be presented in the form

$$f(x) = a_0 + a_1 x + \cdots + a_n x^n \tag{5.2}$$

where a_i are coefficients.

The polynomial passes through every single point. Indeed, $L_k(x_j) = 0$ for $j \neq k$ and $L_k(x_j) = 1$ for $j = k$. Therefore, $f(x_j) = y_j$.

Although the polynomial of degree 5 provides a perfect fit to all data points in Figure 5.4, it is a model with poor predictive properties both within the data range and outside the data range. A simple polynomial $g(x)$

$$g(x) = b_0 + b_1 x + b_2 x^2 \tag{5.3}$$

of second degree may provide a good fit and predictive properties despite that the polynomial may not pass through any single data point.

A robust statistical model means that the output of the model is relatively insensitive to small variations in the input data, for example, system reliability insensitive to small variations in the reliabilities of the components. According to a discussion presented in Chapter 2, in case of a series system including a very large number of components, small errors in the reliability estimates of the individual components gives rise to a large error in the system reliability estimate.

A test on the suitability of a selected model is its approximation of the trivial extreme cases. The inability of a model to approximate correctly some of the trivial extreme cases is a clear indication that the selected model is incorrect.

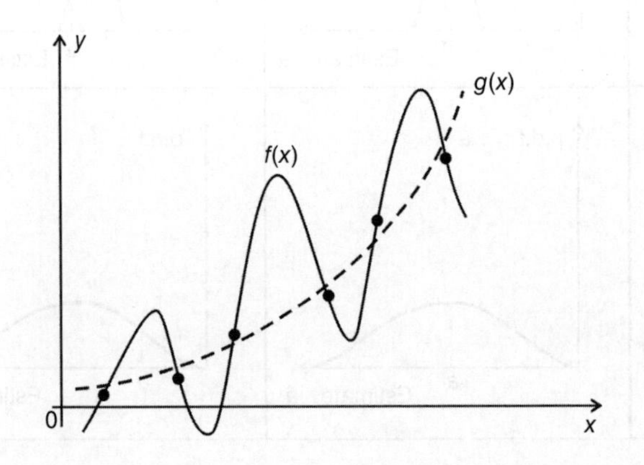

Figure 5.4 As a rule, over-parameterised models have poor predictive capability

A basic final step of model building is the model validation. It involves comparing the predictions of the model under the range of conditions for which it holds with experimental observations (field data).

5.2 Probability Plotting

Probability plotting provides a quick goodness-of-fit evidence whether a data set is consistent with the conjectured model. To construct a probability plot, the observations are first ranked in ascending order $x_{(1)}, x_{(2)}, \ldots, x_{(n)}$. This can be done by sorting the data values $x_1, x_2, \ldots, x_n$. If the number n of data values is large, sorting can be performed by some of the well-known methods (Heapsort, Quicksort, Bubble sort, Insertion, Shell, etc.), covered comprehensively in literature dealing with computer algorithms (Cormen *et al.*, 2001).

Next, an approximation $\hat{F}_i$ of the cumulative distribution function $F(t)$ is made. The plotting positions $\hat{F}_i = i / n$, where i are the indices of the ordered observations, provide a good choice for most applications. The *empirical cumulative distribution function* $\hat{F}_i = i / n$ gives the probability that the random variable t will be smaller than or equal to the value $x_{(i)}$ of the ith observation.

If the probability plot is subsequently used for estimating the parameters of the conjectured distribution, the best plotting positions will depend on the assumed model (Meeker and Escobar, 1998). A basic result in the probability theory states that the empirical cumulative distribution function $\hat{F}_i(x)$ converges to the true cumulative distribution function $F(x)$, in a probabilistic sense, as n grows (DeGroot, 1989). In other words, for a large sample, the empirical cumulative distribution is close to the true cumulative distribution with high probability.

At each value x_i, the empirical cumulative distribution function has a jump of height $1/n$, where n is the number of experimental observations (Figure 5.5). Building an empirical cumulative distribution can be done for the Charpy impact energy from the microstructural zones characterising multi-pass C–Mn welds (Figure 5.6).

The microstructure of the C–Mn multi-run welds is markedly inhomogeneous, consisting roughly of As-deposited metal (the dark zones in Figure 5.6) and reheated zones (the light zones) representing recrystallised weld metal. Each subsequent weld bead grain refines (normalises) part of the previous weld metal underneath, and refinements in microstructure result in improved toughness compared to the microstructure not affected by reheating.

The microstructural zone from which the sample is taken (in which the Charpy V-notch is cut (Figure 5.6b)) has a very strong influence on the distribution of the impact energy in the transition ductile-to-brittle region (Todinov *et al.*, 2000). This is illustrated by the separated distributions of the impact energy from the microstructural zones characterising the ductile-to-brittle transition region of multi-pass C–Mn welds (Figure 5.7).

The empirical distributions in Figure 5.7 provide strong evidence that the separate microstructural zones in multi-run C–Mn weld are characterised by distinct distributions of the Charpy impact energy.

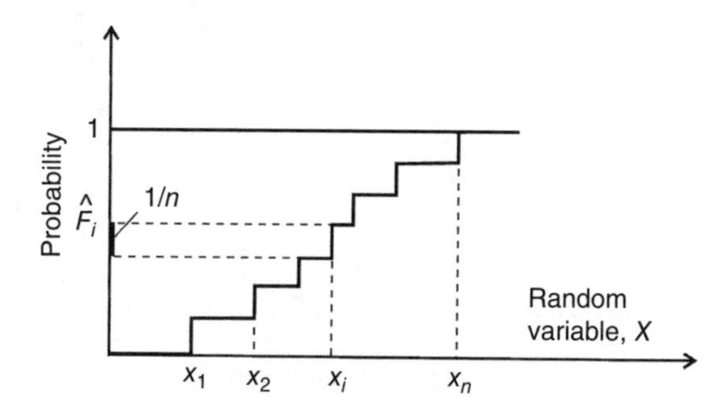

Figure 5.5 Building the empirical cumulative distribution of a random variable

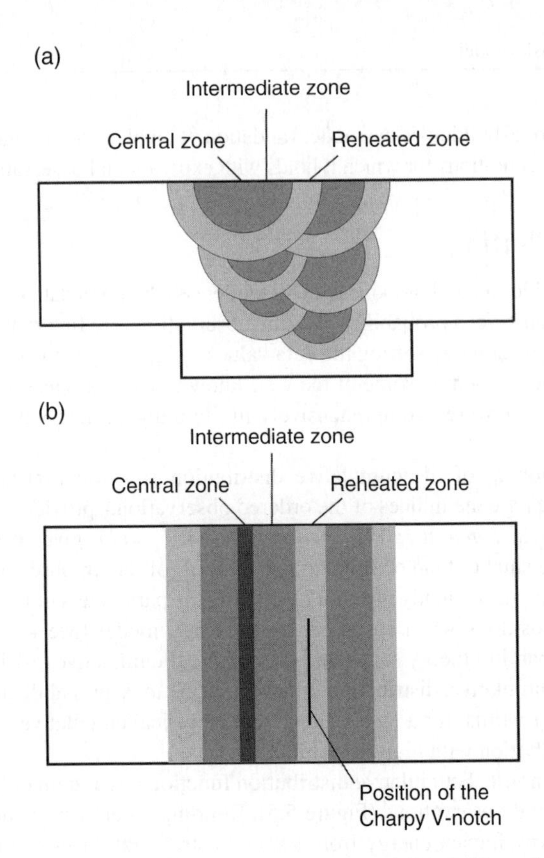

Figure 5.6 (a) A typical inhomogeneous microstructure characterising C–Mn multi-run welds. (b) Position of the Charpy V-notch

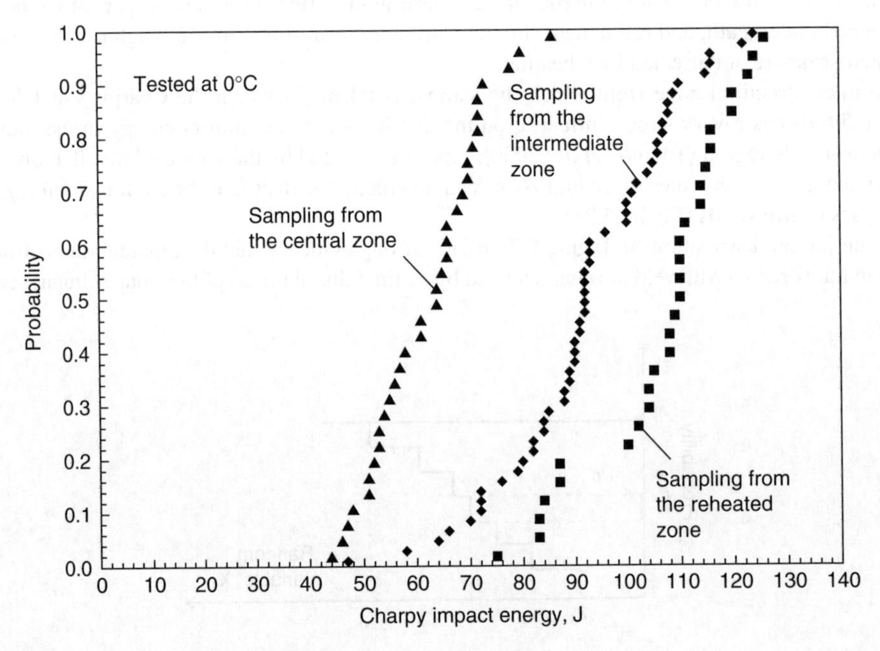

Figure 5.7 Cumulative empirical distribution of the Charpy impact energy characterising sampling from the separate microstructural zones of C–Mn multi-run welds (Todinov *et al.*, 2000)

The central zone is characterised by a poor impact toughness, the reheated zone is characterised by a relatively high impact toughness and the intermediate zone is characterised by an intermediate impact toughness (Figure 5.7).

The cumulative distribution function *F*(*t*) can be linearised by applying an appropriate transformation of the coordinate axes *t* and *F*(*t*). Let the linear transformation be defined by $z = kx + c$, where $z = \Omega(F(t))$ and $x = \Psi(t)$ are functions of *F*(*t*) and *t*.

If $z_i = \Omega(\hat{F}(t_i))$ are now plotted against $x_i = \Psi(t_i)$, the plotted points will fall approximately along a straight line if the data set is consistent with the conjectured statistical distribution *F*(*t*). If the plotted points deviate systematically from a straight line, the conjectured distribution is not consistent with the observations in the data set. A linear transformation is used because even slight deviations from a straight line indicating a departure from the conjectured model are easily identifiable. Fitting a straight line to the transformed data can be done using linear regression.

Once a candidate distribution has been identified, the next step is to estimate its parameters. Usually, the parameter estimates are obtained using the *method of maximum likelihood*. Probability plots, however, can also be used for estimating model parameters.

5.2.1 Testing for Consistency with the Uniform Distribution Model

Let the ordered sample values $x_{(1)}, x_{(2)}, \ldots, x_{(n)}$ are believed to be realizations of a uniformly distributed random variable *X* with probability density function defined by $1/(b-a)$ in the interval [*a*, *b*] and 0 elsewhere. The conjectured cumulative distribution function $F(x) = (x-a)/(b-a)$ of the random variable *X* can be presented as $z = kx + c$, where $z = F(x)$, $k = 1/(b-a)$ and $c = -a/(b-a)$.

Next, the cumulative distribution function has been approximated by $\hat{F}_i = i/(n+1)$, where *n* is the number of observations. The expression $\hat{F}_i = i/(n+1)$ estimates the probability that the random variable *X* will not exceed the observed value $x_{(i)}$. Next, the values $z_i = i/(n+1)$ are plotted against the ordered observations $x_{(1)}, x_{(2)}, \ldots, x_{(n)}$. If the ordered observations do indeed come from a uniform distribution, the points will fall approximately along a straight line. Estimates $\hat{a} = -c^*/k^*$ and $\hat{b} = (1-c^*)/k^*$ of the true values of the parameters *a* and *b* can be obtained from the slope k^* and the intercept c^* of the best-fit straight line. The best-fit straight line is usually determined using the method of least squares.

5.2.2 Testing for Consistency with the Exponential Model

In order to check whether a set of observations is consistent with the negative exponential distribution of a particular random variable *T*, an exponential probability plot can be made using an appropriate linear transformation of the axes. Since the conjectured cumulative distribution function of the random variable *T* is $F(t) = 1 - \exp(-\lambda t)$, by taking a logarithm, the negative exponential distribution can be transformed into an equation of a straight line:

$$z = -\ln\left(1 - F(t)\right) = \lambda t \tag{5.4}$$

Next, the data points are ranked in ascending order, $t_{(1)}, t_{(2)}, \ldots, t_{(n)}$, and the plotting positions are obtained from $\hat{F}_i = i/(n+1)$, which estimates the probability that the random variable *T* will be smaller than the *i*th observed value $t_{(i)}$. If the data set does come from an exponential distribution, the values $z_i = -\ln(1-\hat{F}_i)$ plotted against $t_{(i)}$, $i = 1, 2, \ldots, n$ should be close to a straight line. The slope $\hat{\lambda}$ of the best-fit straight line is an estimate of the unknown parameter λ in the negative exponential distribution.

5.2.3 Testing for Consistency with the Weibull Distribution

Weibull analysis may be carried out by using a very small number of observations (Abernethy, 1994; Dodson, 1994). The Weibull distribution $F(t) = 1 - \exp(-t/\eta)^m$ is linearised first, by taking a double logarithm:

$$\ln\left[\ln\left(\frac{1}{1-F(t)}\right)\right] = m\ln t - m\ln\eta \tag{5.5}$$

This is an equation of a straight line:

$$z = mx + c,$$

where

$$z = \ln\left[\ln\left(\frac{1}{1-F(t)}\right)\right], \quad x = \ln(t) \text{ and } c = -m\ln\eta$$

The ordered observations $t_{(1)} < t_{(2)} < \cdots < t_{(n)}$, are believed to be realizations of a Weibull-distributed random variable T. The median rank approximations for the Weibull cumulative distribution function are $\hat{F}_i = (i-0.3)/(n+0.4)$ (Abernethy, 1994; Dodson, 1994), where $i = 1,2,\ldots,n$ are the indices of the ordered observations and n is the number of observations. $\hat{F}_i$ gives the probability that the random variable T does not exceed $t_{(i)}$. From the estimated $\hat{F}_i$ values, the values $z_i = \ln[\ln(1/(1-\hat{F}_i))]$ are determined. Using the method of least squares, the slope $\hat{m}$ and the intercept $\hat{c}$ of the best-fit straight line (Figure 5.8) are obtained (Draper and Smith, 1981):

$$\hat{m} = \frac{\sum_{i=1}^{n} z_i(x_i - \bar{x})}{\sum_{i=1}^{n} (x_i - \bar{x})^2} \tag{5.6}$$

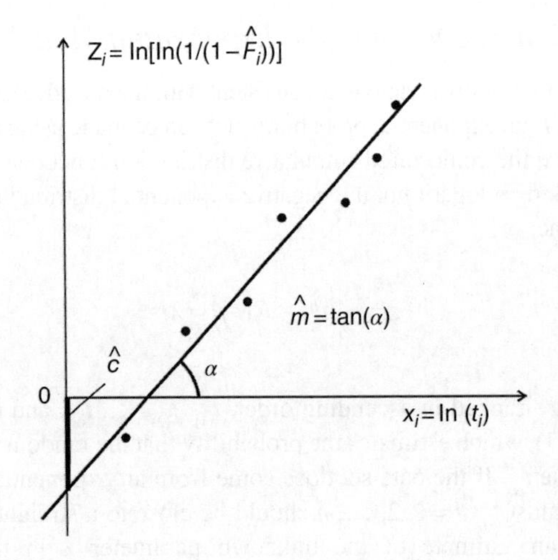

Figure 5.8 A probability plot for a two-parameter Weibull distribution

and

$$\hat{c} = \bar{z} - \hat{m}\,\bar{x} \tag{5.7}$$

where $\bar{x} = (1/n)\sum_{i=1}^{n} \ln t_i$, $\bar{z} = (1/n)\sum_{i=1}^{n} z_i$ and $x_i = \ln(t_i)$. The value $\hat{m}$ is an estimate of the exponent m in the Weibull distribution and $\hat{\eta} = \exp(-\hat{c}/\hat{m})$ is an estimate of the characteristic life η.

5.2.4 *Testing for Consistency with the Type I Extreme Value Distribution*

Suppose that the ordered observations are $x_{(1)} < x_{(2)} < \cdots < x_{(n)}$. The median rank approximations of the cumulative distribution function of the type I extreme value distribution are $\hat{F}_i = (i-0.4)/(n+0.2)$. If the extreme value distribution $F(x) = \exp[-\exp(-(x-\xi)/\theta)]$ is approximated by $\hat{F}_i = \exp[-\exp(-(x_{(i)}-\xi)/\theta)]$, taking natural logarithms twice gives

$$x_{(i)} = \xi + \theta\left[-\ln\left(-\ln \hat{F}_i\right)\right] \tag{5.8}$$

After setting $z_i = -\ln(-\ln \hat{F}_i)$, the values $x_{(i)}$ are plotted against z_i. If the plotted points fall approximately along a straight line, the observed values are consistent with the maximum extreme value model. In case of a good fit, the slope and the intercept of the best-fit line serve as estimates of the unknown parameters θ and ξ, respectively.

5.2.5 *Testing for Consistency with the Normal Distribution*

Let the ordered sample values be $x_{(1)} \le x_{(2)} \le \cdots \le x_{(n)}$, believed to be realizations of a random variable X following a normal distribution. If the sample $x_{(1)}, x_{(2)}, \ldots, x_{(n)}$ comes from a normal distribution with mean μ and standard deviation σ, $z_i = (x_{(i)} - \mu)/\sigma$ will follow the standard normal distribution. Estimates of z_i for n observations can be obtained from the standardised normal scores, which satisfy

$$\frac{i-0.5}{n} = \Phi\left(z_i\right) \tag{5.9}$$

where $\Phi(\bullet)$ is the cumulative distribution function of the standard normal distribution. $\Phi(z_i)$ from Equation 5.9 is equal to the probability $i-0.5/n$ that the random variable X will not exceed the observed value $x_{(i)}$ ($P(X \le x_{(i)})$). From this relationship, the normal scores z_i corresponding to the n observed values can be determined from $z_i = \Phi^{-1}((i-0.5)/n)$, where $\Phi^{-1}(\bullet)$ is the inverse function of $\Phi(\bullet)$. Values of $\Phi^{-1}(\bullet)$ are obtained from tables or by using numerical methods. The normal probability plot is constructed by plotting the obtained n standardised normal scores z_i ($i = 1, 2, \ldots, n$) against the ordered observations $x_{(i)}$.

If the data $x_{(i)}$ do come from a normal distribution, plotting $x_{(i)}$ versus the standard normal scores $z_i = \Phi^{-1}((i-0.5)/n)$ produces points falling approximately along a straight line $x_{(i)} = z_i\sigma + \mu$ with slope σ and intercept μ. A good linear fit indicates that the data are consistent with a normal distribution. In this case, the slope $\hat{\sigma}$ and the intercept $\hat{\mu}$ of the best-fit straight line are estimates of the standard deviation σ and mean μ of the normal distribution from which the data come from.

In Figure 5.9, the normal probability plotting technique has been illustrated with Charpy impact energy data related to the case where the Charpy V-notch (see Figure 5.6) has sampled from different microstructural zones of a multi-run weld (mixed sampling). In Figure 5.9, the systematic deviations of the plotted points from a straight line indicate that the normal distribution is not an appropriate

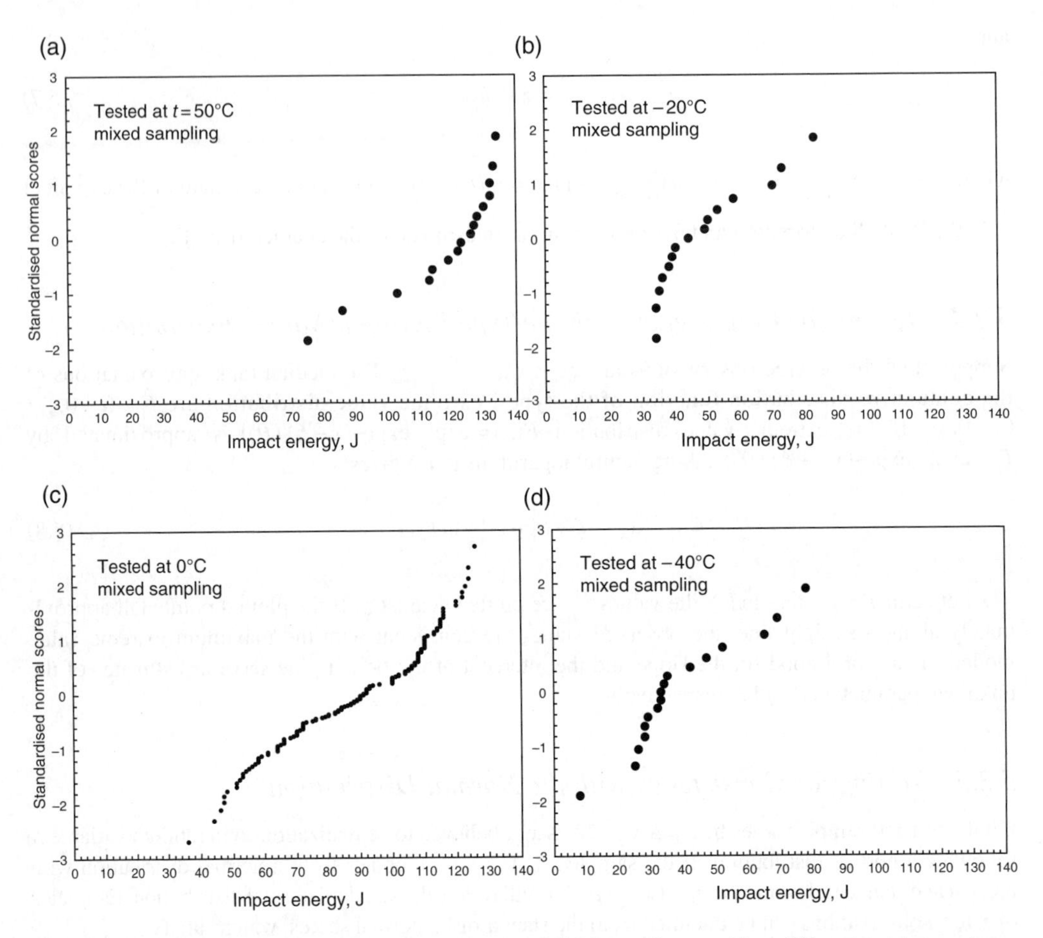

Figure 5.9 Normal plots of the Charpy impact energy of C–Mn multi-run welds at different test temperatures: (a) 50°C; (b) −20°C; (c) 0°C and (d) −40°C (Todinov *et al.*, 2000)

model for the Charpy impact energy of inhomogeneous multi-run welds (Todinov *et al.*, 2000). Consequently, for multi-run welds, a model of the variation of the Charpy impact energy assuming a Gaussian distribution is flawed. The variation of the Charpy impact energy of inhomogeneous welds in the case where different microstructural zones are sampled does not follow a Gaussian distribution. It is a mixture of distributions. The distribution of the impact toughness from sampling all three microstructural zones can be modelled by the distribution mixture

$$F(x) = p_1F_1(x) + p_2F_2(x) + p_3F_3(x) \tag{5.10}$$

where p_1, p_2 and p_3 are the probabilities of sampling the (i) central, (ii) intermediate and (iii) reheated zone (Figure 5.6) and $F_1(x)$, $F_2(x)$ and $F_3(x)$ are the cumulative distribution functions characterising the Charpy impact energy of the three microstructural zones. The mean of the Charpy impact toughness is

$$\mu_0 = p_1\mu_1 + p_2\mu_2 + p_3\mu_3 \tag{5.11}$$

where μ_1, μ_2 and μ_3 are the means of the Charpy impact energy characterising the separate microstructural zones. The variance of the Charpy impact energy from sampling all three microstructural zones is given by

$$V_0 = p_1\sigma_1^2 + p_2\sigma_2^2 + p_3\sigma_3^2 + p_1p_2(\mu_1 - \mu_2)^2 + p_2p_3(\mu_2 - \mu_3)^2 + p_1p_3(\mu_1 - \mu_3)^2$$

where σ_1, σ_2 and σ_3 are the standard deviations of the Charpy impact energy associated with the separate microstructural zones.

This example shows the dangers of assuming automatically Gaussian distribution for data obtained from pooling measurements from different sources even if the distribution of properties characterising each individual source is a Gaussian distribution. This point will be illustrated by the next example.

Example

The strength of components in each of n batches, each containing m_i components, follows a normal distribution with mean μ_i and a standard deviation σ_i, $i = 1, \ldots, n$. All batches have been pooled together in a single, large batch. What is the distribution of the strength of components in the large batch?

Give an expression for the mean and the standard deviation of the strength of components in the large batch.

Solution

The distribution of the strength of components in the single large batch is a mixture of n normal distributions, where $p_i = m_i / \sum_{k=1}^{n} m_k$ is the probability of sampling the ith batch. The distribution *is not* a Gaussian distribution. The mean of the distribution is

$$\mu = \sum_{i=1}^{n} p_i \mu_i$$

The standard error of the distribution is $\sigma = \sqrt{V}$, where

$$V = \sum_{i=1}^{n} p_i \left[\sigma^2 + (\mu_k - \mu)^2\right]$$

is the expression for determining the variance of a distribution mixture.

5.3 Estimating Model Parameters Using the Method of Maximum Likelihood

Suppose that the conjectured model (statistical distribution) $f(x, a)$ depends on a single parameter a. If the *likelihood function* $L(x_1,x_2,\ldots,x_n,a) = \prod_{i=1}^{n} f(x_i,a)$ has been defined, the maximum likelihood estimate $\hat{a}$ of the true parameter a is the value for which the likelihood function attains a maximum:

$$L(x_1,x_2,\ldots,x_n,\hat{a}) = \max_a \left(\prod_{i=1}^{n} f(x_i,a) \right) \quad (5.12)$$

Maximising the logarithm of the likelihood function yields the same maximum likelihood estimate $\hat{a}$:

$$\max_a L(x_1,x_2,\ldots,x_n,a) = \max_a \ln\left[L(x_1,x_2,\ldots,x_n,a)\right]$$

Setting the log-likelihood derivative to zero

$$\frac{d\ln L(x_1,x_2,\ldots,x_n,a)}{da}_{|a=\hat{a}} = 0$$

and solving the equation analytically with respect to a to find a potential candidate for a global maximum, or maximising the log-likelihood function numerically to find a global maximum, yields the estimate $\hat{a}$.

The method of maximum likelihood will be illustrated by estimating the parameter λ of the negative exponential distribution. Suppose that probability plotting has indicated that the data sample $(x_1, x_2, \ldots, x_n)$ is compatible with the negative exponential distribution $f(x) = \lambda\exp(-\lambda x)$. In order to obtain a maximum likelihood estimate of the only model parameter λ, the log-likelihood function is constructed first:

$$\ln L = \sum_{i=1}^{n} \ln(\lambda\exp(-\lambda x_i)) = n\ln\lambda - \lambda\sum_{i=1}^{n} x_i \tag{5.13}$$

Differentiating the log-likelihood function with respect to λ gives $d\ln L / d\lambda = (n/\lambda) - \sum_{i=1}^{n} x_i$. Setting the derivative to 0 and solving

$$\left(\frac{n}{\lambda}\right) - \sum_{i=1}^{n} x_i = 0$$

with respect to λ results in

$$\hat{\lambda} = \frac{n}{\sum_{i=1}^{n} x_i} \tag{5.14}$$

For this value, the second derivative $d^2\ln L / d\lambda^2 = -n/\lambda^2$ is negative; therefore, $\hat{\lambda}$ corresponds to a local maximum which is also the global maximum because the first derivative is positive and decreasing for all $\lambda < \hat{\lambda}$ and negative and decreasing for all $\lambda > \hat{\lambda}$. The value $\hat{\lambda}$ given by Equation 5.14 is the maximum likelihood estimate of the unknown parameter λ. Estimating the parameters of the rest of the models discussed earlier can also be done using the method of the maximum likelihood. Usually, the log-likelihood function cannot be maximised analytically except in cases involving a very small number of model parameters. Numerical optimisation methods are widely used as an alternative.

5.4 Estimating the Parameters of a Three-Parameter Power Law

From a set of measurements y_i, for the values x_i, $i = 1,2,\ldots,N$ of the controlled variable, we would like to obtain unbiased and efficient estimates of the parameters of the power law

$$y = k(x - x_0)^m \tag{5.15}$$

which appears in a number of applications.

A method based on correlation has been proposed in Todinov (2001b). An estimate of the unknown exponent m in Equation 5.15 is obtained by first transforming Equation 5.15 into

$$y^{1/m} = k^{1/m}(x - x_0) \tag{5.16}$$

Let us denote $z = y^{1/m}$. The values of the unknown exponent m are varied in the interval $0 < m \le m_{max}$, where m_{max} is an appropriately selected upper bound. Denoting $z_i = y_i^{1/m}$, the values z_i are plotted versus the observations x_i, and for each value of m, the correlation coefficient

$$\rho(m) = \frac{\sum_{i=1}^{N}(z_i - \bar{z})(x_i - \bar{x})}{s_z s_x} = \frac{\sum_{i=1}^{N} z_i(x_i - \bar{x})}{s_z s_x} \tag{5.17}$$

is determined, where

$$s_z = \sqrt{\sum_{i=1}^{N}(z_i - \bar{z})^2} \tag{5.18}$$

and

$$s_x = \sqrt{\sum_{i=1}^{N}(x_i - \bar{x})^2} \tag{5.19}$$

In Equations (5.17)-(5.18), x-bar and z-bar are the mean values of x_i and z_i ($i=1, ..., n$).

The value $\hat{m}$, for which the maximum value of the correlation coefficient of z_i versus x_i is attained, is an estimate of the unknown exponent m.

Next, the values z_i corresponding to the best estimate $\hat{m}$ are determined from $z_i = y_i^{1/\hat{m}}, i = 1, N$, where y_i are the measured values corresponding to x_i, ($i=1, ..., n$). From the plot z_i versus x_i, the slope

$$p = \frac{\sum_{i=1}^{N} z_i(x_i - \bar{x})}{\sum_{i=1}^{N}(x_i - \bar{x})^2} \tag{5.20}$$

and the intercept

$$c = \bar{z} - p\bar{x} \tag{5.21}$$

of the best-fit straight line are determined. The slope p is an estimate of $k^{1/\hat{m}}$ in Equation 5.16, and the intercept c is an estimate of $-k^{1/\hat{m}}x_0$ in Equation 5.16:

$$p = k^{1/\hat{m}}, \quad c = -k^{1/\hat{m}}x_0$$

The parameters k and x_0 in Equation 5.15 are therefore estimated from

$$\hat{k} = p^{\hat{m}} \tag{5.22}$$

and

$$\hat{x}_0 = -\frac{c}{p} \tag{5.23}$$

It is important to point out that the true value of the location parameter x_0 does not affect the estimate $\hat{m}$ of the shape parameter m. Indeed, let x_i be presented as $x_i = x_0 + \Delta x_i$ where Δx_i is the distance between x_i and x_0. The value Δx_i does not depend on x_0. The expressions $\sum_i (x_i - \bar{x})$ and $\sum_i (x_i - \bar{x})^2$ entering in Equation 5.17 for the correlation coefficient then become $\sum_i (\Delta x_i - \Delta\bar{x})$ and $\sum_i (\Delta x_i - \Delta\bar{x})^2$ $(\Delta\bar{x} = (1/N)\sum_i \Delta x_i)$.

Since neither of these depends on x_0, the value of the location parameter x_0 does not affect the estimate of the unknown exponent m. This constitutes the strength of the proposed method: *estimating the shape of the power curve defined by* Equation 5.15 *is separated from the estimation of its location along the x-axis*. The method yields practically unbiased estimates for the parameters (Todinov, 2001b).

Using appropriate transformations, a number of three-parameter models can be reduced to the power law (5.15) and the unknown parameters estimated using the proposed technique. Such is, for example, the three-parameter Weibull model:

$$F(x) = 1 - \exp\left[-k(x - x_0)^m\right] \tag{5.24}$$

which can be reduced to the three-parameter power law (5.15) by presenting it as $\exp[-k(x - x_0)^m] = 1 - F(x)$ and taking a logarithm from both sides. As a result, Equation 5.15 is obtained, where $y = -\ln[1 - F(x)]$.

5.4.1 Some Applications of the Three-Parameter Power Law

The three-parameter power law can, for example, be used to describe the corrosion kinetics of structural components. Thus, the model

$$d = k(t - t_0)^m \tag{5.25}$$

has been adopted by Dianqing *et al.* (2004) to describe a corrosion of ship structures. In this model, d is the corroded thickness, t is the time, t_0 is the coating life and k and m are constants. If N measurements d_i for the thickness of the corroded plate have been collected at N different times t_i, $i = 1, 2, \ldots, N$, using the technique from the previous section, the unknown parameters in Equation 5.25 can be estimated.

The model (5.15) has also been used to fit the systematic variation of the ductile-to-brittle transition temperature of steels (Todinov, 2001b). The systematic variation $E(x)$ of the Charpy impact energy (Figure 5.10) was modelled by

$$E(x) = E_L + (E_U - E_L)F(x) \tag{5.26}$$

where x denotes the temperature, E_L and E_U are the lower and the upper shelf Charpy impact energies (estimated from experimental data (Todinov *et al.*, 2000)) and $F(x)$ is the normalised Charpy impact energy $[E(x) - E_L]/(E_U - E_L)$ modelled by Equation 5.24.

If x in Equation 5.24 denotes 'time', the equation can be used for modelling the time evolution of transformed phase produced by nucleation and growth. For example, the quantity of transformed phase from a phase transformation with constant nucleation rate and radial growth rate can be approximated by the Kolmogorov–Johnson–Mehl–Avrami (KJMA) equation (Christian, 1965), whose functional form is Equation 5.24.

If x in Equation 5.24 denotes 'time', the equation can also be used for modelling the time evolution of sticking/jamming forces between sliding surfaces. Pilot experiments have indicated that

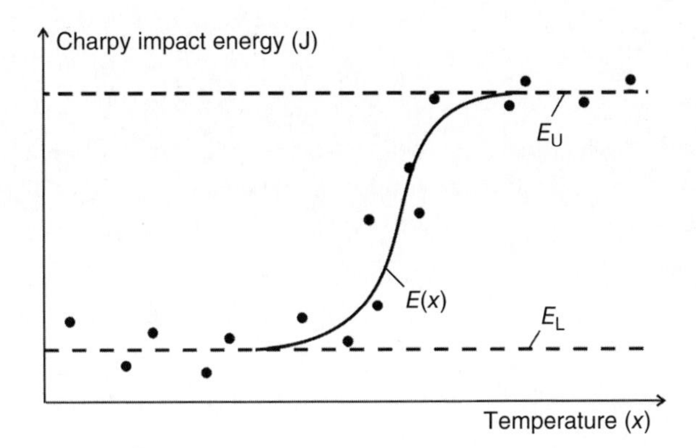

Figure 5.10 Systematic variation of the Charpy impact energy with temperature

Equation 5.26 could be used to describe the time evolution of the force necessary to separate by shear two jammed sliding surfaces. In this case, E_L in Equation 5.26 corresponds to the minimum level of the jamming force, and E_U corresponds to a 100% jamming attained after a significant amount of time; $F(x)$ is given by Equation 5.24.

6

Load–Strength (Demand-Capacity) Models

6.1 A General Reliability Model

Reliability is often derived from the probability of violating a limit (failure) state. A limited number of random variables $X_1, X_2, \ldots, X_n$ usually control the reliability of the system. Such are, for example, the critical design parameters *material properties*, *dimensions* and *loads*. Particular examples of random variables controlling reliability are the fracture toughness, the number density and the size of the flaws, the stress and strain magnitude, the magnitude of electrical power used by a consumer and the magnitude of electrical power produced by a generator. Random variables may not necessarily be statistically independent. Any set $x_1, x_2, \ldots, x_n$ of values for the controlling random variables can be presented as a point in the n-dimensional variable space. The *failure region F* contains all realisations (combinations of values for the controlling variables) that result in failure (*failure states*) as opposed to the *safe region S* containing all realisations that do not result in failure.

The surface $L(x_1, x_2, \ldots, x_n) = 0$ which divides the variable space into a failure and safe region is referred to as *failure surface*. The points on the failure surface are failure states.

Suppose that the controlling random variables are characterised by a joint probability density function $f(x_1, x_2, \ldots, x_n)$. The reliability on demand is then given by the integral

$$R = \int\int_{x_1, \ldots, x_n \in S} \cdots \int f(x_1, x_2, \ldots, x_n)\, dx_1\, dx_2 \ldots dx_n \tag{6.1}$$

where integration is carried out only over the safe region S (Melchers, 1999). Because the integration of the probability density function is performed only at points where no failure can occur, the sum of the probabilities over the safe region must necessarily equal the reliability on demand. If the controlling random variables are statistically independent, their joint probability density function $f(x_1, x_2, \ldots, x_n)$ can be factorised:

$$f(x_1, x_2, \ldots, x_n) = f(x_1) \times f(x_2) \times \cdots \times f(x_n)$$

Reliability and Risk Models: Setting Reliability Requirements, Second Edition. Michael Todinov.

and the reliability integral (6.1) becomes

$$R = \int_{x_1,\ldots,x_n \in S} \int \cdots \int f(x_1) f(x_2),\ldots, f(x_n) dx_1 dx_2 \ldots dx_n \quad (6.2)$$

where $f(x_1)$, $f(x_2)$, …, $f(x_n)$ are the marginal probability densities of the controlling random variables. Significant difficulties arise in solving the general reliability models (6.1)–(6.2), some of which can be summarised as follows:

1. Insufficient amount of data to define the joint probability distribution of the controlling random variables.
2. Difficulties in defining the integration domain (the safe region) S.
3. The multidimensional numerical integration is extremely time-consuming.

These difficulties no longer exist if Monte Carlo simulation is employed to evaluate the reliability integral.

6.2 The Load–Strength Interference Model

A common framework for predicting structural/mechanical reliability at a component level is the load–strength interference model (Freudenthal, 1954) which is related to the interaction of the load distribution and the strength distribution. If strength and load were constant as shown in Figure 6.1, no failure would occur (O'Connor, 2003).

Because of the variability of the load and strength and the interference (overlap) between the load distribution and the strength distribution, the reliability on demand is smaller than 100%.

The reliability on demand is determined by the probability of a relative configuration of the load and strength in which load is smaller than strength ($L < S$). Reliability on demand is controlled by two random variables *load* (L) and *strength* (S) characterised by distinct distributions (Carter, 1986; Freudenthal, 1954). A common application of these concepts is in mechanical overstress failures which occur whenever mechanical load exceeds mechanical strength or local stress exceeds local fracture toughness. There are cases however where failure occurs if the mechanical load cannot overcome mechanical resistance, for example, if a force is applied to open or close a safety-critical valve. In this case, failure occurs if the applied force cannot exceed the jamming (resisting) force.

Load and strength are much broader concepts than their mechanical interpretation. Any two interacting random parameters can be interpreted as 'load' and 'strength'. Load and strength, for

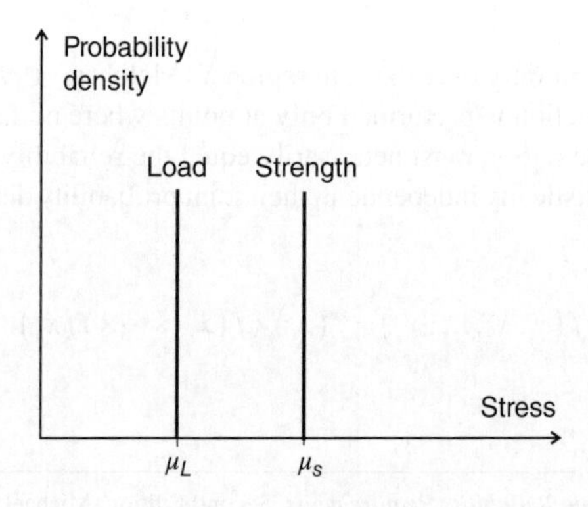

Figure 6.1 Constant values for the load and strength

example, can stand for demand and supply, rate of damage and rate of recovery, stress intensity and fracture toughness, etc. (Figure 6.2). Load and strength can even stand for identical in nature random entities. Such is, for example, the case where load stands for a particular critical parameter (temperature, potential, pressure, concentration, etc.) at a particular point *A* and strength stands for the same critical parameter (temperature, potential, pressure, concentration, etc.) at another point *B*. In this case, failure occurs if the value of the parameter at point *A* exceeds that at point *B* by a particular quantity. The roles of load and strength are reversed, if failure occurs also if the value of the parameter at point *B* exceeds the value of the parameter at point *A*.

The load–strength interference model is a special case of the general reliability model, where only two controlling random variables are present: X_1 (*strength*) and X_2 (*load*). In this case, the failure surface is $L(X_1, X_2) \equiv X_1 - X_2 = 0$. Suppose that the load–strength joint probability density function is $f(x_1, x_2)$ and the strength and load are distributed in the domain defined by $x_{1,\min} \le x_1 \le x_{1,\max}$ and $x_{2,\min} \le x_2 \le x_{2,\max}$, respectively (Figure 6.3). The values $x_{1,\min}$ and $x_{1,\max}$ are the lower and the upper limit of strength, while the values $x_{2,\min}$ and $x_{2,\max}$ are the lower and the upper limit of load.

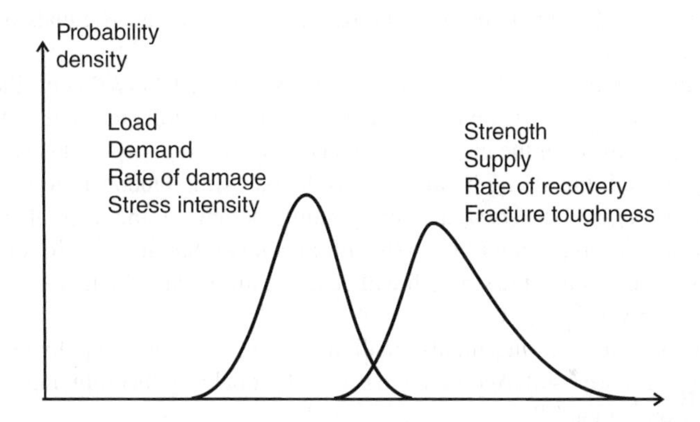

Figure 6.2 The universal nature of the load–strength interference

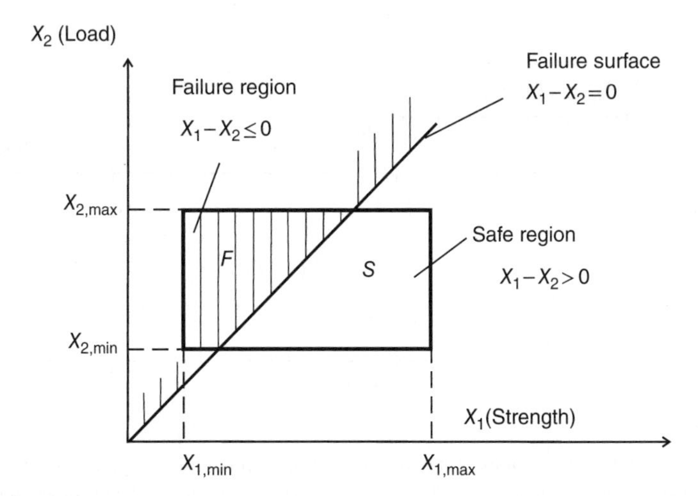

Figure 6.3 Failure region, safe region and failure surface in load–strength interference

According to the general reliability model (6.1), reliability on demand is given by the integral

$$R = \int \int_{x_1, x_2 \in S} f(x_1, x_2) dx_1 dx_2 \tag{6.3}$$

over the safe region S from the rectangular area in Figure 6.3 where $X_1 - X_2 > 0$. If strength and load are statistically independent random variables, the reliability integral becomes

$$R = \int \int_{x_1, x_2 \in S} f(x_1) f(x_2) dx_1 dx_2 \tag{6.4}$$

6.3 Load–Strength (Demand-Capacity) Integrals

In the derivations to follow, load and strength will be denoted by indices L and S. Their probability density functions will be denoted by $f_L(x)$ and $f_S(x)$; the cumulative distribution functions will be denoted by $F_L(x)$ and $F_S(x)$, respectively.

If a *load–strength interference* is present, the reliability on demand R equals the probability that strength will be greater than load.

For the lower and the upper bounds S_{min} and S_{max} of the strength, $f_S(x) = 0$ is fulfilled for $x \le S_{min}$ and $x \ge S_{max}$. If the interval (S_{min}, S_{max}) is divided in non-intersecting elementary intervals Δ_i, the events A_i denoting that strength will lie in the elementary interval Δ_i are mutually exclusive ($A_i \cap A_j = \varnothing$) and exhaustive events. Indeed, the strength value cannot belong simultaneously to two non-intersecting intervals Δ_i and Δ_j $(i \neq j)$, and the strength value certainly belongs to the interval $[S_{min}, S_{max}]$.

The probability that the strength will be in the infinitesimal interval $x, x + dx$ is $f_S(x)\,dx$. The probability of the compound event that the strength will be in the infinitesimal interval $x, x + dx$ and the load will not be greater than x is $F_L(x) f_S(x)\,dx$.

The strength can be in any infinitesimally small interval $x, x + dx$ between the lower bound S_{min} and the upper bound S_{max} (Figure 6.4a). According to the total probability theorem, integrating $F_L(x) f_S(x)\,dx$ gives the reliability on demand R:

$$R = \int_{S_{min}}^{S_{max}} f_S(x) F_L(x) dx \tag{6.5}$$

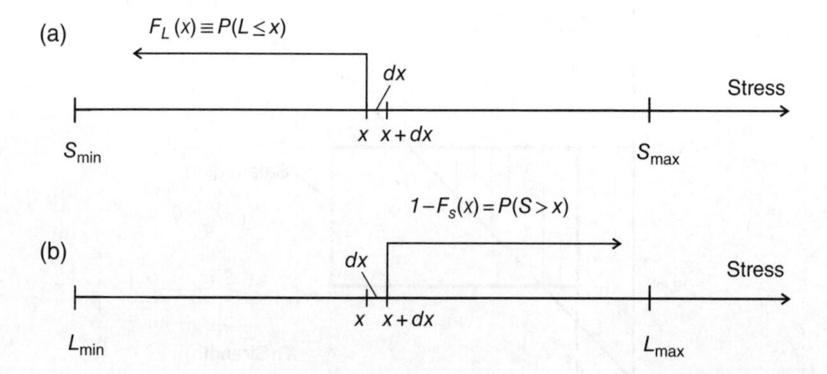

Figure 6.4 Possible safe configurations of the load and strength. Integration is performed between (a) S_{min} and S_{max} – the lower and the upper bound of the strength; (b) between L_{min} and L_{max} – the lower and the upper bound of the load

An alternative form of the *load–strength interference integral* can be derived if integration is performed between the lower and the upper bound of the load (Figure 6.4b).

For the lower and upper bounds L_{min} and L_{max} of the load, $f_L(x)=0$ is fulfilled for $x \le L_{min}$ and $x \ge L_{max}$.

If the interval (L_{min}, L_{max}) is divided in non-intersecting elementary intervals Δ_i, the events A_i denoting that load will lie in the elementary interval Δ_i are mutually exclusive ($A_i \cap A_j = \emptyset$) and exhaustive events. Indeed, the load magnitude cannot possibly belong simultaneously to two non-intersecting intervals Δ_i and Δ_j $(i \ne j)$, and the load magnitude certainly belongs to the interval (L_{min}, L_{max}).

The probability that the load will be in the infinitesimal interval $x, x+dx$ is $f_L(x)\,dx$. The probability of the compound event that the load will be in the infinitesimal interval $x, x+dx$ and the strength will be greater than $x+dx$ is $[1-F_S(x)]\,f_L(x)dx$. Because the load can be in any infinitesimal interval $x, x+dx$ between its lower bound L_{min} and upper bound L_{max}, according to the total probability theorem, integrating $[1-F_S(x)]f_L(x)\,dx$ gives the reliability on demand:

$$R=\int_{L_{min}}^{L_{max}} f_L(x)\left[1-F_S(x)\right]dx = 1-\int_{L_{min}}^{L_{max}} f_L(x)F_S(x)dx \tag{6.6}$$

Suppose that the component is subjected to statistically independent and identically distributed loads. For n load applications, the probability that the loads $L_1, L_2, \ldots, L_n$ will be smaller than a particular value x of the strength is

$$P\left(L_1 \le x \text{ and } L_2 \le x \ldots \text{and } L_n \le x\right) = F_L^n(x) \tag{6.7}$$

where $F_L(x)$ is the cumulative distribution of the load. The probability of the compound event that strength will be in the infinitesimal interval $x, x+dx$ and the loads from all load applications will be smaller than x is $F_L^n(x)f_S(x)\,dx$. Because the strength can be in any infinitesimally small interval $x, x+dx$ between S_{min} and S_{max}, according to the total probability theorem, integrating $F_L^n(x)f_S(x)\,dx$ gives

$$R=\int_{S_{min}}^{S_{max}} f_S(x)F_L^n(x)dx \tag{6.8}$$

for the reliability on demand associated with n statistically independent load applications.

Often, particularly in problems related to demand and supply, the question of interest is the probability that supply exceeds demand by a value greater than a (Figure 6.5). In what follows, supply and demand will be denoted by indices S and D. Their probability density distributions will be denoted by $f_S(x)$ and $f_D(x)$; their cumulative density distributions will be denoted by $F_S(x)$ and $F_D(x)$, respectively. The probability of the compound event that the supply will be in the infinitesimal interval $x, x+dx$ and the demand will be smaller than or equal to $x-a$ is $F_D(x-a)f_S(x)\,dx$.

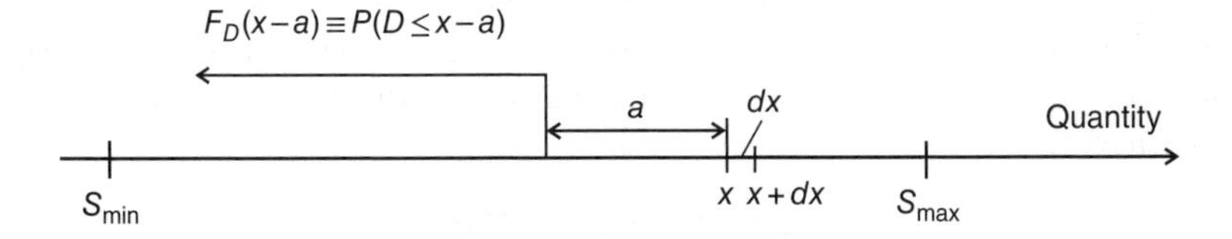

Figure 6.5 Relative configurations of the supply and demand which guarantee that supply exceeds demand by a quantity larger than a

According to the total probability theorem, integrating $F_D(x-a)\, f_S(x)\, dx$ yields

$$R = \int_{S_{\min}}^{S_{\max}} f_S(x)\, F_D(x-a)\, dx \tag{6.9}$$

for the probability that supply will exceed demand by a quantity larger than a. $S_{\min}$ and $S_{\max}$ are the lower and upper bounds of the supply, where $f_S(x) = 0$ for $x \le S_{\min}$ and $x \ge S_{\max}$.

6.4 Evaluating the Load–Strength Integral Using Numerical Methods

The maximum extreme value distribution is often a good model for the maximum load. If the cumulative distribution function $F_L(x)$ of the loading stress is given by

$$F_L(x) = \exp\left[-\exp\left(-\frac{x-\xi}{\theta}\right)\right] \tag{6.10}$$

and the probability density distribution $f_S(x)$ of the strength has been approximated by the three-parameter Weibull distribution

$$f_S(x) = \frac{m}{\eta}\left(\frac{x-x_0}{\eta}\right)^{m-1} \exp\left[-\left(\frac{x-x_0}{\eta}\right)^{m}\right] \tag{6.11}$$

the reliability on demand can be determined from

$$R = \int_{S_{\min}}^{S_{\max}} F_L(x) f_S(x)\, dx \tag{6.12}$$

This integral can be solved numerically using, for example, the *Simpson's method*. This approach will be illustrated by a numerical example related to calculating the risk of failure of a critical component.

Example

A data set is given, regarding the strength of a component (yield strength, fracture toughness, bending strength, fatigue strength, etc.). By using the methods from Chapter 5, the strength has been approximated by the three-parameter Weibull distribution (6.11) with parameters:

$$m = 3.9;\ \eta = 297.3 \text{ and } x_0 = 200 \text{ MPa}$$

A data set is also given, regarding the maximum load over a number of consecutive time intervals (e.g. days, months or years). The measurements have been transformed into a set of calculated maximum loading stress ranges over the specified time intervals. By using the methods from Chapter 5, the appropriate model for the load was found to be the maximum extreme value distribution (6.10) with estimated parameters:

$$\xi = 119.0 \text{ and } \theta = 73.64$$

Find the probability of failure of the critical component.

Solution

Using the numerical values of the parameters, and a sufficiently large value $S_{max} = 3000\,\text{MPa}$ for the upper integration limit of the strength, the value 0.985 is calculated for the reliability on demand:

$$\int_{200}^{3000} \exp\left[-\exp\left(-\frac{x-119}{73.64}\right)\right] \times \frac{3.9}{297.3}\left(\frac{x-200}{297.3}\right)^{3.9-1} \exp\left[-\left(\frac{x-200}{297.3}\right)^{3.9}\right] dx = 0.985$$

The required probability of failure is $1-0.985 = 0.015$. In Chapter 9, this probability will be confirmed by a Monte Carlo simulation.

6.5 Normally Distributed and Statistically Independent Load and Strength

Load (L) and strength (S) are assumed to be statistically independent, *normally distributed* random variables $N\left(\mu_L, \sigma_L^2\right)$ and $N\left(\mu_S, \sigma_S^2\right)$, where μ_L and μ_S are the means and σ_L and σ_S are the standard deviations of the load and strength distribution, respectively.

The random variable $y = S - L$ is normally distributed because y is a sum of normally distributed random variables. Its expected value is

$$E(y) = E(S) - E(L) = \mu_y = \mu_S - \mu_L \tag{6.13}$$

Note that the variance of a random variable X multiplied by a constant c is given by

$$V(cX) = c^2 V(X)$$

Consequently, the variance $V(y) = V(S - L)$ becomes

$$V(y) = \sigma_y^2 = V(S) + V(-L) = V(S) + (-1)^2 V(L) = \sigma_S^2 + \sigma_L^2 \tag{6.14}$$

Because y is normally distributed, with mean μ_y and standard deviation σ_y (Figure 6.6), the probability of failure $P(y \le 0)$ can be found using the linear transformation $z = (0 - \mu_y)/\sigma_y$. This is needed for calculating the probability $P(y \le 0)$ using the standard normal distribution: $P(y \le 0) = \Phi(z)$.

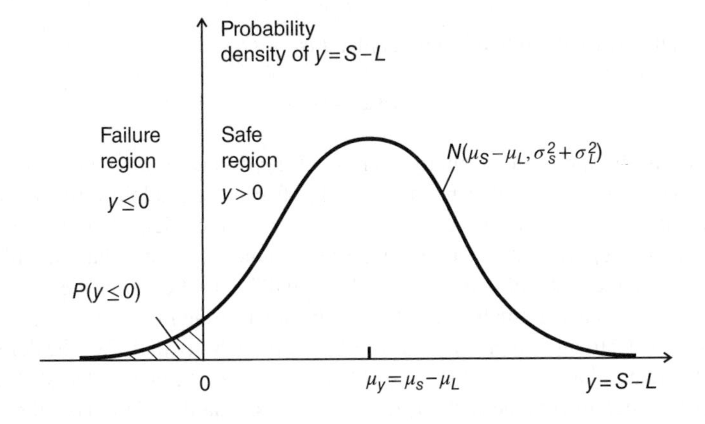

Figure 6.6 A normal probability density distribution of the difference $y = S - L$

Setting

$$\beta = \frac{\mu_y}{\sigma_y} = \frac{\mu_S - \mu_L}{\sqrt{\sigma_S^2 + \sigma_L^2}} \tag{6.15}$$

gives

$$P(y \le 0) = \Phi(-\beta)$$

for the probability of failure, where Φ(•) is the cumulative distribution function of the standard normal distribution.

In Equation 6.15, the quantity β, also known as the *reliability index* or *safety margin*, is an important reliability parameter which measures the relative separation of load and strength. Reliability on demand is determined from

$$R = 1 - \Phi(-\beta) = 1 - [1 - \Phi(\beta)] = \Phi(\beta) \tag{6.16}$$

A table containing the area under the standard normal probability density function is given in Appendix C.

Here, it needs to be pointed out immediately that *the safety margin has a meaning only for load and strength following the Gaussian distribution.* Later, it will be shown that in the general case, a low safety margin does not necessarily mean low reliability and vice versa.

Example

The strength of a structural component is normally distributed with mean 800 MPa and standard deviation 40 MPa. The load is also normally distributed with mean 700 MPa and standard deviation 30 MPa. Calculate the reliability on demand for the component if the load and strength are statistically independent.

Solution

For statistically independent load and strength, the reliability index is

$$\beta = \frac{\mu_S - \mu_L}{\sqrt{\sigma_S^2 + \sigma_L^2}} = \frac{800 - 700}{\sqrt{40^2 + 30^2}} = 2$$

According to Equation 6.16, the reliability on demand is

$$R = \Phi(2) = 0.977$$

Equation 6.16 can also be applied to normally distributed *supply* and *demand*. Demand (D) and supply (S) are assumed to be statistically independent, normally distributed random variables $N(\mu_D, \sigma_D^2)$ and $N(\mu_S, \sigma_S^2)$, where μ_D, μ_S are their means and σ_D, σ_S are their standard deviations, respectively.

The probability that supply will exceed demand by an amount greater than a, called the reserve (Figure 6.7), is equal to the probability that $y = S - D > a$ will be fulfilled. Considering that the random variable $y = S - D - a$ is also normally distributed, the load–strength interference formula can be applied. Using the load–strength interference formula, the probability that after the demand there will remain a quantity greater than or equal to a ($S - D \ge a$ or $S \ge D + a$) can be calculated. $D + a$ is also a normally distributed random variable with mean $\mu_D + a$ and standard deviation σ_D (the variance of the constant a is zero). In other words, by increasing the demand with the required reserve a, the problem

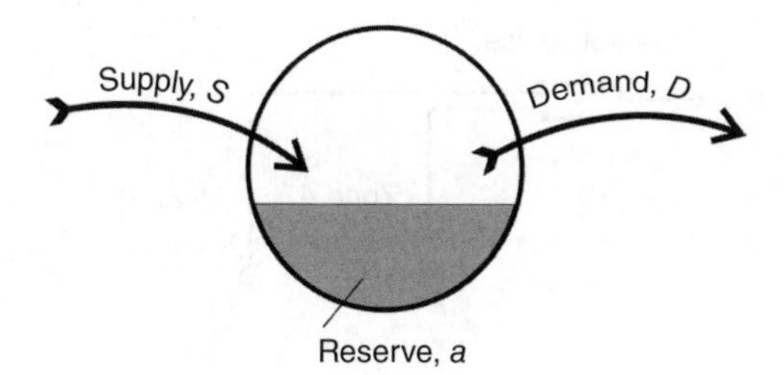

Figure 6.7 Supply, demand and reserve

is reduced to the familiar load–strength interference problem. Consequently, the probability that supply will exceed demand by at least a becomes

$$P(S-D>a)=\Phi\left(\frac{\mu_S-(\mu_D+a)}{\sqrt{\sigma_S^2+\sigma_D^2}}\right) \tag{6.17}$$

Assume that the supply is composed of contributions from M suppliers and the demand is composed of the consumption from N consumers. All supplied and consumed quantities are assumed to be normally distributed or constants. A constant supply or demand of magnitude c can formally be interpreted as 'normally distributed', with mean c and standard deviation zero. The total supply in this case is normally distributed, with mean $\mu_S=\sum_{i=1}^{M}\mu_{Si}$ and variance $\sigma_S^2=\sum_{i=1}^{M}\sigma_{Si}^2$ where μ_{Si} and σ_{Si} are the mean and the standard deviation associated with the ith supplier. The total demand is also normally distributed, with mean $\mu_D=\sum_{i=1}^{N}\mu_{Di}$ and variance $\sigma_D^2=\sum_{i=1}^{N}\sigma_{Di}^2$ where μ_{Di} and σ_{Di} are the mean and the standard deviation of the ith consumer. Substituting these in Equation 6.17 gives

$$P(S-D\geq a)=\Phi\left(\frac{\sum_{i=1}^{M}\mu_{Si}-\left(\sum_{i=1}^{N}\mu_{Di}+a\right)}{\sqrt{\sum_{i=1}^{M}\sigma_{Si}^2+\sum_{i=1}^{N}\sigma_{Di}^2}}\right) \tag{6.18}$$

The next examples illustrate the wide application area of the load–strength interference models.

Example

The magnitude of the temperature is measured continuously by two thermocouples working independently from each other in two different zones of a system, as shown in Figure 6.8. The measured temperatures in the two zones follow normal distribution and are compared by a control device which issues a signal for a system shutdown if the absolute value of the difference between the measured temperatures in the two zones exceeds 25°C. The mean value of the measured temperature in zone A is 80°C, with a standard deviation 16°C. The mean value of the measured temperature in zone B is 84°C with a standard deviation 8°C. Calculate the probability of a shutdown signal because of excessive temperature difference between the two zones.

Solution

Let us denote the two zones by A and B. Let event $E_1\equiv a>b+25$ be 'the temperature value a measured in section A exceeds the temperature value b measured in zone B by more than 25°'. Let event $E_2\equiv b>a+25$ be 'the temperature value b measured in section B exceeds the temperature value a measured in zone A by more than 25 °C'.

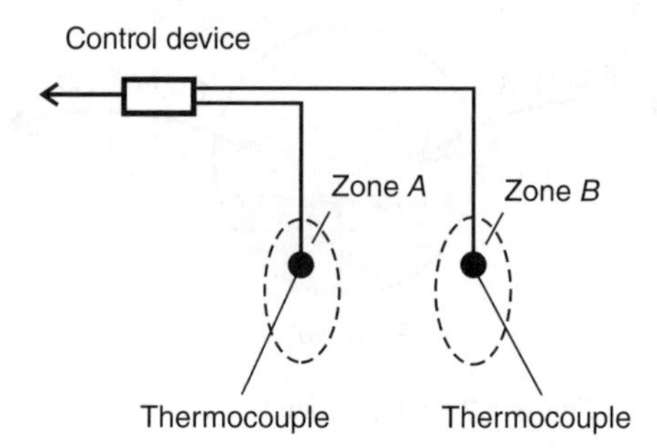

Figure 6.8 Two thermocouples measuring temperature independently from each other in two different zones of a system

The probability $P(E_1)$ of event E_1 can be determined by using the load–strength interference formula, where μ_A, μ_B are the means and σ_A, σ_B are the standard deviations of the temperatures measured in zones A and B:

$$P(a > b+25) = \Phi\left(\frac{\mu_A - (\mu_B + 25)}{\sqrt{\sigma_A^2 + \sigma_B^2}}\right) = \Phi\left(\frac{80-84-25}{\sqrt{16^2+8^2}}\right) = \Phi(-1.62) \approx 0.053$$

Similarly, the probability $P(E_2)$ of event E_2 can be determined by using the load–strength interference formula, where again μ_A, μ_B are the means and σ_A, σ_B are the standard deviations of the temperatures measured in zones A and B:

$$P(b > a+25) = \Phi\left(\frac{\mu_B - (\mu_A + 25)}{\sqrt{\sigma_A^2 + \sigma_B^2}}\right) = \Phi\left(\frac{84-80-25}{\sqrt{16^2+8^2}}\right) = \Phi(-1.17) \approx 0.12$$

The absolute value of the temperature difference between the two zones will be larger than 25°C when either event E_1 or event E_2 occurs. Because events E_1 and E_2 are mutually exclusive events, the probability of their union is a sum of the probabilities of the separate events:

$$P(E_1 \cup E_2) = P(E_1) + P(E_2)$$

The probability that the absolute value of the temperature difference will be larger than 25°C is

$$P(|a-b| > 25) = P(a > b+25) + P(b > a+25) = 0.053 + 0.12 = 0.173$$

Example

In the assembly presented in Figure 6.9, a slider with width b moves into a groove with width B. The width B of the groove is normally distributed, with mean 80 mm and standard deviation 0.5 mm; the width of the slider is also normally distributed, with mean b=79 mm and standard deviation 0.3 mm. In order to avoid jamming, the clearance defined as $\Delta = B - b$ must be at least 0.4 mm. In order to avoid loss of precision, the clearance $\Delta = B - b$ must not exceed 1.5 mm.

Calculate the percentage of faulty assemblies for which either jamming or loss of precision is present.

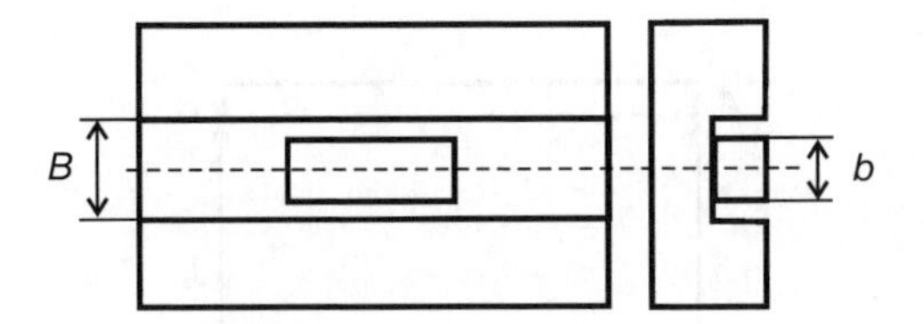

Figure 6.9 A slider assembly subject to two failure modes - loss of precision and jamming

Solution

The probability that the clearance $\Delta = B - b$ will be at least 0.4 mm is essentially the probability that the width of the groove B will be larger than the width of the moving part (the slider) by at least 0.4 mm.

According to the load–strength interference theory, the probability that no jamming will occur is

$$P(B > b + 0.4) = \Phi\left(\frac{80 - (79 + 0.4)}{\sqrt{0.5^2 + 0.3^2}}\right) = \Phi(1.03) = 0.848$$

The probability of jamming is therefore

$$P(B - b < 0.4) = 1 - 0.848 = 0.152$$

Next, according to the load–strength interference model, the probability of 'loss of precision' is

$$P(B > b + 1.5) = \Phi\left(\frac{80 - (79 + 1.5)}{\sqrt{0.5^2 + 0.3^2}}\right) = \Phi(-0.857) = 1 - \Phi(0.857) = 0.196$$

Because 'jamming' and 'loss of precision' are mutually exclusive events, the probability of a faulty assembly ('jamming' or 'loss of precision') is equal to the sum of the probabilities of the events 'jamming' and 'loss of precision'.

The probability of a faulty assembly is therefore $0.152 + 0.196 = 0.348$.

Consequently, approximately 35% of the assemblies will be faulty.

Example

Two key chemical ingredients A and B are produced by two different teams in a plant which works continuously, in three shifts. Both ingredients go into a final product manufactured by the plant and must be added simultaneously into the product. The time of the team producing ingredient A is normally distributed, with mean 120 minutes and standard deviation $\sigma_A = 15$ minutes. The time of the team producing ingredient B is also normally distributed, with mean 120 minutes and standard deviation $\sigma_B = 20$ minutes.

Ingredient A deteriorates if it stays for more than 30 minutes before being added to the final product. Ingredient B deteriorates if it stays for more than 40 minutes before being added to the final product.

Calculate the probability that the plant will fail to manufacture the final product because one of the key ingredients will have deteriorated before the other key ingredient is produced.

Example

The ends A and B of a part with length $L = 150$ mm are fixed at the same distance h from the baseline C (Figure 6.10) so that the part is parallel to the baseline C. Because of imprecision associated with fixing the ends of the part, each of the distances h varies independently of the other, by following a normal distribution with mean 250 mm and standard deviation 1 mm. As a result of this variation, the part is actually inclined at a certain angle with respect to the baseline C. If the acute angle which the part subtends with the baseline C exceeds 1° (≈0.01745 rad), the part cannot perform its function.

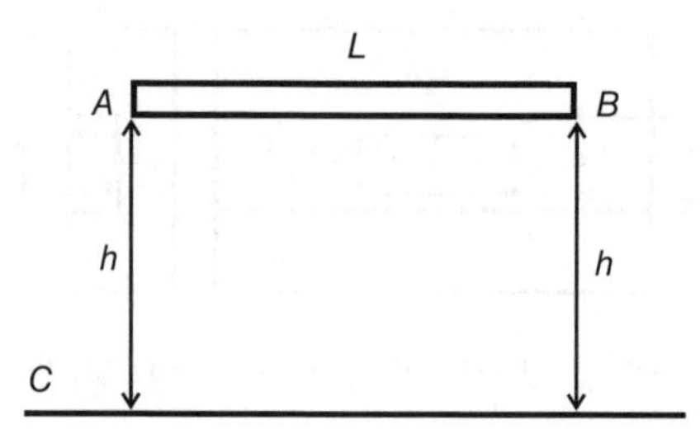

Figure 6.10 Fixing a part parallel to the baseline C is always associated with inaccuracy. As a result, the part may not be capable of performing its function

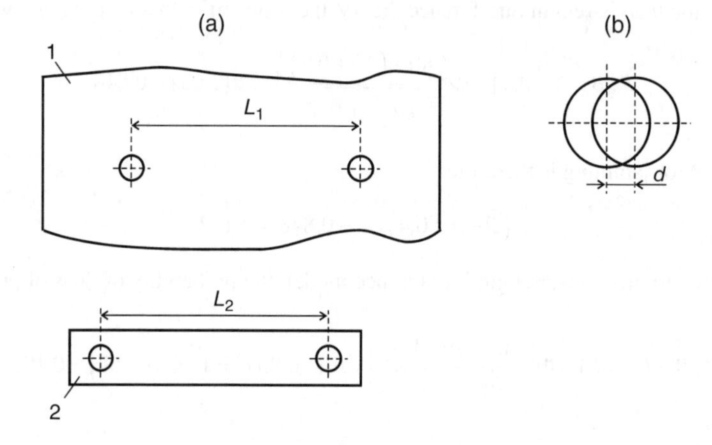

Figure 6.11 (a) Two plates with bolt holes of the same diameter and at the same distances L_1 and L_2; (b) because of variations of the distances L_1 and L_2 the centers of the bolt holes do not coincide

Calculate the probability that the part will not be capable of performing its function because of excessive inclination.

Example

The bolt connection between the plate 1 and the plate 2 in Figure 6.11 can only be made, if the distance d between the centres of the matching holes on plate 1 and plate 2 does not exceed 0.3 mm (Figure 6.11b). The bolt holes in plate 1 and plate 2 are machined independently with mean distances $L_1 = L_2 = 200$ mm between the centres of the bolt holes. Both distances L_1 and L_2 are normally distributed and characterised by the same standard deviation of 0.5 mm. Calculate the percentage of defective assemblies.

6.6 Reliability and Risk Analysis Based on the Load–Strength Interference Approach

6.6.1 Influence of Strength Variability on Reliability

Figure 6.12 illustrates a case where low reliability is a result of large variability of the strength. Large variability of strength is caused by the presence of weak (substandard) items due to poor material properties, manufacturing, assembly and quality control.

The large variability of strength leads to a large overlap of the lower tail of the strength distribution and the upper tail of the load distribution. A large overlap causes low reliability and can be decreased by a high-stress burn-in or proof testing which cause weak items to fail. The resultant distributions (Figure 6.13) are characterised by a small or no overlap (Carter, 1986; O'Connor, 2003).

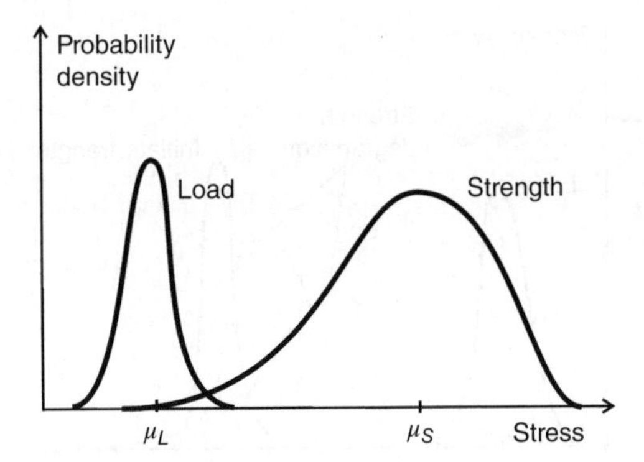

Figure 6.12 Low reliability caused by a large variability of the strength

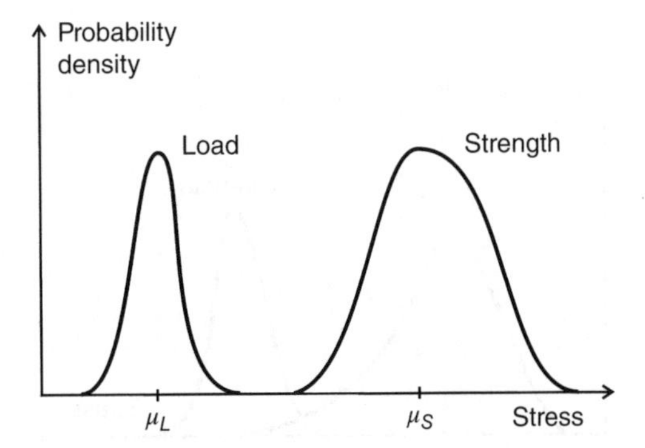

Figure 6.13 Increased reliability after a burn-in

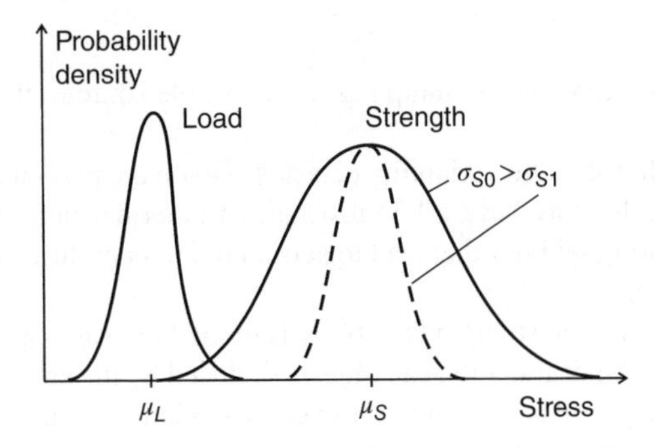

Figure 6.14 The importance of the strength variation to the probability of failure

Strength variability caused by variability of material properties is one of the major reasons for an increased interference with the load distribution which results in increased probability of failure (Figure 6.14).

Here, this point is discussed in some detail. Assume for simplicity that the load and strength follow Gaussian distributions. Since the reliability on demand is $R_0 = \Phi\left[(\mu_S - \mu_L)/\sqrt{\sigma_{S0}^2 + \sigma_L^2}\right]$, decreasing the variability of the strength to $\sigma_{S1} < \sigma_{S0}$ (Figure 6.14) increases the reliability on demand to $R_1 = \Phi\left[(\mu_S - \mu_L)/\sqrt{\sigma_{S1}^2 + \sigma_L^2}\right] > R_0$. If, for example, variability of strength is due to sampling from

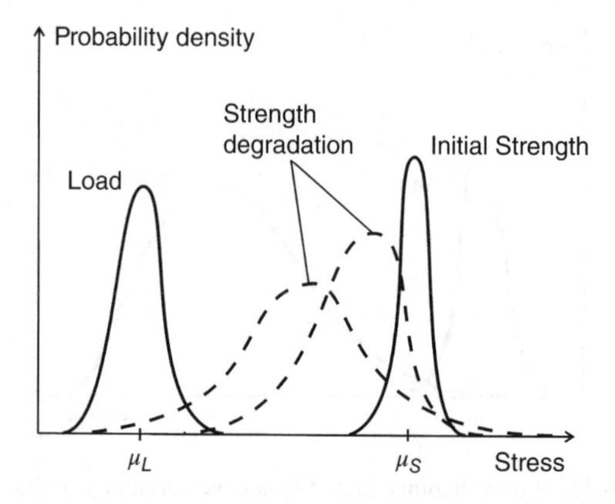

Figure 6.15 Decreased reliability due to strength degradation

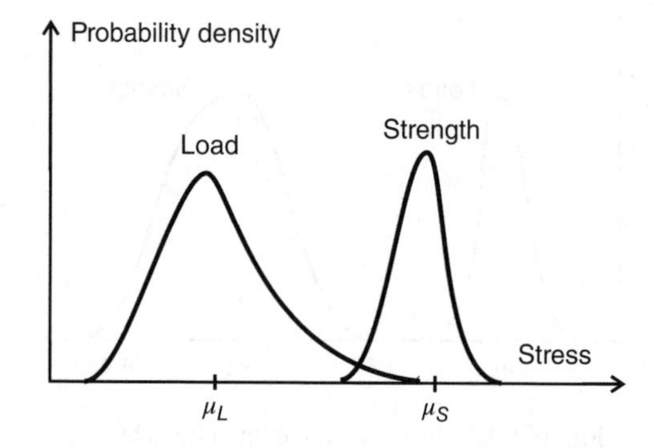

Figure 6.16 Increased load–strength interference due to a large variability of the load (rough loading)

multiple sources, it can be decreased by sampling from a single source - the source characterised by the smallest variance.

It must be pointed out that strength variability depends also on the particular design solution. A particular material, for example, may have a low resistance to thermal fatigue. If the design solution, however, eliminates operating regimes that lead to substantial temperature variations, thermal fatigue will no longer be a problem.

Low reliability due to increased strength variability is often due to ageing and the associated with it material degradation. Material degradation can often be induced by the environment, for example, due to corrosion and ageing. A typical feature of the strength degradation is an increase of the variance and a decrease of the mean of the strength distribution (Figure 6.15).

Low reliability is often due to excessive variability of the load. If variability of the load is large (rough loading), the probability of an overstress failure is significant (Figure 6.16). Mechanical equipment is usually characterised by a rough loading as opposed to electronic equipment which is characterised by a smooth loading (Figure 6.17). A common example of smooth loading is the power supply of electronic equipment through an anti-surge protector or the use of voltage regulators.

Here are some possible options for increasing the reliability on demand:

1. Decreasing the overall variability of strength
2. Altering the lower tail of the strength distribution (e.g. by a burn-in operation)

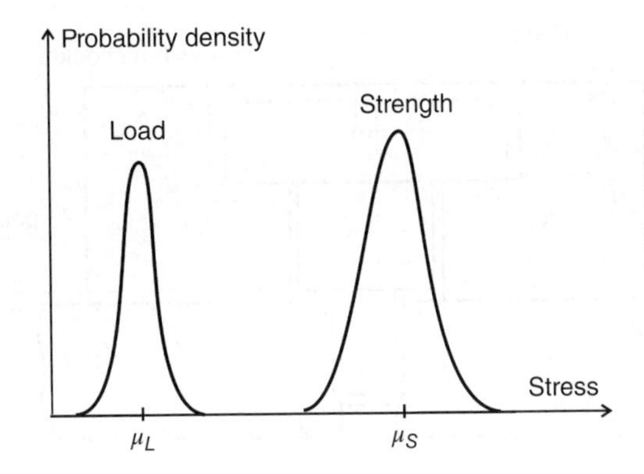

Figure 6.17 High reliability achieved by smooth loading

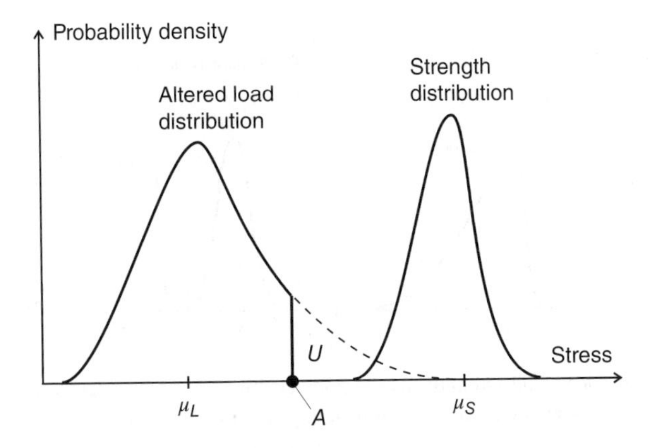

Figure 6.18 High reliability achieved by altering the upper tail of the load distribution using a stress limiter

3. Increasing the mean strength
4. Decreasing the mean load
5. Decreasing the overall variability of the load and obtaining a smooth loading
6. Altering the upper tail of the load distribution (e.g. by truncating it with stress limiters)

Altering the upper tail of the load distribution by using stress limiters is equivalent to concentrating the probability mass beneath the upper tail of the load distribution (the area marked *U* in Figure 6.18) into the truncation point *A*. Typical examples of stress limiters are the safety pressure valves, fuses and switches, activated when pressure or current reaches critical values. The specially designed shoulder on the screw in Figure 6.19 is an example of a stress limiter (Erhard, 2006). The shoulder prevents over-tightening the screw and damaging the plastic component.

Figure 6.20a shows a pair of load and strength distributions yielding a very high reliability. Such a high reliability is sometimes justified by the very high cost of failure of the component, particularly in cases where human fatalities, damaged infrastructure and pollution of the environment are involved. There are a number of cases however where no such high-cost consequences are present. Design to a very high reliability often means purchasing materials with controlled microstructure, free from inclusions and other impurities, small tolerances in the dimensions achieved through a large number expensive

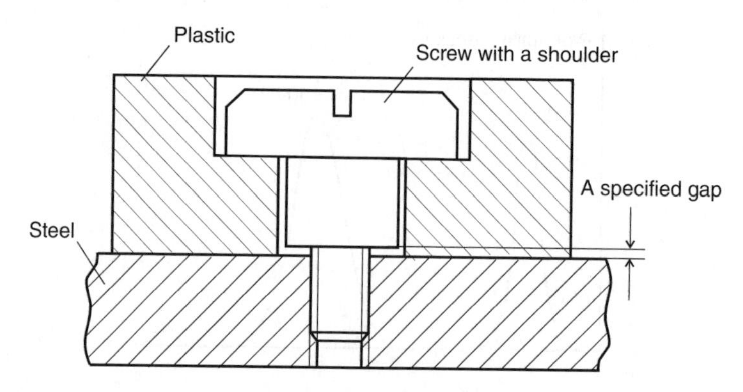

Figure 6.19 An example of a stress limiter: eliminating the risk of damaging the plastic part by a special design of the screw

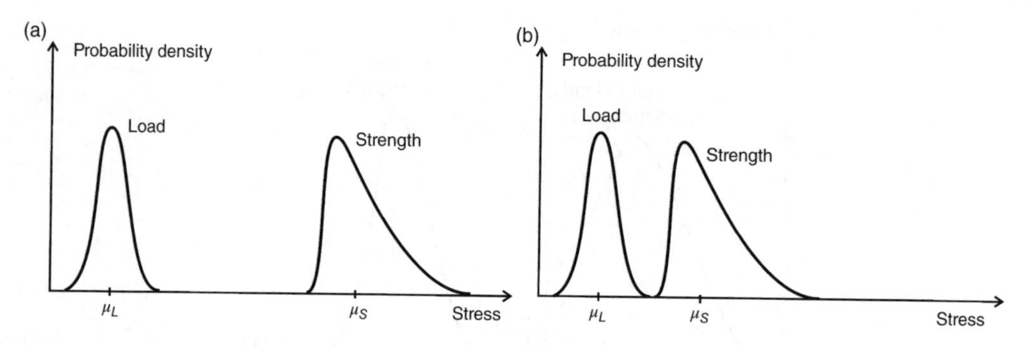

Figure 6.20 Tails of the load and strength distribution resulting in (a) an expensive design with very high reliability; (b) inexpensive design tolerating some probability of failure

machining operations, small tolerances in electronic components achieved through expensive manufacturing processes, etc. The excessive high strength achieved by expensive materials and operations will make it difficult for the company to keep its production costs down. In cases where no high cost of failure is present, an inexpensive design where an acceptable level of probability of failure is tolerated is more preferable (Figure 6.20b).

6.6.2 Critical Weaknesses of the Traditional Reliability Measures 'Safety Margin' and 'Loading Roughness'

For load and strength that do not follow the normal distribution, the traditional reliability measures safety margin and loading roughness can be misleading.

Consider the load and strength distributions from Figure 6.21a. The figure shows a case where a low safety margin $\beta = (\mu_S - \mu_L)/\sqrt{\sigma_S^2 + \sigma_L^2}$ exists ($\mu_S - \mu_L$ is small and $\sigma_S^2 + \sigma_L^2$ is large) yet reliability is high. In Figure 6.21a, μ_S and μ_L are the mean values of the strength and load and σ_S and σ_L are the corresponding standard deviations. Now, consider Figure 6.21b which has been obtained by reflecting symmetrically the distributions from Figure 6.21a with respect to axes r_1 and r_2, parallel to the probability density axis. Since the reflections do not change the variances of the distributions, the only

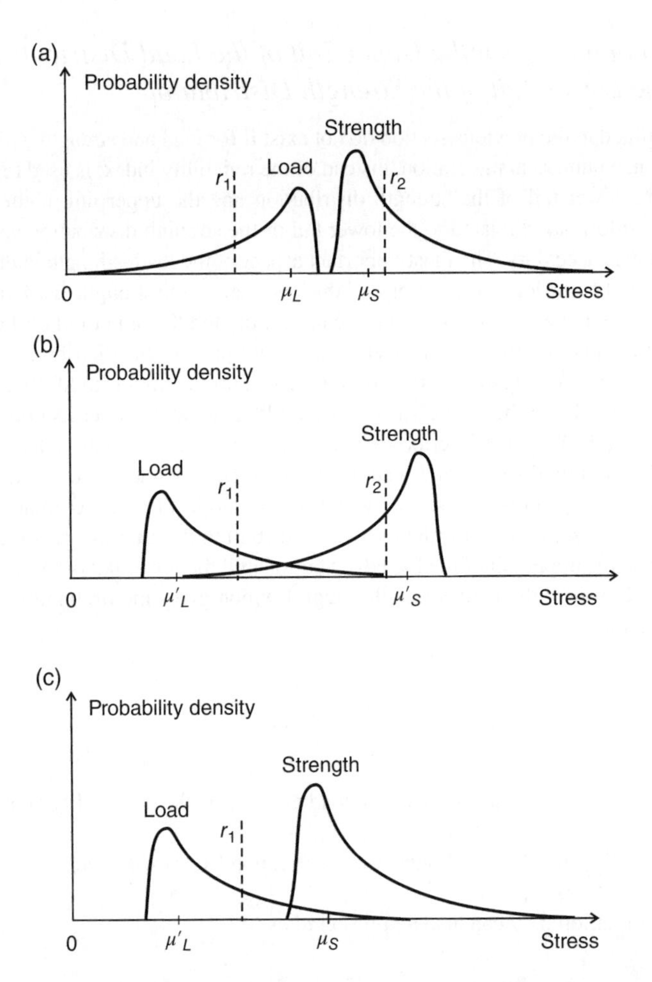

Figure 6.21 A counterexample showing that for (a) skewed load and strength distribution, the traditional reliability measures (b) 'reliability index' and (c) 'loading roughness' are very misleading

difference is the larger difference of the means $\mu_S' - \mu_L' > \mu_S - \mu_L$ (Figure 6.21b). Despite the larger new safety margin

$$\beta' = \frac{\mu_S' - \mu_L'}{\sqrt{\sigma_S^2 + \sigma_L^2}} > \beta = \frac{\mu_S - \mu_L}{\sqrt{\sigma_S^2 + \sigma_L^2}}$$

the reliability on demand related to the distributions in Figure 6.21b is smaller than that related to the distributions on Figure 6.21a. Clearly, *the safety margin concept applied without considering the shape of the interacting distribution tails can be very misleading.*

Similar considerations are valid regarding the parameter *loading roughness* $\sigma_L/\sqrt{\sigma_L^2 + \sigma_S^2}$ introduced by Carter (1986, 1997). If only the load in Figure 6.21a is reflected symmetrically regarding the axis r_1, the loading in Figure 6.21c is obtained. Since the standard deviation σ_L of the load has not been affected by the reflection, the loading roughness in Figure 6.21c, calculated from $\sigma_L/\sqrt{\sigma_L^2 + \sigma_S^2}$, is the same as in Figure 6.21a, despite the much more severe type of loading.

6.6.3 Interaction between the Upper Tail of the Load Distribution and the Lower Tail of the Strength Distribution

The problems outlined in the previous section do not exist if for load and strength which do not follow a normal distribution, a numerical integration, instead of the reliability index, is used to quantify the interaction between the lower tail of the strength distribution and the upper tail of the load distribution. Furthermore, only information related to the lower tail of the strength distribution and the upper tail of the load distribution is necessary. The most important aspect of the load–strength interaction is the interaction of the upper tail of the load distribution and the lower tail of the strength distribution (Figure 6.22).

The values from the lower tail of the strength distribution and the upper tail of the load distribution usually control reliability, not the values covering the other parts of the distributions (Figure 6.22).

Consequently, an adequate model of the strength distribution should faithfully represent its lower tail and an adequate model of the load distribution should adequately cover its upper tail. The normal distribution, for example, may not describe satisfactorily the strength variation in the distribution tails, mainly because the strength distribution is usually asymmetric, bounded on the left. In some cases, the Weibull model and the log-normal model may be suitable models for approximating the variation of material properties, but in many cases, the strength distribution is a mixture of several distributions.

The interaction of the upper tail of the load distribution and the lower tail of the strength distribution can be quantified. Consider the load–strength integral which gives the probability of failure p_f for a single load application:

$$p_f = \int_{S_{min}}^{S_{max}} \left[1 - F_L(x)\right] f_S(x) dx \tag{6.19}$$

where $F_L(x)$ is the cumulative distribution of the load and $f_S(x)$ is the probability density distribution of the strength.

Suppose that the S_{min} and S_{max} in Figure 6.23 correspond to stress levels for which $f_S(x) = 0$ if $x < S_{min}$ or $x > S_{max}$.

The integral in Equation 6.19 can also be presented as

$$p_f = \int_{S_{min}}^{L_{max}} \left[1 - F_L(x)\right] f_S(x) dx + \int_{L_{max}}^{S_{max}} \left[1 - F_L(x)\right] f_S(x) dx \tag{6.20}$$

For $x > L_{max}$, $F_L(x) \approx 1$ holds for the cumulative distribution of the load (Figure 6.23), and the second integral in Equation 6.20 becomes zero $\left(\int_{L_{max}}^{S_{max}} [1 - F_L(x)] f_S(x) dx \approx 0\right)$. Consequently, the probability of failure becomes

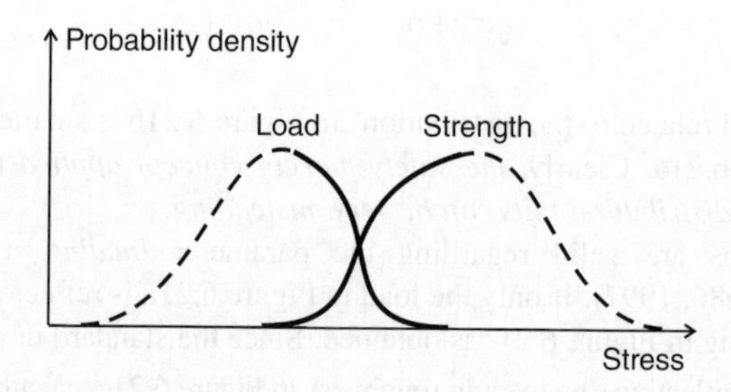

Figure 6.22 Reliability is determined by the interaction of the upper tail of the load distribution and the lower tail of the strength distribution

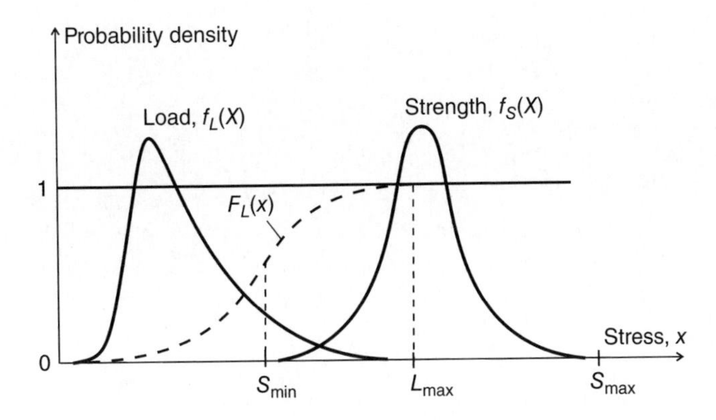

Figure 6.23 Deriving the reliability on demand, by integrating within the interval (S_{min}, L_{max}) including only the upper tail of the load distribution and the lower tail of the strength distribution

$$p_f = \int_{S_{min}}^{L_{max}} \left[1 - F_L(x)\right] f_S(x)\, dx \tag{6.21}$$

Finally, for the reliability on demand, we get

$$R = 1 - \int_{S_{min}}^{L_{max}} \left[1 - F_L(x)\right] f_S(x) dx \tag{6.22}$$

The reliability integral (6.22) which quantifies the interaction of the upper tail of the load distribution and the lower tail of the strength distribution has a clear advantage: To derive the reliability on demand, data covering the lower tail of the load distribution and the upper tail of the strength distribution are no longer necessary.

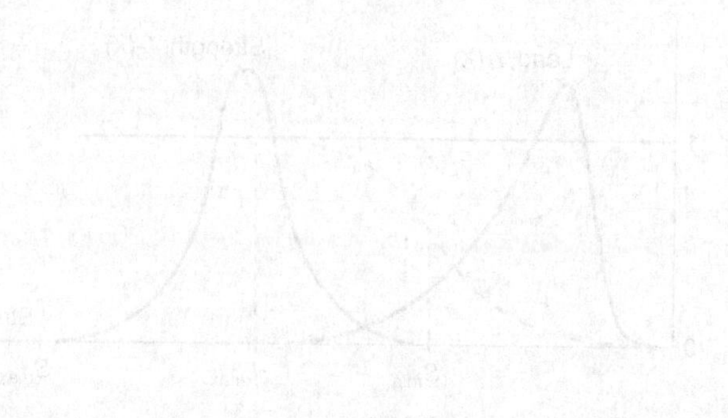

7

Overstress Reliability Integral and Damage Factorisation Law

7.1 Reliability Associated with Overstress Failure Mechanisms

According to the discussions in Chapter 6, overstress failures occur if load exceeds strength. If load is smaller than strength, the load has no permanent effect on the component. In this section, an integral will be presented related to reliability associated with all overstress failure mechanisms.

Suppose that a random load characterised by a cumulative distribution function $F(x)$ has been applied a number of times during a finite time interval with length t, and the times of load applications follow a non-homogeneous Poisson process with intensity $\rho(t)$.

If the strength is described by a probability density distribution $f_S(x)$, the probability R of surviving n load applications is given by the classical load–strength interference model:

$$R = \int_{S_{\min}}^{S_{\max}} F_L^n(x) f_S(x) dx \tag{7.1}$$

where $S_{\min}$ and $S_{\max}$ are the lower and the upper limit of strength. The probability of no failure (reliability) associated with the finite time interval with length t can be calculated from the following probabilistic argument. According to the total probability theorem, the probability of no failure is a sum of the probabilities of the following mutually exclusive and exhaustive events: the probability of no failure if no load has been applied during the time interval of length t, the probability of no failure associated with exactly one, two, three, …, k load applications, etc. Because the probability of k load applications during the time interval 0, t is given by the Poisson distribution $p_{(k)} = (\mu^k / k!)\exp(-\mu)$, where

$$\mu = \int_0^t \rho(v) dv \tag{7.2}$$

is the mean number of load applications in the finite time interval with length t, the probability $R_k(t)$ of no failure during the time interval $(0, t)$ for exactly k statistically independent load applications is

$$R_k(t) = \frac{\mu^k}{k!} \int_{S_{\min}}^{S_{\max}} F_L^k(x) f_S(x) dx$$

Reliability and Risk Models: Setting Reliability Requirements, Second Edition. Michael Todinov.

According to the total probability theorem, the probability of no failure associated with the time interval (0, t) is

$$R(t) = \exp(-\mu) \times \left[1 + \mu \int_{S_{\min}}^{S_{\max}} F_L(x) f_S(x) dx + \frac{\mu^2}{2!} \int_{S_{\min}}^{S_{\max}} F_L^2(x) f_S(x) dx + \cdots \right] \tag{7.3}$$

Equation 7.3 can be presented as

$$R(t) = \exp(-\mu) \left\{ \int_{S_{\min}}^{S_{\max}} f_S(x) dx \left[(\mu F_L(x))^0 + \frac{(\mu F_L(x))^1}{1!} + \frac{(\mu F_L(x))^2}{2!} + \cdots \right] \right\} \tag{7.4}$$

which simplifies to

$$R(t) = \exp(-\mu) \left\{ \int_{S_{\min}}^{S_{\max}} f_S(x) dx \exp\left[\mu F_L(x) \right] \right\} \tag{7.5}$$

Finally,

$$R(t) = \int_{S_{\min}}^{S_{\max}} \exp\left[-\mu(1 - F_L(x)) \right] f_S(x) dx \tag{7.6}$$

For the common case of a constant density of load applications in the interval (0, t), $\mu = \int_0^t \rho(v) dv = \rho t$ and Equation 7.6 becomes (Todinov, 2004d)

$$R(t) = \int_{S_{\min}}^{S_{\max}} \exp\left[-\rho t(1 - F_L(x)) \right] f_S(x) dx \tag{7.7}$$

Equations 7.6 and 7.7 are reliability integrals, associated with an overstress failure mechanism (Todinov, 2004d). The *overstress reliability integrals* (7.6) and (7.7) can be regarded as a generalisation of the Freudenthal's load–strength interference integral (Freudenthal, 1954) with the time incorporated. The integrals are valid for any overstress failure mechanism and provide a closed-form expression for the reliability during multiple load applications (e.g. shock loading) as a function of the length of the time interval and the frequency of the load applications.

The term $\exp[-\rho t(1 - F(x))]$ in the overstress reliability integral (7.7) gives the probability that none of the random load magnitudes in the time interval 0, t will exceed strength of magnitude x, while the term $f_S(x)\,dx$ gives the probability that strength will be in the infinitesimal interval $x, x + dx$. The product $\exp[-\rho t(1 - F_L(x))] f_S(x) dx$ gives the probability of the compound event that the strength will be in the interval $x, x + dx$, and none of the random loads will exceed it. When this product is integrated over all possible values x of the strength magnitude, the reliability associated with time interval 0, t is obtained.

The big advantage of the overstress reliability integrals (7.6) and (7.7) is that they incorporate the time, unlike the load–strength integral (7.1) which only describes reliability on demand.

Using Monte Carlo simulations (see Chapter 9), the overstress reliability integral (7.7) has been verified. Thus, for uniformly distributed load and strength in the interval ($S_{\min}$, $S_{\max}$), the cumulative distribution of the load is

$$F_L(x) = \frac{x - S_{\min}}{S_{\max} - S_{\min}} \tag{7.8}$$

and the probability density of the strength is

$$f_S(x) = \frac{1}{S_{\max} - S_{\min}} \tag{7.9}$$

Substituting these in the overstress reliability integral (7.7) yields for the reliability

$$R(t)=\frac{1-\exp(-\rho t)}{\rho t} \tag{7.10}$$

which for $t=100$ months and $\rho=0.017$ months^{-1} gives $R(t)=0.48$.

For the same parameters $t=100$ months and $\rho=0.017$ months^{-1}, Monte Carlo simulations of uniformly distributed load and strength in the interval $S_{\min}=20$, $S_{\max}=90$ yield the empirical probability $R(t)=0.48$.

Exercise

During a finite time interval $t=100$ months, the tensile load applications on a component follow a homogeneous Poisson process with density that is $\rho=0.5$ months^{-1}. The overloading tensile stress follows the maximum extreme value distribution

$$F_L(x)=\exp\left(-\exp\left(-\frac{x-\xi}{\theta}\right)\right)$$

with parameters $\xi=119$, $\theta=73.64$, and the strength of the component has been approximated by the Weibull distribution

$$f_S(x)=\frac{m}{\eta}\left(\frac{x-x_0}{\eta}\right)^{m-1}\exp\left[-\left(\frac{x-x_0}{\eta}\right)^m\right]$$

with parameters $x_0=200$, $\eta=297.3$ and $m=3.9$.
Determine the reliability associated with the time interval (0, 100 months).

Solution

Numerical integration of the right-hand side of Equation 7.7, within integration limits $S_{\min}=x_0=200$, $S_{\max}=2000$, yields $R(t)=0.597$ for the reliability associated with 100 months. The Monte Carlo simulation of the load–strength interference model with multiple load application and the same parameters (see Algorithm 9.3) yielded an empirical reliability $R(t)\approx 0.597$ which demonstrates the validity of the overstress reliability integral (7.7).

Another advantage of the overstress reliability integral is that it can be applied for load and strength following any distribution. In Chapter 6, it was shown that for a load and strength not following a normal distribution, the standard reliability measure 'reliability index' is misleading.

7.1.1 The Link between the Negative Exponential Distribution and the Overstress Reliability Integral

The term $\exp[-\rho t(1-F_L(x))]$ in the overstress reliability integral (7.7) gives the probability that none of the random loads in the time interval 0, t will exceed strength with magnitude x. In the case of a constant strength $x=s=\text{const}$, the integral yields

$$R(t)=\exp\left[-\rho t\left(1-F_L(s)\right)\right] \tag{7.11}$$

for the reliability associated with the time interval 0, t, which decreases with increasing the number density of the load applications.

Equation 7.11 can be presented as

$$R(t)=\exp[-\rho F_c t] \tag{7.12}$$

where $F_c = 1 - F_L(s)$. A load which exceeds the constant strength s will be referred to as a *critical load*. If the number density of all load applications is ρ, the number density of the critical loads is given by ρF_c. As a result, the number density λ_c of the critical load applications becomes $\lambda_c = \rho F_c$, and from Equation 7.12, an expression for the reliability associated with the time interval (0, t) can be obtained:

$$R(t) = \exp[-\lambda_c t] \tag{7.13}$$

As a result, the negative exponential distribution has a fundamental importance because it can be derived from the overstress reliability integral under very general assumptions: constant strength and statistically independent load applications following a homogeneous Poisson process in the time interval (0, t). In this case, the hazard rate $\lambda = \rho F_c$ characterising the component is constant, equal to the product of the number density of the load applications and the probability of failure during a single load application. Equation 7.12 provides an alternative interpretation of the fundamental concept 'hazard rate' and an opportunity for calculating it from the number density of the load applications and the probability of failure associated with a single load application.

It is interesting to investigate the effect on reliability of a smooth (Figure 7.1a) and rough loading (Figure 7.1b). For ideally rough loading, there exists a non-zero probability that load will be greater than any possible value of strength. With increasing the number density of the load applications, reliability approaches asymptotically zero (Figure 7.1b).

In order to guarantee a perfectly smooth loading (the variance of the load is zero), assume that the load is constant, equal to L: The cumulative distribution function of the load then becomes $F_L(x) = 0$, if $x < L$ and $F_L(x) = 1$, if $x \geq L$. The overstress integral (7.7) can then be presented as

$$\begin{aligned} R(t) &= \int_{S_{\min}}^{L-} \exp[-\rho t(1-0)] f_S(x)dx + \int_{L}^{S_{\max}} \exp[-\rho t(1-1)] f_S(x)dx \\ &= \exp(-\rho t) \int_{S_{\min}}^{L-} f_S(x)dx + \int_{L}^{S_{\max}} f_S(x)dx \end{aligned} \tag{7.14}$$

This result shows that, during ideally smooth loading, with increasing the number of load applications, reliability decreases monotonically, approaching the value $R_\infty = \int_L^{S_{\max}} f_S(x)dx$, which is the probability that strength will be larger than the load L (Figure 7.1a). In other words, in case of a very small variation of the load, increasing the number of load applications beyond a particular value has

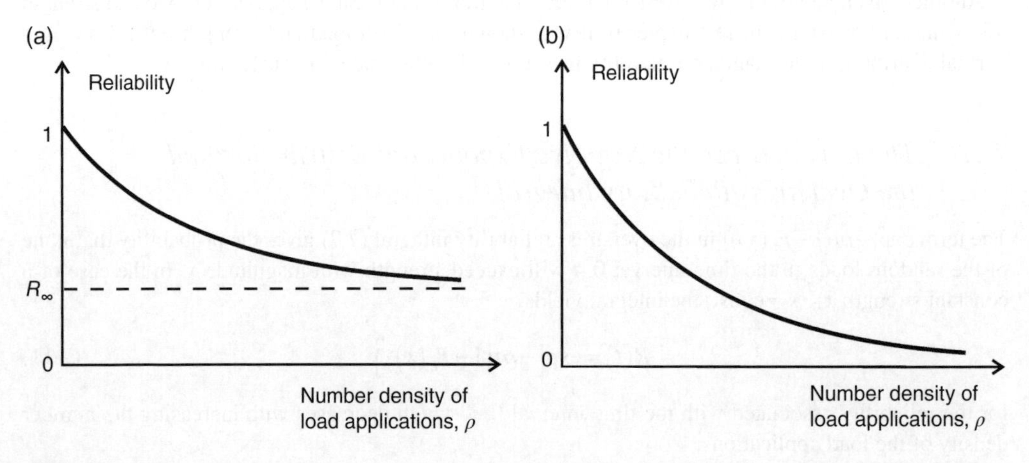

Figure 7.1 Influence of the number density of load applications on reliability: (a) ideally smooth loading; (b) ideally rough loading

no practical effect on reliability. Reliability tends to the probability $R_\infty = \int_L^{S_{\max}} f_S(x)dx$ that strength will be larger than load. Now, suppose that the constant load is so large that strength always remains smaller than load. In this case, $\int_{S_{\min}}^{L-} f_S(x)dx = 1$ and $\int_L^{S_{\max}} f_S(x)dx = 0$ in Equation 7.14 and reliability becomes

$$R(t) = \exp(-\rho t)$$

This is an important reason for the origin of the exponential distribution of the time to failure and the flat region of the bathtub curve. Even if all components undergo wear and deterioration during the time interval 0, t, the time to failure is still given by the negative exponential distribution if failure is controlled by the random load applications and if the times of the load applications follow a homogeneous Poisson process with density ρ.

7.2 Damage Factorisation Law

Suppose that damage due to fatigue, corrosion or any other type of deterioration is a function of time and a particular controlling factor p. During fatigue, for example, the controlling factor can be the stress or strain amplitude. Suppose that a particular component accumulates damage at M different intensity levels $p_1, \ldots, p_M$ of the controlling factor p (Figure 7.2). At each intensity level p_i, the component is exposed to damage for time Δt_i. Suppose that t_i corresponding to constant intensity levels p_i of the controlling factor p denote the times for attaining a critical level of damage a_c, after which the component is considered to have failed (Figure 7.2). It is also assumed that the sequence in which the various levels of the factor p are imposed does not affect the component's life.

The damage factorisation law states that if for a constant level p of the controlling factor, *the rate of damage development can be factorised as a function of the current damage 'a' and a function of the factor level p*,

$$\frac{da}{dt} = F(a)G(p), \tag{7.15}$$

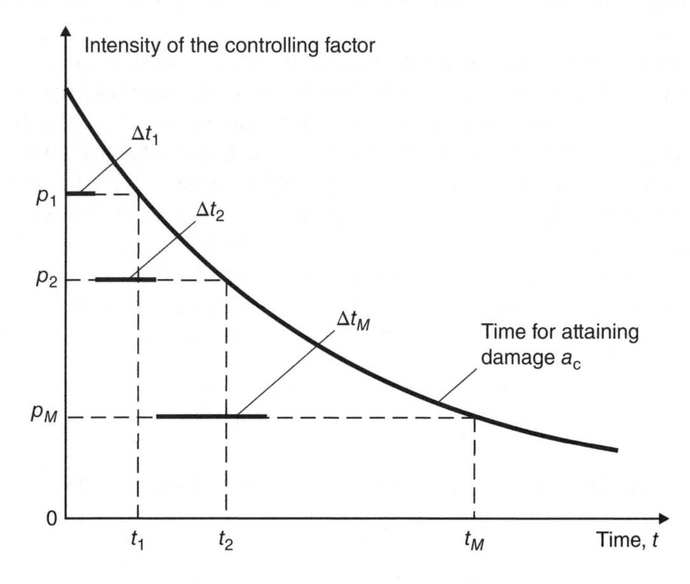

Figure 7.2 Exposure for times Δt_i at different intensity levels p_i of the controlling factor p

the critical level of damage a_c at different levels of the controlling factor p will be attained when the sum

$$\frac{\Delta t_1}{t_1} + \frac{\Delta t_2}{t_2} + \cdots \tag{7.16}$$

becomes unity for some k:

$$\frac{\Delta t_1}{t_1} + \frac{\Delta t_2}{t_2} + \cdots + \frac{\Delta t_k}{t_k} = 1 \tag{7.17}$$

The time t_c to attain the critical level of damage a_c is then equal to

$$t_c = \Delta t_1 + \Delta t_2 + \cdots + \Delta t_k \tag{7.18}$$

Conversely, if the time for obtaining the critical level of damage a_c can be determined using the additivity rule (7.17), the factorisation (7.15) must necessarily hold. *In other words, the damage factorisation law* (7.15) *is a necessary and sufficient condition for the additivity rule* (7.17) (Todinov, 2001a). This means that if the rate of damage law cannot be factorised, the additivity rule (7.17) is not valid and must not be used.

An alternative formulation of the damage factorisation law states (Todinov, 2001a) that if for a constant level p of the controlling factor, the time t for attaining a particular level of damage 'a' can be factorised as a function of damage and the factor level p,

$$t = R(a)\,S(p) \tag{7.19}$$

the time to reach a critical level of damage a_c for different levels of the controlling factor can be determined using the additivity rule (7.17).

Essentially, according to the additivity rule (7.17), the total time t_c required to attain a specified level of damage a_c is obtained by adding the absolute durations Δt_i (Eq. 7.18) spent at each intensity level i of the factor p until the sum of the relative durations $\Delta t_i/t_i$ becomes unity. The fraction $\Delta t_i/t_i$ of accumulated damage at a particular intensity level p_i of the controlling factor p is the ratio of the time Δt_i spent at level p_i and the total time t_i at level p_i needed to attain the specified level a_c of damage (from initial damage zero).

An important application of the additivity rule is the case where damage is caused by fatigue. In this case, the measure of damage is the length a of the fatigue crack. The additivity rule (7.17) also known as the *Palmgren–Miner rule* has been proposed as an empirical rule in case of damage due to fatigue controlled by crack propagation (Miner, 1945). The rule states that in a fatigue test at a constant stress amplitude $\Delta\sigma_i$, damage could be considered to accumulate linearly with the number of cycles. Accordingly, if at a stress amplitude $\Delta\sigma_1$, the component has n_1 cycles of life, which correspond to amount of damage a_c, after Δn_1 cycles at a stress amplitude $\Delta\sigma_1$, the amount of damage will be $(\Delta n_1/n_1)a_c$. After Δn_2 stress cycles spent at a stress amplitude $\Delta\sigma_2$, characterised by a total life of n_2 cycles, the amount of damage will be $(\Delta n_2/n_2)a_c$ and so on. Failure occurs when, at a certain stress amplitude $\Delta\sigma_M$, the sum of partial amounts of damage attains the amount a_c, that is, when

$$\frac{\Delta n_1}{n_1}a_c + \frac{\Delta n_2}{n_2}a_c + \cdots + \frac{\Delta n_M}{n_M}a_c = a_c \tag{7.20}$$

is fulfilled. As a result, the analytical expression of the Palmgren–Miner rule becomes

$$\sum_{i=1}^{M} \frac{\Delta n_i}{n_i} = 1 \tag{7.21}$$

where n_i is the number of cycles needed to attain the specified amount of damage a_c at a constant stress amplitude $\Delta\sigma_i$.

Palmgren–Miner rule is central to reliability calculations, yet no comments are usually made as to whether it is compatible with the damage development laws characterising the different stages of fatigue crack growth. The necessary and sufficient condition for validity of the empirical Palmgren–Miner rule is the possibility to factorise the rate of damage *da*/*dn* as a function of the amount of accumulated damage *a* (the crack length) and the stress or strain amplitude (Δp):

$$\frac{da}{dn} = F(a)G(\Delta p) \tag{7.22}$$

The theoretical derivation of the Palmgren–Miner rule can be found in (Todinov, 2001a). A widely used fatigue crack growth model is the Paris power law (Paris and Erdogan, 1963; Paris *et al.*, 1961):

$$\frac{da(n)}{dn} = C\Delta K^m \tag{7.23}$$

where $\Delta K = Y\Delta\sigma\sqrt{\pi a}$ is the stress intensity factor range, *C* and *m* are material constants and *Y* is a parameter which can be presented as a function of the amount of damage *a* (see Chapter 12). Clearly, the Paris–Erdogan fatigue crack growth law can be factorised as in (7.22), and therefore, it is compatible with the Palmgren–Miner rule. In cases where this factorisation is impossible, the Palmgren–Miner rule does not hold. Such is, for example, the fatigue crack growth law

$$\frac{da}{dn} = B\Delta\gamma a^{\beta} - D \tag{7.24}$$

discussed in Miller (1993), which characterises physically small cracks. In Equation 7.24, *B* and β are material constants, $\Delta\gamma$ is the applied shear strain range, '*a*' is the crack length and *D* is a threshold value.

8

Solving Reliability and Risk Models Using a Monte Carlo Simulation

8.1 Monte Carlo Simulation Algorithms

8.1.1 Monte Carlo Simulation and the Weak Law of Large Numbers

Monte Carlo simulation is a powerful method for determining empirically the probability of an event. If the number of independent trials leading to outcome A is n_A and the total number of trials is n, the probability $P(A)$ of the outcome A can be approximated from

$$P(A) = \frac{n_A}{n} \tag{8.1}$$

for a sufficiently large number of trials n.

It can be shown that

$$\lim_{n \to \infty} P\left(\left|\frac{n_A}{n} - P(A)\right|\right) < \varepsilon = 1 \tag{8.2}$$

for each $\varepsilon > 0$. This fundamental result, which is the foundation of the Monte Carlo method, is known as Bernoulli's *weak law of large numbers*. It was first proved by Jacob Bernoulli in his work *Ars Conjectandi* (1713) (Bernoulli, 1899).

Proof

Let X_i denote the outcome on the ith trial. If the outcome is event A, then $X_i = 1$; otherwise, $X_i = 0$. Because the trials are independent, the random variables X_i are statistically independent and identically distributed. The common mean μ of the random variables X_i can be determined from

$$\mu = 0 \times (1 - P(A)) + 1 \times P(A) = P(A) \tag{8.3}$$

Reliability and Risk Models: Setting Reliability Requirements, Second Edition. Michael Todinov.

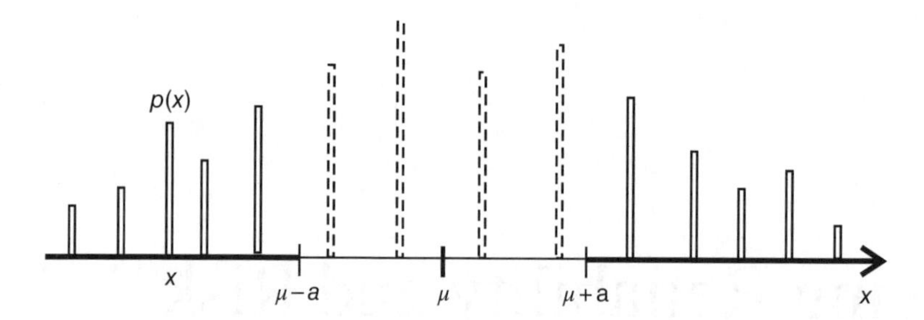

Figure 8.1 Discrete distribution of the random variable X

Because the number of outcomes leading to event A is given by $n_A = X_1 + \cdots + X_n$, to prove expression (8.2), it can be presented as

$$\lim_{n\to\infty} P\left(\left|\bar{X} - \mu\right|\right) < \varepsilon = 1 \tag{8.4}$$

for each $\varepsilon > 0$, where $\bar{X} = (X_1 + \cdots + X_n)/n$.

Before we prove (8.4), the following inequality (known also as *Chebyshev's inequality*) will be proved. If X is a discrete random variable with mean μ and finite variance $\text{Var}(X) = \sigma^2$, then, for each positive number $a > 0$,

$$P\left(\left|X - \mu\right| \ge a\right) \le \frac{\text{Var}(X)}{a^2} \tag{8.5}$$

The inequality states that the probability that a random variable X will deviate by more than a from its mean μ does not exceed the ratio $\text{Var}(X)/a^2$. Let X accept its discrete values x, with probabilities $p(x)\left(\sum_x p(x) = 1\right)$.

Indeed, by definition, $\text{Var}(X) = \sum_x (x-\mu)^2 p(x)$. If the summation is made only over values for which $|x - \mu| \ge a$ (Figure 8.1, the region marked with a thick line), the following inequality holds because both $(x-\mu)^2 \ge 0$ and $p(x) \ge 0$:

$$\text{Var}(X) = \sum_x (x-\mu)^2 p(x) \ge \sum_{|x-\mu|\ge a} (x-\mu)^2 p(x) \tag{8.6}$$

If $(x-\mu)^2$ is replaced by a^2, then

$$\sum_{|x-\mu|\ge a} (x-\mu)^2 p(x) \ge a^2 \sum_{|x-\mu|\ge a} p(x) = a^2 P\left(\left|X - \mu\right| \ge a\right) \tag{8.7}$$

will hold. Considering inequality (8.6), we get

$$\text{Var}(X) \ge a^2 P\left(\left|X - \mu\right| \ge a\right) \tag{8.8}$$

from which inequality (8.5) follows immediately.

Now, going back to (8.4) and applying the inequality (8.5) to $\bar{X} = (X_1 + \cdots + X_n)/n$ yields

$$P\left(\left|\bar{X} - \mu\right| \ge \varepsilon\right) \le \frac{\text{Var}(\bar{X})}{\varepsilon^2} \tag{8.9}$$

From the properties of the variance of a mean: $\text{Var}(\bar{X}) = \sigma^2 / n$, where σ is the standard deviation of the distribution of random variable X. Substituting in (8.9) gives

$$P\left(\left|\bar{X} - \mu\right| \geq \varepsilon\right) \leq \frac{\sigma^2}{n\varepsilon^2}$$

Because σ^2 and ε^2 are finite quantities, with increasing the number of trials n, $\lim_{n\to\infty} P\left(\left|\bar{X} - \mu\right| \geq \varepsilon\right) = 0$. Considering that $\left|\bar{X} - \mu\right| \geq \varepsilon$ and $\left|\bar{X} - \mu\right| < \varepsilon$ are complementary events, finally

$$\lim_{n\to\infty} P\left(\left|\bar{X} - \mu\right| < \varepsilon\right) = 1 \tag{8.10}$$

which proves (8.2).

8.1.2 *Monte Carlo Simulation and the Central Limit Theorem*

Consider an experiment involving n independent trials to determine the mean μ of a random variable characterised by a probability distribution $f(x)$ and standard deviation σ. According to the *central limit theorem*, with increasing the number n of independent trials, the distribution of the mean $\bar{x} = (1/n)\sum_{i=1}^{n} x_i$ approaches a normal distribution with mean μ and standard deviation $\sigma / \sqrt{n}$. The probability that $|\bar{x} - \mu| > k\sigma / \sqrt{n}$ $(k > 0)$ equals $2\Phi(-k)$ where $\Phi(\bullet)$ is the cumulative distribution function of the standard normal distribution. For a fixed k, the equation

$$P\left(\left|\frac{1}{n}\sum_{i=1}^{n} x_i - \mu\right| \leq \frac{k\sigma}{\sqrt{n}}\right) = 1 - 2\Phi(-k) \tag{8.11}$$

holds (Sobol, 1994). The error $\varepsilon = \left|(1/n)\sum_{i=1}^{n} x_i - \mu\right|$ is inversely proportional to the square root of the number of trials ($\varepsilon \propto 1/\sqrt{n}$) and approaches zero as n increases. From Equation 8.11, it follows that reducing the error m times requires increasing the number of Monte Carlo trials by a factor of m^2. In order to improve the efficiency of the Monte Carlo simulation by reducing the variance of the Monte Carlo estimates, a number of techniques such as *stratified sampling* and *importance sampling* can be employed (see, e.g. Ross, 1997; Rubinstein, 1981). The efficiency of the Monte Carlo simulations can also be increased by a better reproduction of the input distribution using *Latin hypercube sampling* (Vose, 2000).

8.1.3 *Adopted Conventions in Describing the Monte Carlo Simulation Algorithms*

In describing the basic Monte Carlo simulation algorithms, a number of conventions will be used.

Thus, the statements in braces {*Statement 1*; *Statement 2*; *Statement 3*; …} separated by semicolons are executed as a single block. The construct

```
for i = 1 to Number_of_trials do
    {
     ....
    }
```

is a loop with a control variable *i*, accepting successive values from one to the number of Monte Carlo trials (`Number_of_trials`). In some cases, it is necessary that the control variable *i* accepts successive decreasing values from `Number_of_trials` to one. The corresponding construct is

```
for i = Number_of_trials downto 1 do
    {
     ....
    }
```

The loops execute the block of statements in the braces `Number_of_trials` number of times. If a statement **`break`** is encountered in the body of a loop, the execution continues with the next statement immediately after the loop (*Statement n*+1 in the next example) skipping all statements between the statement **`break`** and the end of the loop:

```
for i = 1 to Number_of_trials do
    {
    Statement 1;
    ....
    break;
    .....
    Statement n-1;
    Statement n;
    }
    Statement n+1;
```

The construct

```
while (Condition) do  {Statement 1;…;Statement n;}
```

is a loop which executes the block of statements repeatedly as long as the specified condition is true. If the variable `Condition` is false before entering the loop, the block of statements is not executed at all. A similar construct is the loop

```
repeat
  Statement 1;
  ....
  Statement n;
until (Condition);
```

which repeats the execution of all statements between **`repeat`** and **`until`**, until the specified condition becomes true. Unlike the **`while-do`** loop, the **`repeat-until`** loop statements are executed at least once.

The next important construct is the *conditional statement*. In the conditional statement below, the block of statements in the braces is executed only if the specified condition is true:

```
if (Condition) then {Statement 1;…;Statement n;}
```

A *procedure* is a self-contained section to perform a certain task. The procedure is called by including its name ('`proc`' in the next example) in other parts of the algorithm:

```
procedure proc()
{
  Statement 1;
  ...
  Statement n;
}
```

A *function* is also a self-contained section which returns value and which is called by including its name in other parts of the algorithm. Before returning to the point of the function call, a particular value `p` is assigned to the function name ('`fn`' in the next example) with the statement **`return`**:

```
function fn()
{
  Statement 1;
  ...
  Statement n;
  return p;
}
```

Text in italic between the symbols '/*' and '*/' is comments.

8.2 Simulation of Random Variables

8.2.1 Simulation of a Uniformly Distributed Random Variable

A simple and efficient algorithm for generating uniformly distributed pseudorandom numbers is the congruential multiplicative pseudorandom number generator suggested by Lehmer (1951). If an initial value X_0 called *seed* is specified, the random number X_{i+1} in the random sequence with seed X_0 is calculated from the previous value X_i using the formula

$$X_{i+1} = AX_i \bmod M \tag{8.12}$$

where the *multiplier* A and the *modulus* M are positive integers. X_{i+1} is the remainder left when AX_i is divided by M. For a different seed X_0', a different random sequence is obtained. After at most M generated values, the random sequence will repeat itself. Because $0 < X_i < M$, a uniformly distributed pseudorandom number in the interval (0, 1) is obtained from

$$u_i = \frac{X_i}{M} \tag{8.13}$$

Comprehensive discussion on generating random numbers and tests for statistical independence of the generated random numbers is provided in Knuth (1997), Rubinstein (1981), Tuckwell (1988) and L'Ecuyer (1988). A random sequence with very good properties is obtained if $A = 16\,807$ and $M = 2\,147\,483\,647$ are selected for the values of the constants in the recurrence relation (8.12) defining the pseudorandom generator described in (Park and Miller, 1988). The algorithm in pseudocode for simulating a random variable following a uniform distribution in the interval (0, 1) can be presented as follows:

Algorithm 8.1

```
function u_random()
{
  t = A*Seed;
  Seed = mod (t, M);
  u = Seed/M;     /* Generating a random number in the interval (0,1) */
 return u;
}
```

The function **mod** returns the remainder from the division of t and M. Before the first call of the function **u_random()**, the variable Seed is initialised with any number in the range 1, $M-1$. Subsequently, this value is altered in the statement Seed=**mod** (t, M). The value of the variable 'Seed' should be preserved between the calls of the function **u_random()**.

Using the linear transformation $x_i = a + (b-a)u_i$, where u_i is a random number uniformly distributed in the interval (0, 1), a uniformly distributed random value x_i from any specified interval (a, b) can be generated. Uniformly distributed integer numbers in the range $(0, n-1)$ with equal probability of generating any of the numbers $0,1,2,\ldots,n-1$ can be obtained using the expression

$$x_i = [nu_i] \tag{8.14}$$

where $[nu_i]$ denotes the greatest integer which does not exceed nu_i. Consequently, the formula

$$x_i = [nu_i] + 1 \tag{8.15}$$

will generate with equal probability the integer numbers $1,2,\ldots,n$. On the basis of Equation 8.15, a function *Rand(k)* can be constructed which selects with the same probability *1/k* one object out of *k* objects. The algorithm in pseudocode is straightforward:

Algorithm 8.2

```
function Rand(k)
 {
  u = u_random();   /* Generates a uniformly distributed random value
                       in the interval 0,1 */
  x = Int (k*u) +1; /*Generates a uniformly distributed integer value x in the
                      interval 1,…,k */
  return x;
 }
```

Function **Int** (k*u) returns the greatest integer which does not exceed the product k*u.

8.2.2 Generation of a Random Subset

In some applications, it is important to generate a random subset of size k out of n objects. The subset must be generated in such a way that no object in the subset appears twice and every object has an equal chance of being included in the subset. Some of the applications include (i) selecting randomly a set of k test specimens from a batch of n specimens, (ii) selecting randomly a group of k people out of n people, (iii) selecting randomly k components for inspection from a batch containing n components and (iv) a random assignment of n specimens to n treatments. The list can be continued.

Let the n objects be indexed by $1,2,\ldots,n$ and stored in an array *a[n]* of size n. Calling the function *Rand(n)* will select with equal probability *1/n* a random index r $(1 \le r \le n)$ of an array *a[]*. As a result, a random selection of the first object *a[r]* is made. Next, the selected object *a[r]* and the last object *a[n]* are swapped. This means that the object initially stored in the nth cell of the array is now stored in the rth cell and the object from the rth cell has been moved into the nth cell. As a result, all of the objects which have not been selected are now in the first *n-1* cells of the array. The process of random selection of the next object continues with selecting a random object from the first *n-1* cells of the array by calling *Rand(n-1)*. A random object from the first *n-2* cells of the array is selected by calling *Rand(n-2)* and so on. The selection process ends when exactly *k* objects have been selected. The algorithm in pseudocode is as follows:

Algorithm 8.3

```
for i = n downto n-k+1
 {
  k = Rand(i);
  tmp = a[i]; a[i] = a[k]; a[k] = tmp; /* swaps cells a[i] and a[k] */
 }
```

After executing the procedure, all *k* selected objects are stored in the last *k* cells of the array *a[n]* (from $a[n-k+1]$ to $a[n]$). If *k* has been specified to be equal to *n*, the algorithm generates a random permutation from the elements of the array, or in other words, it scrambles the array in a random fashion.

The algorithm is very efficient because it uses exactly *k* random numbers to make a random selection of *k* out of *n* objects. Each of the *k* objects is selected with the same probability 1/*n*.

Proof

Because the numbers generated by the function Rand() have uniform distribution, the probability of selecting the first object is $p_1 = 1/n$. The second object has been selected in the second round. Therefore, no selection of the second object has been made in the first round, the probability of which is $(n-1)/n$, and a selection of the object has been made in the second round, the probability of which is $1/(n-1)$ (in the second selection, the random number generator generates a random number from 1 to $n-1$). The probability of selecting the second object is therefore

$$p_2 = \frac{n-1}{n} \times \frac{1}{n-1} = \frac{1}{n}$$

The third object has been selected in the third round. The probability of selecting the third object is therefore a product of the probabilities of no selection in the first round, in the second round and the probability of a selection in the third round:

$$p_3 = \frac{n-1}{n} \times \frac{n-2}{n-1} \times \frac{1}{n-2} = \frac{1}{n}$$

Continuing this reasoning, it can be found that the probability of selecting the *k*th object is

$$p_k = \frac{n-1}{n} \times \frac{n-2}{n-1} \times \cdots \times \frac{n-(k-1)}{n-(k-2)} \times \frac{1}{n-(k-1)} = \frac{1}{n}$$

This proves the correctness of the algorithm.

8.2.3 *Inverse Transformation Method for Simulation of Continuous Random Variables*

Let *U* be a random variable following a uniform distribution in the interval (0, 1) (Figure 8.2). For any continuous distribution function $F(x)$, if a random variable *X* is defined by $X = F^{-1}(U)$, where F^{-1} denotes the inverse function of $F(x)$, the random variable *X* has a cumulative distribution function $F(x)$.

Indeed, because the cumulative distribution function is monotonically increasing (Figure 8.2), the following chain of equalities holds:

$$P(X \le a) = P(U \le F(a)) = F(a) \tag{8.16}$$

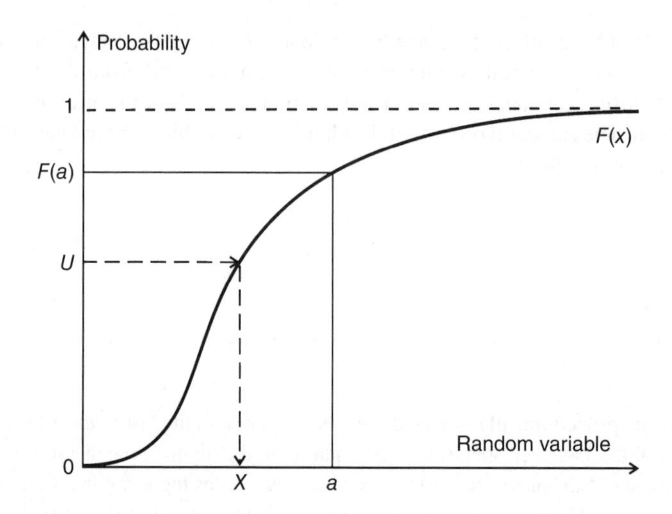

Figure 8.2 Inverse transformation method for generating random numbers

From the first and the last equality, it follows that $P(X \le a) = F(a)$, which means that the random variable X has a cumulative distribution $F(x)$.

8.2.4 *Simulation of a Random Variable following the Negative Exponential Distribution*

The cumulative distribution function of the negative exponential distribution is $F(x) = 1 - \exp(-\lambda x)$ whose inverse is $x = -(1/\lambda)\ln(1-F)$. Replacing $F(x)$ with U, which is a uniformly distributed random variable in the interval (0, 1), gives

$$x = -\frac{1}{\lambda}\ln(1-U) \tag{8.17}$$

which follows the negative exponential distribution. A small improvement of the efficiency of the algorithm can be obtained noticing that $1-U$ is also a uniformly distributed random variable in the range (0, 1), and therefore, $x = -(1/\lambda)\ln(1-U)$ has the same distribution as $x = -(1/\lambda)\ln U$. Finally, generating a uniformly distributed random variable u_i in the interval (0, 1) and substituting it in

$$x_i = -\frac{1}{\lambda}\ln(u_i) \tag{8.18}$$

result in a random variable x_i following the negative exponential distribution.

During the simulation, the uniformly distributed random values u_i are obtained either from a standard built-in or from a specifically designed pseudorandom number generator.

8.2.5 *Simulation of a Random Variable following the Gamma Distribution*

It is not possible to give a closed-form expression for the inverse of the gamma cumulative distribution function. Because the gamma random variable $G(n, 1/\lambda)$ can be presented as a sum of n random variables following the negative exponential distribution with parameter λ, generating a gamma random variable can be based on this property. Each negative exponential random variable is generated from $-(1/\lambda)\ln(u_i)$ $i=1,2,\ldots,n$, where u_i are statistically independent, uniformly distributed random numbers in the interval (0, 1). As a result, the gamma random variable $G(n,1/\lambda)$ can be obtained from the sum

$$G\left(n,\frac{1}{\lambda}\right)=-\frac{1}{\lambda}\ln(u_1)-\frac{1}{\lambda}\ln(u_2)-\cdots-\frac{1}{\lambda}\ln(u_n) \tag{8.19}$$

Equation 8.19 can be reduced to

$$G\left(n,\frac{1}{\lambda}\right)=-\frac{1}{\lambda}\ln(u_1u_2\ldots u_n) \tag{8.20}$$

which is computationally more efficient due to the single logarithm.

8.2.6 *Simulation of a Random Variable following a Homogeneous Poisson Process in a Finite Interval*

Random variables following a homogeneous Poisson process in a finite interval with length a can be generated in the following way. Successive, exponentially distributed random numbers $x_i=-(1/\lambda)\ln(u_i)$ are generated according to the inverse transformation method, where u_i are uniformly distributed random numbers in the interval (0, 1). Subsequent realisations t_i following a homogeneous Poisson process with intensity λ can be obtained from $t_1=x_1$, $t_2=t_1+x_2$, …, $t_n=t_{n-1}+x_n$ ($t_n \le a$). The number of variables n, following a homogeneous Poisson process in the finite time interval, equals the number of generated values t_i within the length a of the interval.

The nth generated value $t_n=-(1/\lambda)\ln(u_1)-(1/\lambda)\ln(u_2)-\cdots-(1/\lambda)\ln(u_n)$ can also be presented as $t_n=(-1/\lambda)\ln(u_1u_2\ldots u_n)$. Generating uniformly distributed random numbers $u_1,\ldots,u_i$ continues while $t_i=(-1/\lambda)\ln(u_1u_2\ldots u_i)\le a$ and stops immediately if $t_i>a$. Because the condition $(-1/\lambda)\ln(u_1u_2\ldots u_i)\le a$ is equivalent to the condition

$$u_1u_2\ldots u_i \ge \exp(-\lambda a) \tag{8.21}$$

generating uniformly distributed random numbers $u_1,\ldots,u_i$ continues while $t_i=u_1u_2\ldots u_i\ge\exp(-\lambda a)$ and stops immediately if $t_i<\exp(-\lambda a)$.

The algorithm in pseudocode for simulating a variable following a homogeneous Poisson process with density λ on the finite interval (0, a) is given next:

Algorithm 8.4

```
function Poisson(λ, a)
{
  Limit = exp(-λa);
  S = u_random();
  k = 0;

  while (S ≥ Limit) do {
                        S = S*u_random();
                        k = k+1;
                        }

  return k;
}
```

At the end, the generated random variable following a homogeneous Poisson process remains in the variable k. Simulating a number of random failures characterised by a constant hazard rate λ in a finite time interval with length a can also be done by the described algorithm.

8.2.7 Simulation of a Discrete Random Variable with a Specified Distribution

A discrete random variable X takes on only discrete values $X = x_1, x_2, \ldots, x_n$, with probabilities $p_1 = f(x_1), p_2 = f(x_2), \ldots, p_n = f(x_n)$ and no other value:

X	x_1	x_2	...	x_n
$P(X=x)$	$f(x_1)$	$f(x_2)$	...	$f(x_n)$

where $f(x) = P(X = x)$ is the probability (mass) function of the random variable $\sum_{i=1}^{n} f(x_i) = 1$.

The algorithm for generating a random variable with the specified distribution consists of the following steps:

Algorithm 8.5

1. Construct the cumulative distribution $P(X \le x_k) \equiv F(x_k) = \sum_{i \le k} f(x_i)$ of the random variable.
2. Generate a uniformly distributed random number u in the interval [0, 1].
3. If $u \le F(x_1)$, the simulated random value is x_1; else, if $F(x_{k-1}) < u \le F(x_k)$, the simulated random value is x_k (Figure 8.3).

A binomial experiment involving n statistically independent trials with probability of success p in each trial can be simulated by generating n random numbers u_i uniformly distributed in the interval (0, 1). If the number of successes X is set to be equal to the number of trials in which $u_i \le p$, the distribution of the number of successes X follows a binomial distribution with parameters n and p. The algorithm in pseudocode is given next:

Algorithm 8.6

```
function Binomial(p,n)
{
  k = 0;
  for i = 1 to n do {
                        S = u_random();
                        if (S ≤ p) then  k = k+1;
                    }
  return k;
}
```

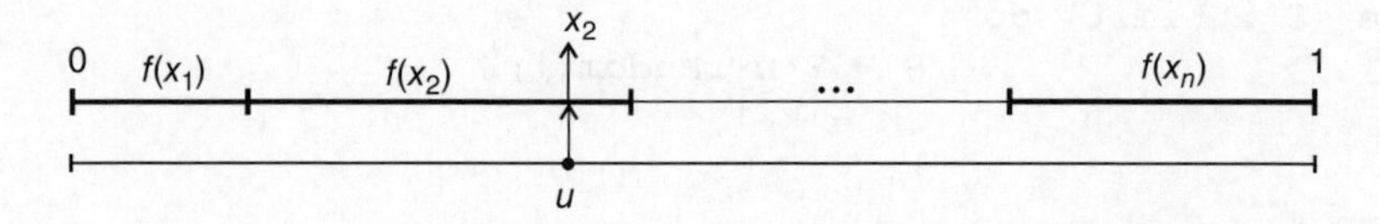

Figure 8.3 Simulating a random variable with a specified discrete distribution

8.2.8 *Selection of a Point at Random in the N-Dimensional Space Region*

In order to select a random point from a bounded three-dimensional region R (Figure 8.4), a *rejection method* can be used. The region R is first surmounted by a rectangular parallelepiped with sides a, b and c. A random point with coordinates au_1, bu_2 and cu_3 is generated in the parallelepiped using three statistically independent random numbers u_1, u_2 and u_3, uniformly distributed in the interval (0, 1). If the generated point belongs to the region R, it is accepted. Otherwise, the point is rejected and a new random point is generated in the parallelepiped and checked whether it belongs to R. The first accepted point is a point randomly selected in the region R.

This approach will be illustrated by a procedure for picking a random point on a circular disc with radius r. The circular disc is inscribed in a square with side $2r$, centred at the origin of the coordinate system (Figure 8.5).

Random points generated in the square but outside the disc are discarded. If the inequality $x^2 + y^2 > r^2$ holds, where (x, y) are the coordinates of the generated random point, the point is outside the disc. If the converse is true ($x^2 + y^2 \leq r^2$), the point is on the disc. The first generated point, for which $x^2 + y^2 \leq r^2$ is fulfilled, is accepted as a point with random location on the circular disc. The corresponding procedure is given next in pseudocode:

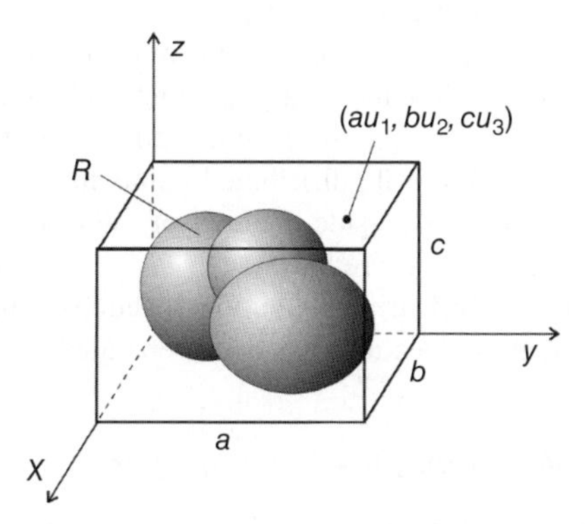

Figure 8.4 Selecting a random point in the region R

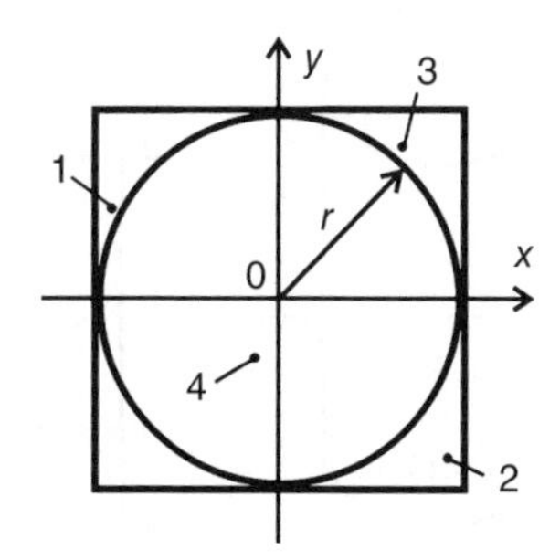

Figure 8.5 Picking a random point from a circular disc by the rejection method

Algorithm 8.7

```
r_sq = r*r;
repeat
  x = -r + 2r * u_random();
  y = -r + 2r * u_random();
until (x*x + y*y > r_sq)
print "The coordinates of the random point are", x,y;
```

8.2.9 Simulation of Random Locations following a Homogeneous Poisson Process in a Finite Domain

Random locations following a homogeneous Poisson process with constant intensity λ in a finite domain (volume, area, interval) can be simulated in two steps. In the first step, using Algorithm 8.4, a number n of random variables is generated following a homogeneous Poisson process with intensity λ. In the second step, n random locations are generated, uniformly distributed in the finite domain.

The method will be illustrated by an algorithm for generating random locations which follow a homogeneous Poisson process with intensity λ, in a cylindrical domain with area of the base S and height H (Figure 8.6). First, the number of locations k in the cylinder is generated according to Algorithm 8.4, where $a = S \times H$ is the volume of the cylinder. Next, k random locations are generated, uniformly distributed in the volume of the cylinder. The second step can be accomplished by generating a uniformly distributed point with coordinates (x, y) across the base of the cylinder, followed by generating the z-coordinate of the location, uniformly distributed along the height H.

A uniformly distributed point across the base is generated by using the rejection method which consists of generating sequentially uniformly distributed points in the circumscribed rectangle containing the base. If a random point falls outside the area of the base, it is rejected. The first point generated inside the base S is accepted (Figure 8.6).

The process of generating uniformly distributed random locations in the cylindrical domain continues until the total number of k random locations has been generated.

8.2.10 Simulation of a Random Direction in Space

Suppose that a random direction needs to be selected from the origin of the coordinate system with axes σ_1, σ_2 and σ_3 as shown in Figure 8.7.

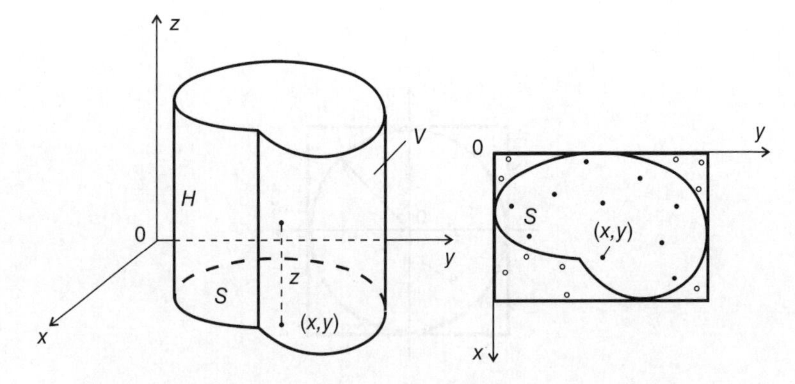

Figure 8.6 Generating uniformly distributed locations in the cylinder V

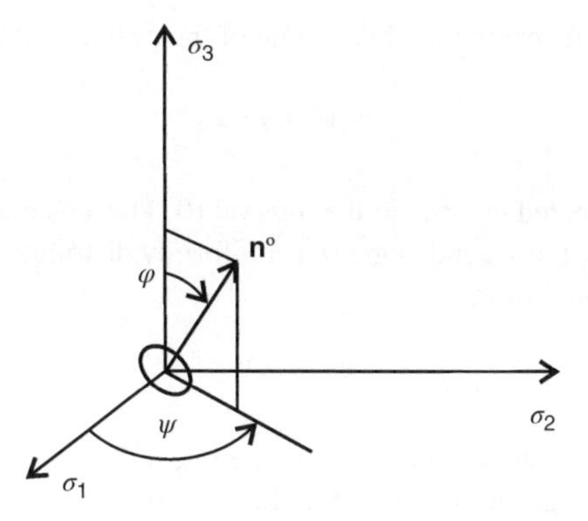

Figure 8.7 A random direction $\mathbf{n}^\circ$ in space defined by angles $0 \le \varphi \le \pi$ and $0 \le \psi \le 2\pi$

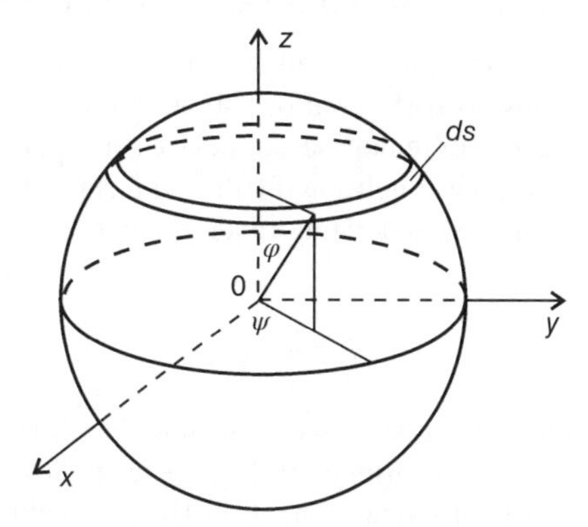

Figure 8.8 Unit sphere and a random direction in space

This problem is common in the applications (see, e.g. Sobol, 1994). Such a problem is present in simulations of brittle fracture triggered by penny-shaped cracks. The orientation of the crack regarding the principal stresses σ_1, σ_2 and σ_3 is specified by the unit normal vector $\mathbf{n}^\circ$ to the crack plane (Figure 8.7).

A random direction in space means that the endpoint of the random direction is uniformly distributed on the surface of the unit sphere in Figure 8.8. The random direction is determined by the two angles $0 \le \psi \le 2\pi$ and $0 \le \varphi \le \pi$ (Figure 8.8).

Let us divide the surface of the sphere with unit radius, by planes perpendicular to one of the axes, into infinitesimally small surface elements ds (Figure 8.8).

Because the endpoints are uniformly distributed on the surface of the sphere, the probability of selection of a particular element ds to which corresponds angle φ should be equal to the ratio of the area of the element ds and the area of the sphere:

$$\frac{ds}{4\pi} = \frac{2\pi \sin(\varphi) d\varphi}{4\pi} = \frac{1}{2}\sin(\varphi) d\varphi$$

The latter expression defines the probability density $(1/2)\sin(\varphi)$ characterising the angle φ. Integrating this probability density gives the cumulative distribution of the angle

$$F(\varphi) = \int_0^{\varphi} \frac{1}{2}\sin(x) dx = \frac{1}{2}[1 - \cos(\varphi)]$$

Applying the inverse transformation method, value of the cosine of the angle φ is generated from

$$\cos(\varphi) = 1 - 2u \quad (8.22)$$

where u is a uniformly distributed number in the interval (0, 1). Considering the axial symmetry with respect to the selected z-axis, the second angle ψ is uniformly distributed in the interval $(0, 2\pi)$, and a value for this angle is generated from

$$\psi = 2\pi v \quad (8.23)$$

where v is another uniformly distributed number in the interval (0, 1), statistically independent from the random number u used for generating the value for $\cos(\varphi)$.

8.2.11 Generating Random Points on a Disc and in a Sphere

The described technique can be used for generating random points inside a disc with radius R. Generating random points on a disc is part of various simulations involving random nucleation and growth in a system, random locations of remote sensor devices on a planar region, etc.

Because of the symmetry, the polar angles φ_i of the uniformly distributed across the area of the disc random points are obtained by generating random numbers φ_i uniformly distributed in the interval $(0, 2\pi)$:

$$\varphi_i = 2\pi v_i \quad (8.24)$$

where v_i are random numbers uniformly distributed in the interval (0, 1).

Let $f(\rho)$ be the probability density distribution of the polar radius which guarantees a uniform distribution of a random point across the area of the disc. The probability that the polar radius of the random point will be within the interval $(\rho, \rho + d\rho)$ will then be given by $f(\rho)d\rho$.

The probability that the polar radius of the random point will be within the interval $(\rho, \rho + d\rho)$ is also equal to the probability that the random point will be located on the elementary area ds enclosed between two circles with radii ρ and $\rho + d\rho$ (Figure 8.9). Because the location of the random point should be uniformly distributed across the area of the disc, this probability should be equal to the ratio of the area of the element ds and the area of the disc:

$$\frac{ds}{\pi R^2} = \frac{2\pi\rho d\rho}{\pi R^2} = \frac{2\rho d\rho}{R^2}$$

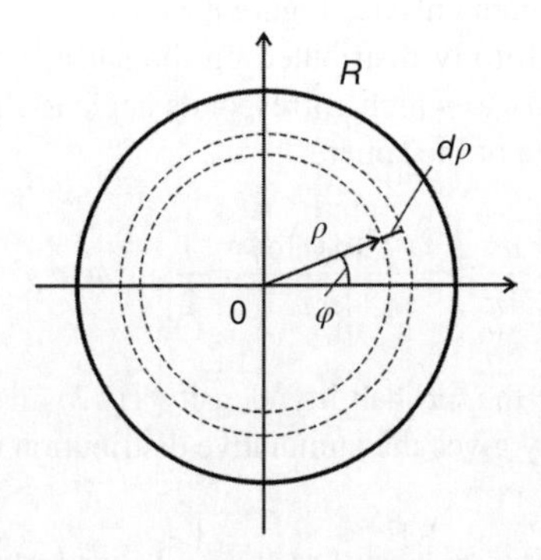

Figure 8.9 Selecting a random point on a disc with radius R

Equating the two probabilities $f(\rho)d\rho = (1/R^2)2\rho\, d\rho$ yields

$$f(\rho) = \frac{2\rho}{R^2}$$

for the probability density distribution of the polar radius ρ of the random point. Integrating this probability density gives the cumulative distribution $F(\rho)$ of the polar radius

$$F(\rho) = \frac{2}{R^2}\int_0^{\rho} u\, du = \frac{\rho^2}{R^2}$$

Applying the inverse transformation method, values of the polar radii ρ_i of the generated random points are obtained from

$$\rho_i = R\sqrt{u_i} \quad (8.25)$$

where u_i are random numbers uniformly distributed in the interval (0, 1).

This method can be used for generating random points inside a sphere with radius R. Random locations inside a spherical region are part of simulations of phase transformations involving nucleation and growth from random centres, simulation of suspended particles, simulation of defects in materials, etc.

The random point is characterised by a polar radius ρ and angles φ and ψ. The coordinates φ and ψ define a random orientation for the selected point with respect to the centre of the sphere and can be determined following the procedure described in the previous section. The polar radius of the random point can be determined by applying a technique very similar to the one used for generating random points on a disc.

Let $f(\rho)$ be the probability density distribution of the polar radius which guarantees a uniform distribution of a random point inside the sphere with radius R. The probability that the polar radius of the random point will be within the interval $(\rho, \rho + d\rho)$ is given by $f(\rho)d\rho$.

This probability is also equal to the probability that the random point will be located within the elementary volume dv enclosed between two spheres with radii ρ and $\rho + d\rho$. Because the location of the random point should be uniformly distributed across the volume of the sphere with radius R, this probability should be equal to the ratio of the volume of the element dv and the volume of the sphere:

$$\frac{dv}{(4/3)\pi R^3} = \frac{4\pi\rho^2\, d\rho}{(4/3)\pi R^3} = \frac{3\rho^2 d\rho}{R^3}$$

Equating the two probabilities $f(\rho)d\rho = (1/R^3)3\rho^2\, d\rho$ yields

$$f(\rho) = \frac{3\rho^2}{R^3}$$

for the probability density distribution of the polar radius ρ of the random point. Integrating this probability density gives the cumulative distribution $F(\rho)$ of the polar radius

$$F(\rho) = \frac{1}{R^3}\int_0^{\rho} 3u^2 du = \frac{\rho^3}{R^3}$$

Applying the inverse transformation method, values of the polar radii ρ_i are generated from

$$\rho_i = R \times \sqrt[3]{u_i} \tag{8.26}$$

where u_i are random numbers uniformly distributed in the interval (0, 1).

8.2.12 *Simulation of a Random Variable following the Three-Parameter Weibull Distribution*

Since the Weibull cumulative distribution function is $F(x) = 1 - \exp\{-[(x - x_0)/\eta]^m\}$, the first step is to construct its inverse $x = x_0 + \eta\{\ln[1/(1 - F(x))]\}^{1/m}$. Next, $F(x)$ is replaced with U, which is a uniformly distributed random variable in the interval (0, 1). As a result, the expression $x = x_0 + \eta\{\ln[1/(1 - U)]\}^{1/m}$ is obtained. Generating uniformly distributed random values u_i in the interval (0, 1) and substituting them in

$$x_i = x_0 + \eta\left[-\ln(u_i)\right]^{1/m} \tag{8.27}$$

yields random values x_i following a three-parameter Weibull distribution.

8.2.13 *Simulation of a Random Variable following the Maximum Extreme Value Distribution*

From the cumulative distribution function of the maximum extreme value distribution $F(x) = \exp\{-\exp[-(x - x_0)/\theta]\}$ the inverse $x = x_0 - \theta \ln[-\ln F(x)]$, is determined. Replacing $F(x)$ with U, which is a uniformly distributed random variable in the interval (0, 1), results in $x = x_0 - \theta \ln(-\ln U)$. Generating uniformly distributed random variables u_i in the interval (0, 1) and substituting them in

$$x_i = x_0 - \theta \ln(-\ln u_i) \tag{8.28}$$

produces values x_i following the maximum extreme value distribution.

8.2.14 *Simulation of a Gaussian Random Variable*

A standard normal variable can be generated easily using the central limit theorem applied to a sum X of n random variables U_i, uniformly distributed in the interval (0, 1). According to the central limit theorem, with increasing n, the sum $X = U_1 + U_2 + \cdots + U_n$ approaches a normal distribution with mean

$$E(X) = E(U_1) + \cdots + E(U_n) = \frac{n}{2}$$

and variance

$$V(X) = V(U_1) + \cdots + V(U_n) = n \times \frac{1}{12}.$$

Selecting $n = 12$ uniformly distributed random variables U_i gives a reasonably good approximation for many practical applications. Thus, the random variable

$$X = U_1 + U_2 + \cdots + U_{12} - 6 \quad (8.29)$$

is approximately normally distributed with mean $E(X) = 12 \times (1/2) - 6 = 0$ and variance $V(X) = 12 \times (1/12) = 1$, or in other words, the random variable X follows the standard normal distribution (Rubinstein, 1981).

Another method for generating a standard normal variable is the Box–Muller method (Box and Muller, 1958). A pair of statistically independent standard normal variables x and y are generated by generating a pair u_1, u_2 of statistically independent, uniformly distributed random numbers in the interval (0, 1). Random variables following the standard normal distribution are obtained from

$$x = \sqrt{-2\ln u_1}\cos(2\pi u_2) \quad (8.30)$$

$$y = \sqrt{-2\ln u_1}\sin(2\pi u_2) \quad (8.31)$$

The derivation of these expressions is given in Appendix 8.1.

From the generated standard normal variable $N(0,1)$ with mean zero and standard deviation unity, a normally distributed random variable $N(\mu,\sigma)$ with mean μ and standard error σ can be obtained by applying the linear transformation

$$N(\mu,\sigma) = \sigma N(0,1) + \mu \quad (8.32)$$

8.2.15 Simulation of a Log-Normal Random Variable

A random variable follows a log-normal distribution if its logarithm follows a normal distribution. Suppose that the mean and the standard deviation of a random variable X are μ and σ, correspondingly. A log-normal random variable can be generated by first generating a normally distributed random variable Y with mean μ and standard deviation σ using

$$Y = \sigma N(0,1) + \mu \quad (8.33)$$

where $N(0,1)$ is a generated standard normal variable (see Eq. 8.29). A log-normal variable X is generated by exponentiating the normal random variable Y:

$$X = e^Y \quad (8.34)$$

It is important to point out that μ and σ in Equation 8.33 are not the mean and variance of the simulated log-normal variable X. The mean and variance of the simulated log-normal random variable X are given by $\mu_X = \exp(\mu + \sigma^2/2)$ and $\sigma_X^2 = \exp(2\mu + \sigma^2) \times [\exp(\sigma^2) - 1]$. Consequently, to generate a log-normal random variable with a specified mean μ_X and standard deviation σ_X, these equations need to be solved with respect to μ and σ which results in

$$\mu = \ln\left(\frac{\mu_X^2}{\sqrt{\sigma_X^2 + \mu_X^2}}\right) \quad (8.35)$$

$$\sigma = \sqrt{\ln\left(1 + \frac{\sigma_X^2}{\mu_X^2}\right)} \quad (8.36)$$

Next, the obtained values μ and σ are used in Equation 8.33 to generate a normally distributed random variable Y, which, after exponentiation (Eq. 8.34), generates a log-normal random variable X with mean μ_X and standard deviation σ_X.

8.2.16 *Conditional Probability Technique for Bivariate Sampling*

This technique is based on presenting the joint probability density function $p(x,y)$ as a product

$$p(x,y) = p(x)p(y \mid x) \tag{8.37}$$

of $p(x)$ – the marginal distribution of X and $p(y \mid x)$ - the conditional distribution of Y, given that $X = x$. Simulating a random number with distribution $p(x,y)$ involves two steps: (i) generating a random number x with distribution $p(x)$ and (ii) for the generated value $X = x$ generating a second random number y, with a conditional probability density $p(y \mid x)$. The obtained pairs (x, y) have a joint distribution $p(x,y)$. A common application of this technique is the random sampling from a *bivariate normal distribution*

$$f(x,y) = \frac{1}{2\pi\sigma_x\sigma_y\sqrt{1-\rho^2}} \times \exp\left\{-\frac{1}{2(1-\rho^2)}\left[\left(\frac{x-\mu_X}{\sigma_X}\right)^2 - 2\rho\left(\frac{x-\mu_X}{\sigma_X}\right)\left(\frac{y-\mu_Y}{\sigma_Y}\right) + \left(\frac{y-\mu_Y}{\sigma_Y}\right)^2\right]\right\} \tag{8.38}$$

where μ_X, μ_Y denote the means and σ_X, σ_Y denote the standard deviation of the μ_X, μ_Y random variables X and Y, respectively. The parameter ρ is the linear correlation coefficient between X and Y, defined by

$$\rho = E\left[\left(\frac{X-\mu_X}{\sigma_X}\right)\left(\frac{Y-\mu_Y}{\sigma_Y}\right)\right] \tag{8.39}$$

An important feature is that the bivariate normal distribution is a natural extension of the normal distribution in the two-dimensional space. If pairs (X, Y) have a bivariate normal distribution, the variables W and Z defined by $W = (X - \mu_X)/\sigma_X$ and $Z = (Y - \mu_Y)/\sigma_Y$ have a standardised bivariate normal distribution with a probability density function

$$f(w,z) = \frac{1}{2\pi\sqrt{1-\rho^2}} \times \exp\left\{-\frac{1}{2(1-\rho^2)}\left[w^2 - 2\rho wz + z^2\right]\right\} \tag{8.40}$$

Given that $X = x$, the conditional distribution of Y, is normal, with mean $\mu_Y + \rho(\sigma_Y/\sigma_X)(x - \mu_X)$ and standard deviation $\sigma_Y\sqrt{1-\rho^2}$ (Miller and Miller, 1999).

A procedure for sampling from the standardised bivariate normal distribution consists of generating two random numbers w and z from the standard normal distribution $N(0,1)$. The random variates x and y following a bivariate normal distribution with specified means μ_X, μ_Y, standard deviations σ_X,σ_Y and a correlation coefficient ρ are obtained from

$$x = \mu_X + \sigma_X w \tag{8.41}$$

$$y = \mu_Y + \rho\frac{\sigma_Y}{\sigma_X}(x - \mu_X) + \sigma_Y\sqrt{1-\rho^2}\,z \tag{8.42}$$

8.2.17 Von Neumann's Method for Sampling Continuous Random Variables

This method, also known as *rejection method*, is convenient in cases where the inverse function of the cumulative distribution function $F(x)$ cannot be expressed in terms of elementary functions or in cases where the probability density function has been specified empirically (e.g. by a histogram).

Suppose that the random variable is defined in the interval (*a*, *b*) and the probability density function $f(x)$ is bounded: $f(x) \le M$ (Figure 8.10).

A value following the specified probability density function $f(x)$ can be generated using the following steps:

1. A uniformly distributed random value $x = a + (b - a)u_1$ in the interval (*a*, *b*) is generated first, where u_1 is a uniformly distributed random number in the interval (0, 1).
2. A random value $y = M \times u_2$ is generated, uniformly distributed in the interval (0, *M*), where U_2 is another uniformly distributed random number in the interval (0, 1).
3. If $y \le f(x)$, the random value x generated on step one is accepted. Otherwise, the random value is rejected, and the process continues with steps (1) and (2) until a generated value x is accepted.

The algorithm of the rejection method in pseudocode is given next:

Algorithm 8.8

```
function rejection_method()
{
 repeat
  x = a + (b - a)*u_random();
  y = M*u_random();
 until (y ≤ f(x));

 return x;
}
```

Indeed, the probability that a generated value x_i will belong to the interval $x_0, x_0 + dx$ is a product of the probability $dx/(b-a)$ that the random value x_i will be generated in the interval $x_0, x_0 + dx$ and the probability $f(x_0)/M$ that it will be accepted. As a result, the probability that a generated value will

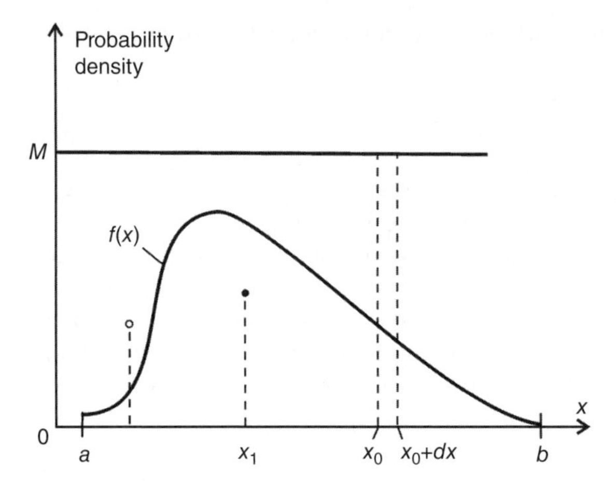

Figure 8.10 Rejection method for generating a continuous random variable with a specified probability density function *f*(*x*)

belong to the interval $x_0, x_0 + dx$ and will be accepted becomes $f(x_0)dx / [(b-a) \times M]$. According to the total probability theorem, the probability of accepting a value is

$$\int_a^b \frac{f(x)dx}{(b-a)M} = \frac{1}{(b-a)M}$$

because $\int_a^b f(x)dx = 1$. Finally, the conditional probability that a value will belong to the interval $x_0, x_0 + dx$ given that it has been accepted is

$$\frac{f(x_0)dx}{(b-a)M} \bigg/ \frac{1}{(b-a)M} = f(x_0)dx$$

which means that the accepted values do follow the specified distribution $f(x)$.

8.2.18 Sampling from a Mixture Distribution

Suppose that a sampling from the distribution mixture

$$F(x) = p_1 F_1(x) + \cdots + p_M F_M(x) \tag{8.43}$$

is required, where $p_i, \sum_{i=1}^{M} p_i = 1$ are the shares of the separate individual distributions $F_i(x)$ in the mixture. Sampling the distribution mixture (8.43) involves two basic steps:

1. Random selection of an individual distribution to be sampled
2. Random sampling from the selected distribution

Random selection of an individual distribution can be done using Algorithm 8.5 for sampling from a discrete distribution with mass function

X	1	2	M
$P(X=x)$	p_1	p_2	p_M

Random sampling from the selected distribution can be performed using, for example, the inverse transformation method.

Appendix 8.1

Indeed, if we denote $r = \sqrt{-2 \ln u_1}$ and $\theta = 2\pi u_2$, then $x = r \times \cos(\theta)$; $y = r \times \sin(\theta)$ and r, θ can be regarded as the polar coordinates of the point (x, y) (Figure 8.11).

The polar angle θ follows a uniform distribution with density $1/2\pi$ in the interval $(0, 2\pi)$. According to the algorithm related to generating a random variable with a negative exponential distribution (see Eq. 8.8), the square $d = r^2$ of the polar radius follows an exponential distribution $(1/2)\exp[-(1/2)d]$ with parameter $\lambda = 1/2$. Because u_1 and u_2 are statistically independent, the polar angle and the polar radius which are functions of u_1 and u_2 are also statistically independent:

$$d = x^2 + y^2 \tag{8.A.1}$$

$$\theta = \arctan\left(\frac{y}{x}\right) \tag{8.A.2}$$

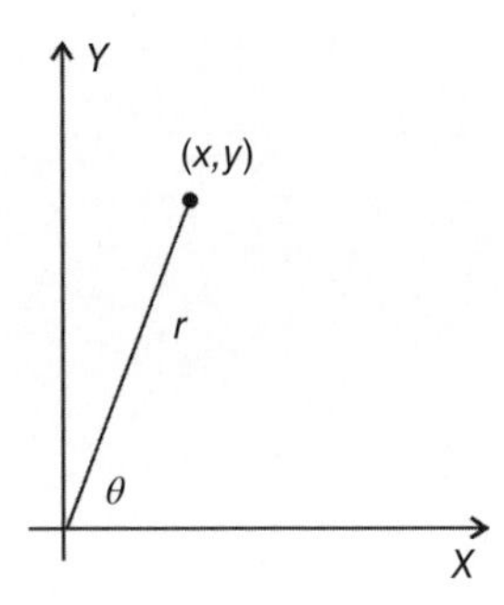

Figure 8.11 Polar coordinates r and θ of point (x, y)

The joint distribution $g(d,\theta)$ of $d = r^2$ and θ is therefore a product of the marginal density distributions of $d = r^2$ and θ (Ross, 1997):

$$g(d,\theta) = \frac{1}{2\pi} \times \frac{1}{2}\exp\left(-\frac{d}{2}\right)$$

The joint distribution $f(x,y)$ of x and y equals the joint distribution of d and θ multiplied by the absolute value of the Jacobian of the transformation (DeGroot, 1989)

$$f(x,y) = g(d,\theta) \times |J|$$

where

$$|J| = \det\begin{vmatrix} \dfrac{\partial d}{\partial x} & \dfrac{\partial d}{\partial y} \\ \dfrac{\partial \theta}{\partial x} & \dfrac{\partial \theta}{\partial y} \end{vmatrix} = 2$$

Finally, for the joint distribution of x and y, we obtain $f(x,y) = (1/2\pi)\exp\left[-(x^2 + y^2)/2\right]$ which can be factorised as

$$f(x,y) = h(x) \times h(y) = \frac{1}{\sqrt{2\pi}}\exp\left(-\frac{x^2}{2}\right) \times \frac{1}{\sqrt{2\pi}}\exp\left(-\frac{y^2}{2}\right)$$

where

$$h(x) = \frac{1}{\sqrt{2\pi}}\exp\left(-\frac{x^2}{2}\right) \quad \text{and} \quad h(y) = \frac{1}{\sqrt{2\pi}}\exp\left(-\frac{y^2}{2}\right)$$

are the probability density functions of two statistically independent standard normal random variables X and Y.

9

Evaluating Reliability and Probability of a Faulty Assembly Using Monte Carlo Simulation

9.1 A General Algorithm for Determining Reliability Controlled by Statistically Independent Random Variables

Suppose that $X_1, X_2, \ldots, X_n$ are statistically independent random variables which control the reliability of a component/system, where $f_1(x_1)$, $f_2(x_2)$, …, $f_n(x_n)$ are their marginal probability densities. The reliability is then given by the integral (Melchers, 1999)

$$R = \iint\limits_{x_1,\ldots,x_n \in S} \cdots \int f_1(x_1) \times f_2(x_2) \times \cdots \times f_n(x_n)\, dx_1 dx_2 \cdots dx_n$$

where the integration is performed within the safe domain S. The safe domain defines all possible combinations of values for the random variables for which no failure occurs. The Monte Carlo algorithm for evaluating the above integral is as follows:

Algorithm 9.1

```
x[n]: /* Global array containing the current values of the n random
         variables */

procedure  Generate_random_variable (j)
 {
     /* Generates a realisation (value) of the jth controlling
        random variable x[j] */
 }

function  Limit_state()
 {
   /* For a particular combination of values of the random
      variables  x[1],…,x[n] returns 1 or 0 depending on whether
      failure state is present or not */
 }
```

Reliability and Risk Models: Setting Reliability Requirements, Second Edition. Michael Todinov.

```
/* Main algorithm */

Failure_counter = 0;
For i = 1 to Number_of_trials do
  {
    /* Generate the ith set of n controlling random variables */
    For j=1 to do
        Generate_random_variable(j);
    Failure = Limit_state(); /* Checks for a limit (failure) state
                                 using the current values of the
                                 random variables in the array x[n] */
    If (Failure=1)  then  Failure_counter = Failure_counter + 1;
  }
Reliability_on_demand = (Number_of_trials - Failure_counter) /
Number_of_trials
```

In the simulation loop controlled by variable *i*, a second nested loop has been defined, controlled by variable *j*, whose purpose is generating instances of all statistically independent controlling random variables. After obtaining a set of values for the random variables, the function **`Limit_state()`** is called to check whether the set of values for the random variables defines a point in the failure region. The function returns '1' if the values of the random variables define failure, otherwise, the function returns zero.

If the set of values defines a point in the failure region, the failure counter is incremented. At the end of the simulation trials, reliability on demand is obtained as a ratio of the number of trials during which no failure occurred and the total number of simulation trials.

9.2 Evaluation of the Reliability Controlled by a Load–Strength Interference

9.2.1 Evaluation of the Reliability on Demand, with No Time Included

Main components of the load–strength interference model are (i) a model for the strength distribution and (ii) a model for the load distribution. A Monte Carlo simulation approach for solving the load–strength reliability integral will be illustrated by providing an alternative solution, by a Monte Carlo simulation, of the example from Chapter 6, related to determining the probability of failure of a component by a numerical integration. For the purposes of comparison with the numerical solution presented in the example from Chapter 6, the distributions and their parameters are the same. Accordingly, the distribution of strength has been approximated by the three-parameter Weibull distribution

$$F_S(x) = 1 - \exp\left[-\left(\frac{x - x_0}{\eta}\right)^m\right]$$

with parameters $m = 3.9, \eta = 297.3$ and $x_0 = 200$ MPa. The load distribution has been approximated by the maximum extreme value distribution

$$F_L = \exp\left[-\exp\left(-\frac{x - \xi}{\theta}\right)\right]$$

with parameters $\xi = 119$ and $\theta = 73.64$.

An algorithm in pseudocode which evaluates the reliability on demand using Monte Carlo simulation is presented next:

Algorithm 9.2

```
function  Weibull_rv()
{/* Generates a Weibull-distributed random variable */}

function  Max_extreme_value_rv()
{/* Generates an Extreme value random variable */}

Failure_counter = 0;
For i = 1 to Number_of_trials do
  {
    /* Generates  the ith pair of random load and random strength */
    Strength = Weibull_rv();
    Load = Max_extreme_value_rv();

    If (Strength <= Load) then  Failure_counter = Failure_counter + 1;
  }
Reliability_on_demand = (Number_of_trials - Failure_counter) /
Number_of_trials;
```

In the Monte Carlo simulation loop controlled by the variable i, instances of random strength and random load are generated in each trial. Their values are subsequently compared in order to check for a safe state. If a failure state is present (strength smaller than or equal to load), the failure counter is incremented. Similar to the previous algorithm, at the end of the Monte Carlo trials, reliability on demand is obtained as a ratio of the number of trials during which no failure occurred and the total number of Monte Carlo trials.

The algorithms of the functions returning random strength following the Weibull distribution and random load following the maximum extreme value distribution have been discussed in Chapter 8.

For the reliability on demand, one million Monte Carlo simulation trials (`Number_of_trials`=1 000 000) yielded 0.985, which coincides with the result from the example in Chapter 6, obtained by a direct numerical integration. Increasing the number of simulations to 10 million does not alter the last significant digit of the result, which shows that a very good convergency of the simulation results is already present at the selected one million trials.

9.2.2 Evaluation of the Reliability Controlled by Random Shocks on a Time Interval

In the important case of random shocks on a time interval, failure of the component occurs when the magnitude of any of the random shocks exceeds the random strength. Suppose that the load and strength distributions from the previous example give the magnitude of the random shocks in the time interval and the strength of the component. It is assumed that the shocks follow a homogeneous Poisson process with density ρ on a time interval with length a. An algorithm in pseudocode for Monte Carlo evaluation of the reliability in the specified finite time interval can be constructed as follows:

Algorithm 9.3

```
function  Weibull_rv()
   { /* Generates a random variable following a Weibull
        distribution with specified parameters */}

function  Max_extreme_value_rv()
   { /* Generates a random variable following the Maximum Extreme
        value distribution with specified parameters */}

function  Generate_number_of_shocks()
   {/* Generates a number of random shocks following a homogeneous
       Poisson process with density ρ, in the finite time interval
       0,a */}

Failure_counter = 0;
for i = 1 to Number_of_trials do
 {
  /* Generates a random strength */
     Strength = Weibull_rv();

  /* Generates the number of random shocks */
     Num_shocks = Generate_number_of_shocks();

  /* Generates the random loads (shocks) and compares each random
     load with the random strength */

  for  k = 1 to  Num_shocks do
   {
     Load = Extreme_value_rv();
     if (Strength <= Load) then
          {Failure_counter = Failure_counter + 1; break;}
   }
  }

Reliability = (Number_of_trials - Failure_counter) /
Number_of_trials;
```

A characteristic feature distinguishing this algorithm from the previous algorithm, dealing only with reliability on demand, is the nested loop with control variable k, accepting values from one to `Num_shocks` (the number of random shocks). A random strength is generated before entering the inner loop (the loop with control variable k) because strength is the same for all load applications (shocks). For each shock, a random load is generated and subsequently compared with the strength. If strength is smaller than or equal to any of the shock magnitudes, failure is registered by incrementing the failure counter (`Failure_counter`), after which the loop with control variable k is exited immediately by the statement **`break`**. By dividing the number of trials during which no failure has occurred, to the total number of trials, the reliability associated with the time interval with length $a = 100$ months is obtained. For the reliability associated with the finite time interval of 100 months and density of the shocks 0.5 shocks/month, a computer programme in C++ based on Algorithm 9.3 and one million simulation trials yields 0.597. This result coincides with the result obtained from a direct integration (see the overstress reliability exercise from Chapter 7). Increasing the number of simulations to 10 million does not alter the last significant digit of the result, which shows that a very good convergency of the simulation results is already present at the selected one million trials.

9.3 A Virtual Testing Method for Determining the Probability of Faulty Assembly

The design parameters related to any product are associated with uncertainty. This is usually caused by variability associated with the external loads acting on the product, the environment where the product operates, the technological processes used in the production of materials, the manufacturing processes and the operation. Variability of the input parameters characterising a particular design transforms into variability of the output properties of the products. Because of the natural variation of design parameters, *particular combinations of parameter values are transformed into undesirable deviations of the output properties from their optimal values* which often constitute a fault. Faults lead to a deteriorated performance and failure.

A typical example of a fault caused by the variation of input design parameters can be given with the *interference fits* (*press* or *shrink fits*) (Vinogradov, 1991) that include a shaft and a hub (Figure 9.1).

This assembly must be capable of carrying torque and axial forces without slippage. All design parameters are associated with a physical variation after manufacturing but the two design parameters that affect most significantly the load-carrying capability of the assembly are the coefficient of friction μ between the hub and the shaft and the interference $\delta = 2\Delta r$ between the diameters of the hub and the shaft (Booker *et al.*, 2001; Vinogradov, 1991). These parameters are important because their variation affects most significantly the friction force f per unit contact area, defined as $f = \mu p$ where p is the contact pressure on the shaft and μ is the coefficient of friction (Figure 9.1). The variations of the coefficient of friction μ affect the friction force directly while the variations of the interference δ affect the friction force through the contact pressure p. If the temperature also varies, the interference is further affected because of the different coefficient of thermal expansion of the hub and the shaft. If, for example, the coefficient of linear expansion of the hub is greater than the coefficient of thermal expansion of the shaft, with increasing temperature, the interference, the contact pressure and the load-carrying capability of the assembly will decrease. Suppose that the friction coefficient varies in the interval $\mu_{\min} \leq \mu \leq \mu_{\max}$, the interference at room temperature varies in the interval $\delta_{\min} \leq \delta \leq \delta_{\max}$ and the temperature varies in the interval $T_{\min} \leq T \leq T_{\max}$. Suppose also that with increasing temperature, the coefficient of friction monotonically decreases. If for a particular interference (press fit) assembly, an elevated temperature is combined with a small coefficient of friction and a small value of the interference δ at a room temperature, the assembly could lose its capability to carry the prescribed torque and axial force.

In the cases where the joint distribution describing the variation of the design parameters is known, the link between the uncertainty in the design parameters and the probability of a faulty assembly due to unfavourable combinations of parameter values can be investigated by a Monte Carlo simulation. In the press fit assembly example, the probability of a faulty assembly can be found by a process called *virtual testing* (Todinov, 2009b). This is essentially a Monte Carlo simulation

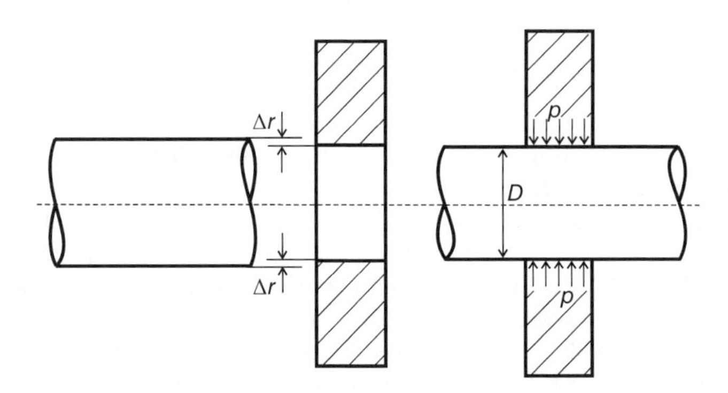

Figure 9.1 The press fit assembly must be capable of carrying torque and axial force without slippage

technique for transforming the uncertainty associated with the input parameters into uncertainty associated with the product's performance.

The simulation involves sampling random values for the friction coefficient, the interference and temperature. After each sampling, a check is performed whether an assembly characterised by the sampled reliability-critical parameters can transmit the required loads without slippage. The ratio of the number of simulations resulting in a faulty assembly and the total number of simulations gives the likelihood of a faulty assembly.

The variation of many design parameters may not have an effect on the reliability of a product. The virtual testing begins with *establishing which design parameters affect the reliability of the component/assembly*. These are the reliability-critical parameters. *In addition, the selected reliability-critical parameters must be associated with uncertainty.*

The algorithm of the virtual testing method can be generalised for an arbitrary number of reliability-critical parameters associated with uncertainty. The description in pseudocode is given next:

Algorithm 9.4

```
x[n]:
/* Global array containing the current values of the n reliability-
   critical design parameters */
Fault_counter = 0;
For i = 1 to Number_of_trials do
 {
/* Generate the ith set of n reliability-critical parameters by
   sampling their joint distribution and placing them in the
   array x[] */
Sample_all_reliability_critical_design_parameters;
/* Check whether the assembly is faulty by using the current values
   of the reliability-critical parameters in the array x[n]. For a
   particular combination x[1],…,x[n] of values, the function Is_
   faulty_assembly() returns 1 or 0 depending on whether a fault is
   present or not */
   Fault = Is_faulty_assembly();
   /* If a faulty assembly is present, then increment the failure
      counter */
   If (Fault=1)  then  Fault_counter = Fault_counter + 1;
 }
Probability_of_faulty_assembly = Fault_counter / Number_of_trials;
```

In the simulation loop controlled by the variable *i*, the procedure ***Sample_all_reliability_critical_design_parameters()*** is called, whose purpose is to generate realisations for all reliability-critical design parameters by sampling their joint distribution. If the reliability-critical parameters are statistically independent, realisations are generated by a sequential sampling of their individual distributions. After obtaining a set of values for the parameters controlling the reliability of the assembly, the function ***Is_faulty_assembly()*** is called to perform design calculations and check whether the set of values for the reliability-critical parameters defines a faulty assembly. If this is so, the fault counter is incremented. At the end of the simulation trials, the probability of a faulty assembly is obtained as a ratio of the number of simulated faulty assemblies and the total number of simulated assemblies.

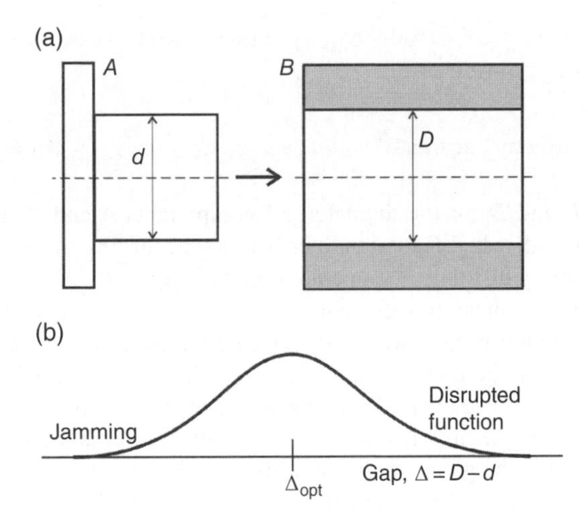

Figure 9.2 (a) Assembly that requires (b) an optimal value for the diameter clearance $\Delta = D - d$

The application of this approach will be illustrated by assessing the probability that an assembly of the type in Figure 9.2 will exhibit a fault during operation. The reliability-critical design parameters associated with uncertainty are the diameters *d* and *D* of components *A* and *B* at room temperature and the working temperature *t*. These three parameters fully determine the magnitude of the diameter clearance Δ, whose excessive deviation from the optimal value Δ_{opt} causes failure.

Suppose that the diameters d_0 and D_0 at temperature $t_0 = 20°C$ follow normal distributions with means $\mu_{d0} = 55\,mm$, $\mu_{D0} = 55.5\,mm$ and standard deviations $\sigma_{d0} = 0.12\,mm$, $\sigma_{D0} = 0.22\,mm$. The optimal diameter clearance of the assembly is $\Delta_{opt} = 0.5\,mm$. Suppose that if the diameter clearance falls below $\Delta_{min} = 0.1\,mm$, jamming occurs and if the clearance exceeds $\Delta_{max} = 0.8\,mm$, precision of operation is lost. The coefficient of thermal linear expansion of component *A* is $\alpha_A = 24\times10^{-6}\,K^{-1}$ and $\alpha_B = 11\times10^{-6}\,K^{-1}$ for component *B*. The working temperature of the assembly can be anywhere in the temperature interval $-55°C \le t \le 400°C$. Consequently, the operating temperature *t* is assumed to be uniformly distributed in this temperature interval.

Since the reliability-critical design parameters associated with uncertainty are statistically independent, sampling from their joint distribution is equivalent to a sequential sampling from their individual distributions. The outlined Algorithm 9.4 transforms into the following algorithm:

Algorithm 9.5

```
Fault_counter = 0;
for i = 1 to Number_of_trials do
 {
        d0 = Sample_ diameter_comp_A();
        D0 = Sample_ diameter_comp_B();
        t = Sample_temperature();
        Δt=t−t0;
        d=d0(1+αAΔt)
        D=D0(1+αBΔt)
```

```
        if (D−d<Δmin or D−d>Δmax ) then Fault_counter =
            Fault_counter+1;
}
Probability_of_faulty_assembly = Fault_counter / Number_of_trials;
```

Initially, instances d_0 and D_0 of the diameters of components A and B at $t_0 = 20°C$ are calculated, by sampling their individual Gaussian distributions, according to the methods presented in Chapter 8. Next, a uniformly distributed temperature in the interval ($t_{min} = -55°C \le t \le t_{max} = 400°C$) is generated, by using the linear transformation $t = t_{min} + (t_{max} - t_{min}) \times u$, where u is a random number uniformly distributed in the interval (0, 1). The temperature change is determined from $\Delta t = t - t_0$. After determining the thermal expansions $d = d_0(1 + \alpha_A \Delta t)$ and $D = D_0(1 + \alpha_B \Delta t)$ of the diameters at temperature t, a check is performed whether $D - d < 0.1$ or $D - d > 0.8$. If any of these inequalities is fulfilled, the assembly is faulty and the fault counter is incremented. The probability of a faulty assembly for the given set of input data has been calculated by implementing the outlined algorithm. The empirical probability of a faulty assembly has been determined to be 11%, obtained on the basis of 100 000 simulation trials.

Increasing the number of simulations to 10 million did not alter this result, which shows that a very good convergence of the simulation results is present at the selected 100 000 trials.

A procedure for virtual testing can be incorporated easily in an optimisation routine to determine the optimal mean values of the design parameters which yield a robust design – a design characterised by the least sensitivity to variations of the design parameters.

Example

The maximum contact pressure p_{max} in the case of two contacting spheres with radii R_1 and R_2, subjected to a load F, is given by the expression

$$p_{max} = \frac{3F^{1/3}}{2\pi \times \left[0.375 \times \frac{(1-v_1)/E_1 + (1-v_2)/E_2}{B}\right]^{2/3}}$$

In this expression, B is a geometry parameter dependent on the radii R_1 and R_2 of spheres, which practically does not vary. The material parameters are the Poisson's ratios v_1 and v_2 for the two spheres, uniformly distributed in the intervals $v_{1,min} \le v_1 < v_{1,max}$ and $v_{2,min} \le v_2 < v_{2,max}$, and the Young's moduli E_1 and E_2 for the two spheres, uniformly distributed in the intervals $E_{1,min} \le E_1 < E_{1,max}$ and $E_{2,min} \le E_2 < E_{2,max}$. The magnitude of the load F follows a normal distribution with mean F_m and standard deviation σ_F.

Describe in detail the steps of a procedure for determining the likelihood that the maximum contact pressure p_{max} will exceed a critical value p_c.

Solution

1. Initialise a fault/failure counter: f_counter=0.
2. Sample the material parameters by using uniformly distributed random numbers u_1, u_2, u_3 and u_4, between 0 and 1:

$$v_1 = v_{min} + u_1 \times (v_{1,max} - v_{1,min})$$
$$v_2 = v_{2,min} + u_2 \times (v_{2,max} - v_{2,min})$$
$$E_1 = E_{1,min} + u_3 \times (E_{1,max} - E_{1,min})$$
$$E_2 = E_{2,min} + u_4 \times (E_{2,max} - E_{2,min})$$

3. Sample the loading force F by using a random number following a normal distribution with mean F_m and standard deviation σ_F.
   - Generate a random number u_{sn} following the standard normal distribution (see Chapter 8 for details):

$$u_{sn} = u_{x1} + u_{x2} + u_{x3} + \cdots + u_{x12} - 6$$

   - Sample the loading force following the normal distribution:

$$F = \sigma_F \times u_{sn} + F_m$$

4. Check by substituting the sampled parameter values in the formula, whether the calculated maximum pressure p_{max} is higher than the critical value p_c, and if so, increment the failure counter:

```
if (p_max > p_c) then f_counter=f_counter+1;
```

5. Repeat steps 2, 3 and 4 a large number of times (N) in order to collect a sufficient amount of statistical information.
6. Divide the number of simulation trials where the calculated maximum contact pressure p_{max} has been larger than the critical value p_c, to the total number of trials, in order to estimate the probability p_f of failure:

$$p_f \approx \frac{f\_\text{counter}}{N}$$

Virtual testing can also be used to reveal the uncertainty of properties and reliability of materials. Thus, the uncertainty in the location of the ductile-to-brittle transition region of multi-run welds is strongly dependent on the number of test temperatures, the choice of the test temperatures and the variation of the impact energy at the test temperatures. With increasing the number of test temperatures, the uncertainty associated with the location of the ductile-to-brittle transition region can be reduced significantly. The virtual testing model presented in (Todinov, 2004e) has an important application in determining whether a shift in the location of the ductile-to-brittle transition region in steels indicates material degradation or is due to a statistical variation.

9.4 Optimal Replacement to Minimise the Probability of a System Failure

This application features a specified time interval $(0, a)$ and a number of components n which undergo fast wearout. It is assumed that the components undergoing fast wearout are logically arranged in series. In other words, random failure of each of these components causes a system failure. A single spare component is kept for each of the working components undergoing fast wearout. Sudden system failures are highly undesirable because they cause sudden and uncontrolled shutdown which is dangerous for the system, and the recovery of the system is associated with big costs. In contrast, controlled replacement of any of the working components with a spare component can be done without disrupting the work of the system and any associated problems.

The problem is to find the optimal replacement times for the components which minimise the probability of a random system failure within the specified time interval $(0, a)$.

Because each components can be replaced at any time during the time interval $(0, a)$, any possible combination of replacement times for the components can be represented as a point in a hypercube domain D with side a. For $n=2$ components, for example, the hypercube domain D is a square with side a (Figure 9.3).

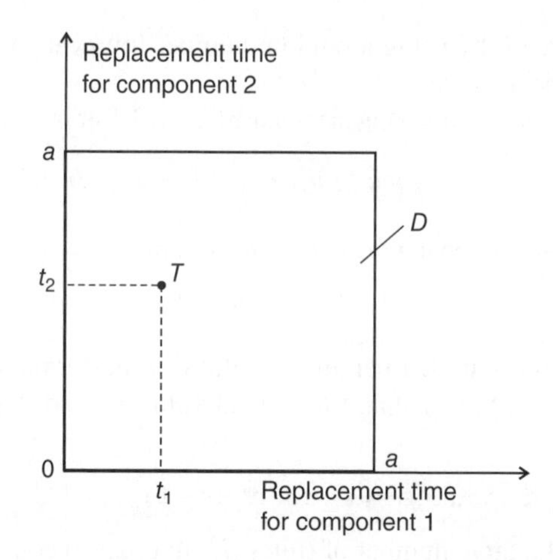

Figure 9.3 All possible combinations of replacement times for two components can be represented by the points of the square domain D

For a specified point T from the domain D, which defines the replacement times for the components, the probability of sudden system failure can be determined by using the following observations:

- The replacement times are always within the time interval $(0, a)$.
- A costly system failure occurs if the time to failure of any of the components is within the time interval $0, a$.
- A costly system failure also occurs if after a replacement of a component within the time interval $(0, a)$, the replacement component fails within the interval $(0, a)$.

Suppose that the replacement times of all components have been specified in the array `repl_time[]`. The algorithm for determining the probability of a costly system failure for the specified replacement times is given next:

Algorithm 9.6

```
/* The replacement times for all components are specified in the
   array repl_time[] */

function Weibull_failure_time(m,eta)
{ /* Returns a random time to failure sampled from a Weibull
    distribution with parameters m (power) and eta (characteristic
    life) */}
function prob_failure()
{
 failure_count=0;
 for i=1 to num_trials do
 {
  for j=1 to n do
    {
    ttf[j] = Weibull_failure_time(m[j], eta[j]);
    if ( repl_time[j] >= ttf[j] ) then { failure_count =
    failure_count+1; break;}
```

```
        else {
                new_ttf[j] = Weibull_failure_time(m[j], eta[j]);
                if (new_ttf[j] + repl_time[j] < a)
                                            then {
                                                    failure_count = failure_
                                                    count+1; break;
                                                  }
               }
          }
      }
      return  failure_count / n;
}
```

The variable 'failure_count' counts the number of simulations for which a system failure occurs. In the inner loop with control variable '*j*', for each component '*j*', a time to failure ttf[*j*] is generated by sampling the Weibull distribution with parameters `m[j]`, `eta[j]` which correspond to the *j*th component. If the generated time to failure ttf[*j*] is smaller than the selected replacement time `repl_time[j]` for the *j*th component, a costly system failure occurs, the failure counter is incremented and the *j*-loop is exited immediately with the statement '`break`'. If the generated failure time `ttf[j]` is larger than the selected replacement time for the *j*th component, a replacement of the failed component is initiated. To determine the time to failure `new_ttf[j]` of the new component which is replacing the *j*th component, the Weibull distribution is sampled again. Next, a check is performed whether the time to failure of the new component plus the time that has elapsed until replacement is smaller than the length of the time interval (0, *a*). If this is so, the new component replacing the *j*th component will fail within the specified time interval (0, *a*). In this case, a system failure will occur; the failure counter is incremented and the *j*-loop is exited immediately with the statement '`break`'.

A standard method for a global optimisation such as *random search, tabu search, simulated annealing, particle swarm optimisation* and *genetic algorithms* (Schneider and Kirkpatrick, 2006) can be used to determine the replacement times `repl_time[]` for the components, which correspond to the smallest value for the probability of system failure. Details about the implementation of a standard algorithm for global optimisation have been omitted.

Example

Given is an operation interval with length $a = 2.5$ years. For a system containing two components with times to failure following the Weibull distributions

$$F_1(t) = 1 - \exp\left[-\left(\frac{t}{eta_1}\right)^{m1}\right]; \; m_1 = 3.6; \; eta_1 = 3$$

and

$$F_2(t) = 1 - \exp\left[-\left(\frac{t}{eta_2}\right)^{m2}\right]; \; m_2 = 1.8; \; eta_2 = 2.7,$$

the procedure for global optimisation yielded an optimal replacement time of 1.25 years for each component. The probability of a system failure within 2.5 years, corresponding to these replacement times, is 0.44.

Indeed, because of the strictly increasing hazard rate function h(t) characterizing each component, it can be shown by a differentiation that a replacement at $x=a/2$ at half of the operational interval minimizes the expected number of failures

$$\mathrm{H}(x)=\int_0^x \mathrm{h}(t)\mathrm{d}t+\int_0^{a-x} \mathrm{h}(t)\mathrm{d}t$$

of the component during the time interval (0, a). Consequently, $x = \mathrm{a}/2$ maximises the probability exp(–H(x)) of surviving the operational interval (0, a) for the component. By induction, it can be proved easily that if n spare parts are available for each component, characterised by a strictly increasing hazard rate, the optimal replacement intervals with length $a/(n+1)$ maximise the probability of surviving the operational time interval (0, a) for each components and for the system.

10

Evaluating the Reliability of Complex Systems and Virtual Accelerated Life Testing Using Monte Carlo Simulation

10.1 Evaluating the Reliability of Complex Systems

Reliability networks can be modelled conveniently by graphs. The nodes are notional and perfectly reliable, while the components are represented by the edges of the graph. Each component is connected to exactly two nodes.

A central question posed for reliability networks is how to calculate the probability of surviving a particular time period of operation a, given the time to failure distribution $F_{ij}(t)$ of each component (edge) (i, j). The answer to this question is given by the methods for system reliability analysis discussed in the reliability literature. There exist a number of methods for system reliability analysis oriented mainly towards small-size systems or systems with simple topology. Such are, for example, the method of network reduction and the decomposition method (Hoyland and Rausand, 1994; Ramakumar, 1993). These methods have serious limitations. For example, the network reduction method is not suitable for topologically complex systems, while the decomposition method is not suitable for large systems.

Methods based on minimal path sets and minimal cut sets are also very common (Billinton and Allan, 1992; Hoyland and Rausand, 1994; Kuo *et al.*, 2001; Ramakumar, 1993).

A path is a set of components which, when working, connect the start node with the end node in the reliability network through working components, thereby guaranteeing that the system is in working state.

A *minimal path* is a path from which no component can be removed without disconnecting the link it creates between the start node and the end node in the reliability network. Consequently, minimal paths are free of loops. In other words, in each minimal path, a particular node may appear only once.

The system reliability can be determined as *the probability of the union of all minimal paths* through the inclusion–exclusion expansion expression (Ebeling, 1997).

A *cut set* is a set of components whose failures cause the system to fail. A *minimal cut set* is a cut set for which no component can be returned in working state without returning the

Reliability and Risk Models: Setting Reliability Requirements, Second Edition. Michael Todinov.

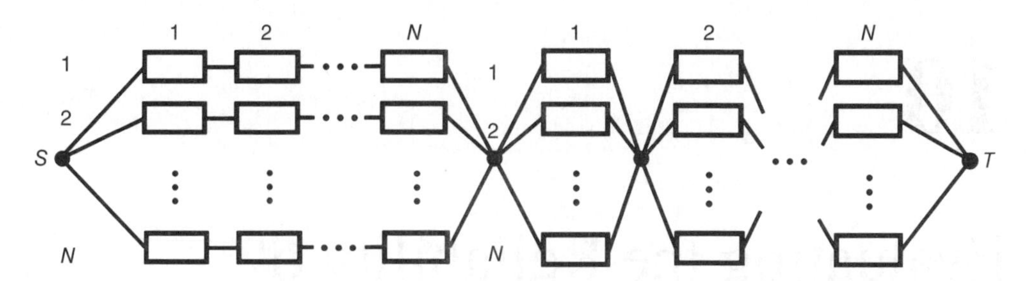

Figure 10.1 An example of a reliability network where the number of minimal paths and minimal cut sets increases exponentially with increasing the size of the network

system into working state. The probability of system failure is determined as *the probability of the union of all minimal cut sets in the system through the inclusion–exclusion expansion expression* (Ebeling, 1997).

Both the minimal paths method and the minimal cut set method *require all paths or cut sets in the system to be known in advance.*

Finding all minimal paths or cut sets, however, is an NP-hard problem (Colbourn, 1987).

Although, for small-size networks, an approach based on minimal paths or minimal cut sets is acceptable, with increasing the size of the network, the number of minimal paths and cut sets increases exponentially, and this approach is no longer feasible. This point can be illustrated immediately with the example in Figure 10.1. The reliability network in the figure has $N^N + N$ minimal cut sets and N^{N+1} minimal paths. Even for the moderate $N = 10$, the storage and manipulation of the minimal paths and cut sets is impossible. As a result, *an algorithm based on determining all minimal paths or cut sets is very inefficient because it will run in exponential time.*

This constitutes the main drawback of methods based on minimal paths and minimal cut sets. Although they work well for small-size systems, for moderately large and large systems, they are not feasible.

System reliability, however, can be defined as a 'probability of existence of a path/connection through working components from the start node to each of the terminal nodes of the reliability network, at the end of the specified time interval of operation'. The analysis of reliability networks can be simplified by noticing that reliability of complex systems can simply be determined as *probability of existence of a connection from the source node to each of the terminal nodes in the reliability network, at the end of the specified time interval* (Todinov, 2006c, 2007).

According to the discussions in Chapter 1, a valid path in a reliability network connecting the start node with any of the terminal nodes can either have forward edges or undirected edges, but it cannot have backward edges.

The probability of existence of a valid path from the start node to each of the terminal nodes is at the heart of the Monte Carlo simulation algorithm for determining the system reliability.

The next algorithm in pseudocode determines the reliability of a complex system with k terminal nodes:

Algorithm 10.1

```
function paths_to_all_terminal_nodes();
function real_random();
function system_reliability();
{
  success_cnt=0;
  for i=1 to num_trials do
  {
```

```
  for j=1 to m do
      {
        tmp = real_random();
        if (tmp > p[j]) then mark edge j as failed;
      }
  paths_exists = paths_to_all_terminal_nodes();
  if (paths_exists=1) then  success_cnt = success_cnt+1;
  Restore the failed edges in the network;
}
  return success_cnt/num_trials;
}
```
■

The array *p*[] contains the reliabilities of the separate edges, which are equal to the probabilities that the separate edges will be in working state at the end of a specified time interval with length *a*. If the time to failure distribution of the *i*th edge is given by $F_i(t)$, then $p[i] = 1 - F_i(a)$.

In a nested loop controlled by the variable '*j*', the state of the separate edges (working/failed) is determined. The state of the *j*th edge is tested by generating a uniformly distributed random number between 0 and 1 from the statement '`tmp = real_random()`' and comparing it with the probability *p*[*j*] that the *j*th edge will be in working state. The number of edges is *m*.

All reliabilities p_j ($j = 1, 2, \ldots, m$) characterising the edges are pre-calculated and stored in the array *p[]*. If the generated random number 'tmp', uniformly distributed between 0 and 1, is greater than $p[j]$, then the *j*th edge is in a failed state and is marked as failed. If the converse is true, the edge remains in working state. A failed edge no longer provides connection between its corresponding nodes and is essentially excluded from the reliability network.

After determining the state of all edges, the function `paths_to_all_terminal_nodes()` establishes whether there exist connections from the start node to each terminal node. The ratio of the number of trials for which a connection from the start node to each of the terminal nodes exists and the total number of simulation trials is an estimate of the reliability of the system. At the end of each simulation trial, all edges marked as 'failed' are restored as working edges. A failed edge which has been restored resumes the connection between the corresponding nodes of the edge and is essentially included in the reliability network.

In the case of very small reliabilities characterising the edges, the precision of the presented Monte Carlo crude sampling can be increased by applying *stratified sampling without replacement*.

Structures for efficient representation of reliability networks and algorithms with linear running time for determining valid paths connecting the start node with a terminal node in networks have been discussed in detail in Todinov (2007, 2013a). The detailed algorithms and data structures related to representing and finding paths in complex reliability networks are beyond the scope of this book, which focuses on reliability and risk models, and will not be included here.

10.2 Virtual Accelerated Life Testing of Complex Systems

10.2.1 Acceleration Stresses and Their Impact on the Time to Failure of Components

The environment has a significant impact on the hazard rates and the times to failure of components. The impact of the environment is manifested through the acceleration stress, which is anything that leads to accumulation of damage and faster wearout. Examples of acceleration stresses are the temperature, humidity, vibration, pressure, voltage, current, concentration of particular ions, etc. This list is only a sample of possible acceleration stresses and can be extended significantly. Because acceleration stresses lead to a faster wearout, they entail a higher failure rate for groups of components. Components affected by an acceleration stress acting as a *common cause* are more likely to fail, which reduces the overall system reliability.

A typical example of this type of *common cause failures* is the high temperature, which increases the susceptibility to deterioration of electronic components or mechanical components (e.g. seals). By simultaneously increasing the hazard rates of the affected components, deterioration due to a high temperature increases the probability of system failure. Humidity, corrosion or vibrations also affect all exposed components.

A common cause failure is usually due to a single cause with multiple failure effects which are not consequences from one another (Billinton and Allan, 1992). Acceleration stresses acting as common causes increase the joint probability of failure for groups of components or for all components in a complex system. Even in blocks with a high level of built-in redundancy, in case of a common cause, all redundant components in the block can fail within a short period of time, and the advantage from the built-in redundancy is lost.

Failure to account for the acceleration stresses acting as common causes usually leads to optimistic reliability predictions – the actual reliability is smaller than the predicted.

For a number of common engineering components, accelerated life models already exist. They have been built by using a well-documented methodology (Kececioglu and Jacks, 1984; Nelson, 2004; Porter, 2004).

Building an accelerated life model for a component starts with the time to failure model for the component. The time to failure model gives the distribution of the time to failure of each component in the system (Nelson, 2004; Porter, 2004). The most common time to failure model is the Weibull distribution

$$F(t) = 1 - \exp\left[-\left(\frac{t}{\eta}\right)^{\beta}\right] \tag{10.1}$$

where $F(t)$ is the cumulative distribution of the time to failure and β (*shape parameter*) and η (*characteristic life/scale parameter*) are constants determined from experimental data. This model is commonly used in the case where the hazard rate depends on the age of the component.

Another common time to failure model is the negative exponential distribution

$$F(t) = 1 - \exp\left(-\frac{t}{\text{MTTF}}\right) \tag{10.2}$$

where $F(t)$ is the cumulative distribution of the time to failure and MTTF is the mean time to failure. The negative exponential distribution can be obtained as a special case from the Weibull distribution for $\beta = 1$ and is used in cases where the hazard rate characterising the component does not practically depend on its age.

The scale parameter η in the Weibull distribution and the MTTF in the negative exponential distribution depend on the acceleration stresses through the *stress–life relationships* (Kececioglu and Jacks, 1984; Nelson, 2004; Porter, 2004; ReliaSoft, 2007). When the stress–life dependence is substituted in Equations 10.1 and 10.2, the acceleration time to failure model for the component is obtained. The acceleration time to failure model gives *the time to failure model at particular levels of the acceleration stresses*.

10.2.2 Arrhenius Stress–Life Relationship and Arrhenius-Type Acceleration Life Models

For this type of accelerated life model, the relationship between the life and the level V of the acceleration stress is

$$L(V) = C \times \exp\left(\frac{B}{V}\right) \tag{10.3}$$

where $L(V)$ is a quantifiable life measure and C and B are constants obtained from experimental measurements. The *Arrhenius stress–life relationship* is appropriate in cases where the acceleration stress is thermal, for example, 'temperature'. The temperature values must be in absolute units [K].

In the case of a Weibull time to failure model, $L(V) \equiv \eta = C \times \exp(B/V)$, where η is the characteristic life (scale parameter) calculated in *years*. Substituting this in the Weibull time to failure model (10.1) yields the *Arrhenius–Weibull time to failure accelerated life model*:

$$F(t,V) = 1 - \exp\left[-\left(\frac{t}{C\exp(B/V)}\right)^{\beta}\right] \tag{10.4}$$

10.2.3 Inverse Power Law Relationship and Inverse Power Law-Type Acceleration Life Models

The relationship between the life of the component and the level V of the acceleration stress is

$$L(V) = \frac{1}{KV^n} \tag{10.5}$$

where $L(V)$ is a quantifiable life measure; K and n are constants obtained from experimental measurements. The inverse power law *(IPL) stress–life relationship* is appropriate for non-thermal acceleration stresses like 'load', 'pressure' and 'contact stress'. It can also be applied in cases where V is a stress range or even in cases where V is a temperature range (in case of fatigue caused by thermal cycling).

In the case of a Weibull time to failure model, the life measure is assumed to be the characteristic life $L(V) \equiv \eta = 1/(KV^n)$, where η is the characteristic life (scale parameter). Substituting this in the Weibull time to failure model (10.1) yields the *IPL-Weibull accelerated life model*:

$$F(t,V) = 1 - \exp\left[-\left(tKV^n\right)^{\beta}\right] \tag{10.6}$$

10.2.4 Eyring Stress–Life Relationship and Eyring-Type Acceleration Life Models

The relationship between the life and the acceleration stress level V is

$$L(V) = \frac{1}{V}\exp\left[-\left(A - \frac{B}{V}\right)\right] \tag{10.7}$$

where $L(V)$ is a quantifiable life measure; A and B are constants obtained from experimental measurements. Similar to the Arrhenius stress–life relationship, the *Eyring stress–life relationship* is appropriate in the case of thermal acceleration stresses. However, it can also be used for non-thermal acceleration stresses such as humidity. In the case of a Weibull time to failure model, $L(V) \equiv \eta = 1/V \exp[-(A - B/V)]$, where η is the characteristic life (scale parameter, calculated in *years*). Substituting this in the Weibull time to failure model (10.1) yields the *Eyring–Weibull accelerated life model*:

$$F(t,V) = 1 - \exp\left\{-\left[tV\exp\left(A - \frac{B}{V}\right)\right]^{\beta}\right\} \tag{10.8}$$

There exist also stress–life models involving simultaneously two acceleration stresses, for example, temperature and humidity (ReliaSoft, 2007). Such are the *temperature–humidity (TH) relationship* and *TH-type acceleration life models* and *temperature–non-thermal relationship (T-NT)* and *T-NT-type acceleration life models*.

The effect of the acceleration stresses on a complex system can be revealed if an accelerated life model for the system is built from the accelerated life models of its components (Todinov, 2011). Apart from revealing the impact of the acceleration stresses on system's performance, building an accelerated life model for a complex system has another significant advantage. During life testing of complex systems, estimating the system reliability under normal operating conditions requires special test rigs and a large amount of time and resources and can be a very complex and expensive task. This task, however, does not have to be addressed if a method is developed for building an accelerated life model of a complex system from the accelerated life models of its components. Deriving the time to failure distribution of the complex system under normal operating conditions from the accelerated life models of its components will be referred to as 'virtual accelerated life testing of a complex system'. The significant advantages of the virtual accelerated life testing can be summarised as follows:

- *The virtual accelerated life testing does not require building test rigs for the various engineering systems, which is an expensive and difficult task. In cases of very large systems, building such test rigs is impossible.*
- *The virtual accelerated life testing reduces drastically the amount of time and resources needed for accelerated life testing of complex engineering systems.*
- *The virtual accelerated life testing permits testing a large variety of systems built with components whose accelerated life models are known.*

The virtual accelerated life testing offers big flexibility in specifying various levels for the acceleration stresses. An efficient algorithm and software tool for building an accelerated life model of a complex system from the accelerated life models of its components has been proposed in Todinov (2011). The software tool has the capability of extrapolating the system's life under normal operating conditions.

The reliability network in Figure 10.2 corresponds to the dual power supply system from Figure 1.15b from Chapter 1. Each of the electromechanical devices (marked by circles with numbers 12–15) receives power from two channels, only one of which is sufficient to maintain the device working. A system failure occurs if an electromechanical device fails or if both power channels have been lost because of failures of components along the paths of the power channels.

The reliability network has been modelled by a set of *nodes* (the filled small circles in Figure 10.2 numbered from n_1 to n_{12}) and components (1, 2, 3, …, 15) connecting them. The system works only if

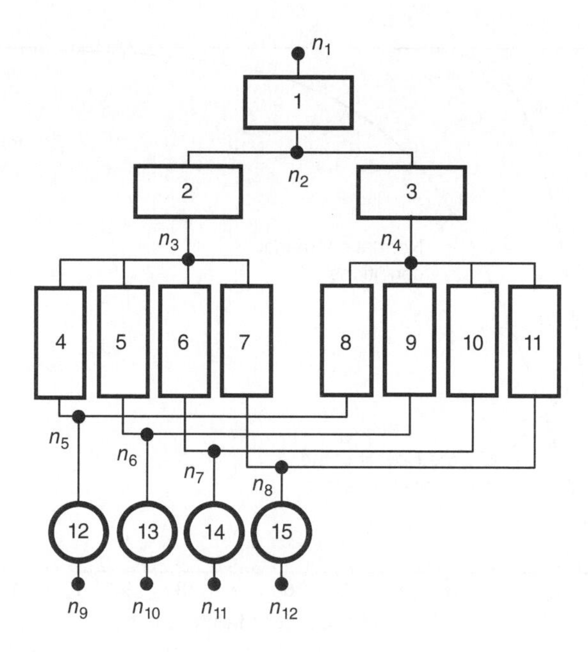

Figure 10.2 Reliability network of the dual control system from Figure 1.17

there exist paths through working components between the *start node* 'n_1' and each of the *end (terminal) nodes* marked by n_9, n_{10}, n_{11} and n_{12}.

As a result, the reliability network in Figure 10.2 can be modelled conveniently by undirected graph. The nodes are the *vertices* and the components that connect the nodes are the *edges* of the graph. Details regarding coding the network topology and algorithms for determining valid paths from the start node to each of the terminal nodes can be found in Todinov (2011, 2013a).

Non-existence of a path to even a single terminal node indicates a system failure. The reliability of the system, associated with a specified time interval 'a' is calculated by counting in the simulation loop the number of trials for which the system failure has not occurred during the specified time interval (0, a) and dividing this number to the total number of simulation trials. At the end of the simulations, the times to failure of the system are sorted in ascending order.

Example

The test example which illustrates the algorithm is based on the dual power supply system in Figure 10.2. Component 1 (Figure 10.2) is characterised by Arrhenius stress–life model (10.3) where the acceleration stress is temperature, set at a level $V = 333\,\text{K}$ (see Eq. 10.3), and the constants in the equation are $B = 461.64$ and $C = 2$. Components 2 and 3 are also characterised by Arrhenius stress–life models with constants $B = 118.77$ and $C = 1.4$, where the acceleration stress is temperature set also at a level $V = 333\,\text{K}$.

Components 4–11 in Figure 10.2 are characterised by Eyring stress–life relationship (10.7) with constants $A = 1.8$ and $B = 3684.8$, where the acceleration stress is temperature, set at a level $V = 413\,\text{K}$. Finally, components 12–15 in Figure 10.2 are characterised by inverse power law stress–life relationship (10.5) with constants $K = 9e-5$ and $n = 1.7$, where the acceleration stress is 'pressure', set at a level $V = 60\,\text{MPa}$.

All components are characterised by negative exponential time to failure distribution. The duration of the time interval for which reliability has been calculated was $a = 1.5$ years. The execution of the programme yielded system reliability equal to 0.162.

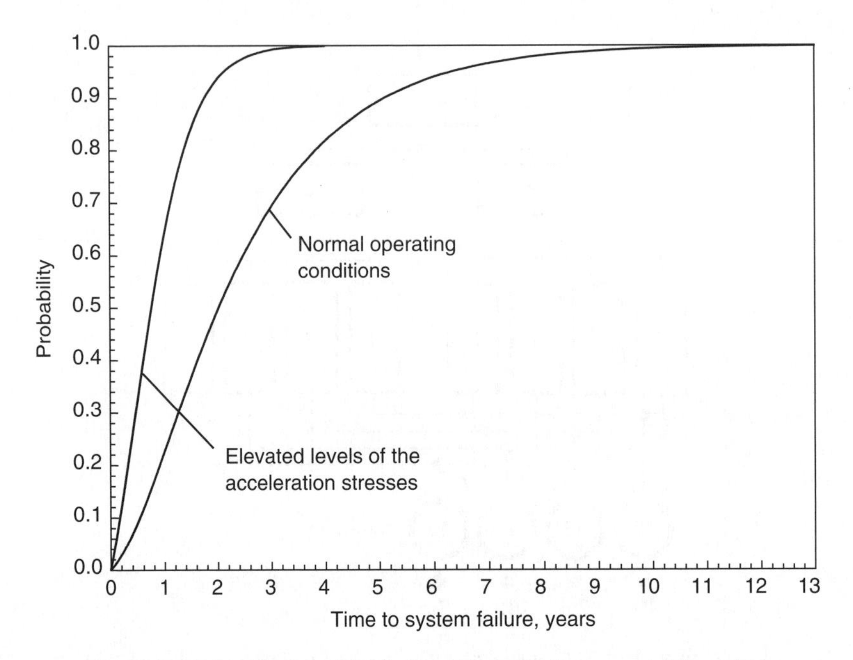

Figure 10.3 Distribution of the times to failure for the system in Figure 10.2

The computational speed is very high; 100 000 simulations have been performed within 1.03 seconds on a laptop with processor *Intel(R) T7200 @ 2.00 GHz*.

The distribution of the times to failure for the system in Figure 10.2 is shown in Figure 10.3 (the curve corresponding to elevated levels of the acceleration stresses). Finally, an extrapolation of the time to failure of the system under normal conditions has been made. This constitutes the main advantage of the developed software tool: *estimating the reliability of a complex system working in normal conditions without allocating time and resources for real testing*. The normal conditions correspond to a temperature 293 K (room temperature) and pressure of 1 MPa. The distribution of the times to failure for the system in Figure 10.2 under normal operating conditions is shown in Figure 10.3.

The execution of the programme yielded system reliability equal to 0.634.

Example

Here is a simple example based on the reliability network from Figure 10.2 revealing the effect of the temperature acting as a common cause acceleration stress. All components with indices 1–11 have been assumed to be identical, characterised by the Arrhenius–Weibull distribution of the time to failure (10.4), where $\beta = 2.3$, $C = 2$, $B = 461$. The time to failure of components 12–15 has been assumed to be the negative exponential distribution, with MTTF = 60 years.

System failure occurs when both control channels to any of the components 12–15 are lost or when any of components 12–15 fail. For an acceleration stress $V = 293$ K, the simulation yielded $R_s = 0.624$ probability that the system from Figure 10.3 will survive 4 years of continuous operation without failure. If, however, the acceleration stress (the temperature) is raised to $V = 523$ K, the simulation yields only $R_s = 0.1$ probability that the system will survive 4 years of continuous operation without failure.

Another advantage of the developed approach consists of the circumstance that it reveals easily the impact of acceleration stresses acting as common causes in the case of topologically complex systems where no simple analytical solution for the system reliability exists.

11

Generic Principles for Reducing Technical Risk

A systematic classification of generic principles for reducing technical risk is crucial to safe operation, developing robust and reliable engineering designs, reliable software and high levels of safety, yet this topic has largely been overlooked in the reliability and risk literature. There is even a view among reliability practitioners that the principal ways of improving reliability of a system are either by improving the reliability of the components/systems or by providing redundancy. Equally unbalanced is the belief that only developing physics of failure models can deliver a real reliability improvement. This view has been fuelled by the failure of some statistical models to predict correctly the life of components and by the lack of failure data to populate statistical models. Another widespread erroneous view is that the quality and utility of reliability models depend strongly on availability of failure data. In Todinov (2009a), it was demonstrated that this view is incorrect. Comparative statistical models can deliver real reliability improvement even in the absence of any failure data. A limited and equally damaging is the pure statistical approach, which disregards the physical principles controlling the operation and failure of engineering systems and components. As can be seen from the compiled principles in this chapter, many of the formulated principles are not routed in the reliability theory and statistics. The statistical modelling is just one of the possible approaches to reliability improvement and risk reduction. These extreme views are an example of unnecessary self-imposed constraints. Increasing reliability can be achieved by using principles which range from pure statistical modelling to pure physics of failure modelling underpinning reliable operation and failure.

Some of the work on improving the reliability of engineering systems has already been done. By distilling already existing approaches to a wide variety of reliability problems and formulating new approaches, the intention is to provide the much needed support for design engineers in their constant efforts for technical risk reduction.

The struggle between the need of increasing efficiency and reducing the weight of components and systems and reliability is a constant source of technical and physical contradictions. Hence, it is no surprise that some of the principles like *altering the shape of components* or *separation* sound similar to some of the inventive principles for resolving technical contradictions formulated by Altshuller in the development of TRIZ methodology for inventive problem solving (Altshuller, 1984, 1996, 1999). Eliminating harmful factors and influences is the purpose of many inventions, and Altshuller's TRIZ system identified a number of useful design principles closely related to eliminating harm.

Reliability and Risk Models: Setting Reliability Requirements, Second Edition. Michael Todinov.

Many of the formulated principles in this chapter however are routed in probabilistic arguments, reliability and risk theory and cannot be deduced from the general inventive principles formulated in TRIZ, which serve as a general guide in developing inventive solutions, as an alternative to the trial-and-error approach. The first principle, for example, stating that the reliability built-in system should be proportional to its cost of failure is routed in the risk theory (Todinov, 2006b). Similarly, the principle '*reliability and risk modelling and optimisation*' requires knowledge of reliability and risk theory in order to apply it and comprehend why it works. Contrary to what some authors stated, optimisation is not necessarily about finding a compromise between several parameters to maximise a particular system output. One of the sections features optimisation reducing the transportation costs by removing parasitic flow loops from networks. In this case, optimisation is based on a new phenomenon and is not done by finding a compromise between several controlling parameters characterising a selected design.

Some principles rely on very specific concepts like 'robust and fault-tolerant design' with reduced sensitivity to the variation of reliability-critical design parameters. Some principles are rooted in the logic of operation of devices and the logic of execution of operations ('failure prevention interlocks'); other principles rely on specific systematic methods for discovery and elimination of failure modes.

The systematic formulation and classification of the generic principles for technical risk reduction were started in Todinov (2007). The principles for reducing the risk of failure have been broadly divided into (i) preventive, reducing mainly the likelihood of failure; (ii) dual, oriented towards reducing both the likelihood of failures and the consequences from failure; and (iii) protective, predominantly reducing the consequences from failure. While protective principles reduce the consequences from failure, preventive principles exclude failures altogether or reduce the possibility of their occurrence. The classification presented in this chapter follows the broad classification from Todinov (2007), but the principles have been updated and a significant number of new principles have been formulated (Todinov 2015).

The set of principles for reducing technical risk aims to suggest efficient methods for reducing technical risk. Prevention is certainly better than cure; hence, preventive principles received a significant emphasis. The first principle, based on the cost of system failure, should be used frequently in cases where the financial impact of failure is big: in cases where failure leads to human fatalities, damage to the infrastructure and the environment or huge financial losses. However, if failure is caused by random factors completely beyond the control of the designer or if the source of failure has not been understood, protective principles should be preferred, which limit the extent of damage. Protective measures are to be preferred also if the reliability of the products is inherently low.

The diverse list of principles prompts research scientists and engineers–designers not to limit themselves within few common familiar ways of improving reliability and reducing risk which often lead to solutions which are far from optimal. Using appropriate combinations of diverse principles often brings a considerably larger effect. The listed principles are very generic, and most of them can be developed further, depending on the specific application area. Thus, the principle of 'thermal design' standing for neutralising the negative effects of temperature in mechanical and electrical devices can be further developed, for example, in the specific area of reducing the thermal stresses. As a result, a number of more specific principles and techniques can be developed, relevant to reducing the thermal stresses only.

The outlined key principles for reducing the risk of failure can be applied with success not only in engineering but in diverse areas of human activity.

For example, the risk reduction principles:

Reducing the likelihood of unfavourable combinations of risk-critical random variables,
Discovering and eliminating a common cause,
Reducing the time of exposure or the space of exposure,
Segmentation,
Designing deliberate weak links,
Self-reinforcement,
Failure prevention interlocks,

are universal principles for reducing technical risk which can be borrowed and applied in diverse areas of the human activity, for example, in environmental sciences, financial engineering, economics, medicine, etc.

11.1 Preventive Principles: Reducing Mainly the Likelihood of Failure

11.1.1 Building in High Reliability in Processes, Components and Systems with Large Failure Consequences

The underlying principle of the risk-based design is that *a process, component or a system whose failure is associated with large losses should have a proportionally high built-in reliability* (Todinov, 2006c). Processes, components and systems should always be designed considering the consequences of their failure. Failure of the cement used for sealing an oil production subsea well for example, causes catastrophic pollution of the environment. Consequently, the cement seal should have a proportionately high built-in reliability. Setting reliability requirements to processes, components and systems with large failure consequences is an important mechanism of reducing risk. Components, processes and operations used in safety-critical applications should be with higher reliability compared to analogous components used in non-critical systems.

According to this principle, no cost savings should be made on components, processes and systems without first analysing the potential impact of these savings on the reliability and safety of the overall system. Thus, a cost saving on the material of a particular component could cause huge losses related to cleaning up polluted environment, human fatalities, damage to health, damaged infrastructure and financial losses.

In manufacturing, cost saving on quality materials has been responsible for big losses related to scraped defective production or expensive system failures. Thus, ordering cheap spring rods characterised by a substantial number of oxide and sulphide inclusions will save the spring company substantial amount of funds on materials. At the same time, this strategy will compromise the fatigue strength of the produced suspension springs and will lead to a loss of market share, loss of customers and ultimately to a loss of business for the company.

This principle can also be illustrated with production systems with hierarchy (Figure 11.1).

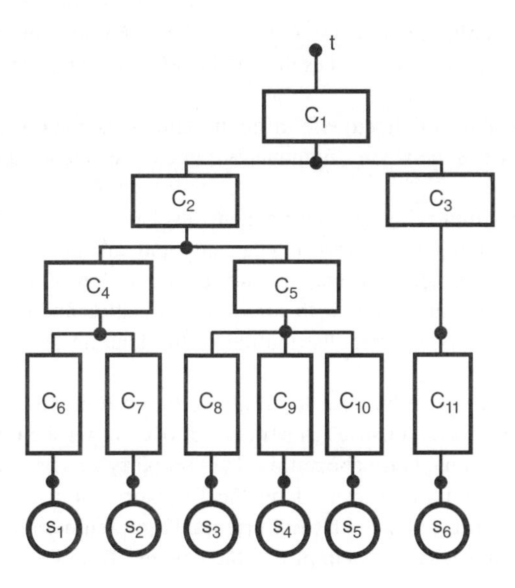

Figure 11.1 A production system with hierarchy based on six production sources

In the production system from Figure 11.1, there are six sources of production (generators, oil and gas wells, pumps, etc.) supplying commodity (electricity, oil, gas and water) to a destination *t*. Despite that components C_2 and C_3 may be identical, failure of component C_3 causes loss of production from source s_6 only, while the failure of component C_2 causes loss of production from five sources (s_1–s_5). The reliability level of component C_2 should be larger than the reliability level of component C_3.

The higher the component in the hierarchy, the more production units will be affected by its failure, the higher the reliability level of the component should be.

11.1.2 Simplifying at a System and Component Level

Simplifying systems and components can be done in various ways: reducing the number of components, simplifying their shape, simplifying their function, reducing the number of functions carried out, etc.

Reducing the number of components is an important way of increasing the reliability of a system. The larger the number of components in a system, the more possibilities for failures exist, the lower is the reliability of the system. Indeed, the reliability of a system composed of *n* components arranged in series is $R = R_1 \times R_2 \times \cdots \times R_n$. If a number of components are removed, the reliability of the initial system can be presented as $R = R' \times R_r$ where R' is the reliability of the simplified system and R_r is the product of reliabilities of the removed components. Since the product of the reliabilities of removed components is a number smaller than unity, for the reliability R' of the simplified system, $R' = R / R_r > R$ holds. In words, the simpler system has a larger reliability.

A powerful way of improving the system reliability is by simplifying the system through eliminating components whose failure modes cause system failures. Often, the functions of the eliminated components can be transferred (integrated) into other components without compromising their reliability and the reliability of the system. Furthermore, the available resources in the environment or in the system can often be used to substitute the functions of the removed component.

Removing components provides the added benefits of (i) reducing the weight of the system, (ii) reducing the cost of the system and (iii) removing failure modes associated with the removed components.

A typical example of simplifying at a system level is the simplification of logic circuits in digital electronic systems. The simplification reduces the number of components in the system and increases its reliability.

Complex designs are often associated with difficult maintenance and small reliability due to the large number of interactions between components which are a constant source of faults and failures.

A typical example of simplifying at a component level is the simplification of component's geometry which results in a larger load-carrying capacity and enhanced reliability. Simplifying the shape of components and interfaces aids manufacturing, creates fewer possibilities for manufacturing faults, reduces the number of regions with stress intensification and improves the load-carrying capacity of the components by a better distribution of the loading stresses in the volume of the components.

Often, physical phenomena can be used to eliminate the introduction of complex control systems. If a ferrite is designed with a Curie temperature of 0°C, it will be magnetic below the water freezing temperature and diamagnetic above this. This property can be used for heating electrical distribution lines to prevent failures caused by the formation of ice at sub-zero temperatures (Altshuller, 1984). As a result, reliance on the Curie transition temperature eliminates the need of introducing a complex control system, simplifies the design of the heating system and improves reliability.

Another way of simplifying components is to simplify their functions which also improves their reliability. Reducing the number of functions reduces the number of possible failure modes. The failure modes characterising a particular component are logically arranged in series (activating any failure mode causes the component to fail), and the effect from reducing the number of functions is similar to the effect from reducing the number of components in a system.

11.1.2.1 Reducing the Number of Moving Parts

An essential part of simplifying at a component and system level is the reduction of the number of moving parts. Moving parts exhibit more failures compared to stationary parts. This is usually due to the increased kinetic energy, wear, fatigue, vibration, heat generation and erosion associated with moving parts. The increased kinetics energy of moving parts (e.g. impellers, fans, turbines, etc.) makes them prone to overstress failures if their motion is suddenly restricted due to lodged foreign objects. Moving parts are also associated with large inertia forces which cause pulsating loading and increased fatigue. If out-of-balance forces are present in the rotating parts and excitation frequencies are reached, the resonance amplitudes are a frequent cause of failure (Collins, 2003).

Vibration is always associated with moving parts and promotes fast wearout and fretting fatigue. Moving parts are sensitive to tolerance faults because they require more precise alignment. The friction and heat generated by moving parts, require lubrication and cooling which make moving parts also sensitive to faults associated with the lubrication or cooling system. As a result, reducing the number of moving parts is an efficient way of improving the reliability of a system.

11.1.3 Root Cause Failure Analysis

The root cause analysis is a solid basis for reducing the hazard rate and for a substantial reliability improvement. Knowledge regarding the circumstances and processes which contribute to the failure events is the starting point for reducing the hazard rate and for a real reliability improvement. The main purpose of the root cause analysis is to identify the factors promoting the failure modes and determine whether related factors are present in other parts of the system.

Identifying the root causes initiates a process of preventing the failure mode from occurring by appropriate modifications of the design, the manufacturing process or the operating procedures.

A typical example of reducing the hazard rate by a root cause analysis can be given with hot-coiled Si–Mn suspension springs suffering from premature fatigue failure. Typically, automotive suspension springs are manufactured by hot winding. The cut-to-length cold-drawn spring rods are austenitised, wound into springs, quenched and tempered. This is followed by warm pre-setting, shot peening, cold pre-setting and painting (Heitmann *et al.*, 1996).

The initial step of the analysis is conducting rig tests inducing fatigue failures to a large batch of suspension springs under various conditions. Fracture surfaces are then preserved, and scanning electron microscopy is used to investigate the fatigue crack initiation sites. If large size inclusions are discovered at the fatigue crack origin, a possible hazard rate reduction measure would involve changing to a supplier of cleaner spring steel.

Optical metallography of the failed springs must also be made in order to make sure that there is no excessive decarburisation. If the depth of the decarburised layer is significant, its fatigue resistance is low and care must be taken to control the carbon potential of the furnace atmosphere in order to avoid excessive decarburisation. Alternatively, the chemical composition of the steel can be altered by micro-alloying in order to make it less susceptible to decarburisation. The grain size at the surface of the spring wire must also be examined because microstructures with excessively large grain size are characterised by a reduced toughness and fatigue resistance. Correspondingly, the austenitisation temperature and the duration of the austenitisation process must guarantee that the grain size remains relatively small.

The spring surface after quenching must also be examined in order to make sure that there are no micro-cracks. Tempering must guarantee an optimal hardness and yield strength which maximise the fatigue life. Finally, after shot peening, the residual stresses at the surface of the spring wire should be measured (e.g. by an X-ray diffractometer (Cullity, 1978)) to make sure that they are of sufficient magnitude and uniformly distributed over the circumference of the spring wire. If, for example, the residual stresses are highly non-uniform or of small magnitude, they would offer little resistance against fatigue crack initiation and propagation. Changes in the shot peening process must be implemented in this case to guarantee a sufficient magnitude and uniformity of the residual stresses.

11.1.4 Identifying and Removing Potential Failure Modes

The risk of failure is reduced by removing potential failure modes, and the design for reliability is largely about preventing failure modes from occurring during the specified lifetime of the product.

Techniques used for identifying potential failure modes have been covered in detail in Chapter 1. These ensure that as many as possible potential failure modes have been identified and their effect on the system performance assessed. The objective is to identify critical areas where design modifications can reduce the probability of failure or the consequences of failure. In this way, potential failure modes and weak spots which need attention are highlighted, and the limited resources for reliability improvement are focused there.

11.1.5 Mitigating the Harmful Effect of the Environment

The environment is a major source of acceleration stresses which lead to accumulation of damage, faster wearout and a significant increase of the hazard rates of components. A typical acceleration stress is the high temperature which increases the susceptibility to deterioration and increases significantly the hazard rates of the affected components. Humidity, corrosion or vibrations also increase the hazard rates of the affected components. Failure to account for the negative impact of the operating environment usually leads to optimistic reliability predictions – the actual reliability is smaller than the predicted.

Reducing the harmful effect of the environment can be done (i) by improving the resistance against the harmful effect, (ii) by modifying the environment in order to reduce its harmful effect and (iii) by replacing it with inert environment.

Modifying the design is often used to reduce the magnitude of a particular acceleration stress generated by the environment. The circuit boards in Figure 11.2a and b contain identical components, yet the reliability of circuit board (b) is higher because of the two additional screws which reduce the amplitude of the vibrations for design (b) and the hazard rates of the affected components (Jais *et al.*, 2013).

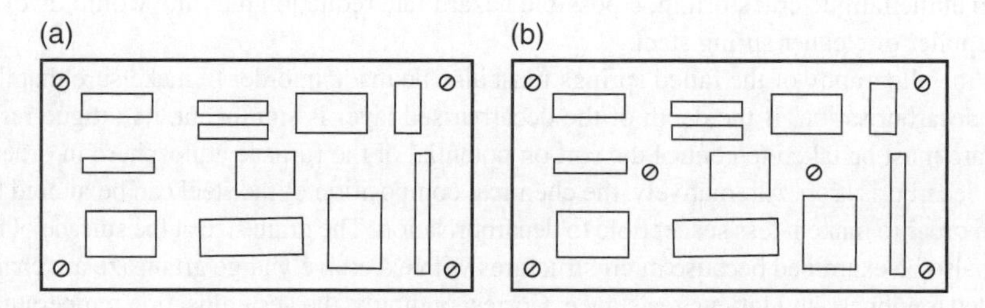

Figure 11.2 (a) Lack of mitigation and (b) mitigation against the acceleration stress 'vibrations'

Arc welding shielded by an inert gas atmosphere such as argon or carbon dioxide is an example of reducing the negative impact of oxygen and improving the reliability of welds. This principle is used, for example, in metal inert gas (MIG) and tungsten inert gas (TIG) welding techniques. Another example is hermetic or plastic encapsulated integrated electronic circuits to protect them from the harmful action of humidity.

Cavitation is generation of cavity bubbles in liquids by rapid pressure changes. When the cavity bubbles implode close to a metal surface, they cause pitting erosion. Typical spots of cavitation damage are suction pipes of pumps and impellers, narrow flow spaces and sudden changes in the flow direction (bends, pipe tees) which cause turbulence.

This type of acceleration stress from the environment can be avoided by designing the flow paths in such a way that sharp pressure drops are avoided (especially below the atmospheric pressure). This can be achieved by designs guaranteeing a multistage pressure drop. Avoidance of turbulence by streamlining the flaw is an important measure decreasing cavitation.

Alternatively, the flow paths can be designed in such a way that the cavitation bubbles implode in the fluid but not next to the metal surface. As a result, cavitation is still present but the metal surfaces are not affected. The susceptibility to cavitation damage can be reduced by using cavitation-resistant materials, welded overlay of metals, sprayed metal coatings or elastomeric coatings.

Corrosion is a name for the degradation of mechanical, microstructural and physical properties of materials due to the harmful effect of the environment. Material degradation due to corrosion is often the root cause of failures entailing loss of life, damage to the environment and huge financial losses. Methods increasing the corrosion resistance include *cathodic protection, corrosion allowance, protective coatings, plastic or cement liners* and use of *corrosion-resistant special alloys.*

The corrosion intensity is a function of the environment. Aggressive environments combined with low corrosion resistance cause expensive early-life corrosion failures.

Corrosion inhibitors are compounds which modify the corrosive environment thereby reducing the rate of corrosion. Corrosion inhibitors are often injected into a pipeline where they mix with the product to reduce corrosion. Corrosion retarding inhibitors are added to the corrosive medium to reduce the intensity of the anode/cathode processes. They can also reduce the corrosion rate by forming barrier films separating the protected surface and the corrosive environment. There are also inhibitors which passivate the protected surface by reacting with it and forming compounds which serve as anti-corrosion barrier. Inorganic inhibitors usually passivate the protected surface, while organic inhibitors usually form a protective film on the surface.

Poor design of the flow paths of fluids containing abrasive material promotes rapid *erosion* which can be minimised by a proper material selection and design. Structural design features promoting rapid erosion (Mattson, 1989) should be avoided. Such are, for example, bends with small radii in pipelines or obstacles promoting turbulent flow. Increasing the pipeline curvature and removing the obstacles result in less turbulent flaw and reduced erosion. Erosion is significantly reduced by appropriate heat treatment increasing the surface hardness.

If the possibility for increasing the curvature of the flow paths is limited, internal coatings resistant to erosion may be considered at the vulnerable spots.

11.1.6 Building in Redundancy

Incorporating redundancy in the design is particularly effective in cases where random failures are present. *Redundancy* is a technique whereby one or more components of a system are replicated in order to increase reliability (Blischke and Murthy, 2000). Since a design fault would usually be common to all redundant components, design-related failures may not be reduced by including redundancy. For active redundancy, all redundant components are in operation and share the load with the main unit from the time the system is put in operation. Full active redundancy is present in cases where the assembly is operational if at least one of the units is operational.

Partial active redundancy (*k-out-of-n redundancy*) is present if the system works *if* and only if at least k out of the n components work.

Active redundancy at a component level (Figure 11.3a), where each component is replicated, yields higher reliability compared to active redundancy where the entire system is replicated (Figure 11.3b). A detailed analysis demonstrating this feature can be found in (Todinov, 2007).

While for an *active redundancy* no switching is required to make the alternative component available, using *passive* (*standby*) *redundancy* requires a switching operation to make redundant component available. In cases of passive redundancy, the redundant components do not share any load with the operating component.

The redundant components are put in use one at a time, after failure of the currently operating component, and the remaining components are kept in reserve (Figure 11.4).

Standby components do not operate until they are sequentially switched in. In contrast to an active redundant system based on n components operating in parallel, the components in the standby system operate one after another. This is why a cold standby redundancy with perfect switching provides a higher reliability compared to an active redundancy (Figure 11.5).

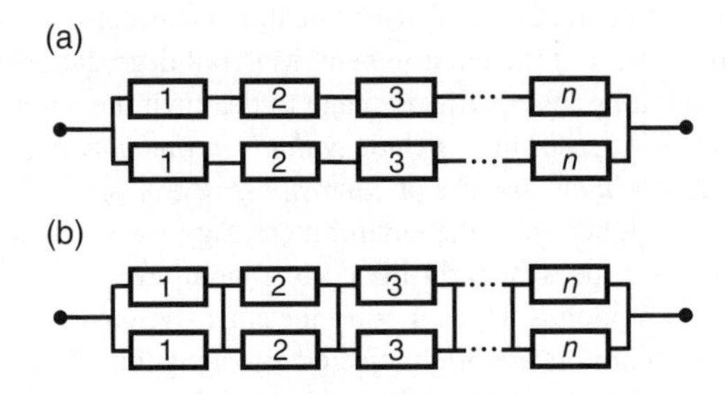

Figure 11.3 Redundant systems (a) with redundancy at a system level and (b) with redundancy at a component level

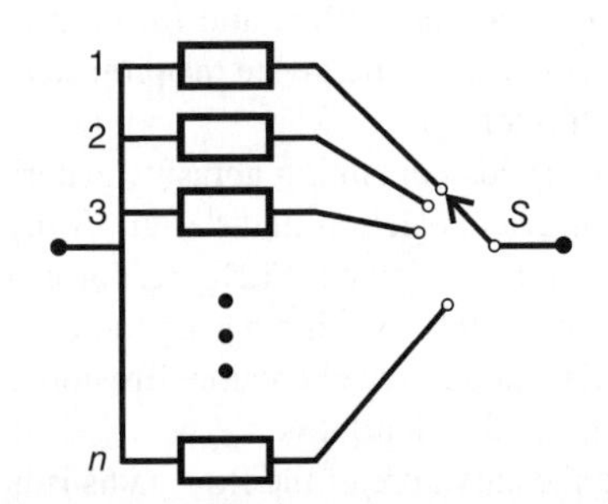

Figure 11.4 A passive (standby) redundancy

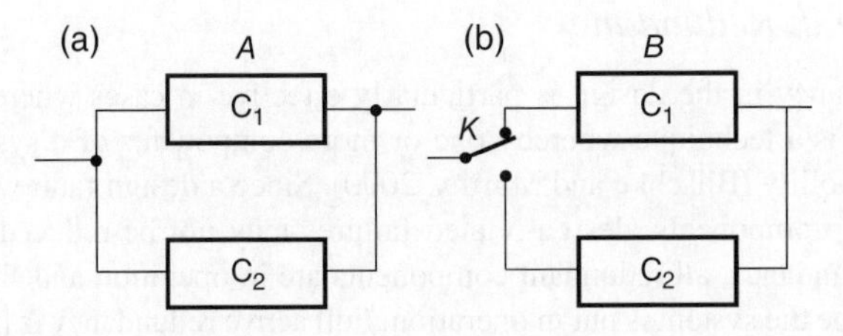

Figure 11.5 Active redundant system (a) and a twofold standby system (b)

If the switch is perfect and the components are identical, the time to failure of the standby system (Figure 11.5b) is a sum of the times to failure of the components ($t_B = t_1 + t_2$). The time to failure of the active redundant system (Figure 11.5a) is equal to the maximal of the times to failure of the components ($t_A = \max\{t_1, t_2\}$). The cold standby system is characterised by a higher reliability ($t_B > t_A$).

Theoretically, by providing a sufficiently large number of standby components, the reliability of a standby system with perfect switching can be made arbitrarily close to 1. Indeed, for the special case of n cold standby components with a constant hazard rate λ, the reliability associated with time t of a standby system with perfect switching is (see Chapter 3)

$$R(t) = \exp(-\lambda t)\left[1 + \frac{(\lambda t)^1}{1!} + \frac{(\lambda t)^2}{2!} + \cdots + \frac{(\lambda t)^{n-1}}{(n-1)!}\right]$$

As can be verified, with increasing the number of components n,

$$\lim_{n\to\infty}\left[1 + \frac{(\lambda t)^1}{1!} + \frac{(\lambda t)^2}{2!} + \cdots\right] = \exp(\lambda t)$$

and, as a result, $\lim_{n\to\infty}[R(t)] = 1$. The number of standby components is limited by constraints such as size, weight and cost. Standby units may not necessarily be identical. An electrical device, for example, can have a hydraulic device for backup.

11.1.7 Reliability and Risk Modelling and Optimisation

11.1.7.1 Building and Analysing Comparative Reliability Models

Building and analysing comparative reliability models is essential in selecting a more reliable solution. If several alternative topologies are available, the selection of the topology which delivers the lowest risk of failure can be done by building and running the system reliability models of the competing solutions and selecting the most reliable solution. This is a powerful strategy for technical risk reduction which can be executed even in the absence of any data (Todinov, 2009a). Here is an example featuring two systems '1' and '2' built of type-*A*, type-*B* and type-*C* components. The reliability networks of the systems are given in Figure 11.6. If a type-*A* component is more reliable than a type-*B*

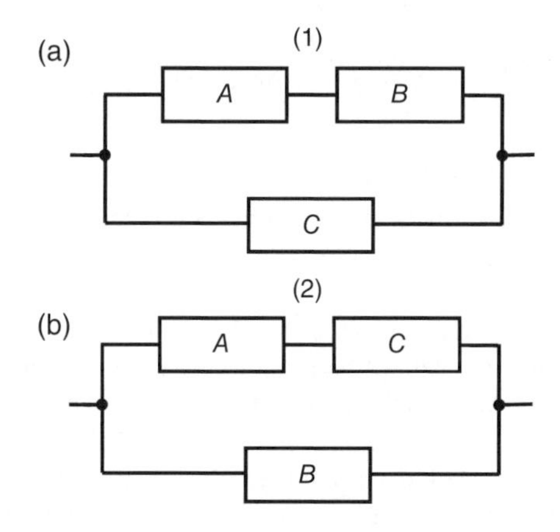

Figure 11.6 Comparing reliabilities of competing systems (a) and (b)

component and a type-B component is more reliable than a type-C component, comparing the reliability of the systems helps to identify the more reliable system.

Indeed, let a, b and c denote the reliabilities of components A, B and C. The reliability of system '1' is $R_1 = 1-(1-ab)(1-c) = c+ab-abc$. The reliability of system '2' is $R_2 = 1-(1-ac)(1-b) = b+ac-abc$.

The difference $R_2 - R_1 = b+ac-c-ab = b-c-a(b-c) = (b-c)(1-a)$. From $b > c$ and $1-a > 0$ $(a < 1)$, it follows that $R_2 - R_1 > 0$. Therefore, the second system is the more reliable system.

As a result, building and comparing the reliability models of the systems helped to identify the more reliable system in the absence of any reliability data.

11.1.7.2 Building and Analysing Physics of Failure Models

One of the big advantages of physics of failure models is that they usually suggest ways of improving the resistance to failure (see Chapter 12).

The classical statistical models of the time to failure require the components to be built and tested before the parameters of the time to failure model can be estimated. In contrast, the physics of failure models are based on first principles, and once they are established, they can be applied to estimate the reliability of new technology with no failure history.

Suppose that one of the important failure modes of a component is 'material degradation', whose instantaneous rate is proportional to the diffusion coefficient of a particular harmful element. The variation of the diffusion coefficient D with temperature, in solids, is given by the Arrhenius' equation:

$$D = D_0 \exp\left(-\frac{E_D}{RT}\right)$$

where T is the absolute temperature (K), E_D is the activation energy for diffusion (J/mol), R is the gas constant in J/(K × mol) and D_0 is a constant. From the analysis of this equation, it can be inferred that the increase in temperature will increase significantly the rate of material degradation. Design measures preventing the increase of temperature therefore will be of critical importance to reducing the likelihood of premature failure.

In another example, suppose that one of the important failure modes is 'fatigue failure initiated from a pre-existing crack' subjected to a varying tensile stress with amplitude $\Delta\sigma$.

A popular model related to the physics of fatigue failure growth from a pre-existing crack is the Paris equation (Paris and Erdogan, 1963; Paris *et al.*, 1961):

$$\frac{da}{dN} = C\left(Y\Delta\sigma\sqrt{\pi a}\right)^m$$

where da/dN is the growth rate of the fatigue crack; $\Delta\sigma = \sigma_{max} - \sigma_{min}$ is the stress range of the loading cycle; C, m are constants depending on the material microstructure, environment, test temperature and load ratio $R = \sigma_{min}/\sigma_{max}$; Y is a geometry factor; a is the crack size; and N is the number of loading cycles.

Suppose that the experimental analysis yields $m \approx 4$ for the constant m. The rate of fatigue crack propagation is then proportional to $(\Delta\sigma)^4$. According to the Paris equation, a reduction of the amplitude of the internal tensile stress will result in a significant increase of the fatigue life and a reduction of the risk of premature failure.

Suppose that a structure is subjected to varying load. Design measures aimed at reducing the amplitude of the internal tensile stress will result in a significant reduction of the risk of premature failure. This example is developed and discussed in detail in Chapter 12.

11.1.7.3 Minimising Technical Risk through Optimisation and Optimal Replacement

The risk of failure can be decreased if a compromise among the design parameters is found which maximises the reliability of the system. This is the purpose of the optimal selection of risk reduction options within available risk budget to achieve a maximum risk reduction (see Chapter 19). The risk of premature failure of a complex system can be reduced significantly by guaranteeing through optimal replacement of components, a high probability of surviving a specified in advance minimum failure-free operating period.

Consider a system where m components, with monotonically increasing hazard rates, undergo intensive wearout and a single spare component is available for each of these components. The time to failure of each of the m components is random, given by the Weibull distribution. The system is operated for a period a. Failure of any of the m components causes system failure. There is an optimal choice of times for replacement of the separate components which minimises the probability of system failure within the operational interval [0,a]. Minimising the probability of system failure is an optimisation problem (see the example at the end of Chapter 9). If the replacements are done too early, the system will fail because of a highly likely premature wearout of the replaced components; if the replacements are done too late, the system is likely to fail before taking advantage of the benefits from the replacement.

Optimisation does not necessarily involve a balance between risk-critical factors within a selected design. It can be based on the selection of an entirely new design.

11.1.7.4 Maximising System Reliability and Availability by Appropriate Permutations of Interchangeable Components

To increase the reliability and availability of systems, *the traditional approach invariably requires investment of resources*. Increased system reliability and availability is commonly achieved by investing resources for purchasing more reliable components, redundancy or building extra connectivity. Another traditional method for maximising system availability is through investment in more efficient maintenance, associated with shorter repair times. Recent research (Todinov, 2014b) on production systems with simple parallel–series arrangement revealed that the availability and reliability of production systems can be improved dramatically, at no extra cost, solely by appropriate permutations of interchangeable components between the parallel branches (see Chapter 19).

11.1.7.5 Maximising the Availability and Throughput Flow Reliability by Altering the Network Topology

Research published in Todinov (2013a) demonstrated that maximising the reliability and availability of flow networks at no extra cost can also be done by altering the network topology. The results indicated that seemingly insignificant alterations of the topology result in a dramatic increase of the system's reliability and availability.

11.1.8 Reducing Variability of Risk-Critical Parameters and Preventing them from Reaching Dangerous Values

Reliability- and risk-critical parameters vary (Carter, 1986, 1997; Haugen, 1980), and this variability can be broadly divided in the following categories: (i) variability associated with material and physical properties, manufacturing and assembly; (ii) variability caused by the product deterioration; (iii) variability associated with the loads the product experiences in service; and (iv) variability associated with the operating environment.

Strength variability caused by production variability and variability of properties (Bergman, 1985) is one of the major reasons for an increased interference of the strength distribution and the load distribution which results in overstress failures. A heavy lower tail of the distribution of properties usually yields a heavy lower tail of the strength distribution, thereby promoting early-life failures. Low values of the material properties exert stronger influence on reliability than do high or intermediate values.

Variability of critical design parameters (e.g. material properties and dimensions) caused by processing, manufacturing and assembly is an important factor promoting early-life failures.

An important way of reducing the lower tail of the material properties distribution is the high-stress burn-in. The result is a substantial decrease of the strength variability and increased reliability on demand due to a reduced interference of the strength distribution and the load distribution.

Due to the inherent variability of the manufacturing processes, however, even items produced by the same manufacturer can be characterised by different properties. Production variability during manufacturing, not guaranteeing the specified tolerances or introducing flaws in the manufactured product, leads to a significant number of failures. Depending on the supplier, the same component of the same material, manufactured to the same specification, is usually characterised by different properties. Between-suppliers variation exists even if the variation of the property values characterising the individual suppliers are small (see Chapter 4). The variability associated with the lower tail of the strength distribution controls the load–stress interference. Stress screening which eliminates substandard items is an efficient way of reducing the variability in the region of the lower tail of the strength distribution and increasing reliability (see the discussion in Chapter 6).

Low reliability is often due to excessive variability of the load. If the load variability is large (rough loading), the probability of an overstress failure is significant. Altering the upper tail of the load distribution is often done by using stress limiters (see Chapter 6). Typical examples of stress limiters are the safety pressure valves, fuses and switches, activated when pressure or current reach critical values. A common example of a stress limiter preventing surges in voltage from reaching dangerous levels is the anti-surge protector used in the power supply of electronic equipment.

Tolerances in geometrical reliability-critical parameters must be controlled during manufacturing. Such control translates into fewer problems and failures during assembly, less possibility for loss of precision, jamming, seizure, poor lubrication and fast wearout.

Often, variability of geometrical parameters causes *fit failures* resulting from interference of solid parts which makes the assembly impossible.

Material quality is positively correlated with the reliability of components. This correlation is particularly strong for highly stressed components. Sources of materials must be controlled strictly without relying on vendor's trade names or past performance. Changes in the processing and manufacturing procedures often result in materials with poor quality.

11.1.9 Altering the Component Geometry

A typical example is the case where the component shape is altered to eliminate stress concentrators. As a result, the stress intensification zones in the component are eliminated, and the risks of fast fracture and fatigue fracture are greatly reduced.

Another example is altering the component shape to achieve thermal design. In order to increase the heat dissipation, the components' surface is often increased by flattening or by introducing cooling ribs which increase the surface-to-volume ratio. Conversely, in cases where heat conduction is unwanted, the shape is made spheroidal which decreases the surface-to-volume ratio. Thus, to reduce erosion of the cladding in blast furnaces due to interaction with molten metal, the cladding components are often made spheroidal. In structures spanning spaces without supporting columns, arches have been used through the entire human history, to eliminate tensile stresses and increase reliability. If an abrasive belt is made with the shape of Moebius ribbon, its working time (durability) can be increased twice at the same length (Altshuller, 1996).

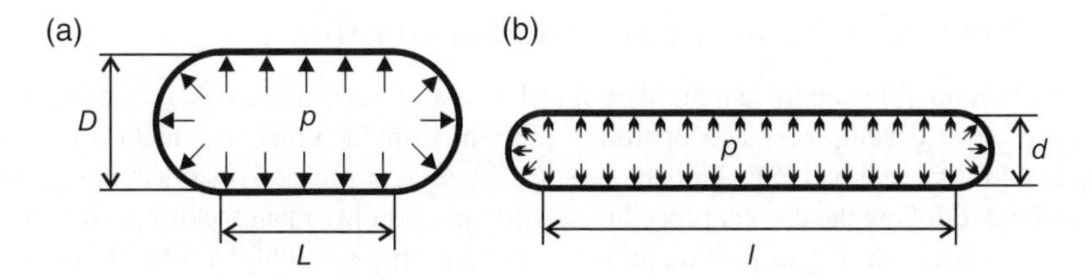

Figure 11.7 (a) A pressure vessel whose shape has been modified (b) by keeping the same volume

Consider the pressure vessel in Figure 11.7a with diameter D, length L and thickness s of the shell. The vessel contains fluid exerting pressure p on the inside of the shell (Figure 11.7a). Altering the shape to the one in Figure 11.7b by keeping the same volume

$$V_1 = \pi D^2\left(\frac{D}{6}+\frac{L}{4}\right) = V_2 = \pi d^2\left(\frac{d}{6}+\frac{l}{4}\right)$$

reduces significantly the hoop stress, which is the largest principal tensile stress acting on an element from the shell (see Chapter 12). The axial principal tensile stress is also reduced.

Thus, by modifying the shape, the hoop stress decreases from $\sigma_{H1} = pD/(2s)$, Figure 11.7a, to $\sigma_{H2} = pd/(2s)$, Figure 11.7b.

11.1.10 Strengthening or Eliminating Weak Links

Consider again a common example of a system with n components, logically arranged in series with reliabilities $R_1, R_2, \ldots, R_n$. The system contains a weak link with reliability r, logically arranged in series with the rest of the components. In other words, $r < R_1, r < R_2, \ldots, r < R_n$ are fulfilled, and the reliability of the system $R_{\text{sys}} = R_1 \times R_2 \times \cdots \times R_n \times r$ is smaller than the reliability of the weakest link. (Indeed, since $R' = R_1 \times R_2 \times \cdots \times R_n < 1$, then $R_{\text{sys}} = R' \times r < r$.)

Interfaces often appear as weak links in the chain, thereby limiting the overall reliability of a system. Consider a common practical example related to two very reliable components with high reliabilities $R_c \approx 1$ connected by an interface with a relatively low reliability $r << R_c$. The reliability of the system is smaller than the reliability r of the interface, and in order to improve the reliability of the system, the reliability of the interface must be increased. One of the reasons why so many failures occur at interfaces, despite that the interfaced components are usually very reliable, is the fact that *often interfaces are not produced to match the reliability of the corresponding components*. Seals in mechanical components and connectors in electrical devices, for example, commonly appear as weak links.

A weak link can be strengthened by improving its reliability or including redundancy. The weak link should be strengthened *sufficiently, but not excessively*. Strengthening a weak link excessively, more than other components, would add cost and weight without increasing the overall reliability of the system.

Strengthening the weak links to avoid failures and improve performance is a truly universal concept. The human body, for example, can also be considered as a chain of individual elements. A weak link in an athlete, for example, might be faulty biomechanics or lack of joint mobility or joint stability or any physical limitation which could result in injury or prevents the athlete from achieving peak performance. To prevent injuries and maximise performance, weak links must be identified by qualified professionals and strengthened by specially prescribed corrective exercises.

11.1.11 Eliminating Factors Promoting Human Errors

Human errors account for a significant number of technical failures. They are an inevitable part of each stage of the product development and operation: design, manufacturing, installation and operation. Following Dhillon and Singh (1981), human errors can be categorised as (i) errors in design, (ii) operator errors (failure to follow the correct procedures), (iii) errors during manufacturing, (iv) errors during maintenance, (v) errors during inspection and (vi) errors during assembly and handling. A thorough analysis on the root causes, conditions and factors promoting human errors is an important step towards reducing them. Some common factors promoting human errors are listed below:

- Time pressure and stress
- Overload and fatigue
- Distractions and high noise levels
- Poor work skills and lack of experience
- Unfamiliarity with the necessary procedures and equipment
- Inadequate information, specification and documentation
- Poor health
- Poor organisation, time management and discipline
- Inattention and lack of concentration
- Making unwarranted assumptions and building a false picture of the reality
- Negative emotional states and disempowering beliefs
- Low confidence
- Poor motivation
- Poor communication
- Poor relationships with the other members of the team
- Poor safety culture

Instructions and procedures must be clearly written, easy to follow and well justified. The procedures must also reflect and incorporate the input from people who are expected to follow them. It must always be remembered that human beings are prone to forgetting, misjudgement, lack of attention, creating false pictures of the real situation, etc. – conditions which are rather difficult to manage. Hardware systems and procedures are much easier to manage and change than human behaviour. Therefore, the efforts should always concentrate on adapting the hardware to humans rather than humans adapting to the hardware.

Learning from past failures and making available the information about past human errors which have caused failures are powerful preventive tools. In this respect, compiling formal databases containing descriptions of failures and lessons learned and making them available to designers, manufacturers and operators are activities of significant value.

Frequent reviews, checks and tests of designs, software codes, calculations, written documents, operations or other products heavily involving people are important tools for preventing human errors. In this respect, *double checking* of the validity of calculations, derivations or a software code is invaluable in preventing human errors. To eliminate common cause errors associated with models and problem solutions, *double checking based on two conceptually distinct approaches* is particularly helpful. Such is, for example, the approach based on creating both an analytical solution and Monte Carlo simulation solution to a probabilistic problem. Obtaining very close results provides a strong support that both the analytical model and the Monte Carlo simulation programme are correct.

A number of human errors arise in situations where a successful operation or assembly is overly dependent on human judgement. Human errors of this type can be avoided by using tools/devices which rely less on a correct human judgement. Poka-Yoke design features and special recording and marking techniques could be used to prevent assembling parts incorrectly. Blocking against common cause maintenance errors could be achieved by avoiding situations where a single person is responsible for all pieces of equipment.

A thorough task analysis often reveals weaknesses in the timing and the sequences of the separate operations and is a key factor for improving their reliability. Additional training has a great impact on the probability of successfully accomplishing a task. Improving the efficiency in accomplishing various required tasks (working smarter) improves the management of the work load, reduces overload and fatigue and the associated with them human errors.

11.1.12 Reducing Risk by Introducing Inverse States

11.1.12.1 Inverse States Cancelling the Anticipated State with a Negative Impact

An inverse state of anticipated states with negative impact can be used to compensate the negative effect. The two states superpose and the result is an absence or a significantly attenuated negative effect.

In acoustics, this principle works in noise-cancellation headphones for reducing the risk of hearing damage caused by noise. A sound wave is emitted with the same amplitude but with inverted phase to the noise. The result is a significant attenuation of the harmful noise and reduced risk of hearing damage.

This principle also underlies active methods of controlling vibration. The active vibration control involves suitable vibration sensors (e.g. accelerometers), controllers and actuators for vibration control. The signal from the vibration sensors is fed to a controller, and through an actuator, a spectrum of cancellation vibrations are generated in response. The advances in the sensor, actuator and computer technology made active methods of control cost-effective and affordable.

11.1.12.2 Inverse States Buffering the Anticipated State with a Negative Impact

Introducing an inverse state as a buffer can be done in many cases where a negative effect has been anticipated and the inverse state is provided for buferring the impact of the anticipated negative effect.

This underlies reducing the risk of failure of zones generating heat. Components working in close contact (e.g. piston cylinder) and moving relative to each other generate heat which, if not dissipated, causes intensive wear, reduced strength and deformations. The risk of failure of such an assembly is reduced significantly if one of the parts (e.g. the cylinder) is cooled to dissipate the released heat which reduces the friction and wear.

The *cold expansion* used in aviation for creating compressive stresses at the surface of fastener holes (Figure 11.8) is another example of using buffering inverse states.

This is done by passing a tapered mandrel through the hole. The inverse state created in the vicinity of the hole (compressive residual stress field) counters the tensile loading stresses during operation and impedes the formation of fatigue cracks at the edge of the hole and their propagation which reduces the risk of fatigue failure.

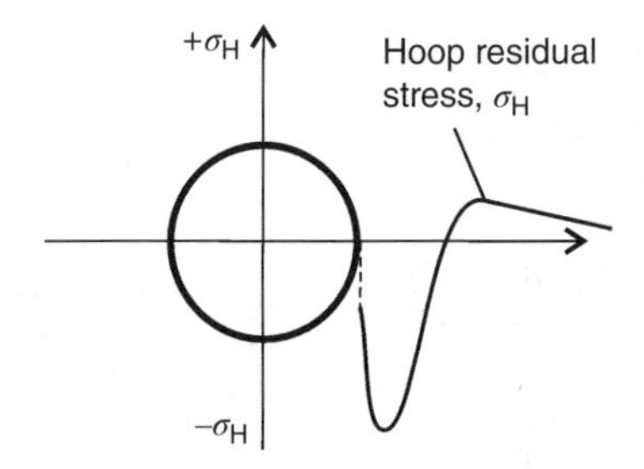

Figure 11.8 Countering the stress concentration effect of a hole by creating compressive stresses through cold expansion

In order to counter the tensile stresses from loading at the surface and improve fatigue resistance, *shot peening*, introducing compressive stresses at the surface, has been used as an important element of the manufacturing technology (Bird and Saynor, 1984; Niku-Lari, 1981). As a result of this operation, the fatigue life of leaf springs, for example, can be increased many times.

Buffering the negative effect by introducing an inverse state is often used in the construction industry where the tensile stresses from bending of concrete beams can be reduced if *preloaded in tension steel bars* are inserted in the beam. After the concrete sets, the beam is preloaded in compression. The compressive stress from preloading is an inverse state which compensates the tensile loading stresses. Since the tensile stresses from bending superpose with the compressive residual stresses, the effective tensile stress during service is reduced significantly.

An inverse state of compressive residual stresses at the surface, acting as a buffer compensating the tensile service stresses from loading, can also be created by a special heat and thermochemical treatment such as *case hardening*, *gas carburising* and *gas nitriding*. The *corrosion*, *erosion* and *wear allowances* added to the computed sections are inverse states anticipating the loss of wall thickness. They act as buffers compensating for the loss of wall thickness and decrease the risk of failure.

The principle of introducing an inverse state as a buffer has a wide application in many other areas of human activity. In project management, providing time buffers for certain critical tasks reduces the risk of delay should particular risks materialise. Similarly, in managing stock in the presence of random demands, increasing the reserve of a particular critical stock (e.g. particular life-saving medicine) reduces the risk of running out of stock in case of clustering of random demands.

Increasing the financial reserves of a bank or a company makes it less vulnerable to depleting its reserves due to materialised credit and market risks.

11.1.12.3 Inverting the Relative Position of Objects and the Direction of Flows

There are cases where inverting the relative position of objects eliminates a detrimental effect. Drilling vertical blind holes in components by a robot on a manufacturing line is associated with the need for cleaning the drilled holes from metal chips. If a hole is drilled on the component upside-down, the need for cleaning the drilled holes from chips disappears because gravity now helps to clean the holes.

The next example in the area of logistic supply is an unexpected application of this principle.

Figure 11.9a features a logistic supply network where a particular interchangeable commodity is delivered from the three sources s_1, s_2 and s_3 to the destinations t_1, t_2 and t_3.

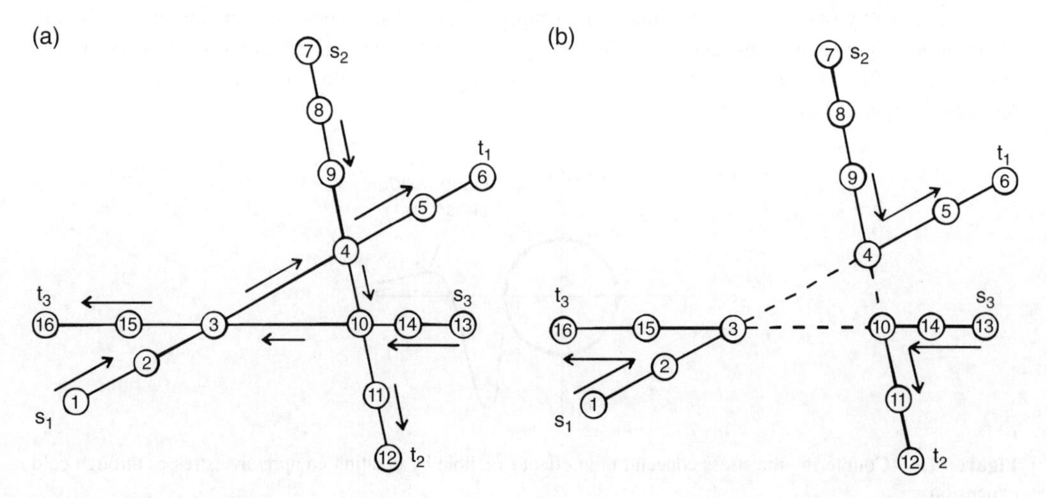

Figure 11.9 (a) Closed parasitic flow loop in a logistic supply network; (b) draining the closed parasitic flow loop

As a result, a closed parasitic flow loop essentially appears between nodes 3,4,10 and 3 despite that the transported interchangeable commodity does not travel along a closed contour. Closed parasitic flow loops essentially are cyclic paths where the flow travels in the direction of traversal (Figure 11.9a, the flow loop 3,4,10,3). By draining the parasitic flow loop by augmenting the cyclic path 3,4,10,3 with flow in opposite direction, the flow loop is eliminated (Figure 11.9b). As a result, value is derived from significantly reducing the transportation losses and risk of congestion without affecting the throughput flow from the sources s_1, s_2 and s_3 to the destinations t_1, t_2 and t_3.

Parasitic flow loops are associated with increased risk of congestion and accidents, big wastage of energy and time and increased levels of pollution to the environment. Parasitic flow loops exist in real transportation networks with a very high probability. Optimising supply networks by draining highly undesirable dominated parasitic flow loops derives significant value by reducing the transportation costs, the risk of congestion and accidents and the environmental pollution (Todinov, 2013a, 2013b, 2014a). The result is billions of dollars saved to the world economy.

The existence of parasitic flow loops in networks remained unnoticed by scientists for nearly 60 years. Ironically, despite the years of intensive research on static flow networks, closed parasitic flow loops appear even in the 'network flow solutions' from all published algorithms (including the famous Ford–Fulkerson algorithm; Ford and Fulkerson, 1956) for maximising the throughput flow in networks, since the creation of the theory of flow networks in 1956.

The parasitic flow loops are not necessarily closed flow loops only. Dominated parasitic flow loops for which more than half of the cyclic path contains flow along a particular direction of traversal are also associated with significant transportation losses and risk of congestion.

In Figure 11.10a, three sources of interchangeable commodity s_1, s_2 and s_3 are supplying three destinations d_1, d_2 and d_3. As a result, a dominated parasitic flow loop 4,5,6,7,2,3,4 appears. The dominated parasitic flow loop can be eliminated by augmenting the cyclic path 4,5,6,7,2,3,4 with flow in the opposite direction of the direction of the dominating flow. As a result, the dominated parasitic flow loop disappears (Figure 11.10b) without affecting the throughput flow from the sources to the destinations (Todinov, 2014a).

11.1.12.4 Inverse State as a Counterbalancing Force

A typical application of an inverse state as a counterbalancing force are the counterweights in cranes which reduce the loading on the lifting motor and improve the balance and stability of the crane. Counterweights are also used on rotating shafts (e.g. on crankshafts in piston engines) to improve balance

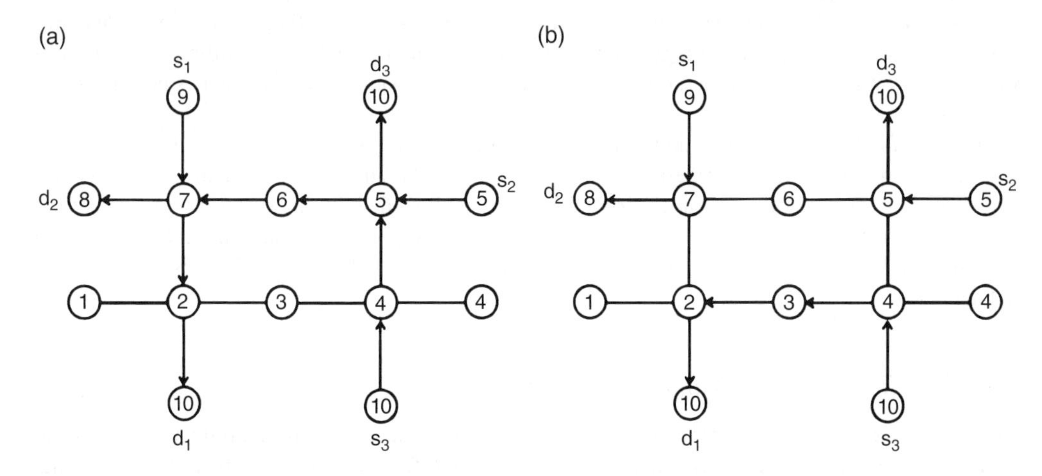

Figure 11.10 Draining a dominated parasitic flow loop (2,7,6,5,4,3,2)

which reduces the magnitudes of the vibrations and the risk of failure. Systems of cables and masts acting as counterbalance are used in architecture to relieve weight and improve the structural reliability. An interesting example of an inverse state as a counterbalancing force can be found in Pahl et al. (2007) where the fast rotating turbine blades are attached to the rotor at an angle. During rotation, the centrifugal force acting on the blade creates a bending moment which opposes the bending moment created by the fluid on the blade. The result is a smaller bending stress at the zone of attachment of the blade to the rotor and increased reliability.

Another example of the application of this principle is the gate valve which is maintained open by a hydraulic pressure acting against a counterbalancing compression spring. Upon failure of the hydraulic system, the counterbalancing spring expands and returns the valve in closed (safe) position.

11.1.13 Failure Prevention Interlocks

Preventing failure modes caused by a wrong sequence or order of actions being taken can be achieved by designing *failure prevention interlocks*. These make the occurrence of failure modes practically impossible.

Physical interlocks are devices and circuits which block against a wrong action or a sequence of actions being taken. Typical physical interlocks are the safety contacts installed in machine guards, which prevent the machine from being switched on before the guard is in place. A physical interlock, for example, will prevent an aeroplane from taking off without setting properly all flight controls or without latching firmly all boarding doors into closed position. If, for example, starting a machine under load will cause failure, a built-in interlock device could make it impossible to start the machine if it is under load. Failures are often caused by exceeding the operational or environmental envelope. Efficient failure prevention interlocks of this type are all circuits which prevent operation during conditions of extreme heat, cold, humidity, vibrations, etc. Such an interlock can be designed for the common fan-cooled device. If the fan fails, the power supply is automatically disconnected in order to prevent an overheating failure of the cooled device.

Poka-Yoke (mistake-proofing) is an effective technique based on either issuing an alert when a mistake is about to be made or preventing the operator from making the mistake. Poka-Yoke designs are often used to prevent operating or assembling a device in the wrong way.

An example of Poka-Yoke application which prevents unintended movement of a car is the interlock which requires the driver to depress the clutch pedal before the engine could start. Another common example of Poka-Yoke technique is the sound signal if an attempt is made to leave the car while the headlights are still on.

Logic interlocks eliminate the occurrence of erroneous actions. Preventing the hand of an operator from being in the cutting area of a guillotine can, for example, be made if the cutting action is activated only by a simultaneous pressure on two separate knobs/handles which engages both hands of the operator.

Time interlocks work by separating tasks and processes in time so that any possibility of collisions or mixing dangerous types of processes and actions is excluded. Suppose that a supply system fails if two or more demands follow within a critical time interval needed for the system to recover. If the operation of the system is resumed only after a built-in delay equal to this minimum critical period, a time interlock will effectively be created which excludes the possibility for overloading from sequential demands.

11.1.14 Reducing the Number of Latent Faults

An efficient way of reliability improvement is the removal of latent faults from products, systems and operations. A *fault* is an incorrect state, or a defect resulting from errors during material processing, design, manufacturing, assembly or operation, with the potential to cause failure or accelerate the occurrence of failure under particular conditions. A software fault is synonymous with bug and is in effect a defect in the code that can be the cause of a software failure.

Fault is not the same as failure. The failure is an 'event' after which the service delivered by the system deviates from the specified system function for the specified operating conditions. The fault is a condition (a state). The latent fault plays a necessary but not necessarily a sufficient role in initiating failure. A system with faults may continue to provide its service, which complies with the specifications until some triggering input condition is encountered which could lead to failure. For example, a software bug allowing the system to read and act on signals from failed sensors is a latent fault. Despite the bug, the system will continue to operate as long as there is no failure of the sensor. The fault will manifestate into a failure only when the sensor fails, but the system continues to read it and act on its false indications as if the sensor was working correctly. In aviation, such a fault may result in unexpected behaviour of an aircraft, causing an accident.

The presence of a software fault/bug does not necessarily result in immediate software failure. Failure will be present only when the logical branch containing the faulty piece of code is executed, which happens only when certain conditions are present.

Typical latent faults in electronic devices include poor solder joints, defective wire bonds, semiconductor impurities, semiconductor defects and component drift.

A fault could lead to a faster accumulation of damage. Such is, for example, the presence of a large defect (e.g. pore) in a stressed component. The component may be operating for some time, but the fatigue crack growing from the defect will cause a premature fatigue failure.

A large building can have as a latent fault improper foundation support which will only manifestate into a catastrophic failure during an earthquake of certain magnitude.

A deviation of a parameter from its safe range is also a latent fault which could lead to failure. For example, deviation of a clearance from its prescribed value could lead to jamming and failure if temperature rises beyond a certain critical level.

Testing and thorough inspection for latent faults are key measures to their removal. In developing software applications, a thorough debugging and operational testing are key to the removal of latent faults. Proper quality management processes must be in place in order to eliminate or minimise the latent faults in the released products.

The objective of *environmental stress screening* (*ESS*) is to simulate expected worst-case service environments. The stress levels used for ESS are aimed at eliminating (screening) the part of the population with faults causing a heavy lower tail of the strength distribution which is the primary reason for many early-life failures.

This process is illustrated in Figure 11.11a where the lower tail of the strength distribution has been altered by stress screening which removes substandard items. In Figure 11.11b, the strength

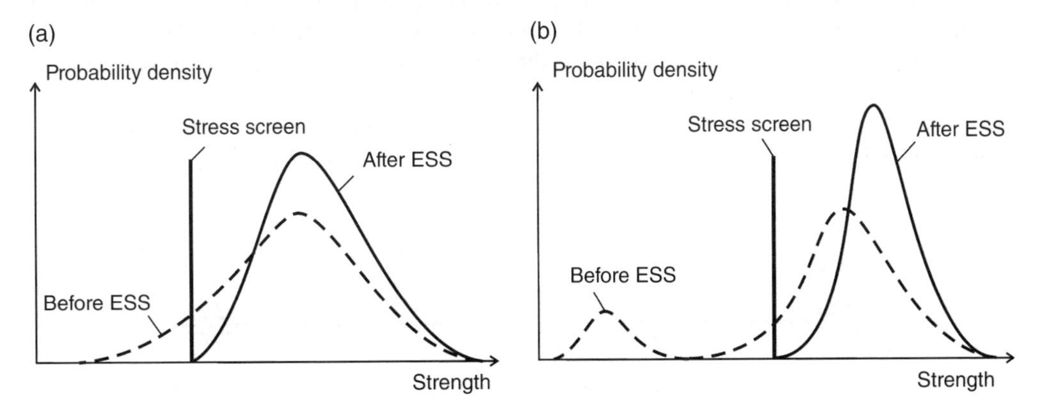

Figure 11.11 Altering the lower tail of the strength distribution by environmental stress screening: (a) unimodal distribution of the strength of items; (b) bimodal distribution of the strength of items

distribution is a mixture of two distributions: a main distribution reflecting the strength of the strong population of items and a distribution characterising the weak population of items with substandard strength. ESS has improved the strength distribution by removing the weak population (Figure 11.11b). By trapping faults and substandard items before they are released to the customer, this operation eliminates early-life failures caused by items with substandard strength. ESS also helps to discover and eliminate sources of faults and weaknesses during design, manufacturing and assembly.

During environment stress screening (or *burn-in*), it is important to find operating and environmental test conditions which permit efficient screening without consuming a substantial part of the life of the components which have passed the screening test. Thermal cycling of integrated circuits, for example, often reveals poor wire bonds, improperly cured plastic packages, poor die bonds, etc.

Particularly useful tests, which could reveal a large number of failure modes and reduce the test time from years to days and hours, are the *highly accelerated life testing* (HALT) and *highly accelerated stress screens* (HASS). The purpose is to expose (precipitate) faults and weaknesses in the design, manufacturing and assembly in order to provide a basis for reliability improvement. The purpose *is not* to simulate the service environment. Precipitation of a fault changes its state from latent/undetected to a detected fault. A latent fault 'poor solder joint' is usually undetectable unless it is extremely poor. Applying vibration, thermal or electrical stress helps to precipitate the fault, conduct failure analysis and perform appropriate corrective action (Hobbs, 2000). The precipitated faults and weaknesses are used as opportunities for improvement of the design and manufacturing in order to avoid expensive failures during service. In this respect, HALT and HASS are particularly useful. The stresses used during HALT and HASS testing are extreme stresses applied for a brief period of time. They include all-axis simultaneous vibration, high-rate broad-range temperature cycling, power cycling, voltage, frequency and humidity variation, etc. (Hobbs, 2000). During HALT and HASS, faults are often exposed with a different type of stress or a stress level than the ones that would be used during service. This is why the focus is not on the stress and the test conditions which precipitate the faults but on the faults themselves.

11.1.15 Increasing the Level of Balancing

Unbalanced forces cause premature wearout, fatigue degradation and failure. As a rule, improving the level of balancing in a system improves the uniformity of the load distribution, reduces the magnitudes of the inertia forces, the loading stresses and increases the reliability of the system. Balancing eliminates unwanted inertia forces and moments in rotating machinery. *Static balancing* guarantees that the mass centre of the rotating parts is on the rotation axis but does not guarantee absence of extra reaction forces in the bearings. Extra bearing reaction forces can still exist due to rotating couples if no dynamic balancing has been performed after the static balancing. A rotating part can be statically balanced and dynamically unbalanced (Figure 11.12).

Improving the distribution of the load across different working parts reduces the stresses acting in the parts and reduces their wearout and deterioration.

The designers' responsibility is to guarantee that the line joining the mass centres of the rotating parts is a straight line coinciding with the rotation axis (Uicker *et al.*, 2003).

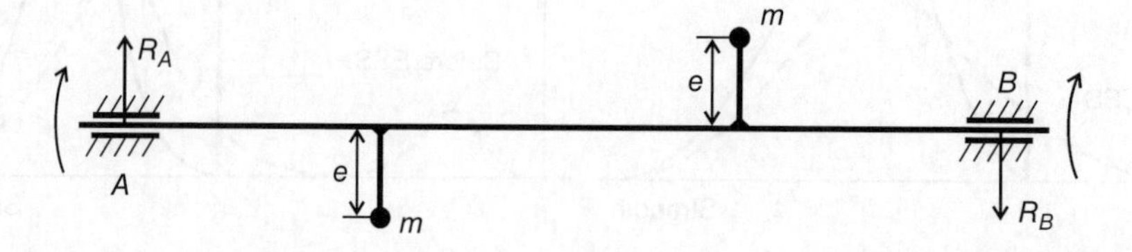

Figure 11.12 A rotating part which is statically balanced and dynamically unbalanced

Another example of the application of the balancing principle in mechanical systems can be found in double helical gears (herringbone gears) where self-balancing of the axial forces is present and symmetrical epicyclic mechanisms where self-balancing of the radial forces acting on the central shaft is taking place.

11.1.16 Reducing the Negative Impact of Temperature by Thermal Design

Temperature variations cause thermal deformations and stresses, degradation of the properties of mechanical components and change in the parameters of electronic components. Above certain temperature, the parameters of electronic components are no longer guaranteed to be within specification.

For mechanical systems, common protection against the development of excessive thermal stresses includes:

(a) ***Reducing the thermal gradients and temperature changes in components***

- *Using materials with large heat conduction coefficients.*
  The use of such materials does not permit the development of large thermal gradients. Thus, the thermal gradient developed in a hot metal pipe during contact with cold environment (e.g. rain) is significantly smaller compared to the thermal stress developed in hot glass pipe or ceramic pipe.
- *Using coatings to mitigate thermal shocks from the environment.*
  Such are the design measures involving coating steel pipes transporting hot fluids (Peng and Peng, 1998). Natural insulation such as burying pipes is also beneficial in reducing the thermal gradients.
- *Avoiding start–stop cycles which introduce thermal gradients.*
- *During heat treatment, conducting quenching in media which guarantees a small heat transfer coefficient.*
- *Reducing the operating temperature.*
  Reducing the operating temperature reduces the temperature differences between working components and does not permit the development of significant thermal stresses.

(b) ***Using materials with special thermal properties***

- *Using materials with small thermal expansion coefficients.*
  Materials with small thermal expansion coefficients do not develop thermal stresses with large magnitudes even for large thermal gradients. The reason is the proportionality of the thermal stress to the thermal expansion coefficient α_t $[K^{-1}]$. For the metal rod with both ends welded to the fixed supports in Figure 11.13a, a change in temperature $\Delta t = t - t_0$ with respect to the ambient temperature t_0 causes tensile thermal stress if $\Delta t = t - t_0 < 0$ (Figure 11.13a) and compressive thermal stress if $\Delta t = t - t_0 > 0$ (Figure 11.13b). (See Chapter 12 for a detailed discussion.)

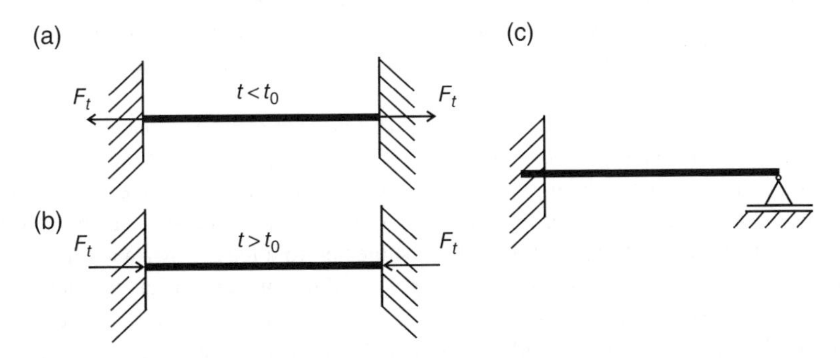

Figure 11.13 (a and b) Design associated with thermal stresses and (c) design free from thermal stresses

No thermal stresses appear if a sliding support is introduced as is shown in Figure 11.13c which provides a compensation for the thermal expansion/contraction. This is an example of reducing thermal stresses by a *relaxation of constraints*.

Selecting material with a small thermal expansion coefficient or reducing the thermal gradient (temperature difference) reduces the thermal stress.

Another way of reducing the thermal stresses is by *using materials with similar thermal expansion coefficients.*

If components with similar coefficients of thermal expansion work together (e.g. bolt and nut made of materials with similar thermal expansion coefficients), the thermal expansion of one of the materials is not severely restricted by the thermal expansion of the other material, and the thermal stresses are with relatively small magnitude. The converse is true for materials with very different thermal expansion coefficients.

The total input energy lost to heat in the system must correspond to the capacity of the components to dissipate heat energy. Otherwise, the result is overheating of particular components which leads to increased thermal stresses, decreased stiffness and strength and decreased fatigue resistance. Consequently, *using materials with large capacity to dissipate heat energy* is an important way of countering the negative impact of heat generation.

(c) ***Altering the component geometry to avoid the development of large thermal stresses***

- *Avoiding stress concentrators.*

  Stress concentrators magnify the thermal stresses which are often sufficient to cause failure or permanent deformation of the components.
- *Avoiding connecting very thick to very thin sections.*

  Because of big differences in volume, thin sections decrease their temperature fast, which is resisted by the thick sections. As a result, thermal stresses with large magnitudes appear which often exceed the material strength.
- *Providing larger expansion gaps towards the zones with higher temperature.*

  This method has an application in designing the shape of the piston in high-performance internal combustion engines. To make allowance for the larger thermal expansion of the piston at higher temperatures, the clearance with the cylinder wall is larger towards the piston head.

(d) ***Relaxation of constraints***

- *Using thermal expansion gaps and expansion offsets.*

  Thermal expansion gaps, offsets and loops are designed to absorb dilatation and are commonly used in railways, buildings and piping systems.
- *Using statically determinate structures.*

  These are free from thermal stresses because they provide compensation for the thermal expansions. A typical example of structures free from thermal stresses is the statically determinate trusses (Barron and Barron, 2012).

(e) ***Using special designs limiting the development of thermal stresses***

- *Using material with intermediate thermal expansion coefficient between two materials with very different thermal expansion coefficients.* Details about this technique can be found in Ishikawa *et al.* (2012).
- *Using material with a very small thermal expansion coefficient between materials with a big mismatch of the thermal expansion coefficients.* This technique has an application in cases where a bolt is clamping a part with larger thermal expansion coefficient (e.g. steel bolt clamping aluminium plate). Because of the different thermal dillatation of the bolt and the clamped plate, with increasing temperature, the bolt will be subjected to a tensile thermal stress. This can be avoided if a sleeve made of material with a very small thermal expansion coefficient is inserted between the bolt head and the clamped part (Pahl *et al.*, 2007). As a result, the thermal dilatation of the bolt becomes comparable with the thermal dilatation of the clamped part, and the thermal stress is reduced.

(f) *Dissipating heat from areas of friction and other heat generation*

- *Using heat sinks.* For electronic devices, common methods for providing thermal protection include heat sinks for components generating considerable amount of heat. Including thermal conduction planes across printed circuit boards redistributes and dissipates heat from components with high heat generation power.
- *Using cooling fluids.* Fans and cooling jackets are often used in high-power devices and engines to remove heat from zones where heat is generated.

11.1.17 Self-Stability

Self-stability in design ensures that the disturbances of the system output produce a stabilising action returning the system in a stable operational state. The self-stabilising design can be found in the coupling of induction motor with torque-speed characteristic 1 and a machine with resisting torque characteristics 2 (Figure 11.14).

An increase in the angular velocity $\omega_{st} + \Delta\omega$ from the point of stable operation ω_{st} causes the driving torque produced by the induction motor to drop below the resisting torque M_{res} ($M < M_{res}$) which slows down the rotor and decreases the angular velocity towards ω_{st}. A decrease $\omega_{st} - \Delta\omega$ in the angular velocity causes the driving torque from the induction motor to increase above the resisting torque ($M > M_{res}$). The result is an increase of the angular velocity towards ω_{st}. Operating at angular velocity ω_{st} results in a stable operation.

A very different behaviour is exhibited if the system is operated at an angular velocity ω_{un}. A decrease in the angular velocity $\omega_{un} - \Delta\omega$ causes the driving torque to further decrease until the induction motor stops. An increase in the angular velocity $\omega_{un} + \Delta\omega$ causes the driving torque to increase which results in a further increase of the angular velocity. The operation at angular velocity ω_{un} is an unstable operation.

Another use of this principle can be found in the negative feedback loops used in stabilising the output of electronic circuits and mechanical systems. A deviation of the system/process from a stable configuration produces a correcting action returning it in a stable state. Commonly, in closed-loop or feedback control systems, the output is measured and fed back to an error detector at the input. A controller is then correcting the parameters of the system/process so that the deviations from the expected output become as close to zero as possible.

Negative feedback is used in audio amplifiers to reduce distortion and in operational amplifier circuits to obtain a predictable transfer function.

The availability of low-cost electrical devices and sensors makes it possible to provide more flexibility and regulate mechanical systems to a finer degree compared to all-mechanical governors.

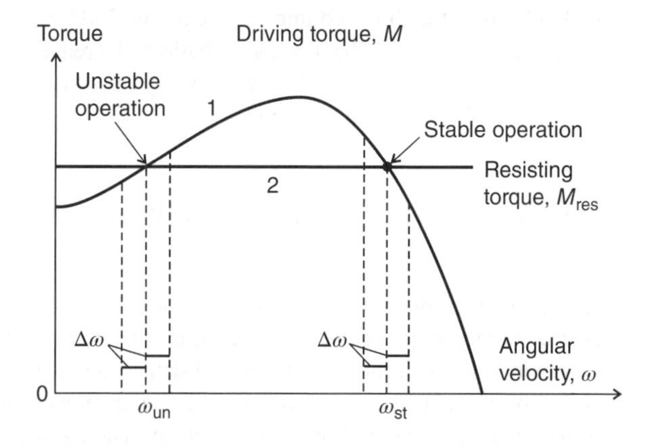

Figure 11.14 Speed-torque characteristic of induction motors

11.1.18 *Maintaining the Continuity of a Working State*

Interrupting the continuity of a working state introduces transient processes and forces associated with failure modes. Maintaining the continuity of a working state avoids high-resistance forces and dynamic transient stresses from start–stop regimes. Thus, the resistance of pressure vessels to thermal fatigue can be enhanced significantly by avoiding start–stop regimes which induce high thermal stresses. The resistance to jamming of sliding surfaces (e.g. stems in valves) can be enhanced by maintaining continuity of motion which prevents the formation of build-ups of corrosion products. Avoiding start–stop regimes of rotating shafts maintains the hydrodynamic lubrication layer in journal bearings, reduces wearout and improves reliability. Avoiding start–stop regimes of electro-motors avoids the high initial currents characterising the start regimes. Avoiding the variation of the speed of rotating shafts avoids operation close to natural excitation frequencies which is associated with excessive stress amplitudes and premature failure.

11.1.19 *Substituting Mechanical Assemblies with Electrical, Optical or Acoustic Assemblies and Software*

As a rule, replacement of mechanical devices reduces significantly the overstress, wear, fatigue corrosion and material degradation, which are major factors leading to failures of mechanical equipment. The replacement of complex mechanical assemblies with electrical, optical or acoustic devices reduces the complexity of design, the number of moving parts, wear and increases precision. During the design of electromechanical components, where possible, the complexity should be transferred to the software. Design should be oriented towards simpler but more refined mechanical components combined with powerful software to guarantee both performance and flexibility (French, 1999).

11.1.20 *Improving the Load Distribution*

Improving the load distribution decreases the loading stresses and improves reliability.

Providing a surface contact instead of line contact and line contact instead of a point contact results in improved load distribution, decreases contact stresses and results in higher reliability. Using conforming contact surfaces increases the contact area and also significantly reduces the contact stresses and wear. Using ribs to distribute loads to supporting walls is another example of improving reliability by an improved load distribution.

A common example of this principle is the use of washers under bolt heads and nuts, which leads to a more uniform load distribution. A similar example is the special design of nuts resulting in a more uniform distribution of the load across the threaded interface (Collins, 2003).

Using more bolts in assemblies increases the load distribution, decreases vulnerability to bolt failures and increases reliability. A balance should be sought however because increasing the number of bolts affects adversely maintainability (Thompson, 1999).

11.1.21 *Reducing the Sensitivity of Designs to the Variation of Design Parameters*

Robustness is an important property of components and systems. Robust designs show small sensitivity of the performance characteristics to variations of the manufacturing quality, drifts in parameter values, operating and maintenance conditions, environmental loads and deterioration with age (Lewis, 1996). Achieving high reliability levels by reducing the variation of geometrical parameters through more precise finishing operations (e.g. grinding, honing and polishing) is not always economically feasible because it leads to escalating costs. Making the design insensitive to the variations of geometrical parameters often achieves a high reliability without excessive costs. A typical example of design, insensitive to variations of geometrical parameters is the involute gear system, where the profiles of the teeth are involutes of a circle. The angular velocity ratio ω_1/ω_2 is insensitive to variations of the distance between the gear axes. It remains constant. Another typical example is the designs of self-adjusting bearing assemblies and couplings which can accommodate geometrical imprecision and misalignment. *Gear couplings*, for example, can compensate simultaneously *radial*, *angular* and *axial* misalignment. In this respect, avoiding a *double fit* is important, where a component is guided at the same time by two surfaces (machined separately). Because of the inevitable variation of tolerances, such components are a source of problems during assembly. They are also a source of problems during operation because of the assembly stresses and the uncertainty regarding the distribution and magnitude of the loading stresses during operation.

In many cases, the reliable work of components and systems occurs under too narrowly specified conditions. Slight variations in the material properties, the quality of manufacturing, the external load or the values of the design parameters are sufficient to induce failures or unacceptable deviations from the expected function/service.

Design solutions requiring fewer parts with simple geometry reduce the susceptibility to manufacturing variability. Designs incorporating appropriate geometry, static determinacy, tolerances and materials with high conductivity reduce the susceptibility to temperature variations and large thermal stresses which are a common cause of failure. For example, making truss structures statically determinate makes them insensitive to temperature variations.

Thus, the statically indeterminate structure in Figure 11.15a is subjected to thermal stresses or assembly-induced stresses if the struts are characterised by different coefficients of thermal expansion, are at different temperatures or are characterised by differences in their lengths. None of these problems exist for the statically determinate structure in Figure 11.15b.

Further examples of robust designs are the sensors for measuring a particular property which are insensitive to variations of other parameters, for example, gas sensors insensitive to the variations of temperature and humidity.

Often, the mean values of reliability-critical parameters are sought which minimise the sensitivity of the performance characteristic to variations of the input parameters.

This approach has been illustrated by the mechanical spring assembly in Figure 11.16a required to provide a constant clamping force of specified magnitude P. The same clamping force can be provided by a stiff spring, with a large spring constant k_1 and a particular initial deflection x_1 ($P = k_1x_1$) or a softer spring, with spring constant $k_2 < k_1$ and larger initial deflection $x_2 > x_1$ (Figure 11.16b). The initial spring deflection is always associated with errors (errors in cutting the spring to exact length, imperfections associated with machining the ends of the spring coil, sagging of the spring with time due to stress relaxation, variations in the length of the spring associated with the pre-setting operation, etc.).

As it can be verified from Figure 11.16b, for the softer spring (with a spring constant k_2), variations of magnitude Δx in the spring deflection cause much smaller variations ΔP_2 in the clamping force compared to the variations ΔP_1 in the clamping force for the stiffer spring, caused by variations of the same magnitude Δx of the spring deflection. Selecting a softer spring results in a more robust design, for which the clamping force P is less sensitive to variations in the spring deflection. For the same amount of stored potential energy $E = (1/2)k\Delta x^2$, for a soft spring with small constant k and large deflection Δx, the variation of the spring constant k causes a smaller variation in the amount of stored

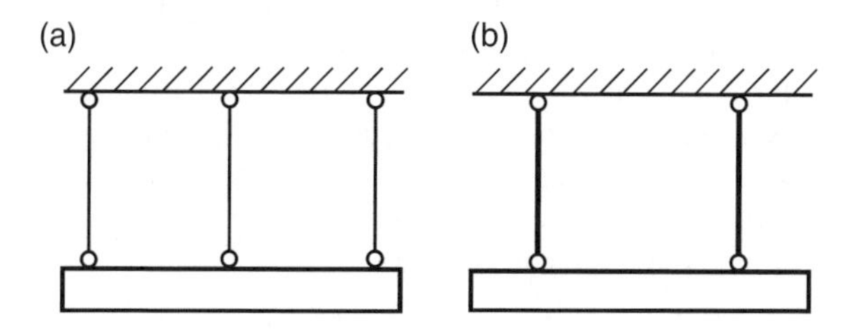

Figure 11.15 (a) Statically indeterminate and (b) statically determinate structure

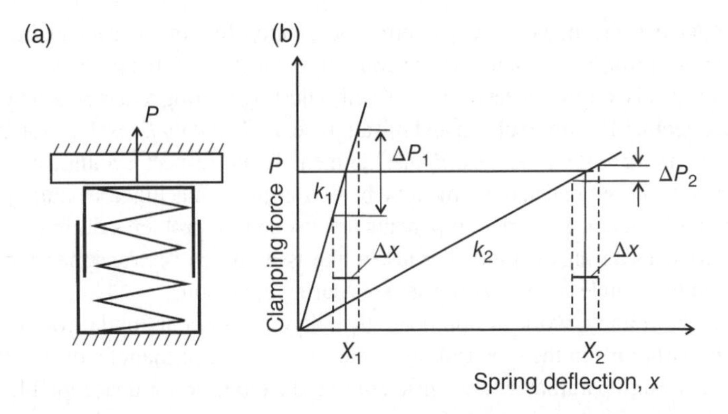

Figure 11.16 (a) Clamping force from a spring; (b) Clamping force variation for a stiff and soft spring

energy E, compared to a stiff spring with large constant k and a small deflection Δx. Due to variation of the shear modulus of the spring material and the diameter of the spring wire, the spring constant k does vary substantially.

In the general case, determining the mean values of the control parameters which minimise the variation of the performance characteristic requires efficient algorithms for constrained non-linear optimisation. For these optimal values of the design parameters, the output performance characteristics will be least sensitive to variations in the design parameters. *Such are for example the flat regions of the output performance characteristic, where the partial derivatives with respect to the separate parameters are small or zero*. In many cases, the relationship between the performance characteristic and the reliability-critical variables cannot be presented in a closed form, or if it exists, it is too complex. Furthermore, the reliability-critical parameters may be interdependent, subjected to complex constraints. In these cases, the simulation-based optimisation is a powerful alternative to other methods. It is universal, handles complex constraints and interdependencies between reliability-critical variables and does not require a closed form function related to the performance characteristics. Furthermore, its algorithm and implementation are relatively simple and straightforward.

Large variations of the internal stresses in components and structures caused by variation of the loading forces are a frequent cause of fatigue failures. Fatigue life is very sensitive to the amplitude of the internal stresses. As a consequence, in regions with large internal stress ranges, the fatigue crack growth rate is significantly increased, and the life of the affected components is short. Design solutions restricting the variations of the internal stresses include but are not limited to stress limiters, appropriate modifications of the component geometry, appropriate modifications in the loading geometry, appropriate stress relieve notches, stress relieve grooves, etc.

For a bolted joint of a lid covering a container where the pressure fluctuates, the force-deformation diagram for the bolt and the flange is shown in Figure 11.17a. The elastic constants of the bolt and the flange are $k_b = \tan(\alpha)$ and $k_f = \tan(\beta)$, correspondingly. The bolted joint is subjected to a preload force of magnitude P. Because of the pressure in the container, the bolted joint is subjected to a pulsating external force with magnitude F, which causes pulsating force of magnitude F_b in the bolt (Figure 11.17a).

With reducing the elastic constant of the flange, the pulsating force F_b in the bolt increases (Figure 11.17b). Increasing the elastic constant of the flange has the opposite effect. With reducing the elastic constant of the bolt, the amplitude F_b of the pulsating force in the bolt is reduced (Figure 11.17c). Increasing the elastic constant of the bolt has the opposite effect. To reduce the amplitude of the pulsating force in the bolt and increase its fatigue life, the appropriate selection is elastic bolt and a stiff flange. This effect can be quantified by the amplitude of the force F_b by expressing the length L_{AB} of the segment AB in two different ways (Figure 11.17a):

$$L_{AB} = \frac{F_b}{\tan(\alpha)} = \frac{F - F_b}{\tan(\beta)}$$

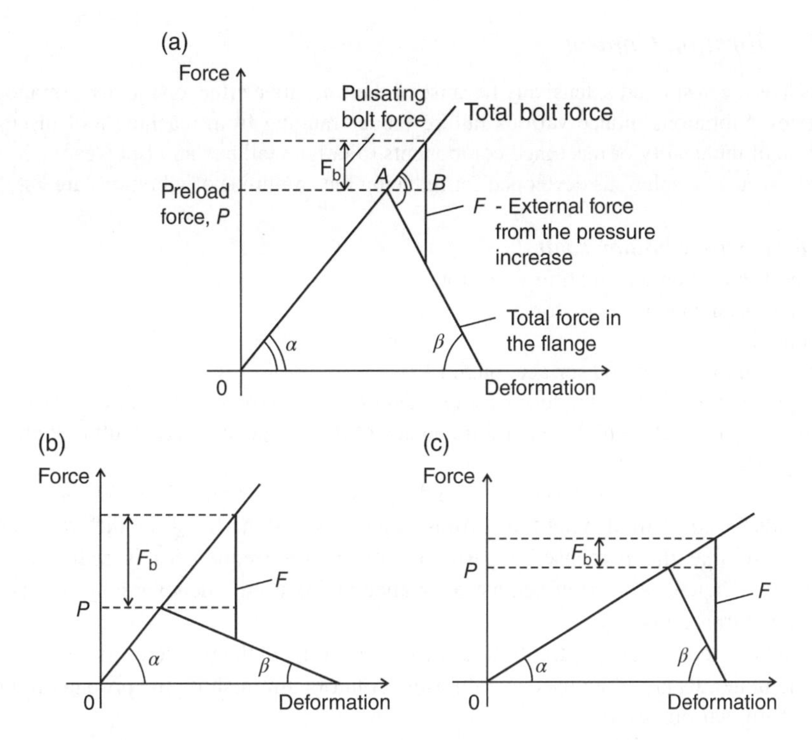

Figure 11.17 (a) Forces acting on a preloaded bolted joint; (b) with a reduced elastic constant of the flange; (c) with a reduced elastic constant of the bolt.

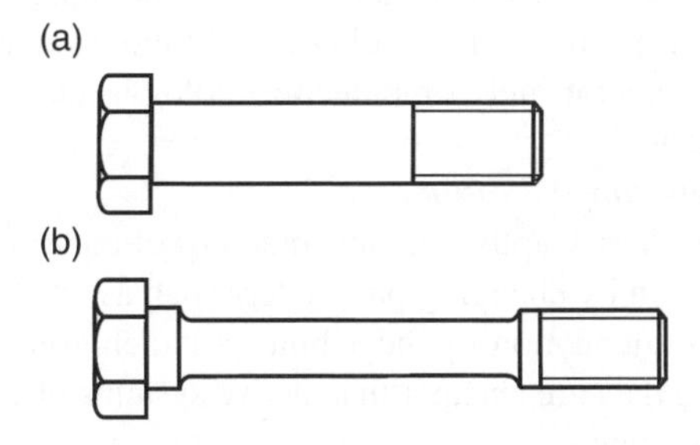

Figure 11.18 (a) A bolt whose elastic constant has been reduced by (b) increasing the length of the bolt and thinning down the shank diameter

Considering that $k_b = \tan(\alpha)$ and $k_f = \tan(\beta)$, the above relationship gives

$$F_b = \frac{k_b}{k_b + k_f} F$$

The first derivative with respect to k_b is positive everywhere which shows that reducing the elastic constant k_b of the bolt, reduces the magnitude of the pulsating force F_b in the bolt and increases its fatigue life. The elastic constant of the bolt is given by $k_b = EF / L$, where L is the length of the bolt, F is the cross-sectional area and E is the Young modulus of the material. Reducing the elastic constant of the bolt can be done by selecting a bolt with a larger length L or reducing the cross-sectional area F by thinning down the shank diameter of the bolt as shown in Figure 11.18.

Other methods for reducing the amplitude of the stresses in components are discussed in detail in Chapter 12.

11.1.22 Vibration Control

Vibrations have been studied extensively because of their negative effect on the performance of technical systems. Vibrations induce various failure modes ranging from machine tool instability and deterioration of the quality of machined components to fatigue failures and fast wear.

The main groups of solutions developed for reducing the amplitude of vibrations are listed next:

Methods for passive vibration control:

1. Methods based on eliminating vibrations:
   - Better balancing of inertia forces.
   - Better lubrication.
   - Reduced tolerances and surface roughness.
   - Improved fluid flow and reduced vortex shedding for flow-induced vibrations. A vortex shedding frequency close to the natural frequency of the component may result in vibrations with very large amplitudes.
   - Reducing parameter variation to reduce the possibility of parametric excitation.
2. Methods based on modifying the stiffness, damping and inertia of the different parts of the system so that the response to particular excitation frequencies is reduced. Damping provides efficient protection against resonance and is used whenever a system is operated near a natural frequency.
3. Methods aimed at isolating the sources of vibrations from the protected system. Rubber and plastic materials are often used as vibration isolators diminishing the propagation of vibrations from various sources.

Active methods of controlling vibration.

The active vibration control involves suitable vibration sensors (e.g. accelerometers), controllers and actuators for vibration control. The signal from the vibration sensors is fed to a controller, and through an actuator, a spectrum of cancellation vibrations are generated in response. The advances in the sensor, actuator and computer technology made active methods of control cost-effective and affordable.

Semi-active methods of controlling vibration.

These methods are also known as adaptive-passive methods of control where the passive element tunes to the vibration system by changing parameters such as 'stiffness' and 'damping coefficient' so that a maximal attenuation of the vibration is achieved. The semi-active methods gained popularity because they are cheaper than active systems of control and offer more flexibility than the passive systems.

11.1.23 Built-In Prevention

This principle in action can be found in the construction industry where reinforcing steel bars in concrete structures are placed in the regions of high tensile stresses. Steel bars have a very good tensile strength and compensate the lack of tensile strength in concrete. The result is a significant reduction of the probability of overstress failure. This principle is also used in the design of components from fibre-reinforced composite materials. The composite material is oriented so that the direction of its fibres is parallel to the direction of the tensile stresses. Because the composite has a much higher tensile strength along the direction of its reinforcing fibres compared to the tensile strength in lateral direction, this layout increases the load-carrying capacity of components and reduces the risk of overstress failure.

An example where the risk of failure is reduced by building in prevention compensating negative effects during service is the allowance for lost wall thickness. The *corrosion*, *erosion* and *wear allowances* added to the computed sections compensate for the loss of wall thickness and decrease the risk of failure.

Components working in close contact (e.g. piston cylinder) and moving relative to each other generate heat which, if not dissipated, causes intensive wear, reduced strength and deformations. The risk of failure of such an assembly can be reduced significantly if one of the parts (e.g. the cylinder) is cooled to dissipate the released heat which reduces friction and wear.

Preloading with the same sign with the loading stresses is often used as a built-in prevention to counter effects increasing the risk of failure. It is frequently applied to bolted joints and flange-and-gasket assemblies. *Tensile preloading* increases the fatigue life of a part subjected to a completely reversed zero-mean alternating stresses. The mean stress is increased, but the equivalent completely reversed cyclic stress is reduced significantly, and as a result, the fatigue life is increased substantially (Collins, 2003).

Built-in prevention can also be based on *using available sources in the environment*. A good example is the design of the cat's eyes on the roads of the United Kingdom. A rubber wiper cleans the reflectors from dirt as they are pushed below the road surface by passing traffic. The rain water collected at the base of the cat's eye also helps this process and makes it more efficient.

11.2 Dual Principles: Reduce Both the Likelihood of Failure and the Magnitude of Consequences

11.2.1 Separating Critical Properties, Functions and Factors

In general, it is difficult to optimise a single component carrying many functions with regard to every single function. Separating critical functions and properties is often the key to improving reliability and reducing technical risk. The separation principle can be discovered in the design of flexible pipes carrying hydrocarbons. The different layers in the flexible pipe are designed for different functions: to protect against external corrosion, to resist tensile loads, to resist radial loads resulting from internal pressure, to make the pipe leak proof and to prevent collapse due to external pressure. It is difficult to optimise a homogeneous pipe with respect to each of these functions. The separate layers building the flexible pipe however can be optimised with respect to the function they carry. The result is increased reliability of the pipe. Separating functions to different components relieves the load on components and reduces the risk of failure.

This principle is very useful in cases where reliability is balanced against weight and cost. Improving reliability only locally, where it matters, saves resources and results in lightweight designs.

Separating critical properties is often present in the design of complex alloys where some of the microstructural components provide resistance to wearout, while other components provide toughness (resistance to crack propagation).

Often, the necessity of reducing the risk of failure requires different properties from the different parts of the component. Guaranteeing different properties to different parts of the components is the underlying principle behind coatings improving the wear resistance and corrosion resistance. The case hardening of components, which consists of a local induction heating of the surface layers followed by quenching, improves the surface resistance to large contact stresses and wear while leaving the core tough which is necessary to withstand impact loads.

This principle is often used in composite materials combining structural constituents with different properties in different directions. Concrete used in the construction industry is a material with good compressive strength but small tensile strength. The steel bars in reinforced concrete are placed in areas loaded in tension where the concrete cannot resist tensile stresses.

The separation principle can also be used for mitigating the consequences from failure because an increased local reliability delays the propagation of damage to the rest of the component/structure. Thus, increasing the reliability of a fire door by additional fireproof coating delays the fire escalation and limits the fire damage.

The separation principle has a wide application. Reliable operation often depends on critical properties, events or factors not being present at the same time or in the same space region. Separating people from hazards is an important measure for reducing risk. A typical example of time separation of risk-critical factors is the traffic lights, preventing collision between intersecting flows of traffic and flows of pedestrians. Two incompatible risk-critical factors can be introduced simultaneously by transforming their action from continuous to periodic and inserting the action of one of the factors in the pauses of the other factor.

A typical example of space separation of risk-critical factors is the separation of intersecting flows of traffic and flows of pedestrians at different levels which eliminates the risk of collisions and accidents.

Limiting the spread of infection by urgent quarantine measures isolating infected individuals is another example of space separation used for risk reduction.

11.2.2 Reducing the Likelihood of Unfavourable Combinations of Risk-Critical Random Variables

Reliable and smooth operation and consequences of failure often depend on not having a particular critical combination of values of the random variables which control the performance of components and systems.

Monte Carlo simulation is a very good technique for revealing the probability of critical combinations of random variables. Some of the applications of the Monte Carlo simulation have been demonstrated in Chapter 9. If, for example, a low value of the diameter of the shaft, a high value for the diameter of the hub and a low value for the coefficient of friction are combined, a press-fit assembly could lose its capability to transmit torque. Design should aim to minimise the probability of unfavourable combinations of risk-critical random variables.

In some applications, it is possible that several redundant components will wear out and fail almost simultaneously which will also lead to a catastrophic failure.

This problem can be avoided by designing the redundant components in such a way that a simultaneous failure of all of them is very unlikely.

Consider a finite time interval during which a number of large consumers connect to a supply system independently and randomly. Suppose that the supply system needs a minimum time interval to recover and stabilise after a demand from a consumer. The supply system is overloaded if two or more demands follow (cluster) within the critical time interval.

Here are some other common examples where reliability depends on the existence of minimum critical distances *between* the occurrences of random events:

- *Users demanding the same resource (e.g. piece of equipment) for a fixed time s.* Collisions between user demands occur if two or more users arrive within the time interval s allocated per each user.
- *A single repairer servicing a number of devices upon failure.* If the repair time is s, clustering of two or more failures within a time interval s will cause delay in some of the repairs.

Decreasing the expected number density of events in the time interval is an efficient way of decreasing the probability of time clustering of reliability-critical events. Thus, reducing the expected number of events from 10 events per year to 5 events per year reduces the probability of time clustering within a week from 77.5 to 34%. The alternative way of decreasing the probability of time clustering, which consists of increasing the length of the time interval, is not as efficient. Doubling the time interval to 2 years by keeping the expected number of demands the same (10 expected number of events per 2 years) results in a probability of time clustering 56.7% within 2 years, which is still very large.

Space clustering of random risk-critical factors may also occur. Clustering of two or more random flaws over a small critical distance decreases dangerously the load-carrying capacity of thin fibres and wires. As a result, a configuration of two or more flaws within a critical distance could cause failure during loading.

11.2.3 Condition Monitoring

Condition monitoring is used for detecting changes or trends in controlling parameters or in the normal operating conditions which indicate the onset of failure. By providing an early problem diagnosis, the underlying idea is to organise in advance the intervention for replacement of components whose failure is imminent thereby avoiding heavy consequences. Condition monitoring is particularly important in cases where the time for mobilisation of repair resources is significant. The early problem diagnosis helps to reduce significantly the downtimes due to unplanned intervention for repair. A planned or opportune intervention is considerably less expensive than unplanned intervention initiated when a critical failure occurs.

Early identification and action upon detection of an incipient failure reduces significantly the risk of environmental pollution, the number of fatalities, the loss of production assets and the losses caused by dependent failures associated with damage to other components and systems. The earlier the warning, the shorter is the response time, the more efficient is the loss prevention, the more valuable is the condition monitoring technique.

11.2.4 Reducing the Time of Exposure or the Space of Exposure

11.2.4.1 Time of Exposure

A typical example of limiting the risk of failure by reducing the time of exposure is reducing the length of operation in order to reduce the probability of encountering an overstress load. Indeed, if the critical overstress load follows a homogeneous Poisson process with density ρ_c and the length of the time interval is a, the probability of encountering an overstress load during the time interval (0, a) is $p_f = 1 - \exp(-\rho_c a)$. This probability can be reduced significantly by reducing the length of the time interval a. If the process or action is conducted within a very small time interval $a \approx 0$, the probability of failure tends to zero. There is simply no time for the hazards to materialise:

$$\lim_{a \to 0}\left[p_f = 1 - \exp(-\rho_c a)\right] = 0$$

For road accidents following a homogeneous Poisson process with density ρ, along a road with length L_1, the probability of no road accident associated with the length L_1 is $p_1 = \exp(-\rho L_1)$. If the road length is increased by a factor of m ($L_2 = mL_1$), the probability of no accident will be $p_2 = \exp(-\rho m L_1)$. Taking the ratio of the logarithms of these probabilities gives

$$\frac{\ln p_1}{\ln p_2} = \frac{1}{m}$$

from which $p_2 = (p_1)^m$. From the last expression, if, for example, the probability of no road accident associated with the road length L_1 is $p_1 = 0.9$, increasing the length of the road four times decreases the probability of no road accident to $0.9^4 = 0.66$. Unlike short journeys, long journeys are likely to be affected by delays caused by road accidents. Consequently, for long journeys, a delay reflecting potential accidents should be calculated in the estimated overall time of the journey.

Suppose that a random load, characterised by a cumulative distribution function $F_L(x)$, is applied a number of times during a finite time interval with length t and the times of load application follow a homogeneous Poisson process with intensity ρ. Suppose that the strength is characterised by a probability density distribution $f_S(x)$. It is also assumed that the load and strength are statistically independent random variables. The probability of no failure (the reliability) associated with the finite time interval (0, t) can be calculated from the overstress reliability integral derived in Chapter 7:

$$R(t)=\int_{S_{\min}}^{S_{\max}} \exp\left[-\rho t\left(1-F_L(x)\right)\right] f_S(x)dx$$

The term $\exp[-\rho t(1-F_L(x))]$ in the *overstress reliability integral* gives the probability that none of the random loads in the time interval (0, t) will exceed strength with magnitude x. With reducing the time of exposure t, the probability $\exp[-\rho t(1-F_L(x))]$ that the load will exceed strength (the probability of failure) decreases significantly which leads to a significant increase in the reliability $R(t)$.

Another example of limiting the risk by reducing the time of exposure is in cases where the accumulated damage is proportional to the time of exposure. In this case, reducing the time of exposure prevents damage from reaching a critical level. Thus, reducing the amount of time spent in a hazardous area is an important measure limiting the damage to health. In some cases, the extent of the damage (e.g. in case of carbon monoxide poisoning or radiation damage) is strongly correlated with the amount of time spent in the hazardous area.

Often, reducing the time of exposure does not permit the negative effect to develop. A typical example is increasing the speed by which a rotating shaft goes through its natural frequencies which does not allow developing large resonance amplitudes. Another example is reducing the time of cutting plastic materials by increasing the speed of cutting faster than the speed with which plastic deformation can spread, eliminates the deformation of the processed plastic material.

11.2.4.2 Length of Exposure and Space of Exposure

A typical example of limiting the risk of failure by reducing the length of exposure is reducing the length of a piece of wire in order to reduce the probability that a critical defect will be present. Indeed, if the critical flaws in the wire follow a homogeneous Poisson process with density λ_c and the length of the wire is L, the probability of encountering a critical flaw along a length (0, L) is $p_f = 1-\exp(-\lambda_c L)$. This probability can be reduced significantly by reducing the length L of the wire.

If the piece of wire is with a very small length $L \approx 0$, the probability of having a critical flaw on the length L tends to zero:

$$\lim_{L\to 0}\left[p_f = 1-\exp(-\lambda_c L)\right]=0$$

Another example is limiting the risk of an error in a long chain of the same type of calculations. Assuming that the errors follow a Poisson distribution, if λ is the number of errors per unit number of calculations, the probability of an error associated with the total number of calculations N is given by $p_f = 1-\exp(-\lambda N)$. Reducing the number of calculations N dramatically reduces the probability of a calculation error p_f.

11.2.5 Discovering and Eliminating a Common Cause: Diversity in Design

A common cause reduces the reliability of a number of components simultaneously. The affected components are then more likely to fail, which reduces the overall system reliability.

Typical conditions promoting common cause failures are common design faults, common manufacturing faults, common installation and assembly faults, common maintenance faults and shared environmental stresses by several components, for example, high temperature, pressure, humidity, erosion, corrosion, vibration, radiation, dust, electromagnetic radiation, impacts and shocks. Common cause may also be due to common power supply, common communication channels, common piece of software, etc. Thus, two programmable devices produced by different manufacturers and assembled and installed by different people can still suffer a common cause if the same faulty piece of software code has been recorded in the devices.

Maintenance and operating actions common to different components are major sources of common cause failures. Software routines written by the same person/team also may exhibit common faults and failures.

Acceleration stresses leading to accumulation of damage and faster wearout are typical examples of common causes. Examples of acceleration stresses are the temperature, humidity, cycling, vibration, speed, pressure, voltage, current, concentration of particular ions, etc. This list is only a sample of possible acceleration stresses and can be extended significantly. Because acceleration stresses lead to a faster wearout, they entail a higher propensity to failure for groups of components which reduces the overall system reliability.

A typical example of this type of common cause failures is the high temperature which increases the susceptibility to deterioration of a group of electronic components. By simultaneously increasing the hazard rates of the affected components, the probability of system failure is increased. Humidity, corrosion or vibrations increase the joint probability of failure of the affected components and shorten the system's life. Even in blocks with a high level of built-in redundancy, in case of a common cause, all redundant components in the block may fail within a short period of time, and the advantage from the built-in redundancy is lost.

Failure to account for the acceleration stresses acting as common causes usually leads to optimistic reliability predictions – the actual reliability is smaller than the predicted.

Designing against common causes of failure in order to reduce the risk of failure can be done by (i) identifying and eliminating sources of common faults and common cause failures, (ii) decreasing the likelihood of the common causes and (iii) reducing the consequences from common cause failures.

Common cause failures are difficult to identify and are frequently overlooked if little attention is paid to the working environment and the possibility of latent faults. Designing out common causes is not always possible, but it should be done if opportunity arises. Simultaneous corrosion of various components in a cooling circuit or a hydraulic circuit, for example, can be eliminated by selecting non-corrosive working fluids. Erosion of various components caused by a production fluid can be reduced if a filter eliminating the abrasive particles is installed. Destruction of all communication lines due to fire, accident or vandalism can be avoided by avoiding placing all of the communication lines in a common conduit.

A typical example of reducing the impact of a common cause is the use of corrosion inhibitors which, when mixed with the cooling agent, reduce significantly the corrosion rate. The impact of a common cause can also be reduced by strengthening the components against it. Such is, for example, the intention behind all corrosion protection coatings. Other examples are the water-tight designs and couplings in underwater installations which reduce the possibility of contamination due to seawater ingress.

Common case failures can also be reduced by *decreasing the likelihood of occurrence of common cause events*, for example, common design, manufacturing and assembly faults, fire and faulty maintenance. Frequent design reviews and strict control of the manufacturing and assembly reduce the likelihood of latent faults which could be a common cause for expensive failures. A strict control of the maintenance operations reduces the maintenance faults which often initiate common failures. Furthermore, providing maintenance of components by two different operators reduces the likelihood of a common cause failure due to faulty maintenance.

Providing diversity in design is a very efficient way of blocking out a common cause and reducing common cause failures. The idea is to prevent several components from being affected by the same common cause. A common cause failure due to a software bug, for example, can be avoided if an alternative algorithm and implementation is provided for the same task or if a different team is involved in developing the same piece of software, independently. If two cooling pumps (a main pump and an emergency pump) participate in cooling down a chemical reactor, failure of both pumps creates an emergency situation. If the two cooling devices are from different manufactures or operate on different principles, common cause faults will be blocked out. For redundant cooling devices, if one of them is powered by electricity and the other uses natural gravitation to operate, the common cause 'absence of power supply' will be eliminated. If in addition, the two cooling devices are serviced/maintained by different operators, the common cause 'faulty maintenance' will also be blocked out. Similarly, a common cause due to an incorrect calibration of measuring instruments due to a human error can be avoided if the calibration is done by different operators. If finally, the cooling devices are separated in different rooms, the common cause failure due to fire will also be blocked out.

Separating the components at distances greater than the radius of influence of the common cause is an efficient way of reducing the risks from a common cause failure.

Thus, separating large fuel containers at safe distances from one another limits the extent of damage from accidental fire. Separating two or more communication centres at distances greater than the radius of destruction of a missile increases the probability of survival of at least one of the centres. Multiple backups of the same vital piece of information kept in different places protect against the loss of information in case of fire, theft or sabotage.

Another implementation of this principle is the separation of vital control components from a component whose failure could inflict damage. A typical example is separating the control lines at safe distances from the aeroplane jet engines. In case of engine explosion, the flight controls will not be lost.

Insulating some of the redundant components from contact with environment characterised by excessive dust, humidity, heat or vibrations.

Avoiding common links which can be affected by a common cause is an efficient way of blocking out common causes. Such are, for example, the common conduits for cables and common location for components and devices with vital functions.

Investing in many unrelated sectors protects against a common cause reducing simultaneously the return from a number of sectors (e.g. agricultural sectors simultaneously affected by bad weather or disease, consumer sectors simultaneously affected by a health scare, investments in a country affected by a political crisis, financial crisis or social unrest, etc.)

Sundararajan (1991) suggests preliminary common cause analysis which consists of identifying all possible common causes to which the system is exposed and their potential effects. The purpose is to alert design engineers to potential problems at the early stages of the designs.

11.2.6 Eliminating Vulnerabilities

A design vulnerability is often present where a single failure of a component leads to a catastrophic failure escalation with serious consequences. Particularly sensitive to a single failure are systems possessing a large amount of energy, whose safe containment depends on the safe operation of one or several components or on a safe sequence of operations. Such systems can be compared to loaded springs accumulating a large amount of potential energy controlled by a single lock. Failure of the lock will release a large amount of stored energy with large destructive power.

Consider a dam built from non-compacted material whose strength depends on the reliable operation of the draining system. Such a dam is vulnerable because if the draining system fails (which can happen if the draining pipes are blocked by silt or debris), the strength of the dam can be eroded quickly which leads to triggering of a catastrophic failure.

Another example of a vulnerable design is the case where preventing the release of toxic substances in the environment depends on the reliable operation of a single control sensor. An accidental damage of the sensor entails grave consequences.

Yet another example of sensitivity to single failures is a system whose safe operation overly depends on the absence of human error. In this case, a human error during a critical operation may trigger a major failure associated with grave consequences. One way of counteracting this type of sensitivity is to build in fail-safe devices or failure prevention systems that make the dangerous operation impossible. Such are the various failure prevention interlocks which do not permit conducting an operation until particular safety conditions are in place.

In computer security, vulnerability is a weakness which allows an attacker to acquire access to a valuable service or data. Failure to include checks for a stack overflow or division by zero, for example, leaves serious vulnerabilities which could be exploited by attackers to execute malicious code.

Monte Carlo simulation is an important method for identifying and assessing the likelihood of vulnerabilities caused by an unfavourable stack up of values of risk-critical parameters.

11.2.7 Self-Reinforcement

Typical applications of the self-reinforcement principle are the self-locking screws and the self-locking wedges. The self-reinforcement principle is also the basis of the design of other self-locking devices: (i) self-locking grips in tensile testing machines, (ii) self-locking plate clamps, (iii) self-locking hooks, (iv) self-locking climbing equipment, (v) self-locking marine cleats, etc.

Another application of this principle can be used in the design of covers for containers under pressure (Figure 11.19).

For the design in Figure 11.19a, the loading stresses in the screws can be reduced, and the reliability of the seal can be increased by selecting the self-reinforcement version in Figure 11.19b. In the design from Figure 11.19b, the pressure helps to form the seal and prevent leakage.

This principle can be also used to limit the consequences in the case of failure. Considering the example in Figure 11.19b, if loss of containment occurs, the leak can be reduced if a flexible screen is inserted in such a way that the pressure causing the leak is now used to seal the leak.

11.2.8 Using Available Local Resources

The essence of this principle is the use of available resources from the environment or products of the system's own operation to achieve a risk reduction.

A typical example is the use of protective earthing. Keeping the exposed conductive surface of electrical devices at earth potential reduces the risk of electrical shock. Earthing is also commonly used for protecting structures from lightning strike.

Another example of this principle is the use of gravity to provide cooling in the case of a power failure. Freezing locally the water in a pipe to provide a temporary plug in order to repair a damaged

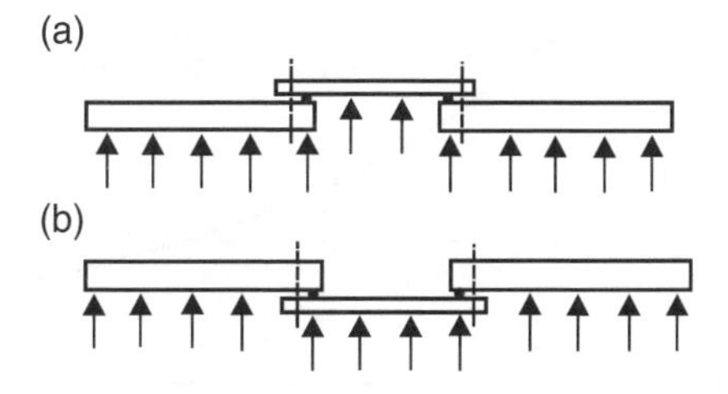

Figure 11.19 Design of a cover for a container under pressure (a) without self-reinforcement (b) with self-reinforcement

pipe section is an example of the use of local resources to reduce the consequences of failure. Yet another example is using the marine water as electrolyte in a circuit for cathodic protection, designed to reduce the corrosion of a pipeline.

11.2.9 Derating

Derating is one of the most powerful tools available to the designer for reducing the likelihood of failures. It is commonly done by reducing the operating stresses below their rated levels. The life of many components and systems increases dramatically if the stress levels are decreased. This makes the components and systems robust against the inevitable variations of the load and strength. The intensity of the wearout and damage accumulation also decreases significantly with reducing the stress magnitude.

The degree of wearout and deterioration is a function of the time and a particular controlling (wearout) factor p (Figure 11.20). As can be seen from Figure 11.20, reducing the intensity level of the wearout factor from p_1 to p_2 enhances the component's life because of the increased time for attaining a critical level of damage, after which the component is considered to have failed. A typical application of this method is the reduction of the stress amplitude, which results in a significant increase of the fatigue life of components.

Derating essentially 'overdesigns' components by separating the strength distribution from the load distribution thereby reducing the interaction between the distribution tails. The smaller the interaction of the distribution tails, the smaller the probability that the load will exceed strength and the smaller the probability of failure. Derating however is associated with inefficient use of the strength capacity of components.

In general, the greater the derating, the longer the life of the component. Voltage and temperature are common derating stresses for electrical and electronic components. The life of a light bulb designed for 220V, for example, can be enhanced enormously, simply by operating it at a voltage below the rated level (e.g. at 110V). For mechanical components, common derating stresses are the operating speed, stress amplitude, temperature, pressure, etc.

Derating also limits the consequences from failure. Thus, decreasing the operating temperature of a gaseous substance not only reduces the likelihood of failure but also limits the consequences given that failure occurs.

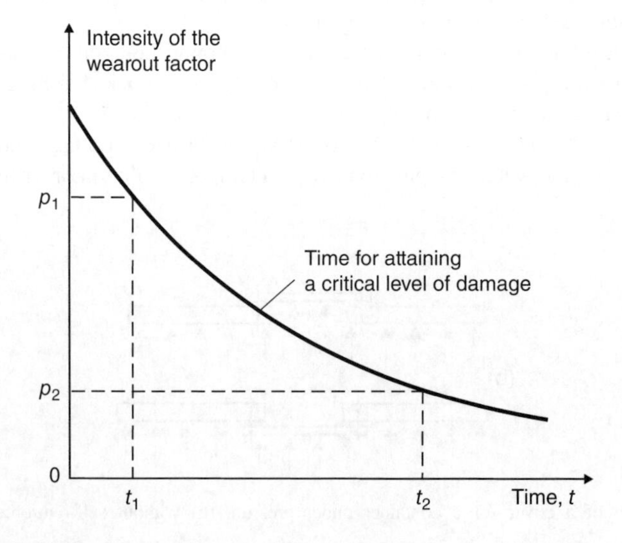

Figure 11.20 Time to failure for different intensity levels of the wearout factor

11.2.10 Selecting Appropriate Materials and Microstructures

Material selection is central to design and guarantees the combination of properties which are needed for the required function. Material selection and selection of appropriate microstructures play a very important role in reducing technical risk. Each particular application requires appropriate combinations of material properties: density, conductivity, Young modulus, shear modulus, toughness, strength, corrosion resistance, thermal resistance, etc. (Ashby, 2005; Ashby and Jones, 2000). Materials which do not comply with these requirements cannot deliver the expected functions and exhibit failure modes.

Consider the common case of tubes transferring corrosive production fluids or tubes working in corrosive environment. Selecting steel instead of appropriate corrosion-resistant polymer or composite will result in costly failures due to corrosion. In turn, selecting polymer instead of metal or composite for components subjected to intensive cyclic loading result in a short fatigue life. In high-temperature aggressive environment, selecting metals instead of ceramics often leads to premature failure.

Composite materials are widely used because they can be tailored to meet specific design requirements and combine various properties. Fibre-reinforced composites combine a very high strength to weight ratio, high fatigue resistance, corrosion resistance, damage tolerance and lightweight and are increasingly replacing plastics and metals in many applications.

Controlling the structure of materials brings unique properties which limit the risk of failure in safety-critical applications. Altering the microstructure of steel, for example, by an appropriate heat treatment, can result in changing the failure mode from brittle to ductile. The brittle failure mode is sudden and proceeds at a high speed, with no additional energy required, and the consequences are much more severe compared to the ductile failure mode.

Selecting appropriate materials also limits the consequences given that failure has occurred. Such are, for example, the materials and structures used for crumpling zones in road vehicles and racing cars. These materials have a pronounced capability to absorb and steadily dissipate impact energy, which reduces the impact load on passengers in the case of collision.

11.2.11 Segmentation

Segmentation reduces risk by (i) improving the load distribution, (ii) reducing the vulnerability to a single failure, (iii) reducing the damage escalation and (iv) limiting the hazard potential.

11.2.11.1 Segmentation Improves the Load Distribution

Consider a flange with very few fasteners. A flange connection with a very small number of fasteners leads to excessive stresses in some of the fasteners. Segmentation involving an increased number of fasteners improves the load distribution and reliability.

A significant reliability increase can be achieved if the load is distributed upon many load-carrying units. Thus, the load capacity of a V-belt cannot be increased by increasing its thickness because of increased bending stresses and big hysteresis losses overheating the belt. The load-carrying capacity and reliability however can be increased significantly by multiple parallel V-belts.

11.2.11.2 Segmentation Reduces the Vulnerability to a Single Failure

Segmentation also decreases vulnerability to a single failure. Consider again a flange with very few fasteners. Failure of a single fastener is very likely to cause a loss of containment. A flange with a larger number of fasteners will not be vulnerable to a single failure or even to several failures.

Failure of one of the multiple parallel V-belts will not cause failure of the transmission system. Similarly, failure of a single wire in a rope built by twisting many wire strands will not normally cause failure of the rope.

11.2.11.3 Segmentation Reduces the Damage Escalation

Segmentation also helps to reduce the damage escalation and the consequences given that failure has occurred. Segmenting a pipe into many separate sealed segments helps to limit the damage from a propagating crack within a single segment only, which reduces significantly the consequences from failure (see Figure 12.11b from the next chapter).

Crack arrestors can be strips or rings made of tougher material (Figure 12.11a). The mechanism of crack arrest consists of reducing the strain energy flow to the crack tip upon encountering a tougher material strip. The crack can also be arrested at the edge of the pipeline section (Figure 12.11b from the next chapter). Segmentation, in this case, does not prevent cracks from becoming unstable; it only limits the extent of damage once the damage has started escalating. In this case, segmentation reduces risk by limiting the consequences of failure.

Another application example of the segmentation principle can be given with buckling of a pipeline subjected to a high external hydrostatic pressure. Buckling could be eliminated by increasing the thickness of the pipeline, but this option is associated with significant costs. Control of buckling propagation achieved by using buckle arrestors is a cheaper and more preferable option. Buckle arrestors are thick steel rings welded to or attached at regular intervals to the pipeline in order to halt the propagating buckle and confine damage to a relatively small section (see Figure 12.11a from the next chapter). In this way, the losses from buckling are limited to the length of the section between two buckle arrestors. In case of failure, only the buckled section will be cut and replaced. The spacing between buckle arrestors can be optimised on the basis of a cost–benefit balance between the cost of installation of the arrestors and the expected cost of intervention and repair.

The segmentation principle can be used for reducing risk in a wide range of applications.

Segmentation is used to increase the resistance of a ship to flooding. The volume of the hull is divided into watertight compartments. If the flooding is localised, only one or very few compartments are affected which allows the ship to retain buoyancy. Segmentation of the corridors in a building, with fireproof doors, protects against the fast escalation of fire.

Segmentation can be used to prevent the spread of infectious diseases. Such is the purpose of preventing formations of large gatherings of people in case of infectious disease.

11.2.11.4 Segmentation Limits the Hazard Potential

Segmentation can be applied with success to limit the amount of energy possessed by hazards which limits their potential to cause harm. Thus, processing small (segmented) volumes of toxic substances at a time, reduces the hazard potential of the substance and the risk of poisoning in case of accidental spillage. Preventing the formation of large build-ups of snow, water, overheated water vapour, etc., reduces both the likelihood of an accident and its destructive power should it occur.

11.2.12 Reducing the Vulnerability of Targets

Vulnerability of humans is reduced by various barriers, guards, rails and by using personal protective equipment. Examples of personal protective equipment are protective clothing, harnesses, breathing devices, hats, goggles, boots, gloves, masks, radiation indicators, toxic gas release detectors, lifting and handling equipment, vaccines, etc. Vulnerability of the equipment and systems is decreased by using protection barriers, housing, encapsulation, anti-corrosion and anti-erosion coatings, use of CCTV surveillance, metal shutters, exclusion zones, security systems for access, etc.

Vulnerability of data is reduced by using security systems and limiting the access to personal records and confidential data.

11.2.13 Making Zones Experiencing High Damage/Failure Rates Replaceable

The life of a system can be extended by making blocks of components experiencing intensive failure rates modular. The failed blocks can then be replaced without replacing the entire system. This principle has been used widely in electronic systems and mechanical systems.

Often, even the life of components can be increased by identifying zones subjected to intensive failure rates and making them replaceable. This avoids replacing the entire component after the failure of such a zone. Thus, the life of a conveyer belt can be increased significantly if the surface zones in contact with the abrasive material are designed as replaceable plates. This principle has been used for a long time in the design of journal bearings and brakes.

11.2.14 Reducing the Hazard Potential

The purpose is to limit the amount of energy possessed by hazards which limits their potential to cause damage. Thus, preventing the formation of large build-ups of snow reduces both the likelihood of an avalanche and its destructive power should it occur. This is an example of reducing the hazard potential by reducing the amount of stored potential energy.

Another related example is reducing the voltage in electrical devices to reduce the consequences from an electric shock.

Instead of investing in safety devices and passive barriers, often, it is much more cost efficient to passivate hazardous wastes or spilled hazardous substances. This eliminates or reduces significantly their hazard potential and with it, the associated risk. There are various methods by which this could be achieved:

- Treatment with chemicals which reduce the chemical activity and toxicity of the hazardous substances.
- Reducing the inherent tendency to ignite or burn (e.g. chemicals which cover spilled fuel and prevent it from catching fire).
- Reducing the capability to evaporate.
- Reducing the possibility of auto-ignition (e.g. by avoiding piles of flammable materials).
- Changing the aggregate state. Solidifying liquid toxic waste, for example, reduces significantly its potential to penetrate through the soil and contaminate underground water.
- Dilution.

11.2.15 Integrated Risk Management

The goal of integrated risk management is to ensure that all major technical risks are identified, assessed, prioritised, aggregated and mitigated where possible.

Prioritisation is an important part of risk management. It may be unnecessary to allocate resources for risks whose impacts are very small.

Consequently, the process of managing technical risk can be summarised by the stages in Figure 11.21.

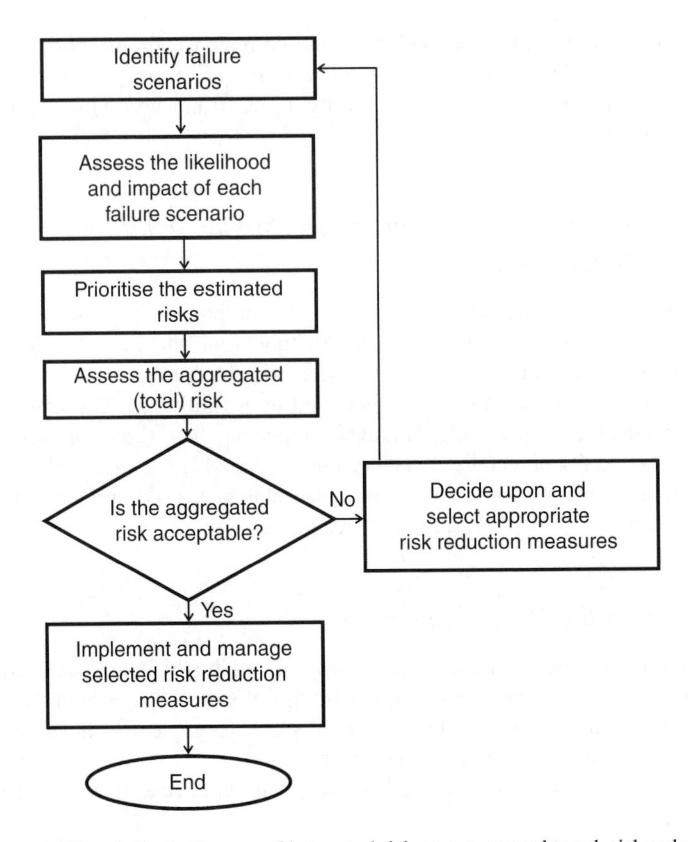

Figure 11.21 A block diagram of integrated risk management through risk reduction

A basic step of the risk management is the identification of as many failure scenarios as possible and assessing their likelihood and impacts. After the aggregated (total) risk associated with the identified failure scenarios has been estimated, the focus is on making a decision. If the aggregated risk is low, the risk is accepted and no further action is taken. Otherwise, the risk must be transferred, spread or reduced.

If the risk can be managed easily by a risk reduction, a large aggregated risk would require selecting and implementing appropriate risk reduction measures.

After assessing the risks corresponding to the separate failure scenarios, the associated risks are ranked in order of magnitude. A Pareto chart can then be built on the basis of this ranking, and from the chart, the failure scenarios accountable for most of the total risk during the specified time interval are identified. Risk reduction efforts are then concentrated on the few critical failure scenarios accountable for most of the aggregated risk.

Appropriate risk reduction measures are also identified which will reduce major risks associated with the critical failure scenarios. Next, new possible failure scenarios are identified, and the aggregated risk is estimated and assessed again. This iterative process continues until the risk assessment procedure indicates that the aggregated risk is acceptable.

Deciding upon and selecting particular risk reduction measures may not necessarily reduce the aggregated (total) risk. Indeed, a common situation during the design of complex systems is present when design modifications to eliminate a particular failure mode create other failure modes. In order to reduce the possibility of introducing new failure modes, each time after deciding upon and selecting appropriate risk reduction measures (e.g. design modifications), possible new failure scenarios are identified and assessed again. Furthermore, risks are often interrelated. Decreasing the risk of a particular failure scenario may increase the risk of other failure scenarios. Thus, building a tourist attraction on a remote place with sunny weather reduces the risk of reduced number of tourists due to a bad weather but simultaneously increases the risk of reduced number of tourists due to higher transportation expenses (Pickford, 2001). An effective protection against interrelated risks is *integrated risk management* which includes assessment of all individual risks and the total risk after deciding upon each risk reduction measure.

Alongside risk, the integrated risk management also incorporates the expected reward (Chapman and Ward, 2003). A particularly important issue for a company is striking the right balance between risk and profitability. Thus, borrowing from banks and investing in projects provide leverage and increase profitability but also increase the risk. Conversely, an increase in the cash position reduces risk but also reduces profitability because of the reduced rate of return.

11.3 Protective Principles: Minimise the Consequences of Failure

11.3.1 Fault-Tolerant System Design

A fault may not cause failure if the component/system is *fault tolerant*. Differentiation between failures and faults is essential for fault-tolerant systems. An example of a fault-tolerant component is a component made out of composite material resistant to cracks, defects and other imperfections. At the other extreme is a component made of material with low toughness (e.g. hardened high-strength steel) sensitive to different type of inclusions and mechanical flaws. At a system level, a system with built-in redundancy is fault tolerant as opposed to a system with no redundancy.

A digital circuit implementing *triple modular redundancy* is a typical example of a fault-tolerant design in digital electronics.

The *k-out-of-n* redundancy discussed in Chapter 3 is a popular type of redundancy because it makes the system fault tolerant. Such is, for example, a system of power supply based on six energy sources where only three working sources at any time are sufficient to guarantee uninterrupted supply. Such a system will be resistant to faults and failures associated with the separate power supply sources.

A power distribution system with radial, tree-like structure is simple but also highly unreliable – a fault along any branch cuts off the power supply to the entire downstream part. An interconnected power grid with mesh-like topology is significantly more fault tolerant, because congestion and faults can be isolated without affecting the other sections of the grid.

Software including internal tests and exception handling routines which set up safe conditions in case of errors is an example of fault-tolerant software.

11.3.2 Preventing Damage Escalation and Reducing the Rate of Deterioration

Implementing appropriate protective barriers can prevent damage from escalating. Protective barriers control an accident by limiting its extent and duration. They can also arrest the evolution of the accident so that subsequent events in the chain never occur. Protective barriers can also prevent particular event sequences and processes which cause damage by blocking the pathways through which damage propagates. The defence against a release of toxic substance, for example, combines:

Passive physical barriers (machine guards, fences, protection equipment, clothing, gloves and respiratory masks)
Active physical barriers (ventilation triggered by a detector)

Immaterial barriers (handling rules minimising the released quantity in case of an accident (e.g. handling a single container with toxic material at a time))
Human actions barrier and *organisational barrier* (evacuation)
Recovery barriers (first aid and medical treatment)

A number of different types of protective barriers reducing the consequences from accidents or failures have been discussed in Todinov (2007).

A common way of preventing damage from escalating is by using *damage arrestors*.

Various examples of damage arrestors have been discussed in Section 12.2 (Chapter 12).

The damage escalation can also be avoided by *avoiding concentration of vulnerable targets in close proximity*.

An example of this principle is the practice of avoiding building large containers for fuel storage in close proximity and leaving safe exclusion zones separating the containers. This measure makes the storage containers invulnerable to domino-type failures and damage escalation and prevents an accidental explosion of a storage container initiating other explosions.

An efficient method of limiting the consequences from an accident or failure is *blocking the pathways through which the damage escalates.* This is done by studying the pathways through which the consequences propagate and where possible, automatically sealing them off in case of an accident. A good example is the urgent quarantine measure of tracking and isolating infected individuals to prevent the spread of infectious disease. *Active protection systems* limit the consequences by blocking automatically the pathways through which the consequences propagate. Such are, for example, the shutdown systems and fail-safe devices which automatically close key valves in case of a critical failure, thereby isolating toxic or flammable production fluids and reducing the consequences from failures. Various stop buttons halting the production cycle in case of failure along a production line are also part of the active protection systems. Other examples of protection devices are the cut-off switches or fuses which disconnect a circuit if the current exceeds a maximum acceptable value.

An example of *delaying the rate of deterioration in case of failure* are the fireproof coatings of steel supporting structures, limiting the consequences should fire breaks out. Without the fireproof protection, in case of fire, the steel loses quickly its strength and may cause the entire structure to yield and collapse.

11.3.3 Using Fail-Safe Designs

The idea behind the fail-safe principle is to establish safe operating conditions for the system *after* the failure occurrence.

Hardware fail-safe devices are often an integral part of protection systems. A fail-safe gate valve, for example, uses the accumulated elastic energy of a spring to return the valve in safe 'closed' position should the hydraulic pressure keeping the valve open suddenly drop.

The fail-safe electrical contacts, part of machine guards, are designed to prevent failure causing a dangerous state of contacts sticking together. Without a fail-safe design, a failure resulting in electrical contacts sticking together permits the machine to be switched on in the absence of a machine guard.

The *leak-before-break concept* (see Section 12.1.3) in designing pressure vessels is an application of the fail-safe principle – the pressure vessel must be able to tolerate without fast fracture a through crack of length twice the thickness of the vessel.

Software fail-safe devices work by using internal programme tests and exception handling routines which set up safe conditions in case of errors. In another example, a control can be set up in 'safe' position and error indicated if an important component or a sensor has failed.

Using devices which permit operation in severely degraded conditions also belongs to this category. A rigid metal disc in a car tyre or a substance which automatically seals the tyre puncture allows the driver to maintain steering control and avoid accidents.

11.3.4 Deliberately Designed Weak Links

The consequences from failure can be decreased if potential failures are channelled into deliberately designed weak links. Should the unfavourable conditions occur, the weak links are the ones to fail and protect the expensive components or the system. Impact attenuators, including *crumpling zones* in road cars and *crash cones* in racing cars are typical examples of deliberately designed weak links which absorb the energy of impact and protect the driver and passengers. In this respect, the honeycomb sandwich panels with carbon fibre skins have found application in motor racing to minimise the consequences from an impact.

Shear pins are mechanical sacrificial parts, analogous to an electrical fuse. They are frequently used in couplings connecting expensive mechanical equipment with driving shafts. Should sudden overload occur, the shear pin shears at a specified torque to prevent damage to the driveshaft or other components.

Shear pins are also used in airplanes to release the engines and prevent fire upon emergency landing.

Rupture discs are another example of deliberate weak links. They are sacrificial membranes which protect equipment from over-pressurisation or from potentially damaging vacuum.

Wire rope fuse provides a warning that the tensile load on the rope has exceeded the safe load.

Blowout panels are intentionally wakened areas which fail in predictable manner and are used in situations where sudden overpressure can occur. The pressure wave is channelled through the weakened area, thereby protecting the rest of the structure and the neighbouring structures from catastrophic damage. An example of a blowout panel is the deliberately weakened wall in a room used to store compressed gas cylinders and the deliberately weakened roof of a bunker for ammunition storage.

Sacrificial galvanic anodes are another example of a deliberate weak link, designed to protect buried metal structures (e.g. pipes) or hulls of ships from corrosion. All electrical fuses are examples of deliberate weak links, whose failure protects expensive electrical equipment.

11.3.5 Built-In Protection

This method is often used for components and systems whose likelihood of failure is large or the consequences of failure are significant.

Typical examples are the crumple zones and airbags in cars, the security marking of valuable items, the safety devices transmitting signals for easy identification of the location of tourists in mountains, etc.

This principle is the reason behind the safety practice of building residential areas beyond the radius of harmful influence of toxic substances from chemical plants, compost production, fuel depots, etc.

Passive protective barriers physically separate the hazards (the energy sources) from targets. Physical barriers isolate and contain the consequences and prevent the escalation of accidents. They provide a passive protection against the spread of fire, radiation, toxic substances or dangerous operating conditions. A blast wall, for example, guards against a blast wave. Increasing the distance between sources of hazards and targets minimises the damage in case of an accident.

Examples of built-in protective barriers are the safeguards protecting workers from flying fragments caused by a disintegration of parts rotating at a high speed, the protective shields around nuclear reactors or containers with radioactive waste, the fireproof partitioning, the double hulls in tankers limiting oil spillage if the integrity of the outer hull is compromised, etc.

- *Using active protective barriers.*

  The consequences from an accident or failure can be mitigated significantly by activating built-in protective systems. Typical examples of active barriers designed to mitigate the consequences from accidents are:

  - The safety devices activating sprinklers for limiting the spread of fire, extraction fans limiting the concentration of toxic gases, surge barriers limiting the consequences from floods, automatic brakes in case of a critical failure, automatic circuit breakers in case of a short cut, etc.

11.3.6 Troubleshooting Procedures and Systems

Computer-based expert troubleshooting systems are a powerful tool for reducing the downtimes in case of failure. Expert systems capture and distribute human expertise related to solving common problems and the appropriate course of action in particular situations. Unlike people, these systems fully retain the knowledge about vast number of situations and problems and the appropriate operating procedures. Furthermore, the troubleshooting prescriptions are objective and not coloured by emotions. Troubleshooting systems can help in training the staff to handle various problems or accidents. They also help counteract the constant loss of expertise as specialists leave or retire (Sutton, 1992).

11.3.7 Simulation of the Consequences from Failure

Given that an accident/failure has occurred, for each identified set of initiating events, an assessment of the possible damage is made. In case of loss of containment, for example, depending on the *release rate* and the *dispersion rate*, the consequences can vary significantly. In case of a leak to the environment, the consequences are a function of the magnitude of the leak and the dispersion rate. For a leak with large magnitude and substantial dispersion rate, a large amount of toxic substance is released for a short period of time before the failure is isolated. Where possible, a distribution of the conditional losses (consequences given failure) should be produced. This distribution gives the likelihood that the consequences given failure will exceed a specified critical threshold.

Suppose that a leak caused by a dropped object penetrating a vessel containing fluid under pressure is initiated and the size of the hole made by the dropped object can vary anywhere within a certain range. The time to discover and repair the leak and the pressure inside the vessel are also random variables, characterised by particular distributions.

Using a simulation, the distributions of the size of the hole, the pressure inside the vessel and the time to repair the leak can be sampled and subsequently, for each combination of sampled parameters, the amount of released toxic substance can be calculated. Repeating this process a large number of times will produce a distribution of the amount of released substance, from which the probability that the released amount will be greater than a critical limit can be estimated easily.

The full spectrum of possible failure scenarios should be analysed. Event trees are often employed to map all possible failure scenarios.

In case of a release of toxic chemical or contaminant in a confined space, for example, depending on the volume released, the concentration in the environment will vary, and the consequences due to exposure will also vary. The number of people exposed to the toxic substance depends on the actual occupancy of the space which varies during the year. Suppose also that given that a person has been exposed to the toxic substance, the probability of developing a particular condition is a function of the concentration of the substance (the released volume) and the duration of the exposure. The distribution of the consequences from the release, in other words, the distribution of the number of fatalities, can then be determined by a simulation. This involves sampling from the distributions related to the released amount, the occupancy of the space and the percentage of people developing the condition. Repeating this calculation over a large number of simulation trials yields the distribution of the number of fatalities. In this way, the conditional losses can be determined as well as the associated uncertainty.

Usually, in case of a large number of different failure scenarios and complex interrelationships between them, Monte Carlo simulation software is needed to determine the distribution of the conditional losses. Consequence modelling tools help to evaluate the consequences from dispersion of toxic gases, smoke, fires, explosions, etc.

From the simulation study, key risk factors will be identified whose alteration reduces the consequences significantly. Subsequently, this knowledge is used for devising an efficient risk mitigation strategy.

11.3.8 Risk Planning and Training

The purpose of risk planning is to specify the most appropriate response should failure scenario occur. Risk planning guarantees that the optimal course of action will be taken for dealing with the consequences from failure. Usually, in the absence of planning, the quickest and the most obvious actions are taken which are rarely the optimal ones.

Risk planning prepares for the unexpected. It yields contingency plans for the course of action in the case of failure or accident. Planning guarantees proactive rather than reactive attitude to risk and provides more time to react. It is closely linked with the research preparation involving a careful study of the system or process, identifying possible sources of risk and training for providing emergency response given that the risks materialise. The time invested in risk planning and training pays off because the response time, the chances of taking the wrong course of action and the chances of providing inadequate response to the materialised risks are reduced significantly. Risk planning and training help avoid panic and hasty actions which could otherwise promote errors aggravating the consequences from failure.

Planning also provides an answer to the important question 'how much resource to allocate now, given the possibility of failure scenarios in the future, in order to minimise the total cost'. In this sense, quantifying the risks associated with the different scenarios and deciding on a mitigating strategy are at the heart of risk planning.

A good accidents response management based on well-established rules and training is a major factor mitigating the consequences from accidents. Evacuation procedures, fast rescue operations, fast response to extreme emergency and fast response to crime reduce significantly the consequences from accidents. Various types of emergency training help reduce casualties and losses should accidents occur. Adequate first-aid training, security training and crime combat training are important factors mitigating the consequences should risks materialise.

12

Physics of Failure Models

Following Dasgupta and Pecht (1991), the mechanisms of failure can be divided broadly into two categories:

1. Overstress failures: (i) brittle fracture, (ii) ductile fracture, (iii) yield, (iv) buckling, etc.
2. Wearout failures: (i) fatigue, (ii) corrosion, (iii) stress-corrosion cracking, (iv) wear, (v) creep, etc.

Overstress failures occur when load exceeds strength. If load is smaller than strength, the load has no permanent effect on the component. Conversely, wearout failures are characterised by a damage which accumulates irreversibly and does not disappear when the load is removed. Once the damage tolerance limit is reached, the component fails (Blischke and Murthy, 2000). One of the most important examples of the overstress and wearout failures is the fast fracture and fatigue failure, and they will be considered in greater detail.

12.1 Fast Fracture

12.1.1 Fast Fracture: Driving Forces behind Fast Fracture

Unlike ductile fracture, *fast fracture* occurs suddenly and proceeds at a high speed, and in order to progress, there is no need for the loading stress to increase (Anderson, 2005; Ewalds and Wanhill, 1984; Hertzberg, 1996). Fast fracture also requires a relatively small amount of accumulated strain energy. These features make fast fracture a dangerous failure mode and require a conservative approach to the design of safety-critical brittle components which are prone to fast fracture. The need for control of fast fracture increases:

1. As the magnitude of the loading increases
2. As factors of safety decrease because of more precise computer designs
3. As the use of lightweight designs increases

Reliability and Risk Models: Setting Reliability Requirements, Second Edition. Michael Todinov.

4. As the use of high-strength welded steels becomes more common compared to lower-strength riveted or bolted steels
5. As the need to make savings by prolonging the service life of components and structures increases

Designers often avoid yielding by using high-strength materials, but these materials also have a low fracture toughness and are prone to fast fracture.

Fast fracture is associated with initiation of an unstable crack, usually triggered by cracking or decohesion of particles. Some of these are inclusions, but in other cases, they may be an integral part of the microstructure. Under overstress, a high stress concentration can occur at these microflaws which results in a nucleation and propagation of cracks. Cleavage fracture is the most common type of fast fracture. It is associated with a relatively small plastic zone ahead of the crack tip and consequently, with a relatively small amount of energy needed for crack propagation. The very small amount of plastic deformation is indicated by the featureless, flat fracture surface. Due to the lack of plastic deformation, no blunting of the sharp cracks occurs, and the local stress ahead of the crack tip can reach very high values which are sufficient to break apart the interatomic bonds. As a result, the crack spreads between adjacent atomic planes yielding a flat cleavage surface.

The severity of the stress field around the crack tip is governed by the *stress intensity factor*. For a tensile mode of loading (mode I), of a surface crack, where the crack flanks are pulled directly apart, the stress intensity factor is (Anderson, 2005; Dowling, 1999)

$$K_{\mathrm{I}} = Y\sigma\sqrt{\pi a} \tag{12.1}$$

where σ is the remote stress applied to component (not to be confused with the local stresses) and Y is *the geometry factor (correction factor, shape factor, calibration factor)* which is a function of the crack size a, width of specimen, etc. Thus, for a through crack in an infinite plate (Figure 12.1a), $Y = 1$, while for an edge crack on an infinite plate (Figure 12.1b), $Y \approx 1.12$.

The stress intensity factor K relates several design variables: the nominal applied stress σ (calculated assuming that no crack is present), the crack size a (Figure 12.1a,b) and the geometry factor Y.

The critical value K_{Ic} of the stress intensity factor that would cause fast fracture is the *fracture toughness* of the material. Fracture toughness is a measure of the material's resistance to crack extension. It is a material property independent of the size and geometry of the cracked body and is determined from laboratory experiments. To provide conservatism in the calculations, the plane-strain fracture toughness K_{Ic} is commonly quoted in reference books.

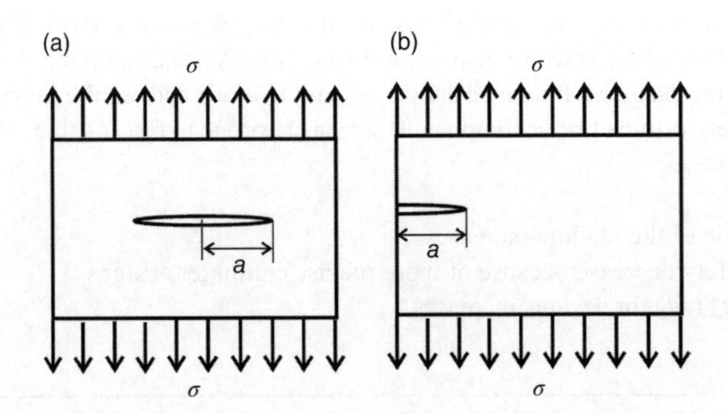

Figure 12.1 Definition of crack size a for (a) central cracks and (b) edge cracks

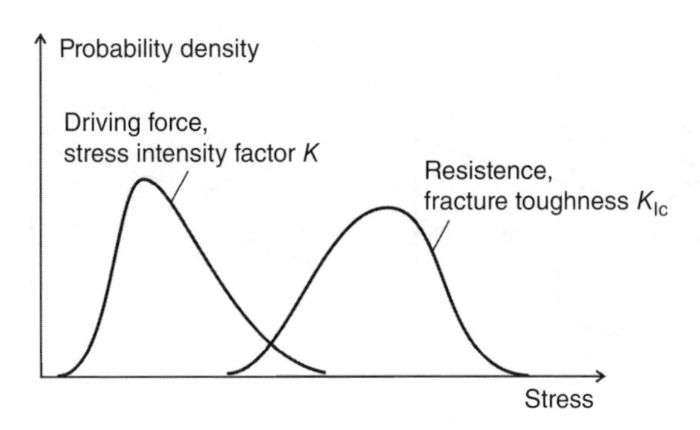

Figure 12.2 In order to avoid fast fracture, the driving force (the stress intensity factor) must be smaller than the resistance (the fracture toughness)

The fundamental design approach to preventing fast fracture in structural materials is to keep the stress intensity factor $K = Y\sigma\sqrt{\pi a}$ (the driving force) below the critical stress intensity factor K_{Ic} (the resistance):

$$Y\sigma\sqrt{\pi a} < K_{Ic} \tag{12.2}$$

For a specified fracture toughness K_{Ic} of the material and shape of the component, the domain preventing fast fracture is given by all combinations of loading stress and crack size for which inequality 12.2 is fulfilled.

Each of these factors can be changed in the designed component to improve the reliability of the design. For example, the welding process can be improved to reduce the number of cracks and their sizes, the component could be redesigned to reduce the stress intensification due to stress raisers, or the material could be selected to have a high fracture toughness.

In reality, both the stress intensity factor and the fracture toughness vary. An overlap between the upper tail of the distribution of the stress intensity factor and the lower tail of the distribution of the fracture toughness is associated with a non-zero probability of fast fracture (Figure 12.2).

A combination of a high-magnitude normal stress and a high-magnitude shear stress is often present for certain orientations of the crack and both, the stress intensity factor K_I characterising the *tensile crack opening mode* (Figure 12.3a) and the stress intensity factor K_{II} characterising the *sliding crack opening mode* (Figure 12.3b) make a significant contribution towards the fast fracture initiation.

In opening mode I or tensile opening mode, the crack faces are pulled apart (Figure 12.3a), while in opening mode II (sliding opening mode or in-plane shear mode), the crack faces slide over each other (Figure 12.3b).

However, the capability of existing design methods to detect vulnerability to brittle failure initiated by flaws is limited. The standard design approach is to place a sharp crack with size equal to the threshold detection limit of the non-destructive inspection technique, in the most dangerous position and in the most dangerous orientation (where the stress takes its maximum value) and to test by using a fracture criterion whether the crack will be unstable. This approach however, works only in cases of components with a simple shape and loading characterised by a one-dimensional or two-dimensional stress state. In these cases, it is relatively easy to identify the most dangerous orientation of the crack. In cases of components with complex shape, loaded in complex fashion, as a rule, the stress state is three-dimensional, and it is near to impossible for researchers and design–engineers to identify the most unfavourable orientation of the crack associated with the highest driving force for crack extension. The space of possible orientations of the crack with respect to a three-dimensional stress tensor

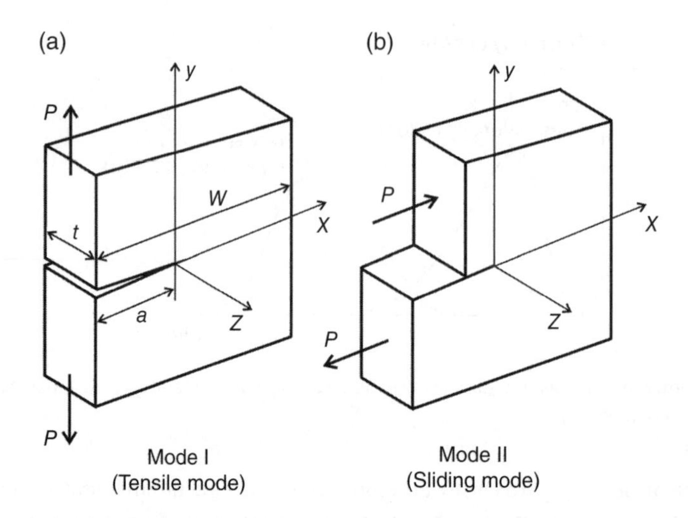

Figure 12.3 (a) The tensile opening mode and (b) the sliding opening mode

is very big, which makes it practically impossible to identify the most unfavourable orientation, even after a substantial number of empirical trials.

This predicament can be resolved with a conservative mixed-mode fast fracture criterion, incorporating the most unfavourable orientation of the crack. According to the adopted conservative mixed-mode criterion, fracture occurs if C in (12.3) exceeds unity for a certain crack orientation:

$$C = \frac{K_{\mathrm{I}}}{K_{\mathrm{Ic}}} + \frac{K_{\mathrm{II}}}{K_{\mathrm{IIc}}} \tag{12.3}$$

In (12.3), K_{Ic}, K_{IIc} is the fracture toughness of the material characterising mode I and mode II crack opening, correspondingly. K_{I} and K_{II} are the stress intensity factors characterising mode I and mode II crack opening. The criterion for fast fracture is given by the condition $\max C > 1$.

It has been shown in (Todinov and Same, 2014) that the critical combinations K_{I}^*, K_{II}^* of the stress intensity factors leading to fast fracture, exhibit a significant scatter. Because of this scatter, mixed-mode criteria of the type $(K_{\mathrm{I}}/K_{\mathrm{Ic}})^m + (K_{\mathrm{II}}/K_{\mathrm{IIc}})^n > 1$ (where $m > 1$, $n > 1$ are constants specific to the material) could lead to fracture occurring in the defined by the fracture criterion safe zone. The locus of stress intensity factors K_{I}^*, K_{II}^* determining fast fracture states is associated with large uncertainty and, consequently, the deterministic concept 'fracture criterion' specified by a particular function cannot guarantee the safety of the designed components. To guarantee a low risk of failure for safety-critical components, it is necessary to specify a conservative safety zone, away from the scatter associated with the locus of stress intensity factors K_{I}^*, K_{II}^* defining fracture states.

The fracture criterion (12.3) defines such a conservative safe zone, well separated from the scatter of the fast fracture states and is strongly recommended for safety-critical calculations.

The most unfavourable orientation of the crack maximises C in Equation 12.3. It can be proved rigorously (Todinov and Same, 2014) that the maximum of C in (12.3) is given by

$$\max C = \frac{\sigma_1 + \sigma_3}{2}\theta + \frac{\sigma_1 - \sigma_3}{2}\sqrt{\theta^2 + \gamma^2}$$

where σ_1, σ_2 and σ_3 are the principal stresses and θ and γ are material constants given by

$$\theta = \frac{Y_{\mathrm{I}}\sqrt{\pi a}}{K_{\mathrm{Ic}}} \tag{12.4}$$

and

$$\gamma = \frac{Y_{\mathrm{II}}\sqrt{\pi a}}{K_{\mathrm{IIc}}} \tag{12.5}$$

depending on the size a of the crack, the fracture toughness of the material characterising mode I and mode II crack opening (K_{Ic}, K_{IIc}) and the geometry factors (Y_{I}, Y_{II}) characterising mode I and mode II crack opening.

Fast fracture occurs if the inequality

$$m = \frac{\sigma_1 + \sigma_3}{2}\theta + \frac{\sigma_1 - \sigma_3}{2}\sqrt{\theta^2 + \gamma^2} \geq 1 \tag{12.6}$$

is fulfilled for certain tested location, characterised by a maximum and minimum principal stresses σ_1 and σ_3.

The analytical criterion (12.6) for crack initiation should be applied in a traditional deterministic design to determine for a flaw of given size, whether an unstable crack will be initiated. A criterion based on the maximum principal tensile stress may lead to a non-conservative design. Given the stress tensor components ($\sigma_x, \sigma_y, \sigma_z, \tau_{xy}, \tau_{yz}, \tau_{zx}$) at the tested location, the procedure for applying the mixed-mode fast fracture criterion involves the following steps.

The directions $\mathbf{x}$ of the principal stresses at the inspected location (the *eigenvectors*) are determined by solving the matrix equation

$$\mathbf{Sx} = \lambda \mathbf{x}$$

and determining its non-trivial solutions $\mathbf{x} \neq 0$, where

$$\mathbf{S} = \begin{pmatrix} \sigma_x & \tau_{xy} & \tau_{xz} \\ \tau_{yx} & \sigma_y & \tau_{yz} \\ \tau_{zx} & \tau_{zy} & \sigma_z \end{pmatrix}$$

is a 3×3 symmetric matrix known as *stress tensor*, and $\mathbf{x} = [x_1, x_2, x_3]^{\mathrm{T}}$ is a 3×1 column vector. The values of λ for which such non-trivial solutions of the matrix equation exist are called principal stresses (*eigenvalues*).

The matrix equation can also be presented as

$$(\mathbf{S} - \lambda \mathbf{I})\mathbf{x} = 0 \tag{12.7}$$

The matrix Equation 12.7 represents a system of homogeneous linear equations which has a non-trivial solution if and only if the determinant $|\mathbf{S} - \lambda \mathbf{I}|$ is equal to zero, which is equivalent to

$$\begin{vmatrix} \sigma_x - \lambda & \tau_{xy} & \tau_{xz} \\ \tau_{yx} & \sigma_y - \lambda & \tau_{yz} \\ \tau_{zx} & \tau_{zy} & \sigma_z - \lambda \end{vmatrix} = 0 \tag{12.8}$$

1. The expansion of the determinant 12.8 yields a polynomial of third degree with respect to λ. The roots $\lambda_1, \lambda_2, \lambda_3$ of the polynomial (for which Eq. 12.8 is fulfilled) are the eigenvalues of the stress tensor S describing the stress state of the location. These eigenvalues coincide with the magnitudes of the principal stresses $\sigma_1, \sigma_2, \sigma_3$ at the selected location. Because the stress tensor is symmetric, it is guaranteed that the eigenvalues (the principal stresses $\sigma_1, \sigma_2, \sigma_3$) will be real numbers.
   The easiest way to determine the eigenvalues is to use the corresponding eigenvalue function in a mathematical software package.
2. The second step involves arranging the obtained eigenvalues in descending order. The largest eigenvalue is the principal stress σ_1, ($\sigma_1 = \max\{\lambda_1, \lambda_2, \lambda_3\}$), the smallest eigenvalue is the principal stress σ_3 ($\sigma_3 = \min\{\lambda_1, \lambda_2, \lambda_3\}$) and the intermediate eigenvalue is the principal stress σ_2.
3. The third step involves evaluating (12.6). A sharp crack is assumed, with size equal to the threshold value of the non-destructive detection technique. If inequality (12.6) is fulfilled, the crack is unstable and will initiate failure.

Example

The stresses in MPa at a particular location in a loaded component are $\sigma_x = -145$, $\sigma_y = 258$, $\sigma_z = 330$, $\tau_{xy} = 225$, $\tau_{xz} = 218$, and $\tau_{yz} = 93$. The ultrasonic inspection technique used cannot detect a crack smaller than 2 mm. If the fracture toughness of the material is $K_{\mathrm{Ic}} = K_{\mathrm{IIc}} = 50\ \mathrm{MPa}\sqrt{m}$, determine whether an undetected crack at this particular location will cause fast fracture. Assume $Y_{\mathrm{I}} = Y_{\mathrm{II}} = 1$.

Solution

The principal stresses are determined as the eigenvalues of the stress tensor

$$\begin{bmatrix} -145 & 225 & 218 \\ 225 & 258 & 93 \\ 218 & 93 & 330 \end{bmatrix}$$

This stress tensor has eigenvalues (–290.7, 198.9, 534.7). The principal stresses are

$$\sigma_1 = 534.7\ \mathrm{MPa},\ \sigma_2 = 198.9\ \mathrm{MPa}\ \text{and}\ \sigma_3 = -290.7\ \mathrm{MPa}$$

The constants θ and γ are calculated next:

$$\theta = \frac{Y_{\mathrm{I}}\sqrt{\pi a}}{K_{\mathrm{Ic}}} = \frac{\sqrt{\pi \times 0.002}}{50} = 0.0016\ \mathrm{MPa}^{-1}$$

$$\gamma = \frac{Y_{\mathrm{II}}\sqrt{\pi a}}{K_{\mathrm{IIc}}} = \frac{\sqrt{\pi \times 0.002}}{50} = 0.0016\ \mathrm{MPa}^{-1}$$

Substituting in the mixed-mode criterion

$$m = \frac{\sigma_1 + \sigma_3}{2}\theta + \frac{\sigma_1 - \sigma_3}{2}\sqrt{\theta^2 + \gamma^2}$$

for unstable crack gives

$$m = \frac{534.7 - 290.7}{2} \times 0.0016 + \frac{534.7 + 290.7}{2} \times 0.0023 = 1.14 > 1$$

Consequently, the crack will be unstable and there will be fast fracture.

If the single-mode (tensile mode) criterion 12.1 was used to test for fast fracture, the obtained stress intensity factor

$$K_{\mathrm{I}} = Y\sigma\sqrt{\pi a} = 534.7\sqrt{\pi \times 0.002} = 42.4\ \text{MPa}\sqrt{m}$$

is smaller than the fracture toughness $K_{\mathrm{Ic}} = 50\ \text{MPa}\sqrt{m}$ of the material which leads to the incorrect conclusion that there would be no fast fracture. *The mixed-mode criterion provides the basis for conservative calculations regarding the occurrence of fast fracture, which makes the design safer.*

The fracture criterion (12.6) for a mixed-mode fast facture can be applied to check the safety of loaded brittle components with complex shape, where fracture is locally initiated by flaws. Ceramics, high-strength steels, glasses, stone, etc. are examples of materials with such failure mechanism. The described model is also valid for components from low-carbon steels undergoing cleavage fracture at low temperature. Cleavage in steels usually propagates from cracked inclusions (McMahon and Cohen, 1965; Rosenfield, 1997). It usually involves a small amount of local plastic deformation to produce dislocation pile-ups and crack initiation from a particle which has cracked during the plastic deformation.

A postprocessor based on the fracture criterion (12.6) can be easily developed for testing loaded safety-critical components with complex shape. For each finite element, only a single computation of the fracture criterion is made. This guarantees a very high computational speed, which makes the postprocessor particularly suitable for testing numerous design variants in an optimisation loop.

Here are the steps of the algorithm:

Algorithm 12.1

1. ***Determine the stress state in the loaded component by FE software.***
2. ***By using a post-processor, extract from the FE solution the maximum and minimum principal stresses characterising each finite element.***
3. `Failure_flag = 0; //Initialises a failure flag`
4. **For** (each finite element of the stressed component) **do**

```
{
    Extract the largest (σ1) and the smallest (σ3) principal
    stress characterising the current finite element;

    m = (σ1 + σ3)/2 θ + (σ1 − σ3)/2 √(θ² + γ²) ≥ 1; // Calculates the fast
                                                    fracture mixed-mode
                                                    criterion
    If (m > 1) then {Failure_flag = 1; break;}
}
If (Failure_flag = 1) then Print ("Unsafe design")
                       else Print ("Safe design").
```

$$m = \frac{\sigma_1 + \sigma_3}{2}\theta + \frac{\sigma_1 - \sigma_3}{2}\sqrt{\theta^2 + \gamma^2} \geq 1$$

The proposed fracture condition is particularly suitable for optimising the shape of brittle components (e.g. the cross section of ceramic beams) in order to increase their resistance to failure locally initiated by flaws.

12.1.2 Reducing the Likelihood of Fast Fracture

Fast fracture is an event associated with a negative impact, characterised by a likelihood and consequences. The risk of fast fracture can be reduced by reducing the likelihood of fast fracture.

One of the big advantages of physics of failure models is that they suggest ways of reducing the likelihood of a particular failure mode. In the case of fast fracture, the physics of failure model (12.2) suggests several ways of reducing the likelihood of fast fracture by modifying the fracture toughness of the material, crack length, loading stress and geometry factor.

12.1.2.1 Basic Ways of Reducing the Likelihood of Fast Fracture

The physics-of-failure relationship (12.2) reveals the roles of the design stress σ and the geometry factor Y, the material processing characterised by the size a of the flaw and the material properties characterised by the fracture toughness K_{Ic}. From this relationship, several generic ways of reducing the likelihood of fast fracture can be inferred immediately:

- *Increasing the fracture toughness of the material* K_{Ic}
- *Reducing the maximum size of the flaws* a
- *Reducing the design stress* σ
- *Reducing the value of the geometry factor* Y
- *A combination of several of the listed basic ways*

Fracture toughness of the material can, for example, be increased by:

- Control over the microstructure (e.g. grain size control)
- Heat treatment (e.g. quenching, tempering, ageing, recrystallisation)
- Reinforcing with particles and fibres
- Alloying
- Eliminating anisotropy
- Eliminating harmful inclusions
- Transformation toughening

The size of the flaws can be controlled by design changes, fabrication and inspection. A crack with size a becomes unstable at the critical applied stress σ_c, for which the stress intensity factor $K_I = Y\sigma\sqrt{\pi a}$ in inequality 12.2 becomes equal to or exceeds the fracture toughness K_{Ic}. Relationship (12.2) illustrates the available design trade-offs.

Given that a surface crack with maximum size a_{max} is present in a particular material, in order to prevent fast fracture, the design stress σ_d must be smaller than the critical stress $\sigma_d < \sigma_c = K_{Ic} / \left[Y\sqrt{\pi a_{max}} \right]$.

This is indeed the case where high strength and light weight are required, and K_{Ic} is usually fixed because of limited number of available materials combining high strength and light weight. The applied stress can be altered by design changes, through the loading scheme or by using stress limiters.

If the loading stress must be kept to a minimum level σ_{min}, because of the need to fully utilise the material, for the maximum flaw size in the material a_{max}, the relationship

$$a_{max} < a_c = \frac{1}{\pi}\left(\frac{K_{Ic}}{Y\sigma_{min}}\right)^2$$

must be fulfilled. In the above expression, a_c is the critical flaw size which becomes unstable at a stress magnitude σ_{min} and causes fast fracture. A material, manufacturing method and heat treatment must be selected in such a way that the maximum size of the flaw a_{max} is smaller than the critical flaw size a_c.

Here, one must be careful that the critical flaw size a_c is large enough to be detected by the non-destructive inspection. Otherwise, fast failure may occur suddenly.

Thus, for an initial crack with initial size a_0 subjected to loading by a stress with magnitude σ, the relationship $Y\sigma\sqrt{\pi a_0} < K_{Ic}$ may be fulfilled. In this case, the crack is stable and will not cause fast fracture. Due to cycling of the loading stress σ however or due to a stress corrosion at a constant stress σ, the crack may grow and reach critical size $a_c > a_0$, at which fast fracture is triggered because $Y\sigma\sqrt{\pi a_c} = K_{Ic}$.

To prevent fast fracture, the inspection technique must be capable of detecting a crack of size a_c. Despite that the combination of a nominal stress σ and initial crack size a_0 lies in the safe domain preventing both plastic collapse and fast fracture, the crack may propagate by a fatigue growth or through a stress-corrosion mechanism, leave the safe domain and cause fast fracture.

The design parameters must be selected in such a way that a crack size that causes fast fracture is detectable by the inspection technique.

In cases where different materials can be selected, selecting a material with higher fracture toughness K_{Ic} will increase the critical stress σ_c of triggering fast fracture and will increase the fast fracture resistance. The geometry factor can be reduced by modifying the shape of the component.

A heat treatment or replacement of an existing low-strength steel grade with a high-strength grade is generally accompanied with a reduction of the fracture toughness (Dowling, 1999).

As can be verified from the equation, $a_c = (1/\pi)[K_{Ic}/(Y\sigma_{min})]^2$, with reducing the fracture toughness K_{Ic}, the critical size of the crack a_c may decrease below the threshold limit of the available non-destrictive testing (NDT) technique which makes the high-strength steel unsafe to use. This point can be illustrated by an example.

Example

To reduce weight, a reduction of the load-carrying cross section of a loaded in tension steel bar is proposed. To withstand the minimum design stress of $\sigma_{min} = 840$ MPa, it is proposed that the existing steel grade with yield strength of 1400 MPa and fracture toughness of 70 MPa$\sqrt{m}$ should be heat treated to a strength 2000 MPa and fracture toughness of 35 MPa$\sqrt{m}$. The detection threshold of the existing NDT technique is a 1 mm long crack. Should this design modification be supported?

Solution

The worst-case scenario of an edge crack with geometry factor $Y = 1.12$ is assumed, in order to assure a conservative estimate of the crack size which causes fast fracture.

The critical flaw size for the two steels is determined from $a_c = (1/\pi)(K_{Ic}/(Y\sigma_{min}))^2$:

$$a_{1,c} = \frac{1}{\pi}\left(\frac{K_{Ic}^{(1)}}{Y\sigma_{min}}\right)^2 = \frac{1}{\pi}\left(\frac{70\times 10^6}{1.12\times 840\times 10^6}\right)^2 \approx 1.8\times 10^{-3}\,\text{m}$$

$$a_{2,c} = \frac{1}{\pi}\left(\frac{K_{Ic}^{(2)}}{Y\sigma_{min}}\right)^2 = \frac{1}{\pi}\left(\frac{35\times 10^6}{1.12\times 840\times 10^6}\right)^2 \approx 0.4\times 10^{-3}\,\text{m}$$

The critical flaw size for the steel with lower yield strength is $a_{1,c} = 1.8$ mm which will be detected by the existing NDT; therefore, this steel is safe to use. The critical crack size for the high-strength steel however is $a_{2,c} = 0.4$ mm. A critical crack of this size cannot be detected by the existing NDT and therefore, the high-strength steel is unsafe to use. The proposed design modification must not be supported.

12.1.2.2 Avoidance of Stress Raisers or Mitigating Their Harmful Effect

Zones in components, characterised by large stress gradients within a small volume, are referred to as *stress concentration zones*. Features such as notches, fillets, holes, threads, steps, grooves, keyways, rough surface finishes, quenching cracks and inclusions, causing stress concentration zones, are referred to as *stress raisers*. Notches, for example, are associated with stress intensification, triaxial tensile stress and geometrical constraint, and as a result, they increase the tendency towards fast fracture.

The stress raisers are reliability-critical design features because they make it easy for a crack to initiate and become unstable.

The stress intensification around a fillet with a small radius and around a hole is shown in Figures 12.4a and 12.5a.

The maximum tensile stress σ_{max} associated with a stress raiser is determined from the equation

$$\sigma_{max} = K_t \times \sigma_{nom}$$

in terms of the stress concentration factor K_t and the nominal tensile stress σ_{nom} (away from the stress raiser).

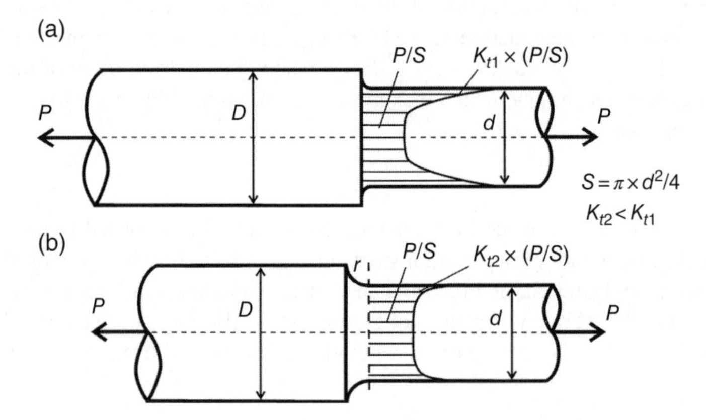

Figure 12.4 (a) Stress intensification around a step with a very small radius. (b) Reducing the stress concentration by introducing a fillet with radius r

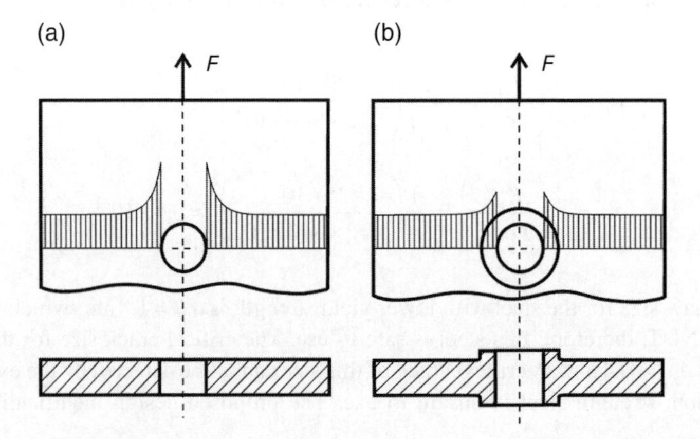

Figure 12.5 (a) Stress intensification around a hole. (b) Reducing the maximum stress in the vicinity of the hole could be achieved by appropriate design (Orlov, 1988)

A common numerical method for determining the stress concentration factors is the *finite element method* (FEM). Obtaining precise results however requires the use of a very fine mesh in the stress intensification zone. Common experimental methods for detecting the stress concentration factors are precision strain gauges and photoelasticity. The results are often presented in stress concentration charts (e.g. Peterson, 1974).

As a result of the stress intensification due to the presence of a stress raiser, the stress σ in inequality 12.2 is increased and becomes $\sigma = K_t \times \sigma_{\text{nom}}$. The result is increased stress intensity factor $K = YK_t\sigma_{\text{nom}}\sqrt{\pi a}$ and increased driving force behind the crack extension. Reducing the risk of fast fracture therefore can be done by avoiding stress raisers ($K_t = 1$) or by reducing the stress concentration factors through appropriate design. Such is, for example, the design of fillets with increased radii (Figure 12.4b) or other design measures (Figure 12.5b).

Other ways of reducing the stress concentration factors and the likelihood of fast fracture are:

- Reducing the stress concentration by using stress-relief notches and stress-relief holes (Budynas, 1999)
- Designing simple shapes
- Specifying clean material without microstructural defects acting as stress raisers (e.g. inclusions, pores, seams, etc.)
- Avoiding shapes and loading resulting in a triaxial tensile stress state
- Avoiding holes too close to walls
- Smoothing edges and avoiding sharp corners. Smoothing sharp corners at the bottom of deep grooves
- Avoiding sharp transitions between thin and thick sections
- Avoiding excessive surface roughness and surface discontinuities

12.1.2.3 Selecting Materials Which Fail in a Ductile Fashion

Ductile fracture is accompanied with a considerable amount of plastic deformation. The crack will not normally extend unless an increased stress is applied. Ductile fracture is associated with a substantial amount of absorbed energy. This is indicated by the large area A beneath the *load–displacement curve* which is numerically equal to the work done to break the component. Indeed, the elementary work dA to extend the component from l to $l + dl$ is given by $dA = Fdl$, where F is the magnitude of the loading force. The total work to fail a component with initial length l_0 to a length l_f at failure is then given by $\int_{l_0}^{l_f} Fdl = A$, where A is the area beneath the load–displacement curve (Figure 12.6a).

Brittle failure is accompanied with little or no plastic deformation. *Once initiated, the crack extends at a high speed without a need for increasing the loading stress. The component fails quickly, without warning.* Brittle fracture is associated with a small amount of energy to break the component which is indicated by the small area A beneath the load–displacement curve (Figure 12.6b). In contrast, ductile fracture is preceded by a large amount of plastic deformation (Figure 12.6a), redistribution and relaxation of the high stresses in the zones of stress concentration. The component deforms before fracture which gives early warning and sufficient time for intervention and repair. In addition, ductile fracture requires more strain energy in order to develop, and in the process of plastic deformation, the strength of the material is enhanced through strain hardening (Figure 12.6a).

Consequently, *in engineering applications, where safety concerns are involved, the materials with ductile behaviour are the obvious choice.*

There are various other ways of reducing the likelihood of brittle fracture.

Avoidance of low-temperature environment. High temperature promotes a ductile fracture while low temperature promotes brittle fracture. This trend is particularly pronounced for metals with body-centred cubic crystal lattice (e.g. steels).

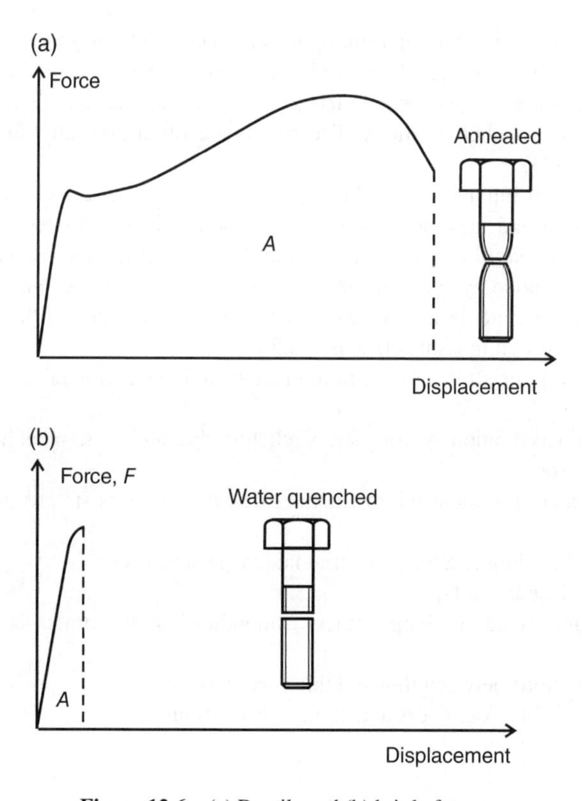

Figure 12.6 (a) Ductile and (b) brittle fracture

Under impact loading, the impact toughness of low-carbon steels (measured by the energy absorbed to break the specimen) is reduced markedly with reducing temperatures (see Figure 12.7). The result is a well-defined ductile-to-brittle transition region. In the upper shelf region, the microscopic fracture mechanism is void nucleation, growth and coalescence, associated with a large amount of absorbed impact energy. With decreasing test temperature, zones fracturing by a distinct cleavage mechanism appear. The latter is associated with brittle fracture which absorbs a relatively small amount of impact energy.

The effect of temperature is assessed by using the Charpy or Izod test, consisting of a weight that moves with a pendulum action. The weight is lifted to a starting position then released so that it hits the specimen with a known energy. The weight continues to move after it has broken the specimen, and the height that the weight attains is a measure of how much energy the specimen has absorbed in fracture (Dowling, 1999).

Avoidance of unfavourable stress states. As a rule, plane stress promotes ductile behaviour while plane strain promotes brittle behaviour and fast fracture. Triaxial tension promotes fast fracture while hydrostatic compression promotes ductile behaviour. Normally, brittle materials may exhibit considerable ductility if the hydrostatic component of the stress tensor is compressive, with a large magnitude. This behaviour is used to process brittle materials (e.g. limestone) in pressurised chambers. *Fracture and yielding are relative* and depend strongly on the stress state.

Geometrical features (e.g. notches) and loading resulting in triaxial tensile stresses promote brittle behaviour and should be avoided.

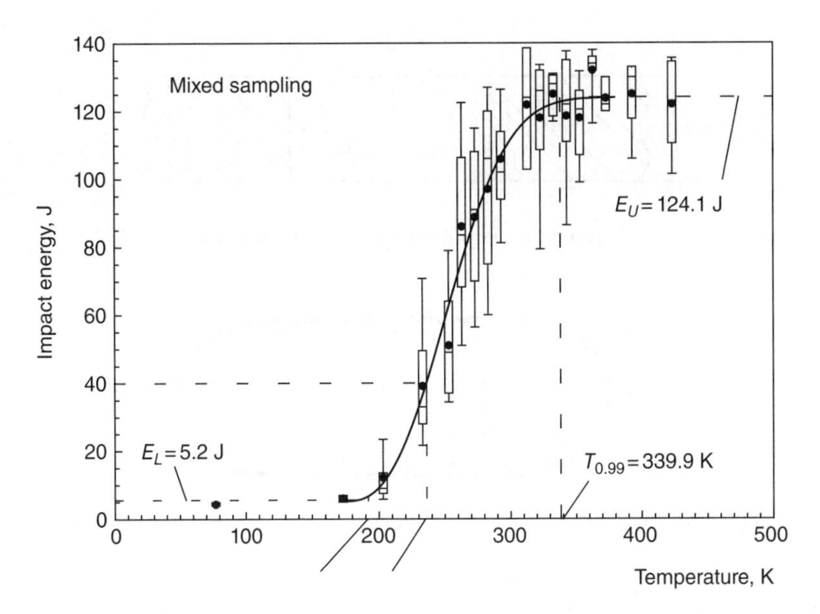

Figure 12.7 Systematic and random component of the Charpy impact energy in the ductile-to-brittle transition region (Todinov *et al.*, 2000)

Avoidance of a high rate of loading. Brittle behaviour and fast fracture are likely when the plastic flow is restricted. High strain rates and impact loads promote brittle fracture. Low rates of loading allow time for the shear to occur and promote ductile behaviour.

Avoidance of a corrosive environment. Corrosive environments promote brittle fracture, particularly intergranular fast fracture. For example, the material of a tank holding ammonia often exhibits intergranular fracture.

Finally, it needs to be pointed out that materials with a low yield strength tend to exhibit ductile behaviour as opposed to materials with a high yield strength which are more prone to fast fracture.

12.1.3 Reducing the Consequences of Fast Fracture

12.1.3.1 By Using Fail-Safe Designs

An important way of reducing the risks associated with fast fracture is the implementation of *fail-safe design*. Fail-safe designs reduce the consequences, given that fast fracture has occurred. Such is the fail-safe design based on the 'leak-before-break' concept. Pressure vessels often carry harmful substances at a high pressure. If cracks go undetected, the associated stress intensity factor may reach the fracture toughness of the material and initiate fast fracture that would cause the pressure vessel to explode. One way of avoiding this is to implement the design concept *leak-before-break* (Figure 12.8)

Leak occurs when the critical crack depth a_c is equal to the thickness of the pressure vessel t. For a leak-before-break design, the pressure vessel must be able to tolerate without fast fracture a through crack of length twice the thickness of the vessel.

Note that the requirement for tolerating a through crack (Figure 12.8) is essential because it provides conservatism in the design calculations.

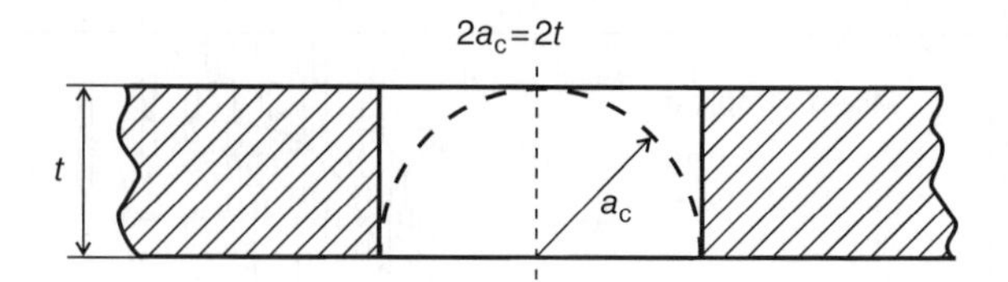

Figure 12.8 Leak-before-break design concept

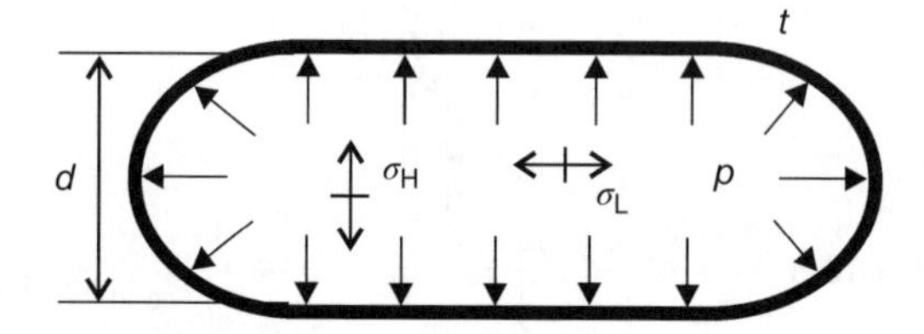

Figure 12.9 Hoop tensile stress σ_H and axial tensile stress σ_L acting in the shell of a vessel under pressure with magnitude p

If there is a through crack and a leak of the substance under pressure, this will be detected by the pressure reduction, the smell, noise, etc. The leak is therefore detectable and repair of the pressure vessel can be initiated without any catastrophic consequences. As a result, the designer should select material, wall thickness, vessel diameter and pressure that ensure that a through crack of size twice the thickness of the pressure vessel will be stable (non-propagating).

Consider the thin-walled pressure vessel in the Figure 12.9 with internal diameter d and thickness of the shell t. The vessel contains fluid exerting pressure p on the inside of the shell.

The largest tensile stress acting in the pressure vessel wall is the hoop stress determined from

$$\sigma_H = \frac{pd}{2t} \tag{12.9}$$

This expression can be easily derived from the equilibrium of the elementary forces along the y-axis in Figure 12.10a. A slice with length L has been taken from the cylinder (Figure 12.10a). The z-axis is perpendicular to the x and y axes.

The pressure is always perpendicular to the wall of the cylinder. The elementary force created by the pressure p on an elementary surface area ds on the inside wall of the cylinder is equal to pds. The component of this elementary force along the y-axis is $(pds) \times \cos\alpha$ where α is the angle which the normal to the elementary surface element subtends with the y-axis (Figure 12.10a). Note that the projection of the elementary force can be written as a product of the pressure p and the projection of the elementary area $ds \times \cos\alpha$ on the x, z plane. The sum of the components of all elementary forces along the y-axis, due to the internal pressure p, is therefore given by the product of the pressure and the total projected area $L \times d$ of the inner surface of the cylindrical slice on the x, z plane, where $d = 2r$. This resultant force must be counterbalanced by the sum of the two forces $tL\sigma_H$ created by the hoop stress σ_H acting in the cylinder wall (Figure 12.10a). From the equilibrium equation

$$2tL\sigma_H = pLd$$

Equation 12.9 is obtained immediately. It is important to emphasise that the hoop stress σ_H is larger than the axial stress σ_L acting in the cylinder wall, along the length of the cylinder (along the z-axis) (Figure 12.9) and perpendicular to the hoop stress σ_H. Indeed, using a very similar reasoning, it follows

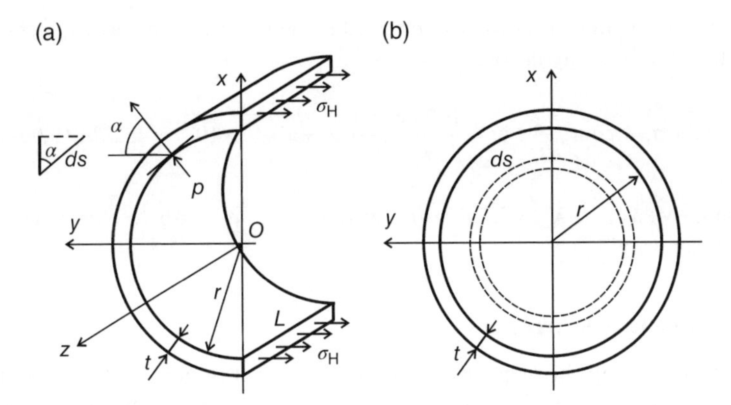

Figure 12.10 Pressure with magnitude p acting on: (a) the cylindrical shell of the pressure vessel and (b) the semi-spherical cap of the vessel

that the sum of the projections along the z-axis of all elementary forces created by the pressure p acting perpendicular to the inside surface of the semi-spherical cap (Figure 12.10b) is equal to the product of the pressure p and the projected area $\pi d^2/4$ of all elementary surface elements ds on the x, y plane ($d = 2r$). This resultant force must be counterbalanced by the sum of the forces $\pi dt\sigma_L$ created by the axial stress σ_L acting perpendicular to the area of the cross-sectional ring (t is the thickness of the spherical cap). From the equilibrium equation

$$p \times \frac{\pi d^2}{4} = \pi dt\sigma_L$$

the expression

$$\sigma_L = \frac{pd}{4t}$$

is obtained.

The magnitude of the axial stress σ_L is half the magnitude of the hoop stress σ_H from Equation 12.9.

For a through crack, the geometry factor is $Y = 1$ and the stress intensity condition for the leak-before-break design becomes

$$\frac{pd}{2t}\sqrt{\pi t} < K_{Ic} \tag{12.10}$$

The leak-before-break design concept will be illustrated by the next example.

Example

An aluminium alloy with fracture toughness of $K_{Ic} = 35\,\text{MPa}\sqrt{m}$ and yield stress 480 MPa has been used for a hydraulic actuator whose cylindrical housing has 60 mm internal diameter and a wall thickness of 12 mm. If the internal pressure is 100 MPa, will there be a leak-before break event?

Solution

The hoop stress in the wall of the hydraulic actuator is given by $\sigma_H = (pd)/(2t)$, where p is the pressure magnitude and d is the internal diameter of the actuator. The hoop stress has a magnitude $\sigma_H = (pd)/(2t) = (100 \times 10^6 \times 60 \times 10^{-3})/(2 \times 12 \times 10^{-3}) = 250$ MPa, which is well below the yield stress of the material.

For a through crack with half-length $a = t$ oriented in the worst possible way (perpendicular to the hoop stress), the stress intensity factor is $K_{\mathrm{I}} = \sigma_{\mathrm{H}}\sqrt{\pi a}$:

$$K_{\mathrm{I}} = \sigma_{\mathrm{H}}\sqrt{\pi a} = \frac{pd}{2t}\sqrt{\pi t} = \frac{100 \times 60 \times 10^{-3}}{2 \times 12 \times 10^{-3}}\sqrt{\pi \times 12 \times 10^{-3}} \approx 48.5\ \mathrm{MPa}\sqrt{m}$$

Since $K_{\mathrm{I}} = 48.5\ \mathrm{MPa}\sqrt{m} > K_{\mathrm{Ic}} = 35\ \mathrm{MPa}\sqrt{m}$, there will be no leak-before-break event. The actuator will fail without warning.

12.1.3.2 By Using Crack Arrestors

Crack arrestors can be strips or rings made of tougher material (Figure 12.11a). The mechanism of crack arrest consists of reducing the strain energy flow to the crack tip upon encountering a tougher material strip. Crack arrestors do not prevent cracks from becoming unstable, they only stop them after they have reached a certain size. Crack arrestors therefore are a protective measure, not a preventive measure. They limit the extent of damage once the damage has started to escalate.

In Figure 12.11b, the crack is arrested at the edge of the pipeline section.

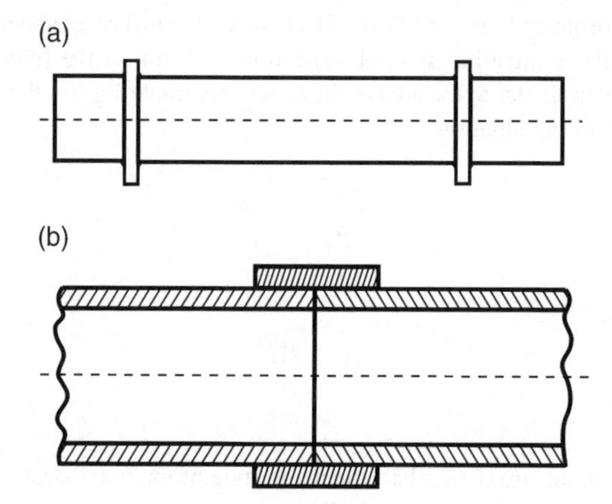

Figure 12.11 A pipeline with crack arresters of types (a) stiffened welded rings and (b) edges of the pipeline sections

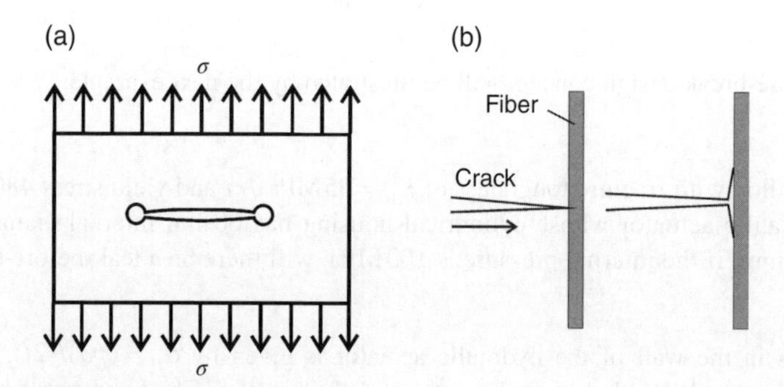

Figure 12.12 Arresting the propagation of an existing crack (a) by drilling holes at the crack tips and (b) by reinforcing with fibres acting as crack deflectors

Arresting the propagation of an existing crack can also be achieved by drilling holes at the crack tips (Figure 12.12a). This reduces significantly the stress magnitude at the crack tip.

The relatively low toughness of materials like resins can be increased by reinforcing them with carbon or glass fibres. In this case, the fibres act as crack deflectors (Figure 12.12b). This explains why the presence of brittle fibres (graphite, glass) into an equally brittle matrix, increases the fracture toughness of the composite.

12.2 Fatigue Fracture

Fatigue failures are wearout failures associated with components experiencing cyclic stresses or strains which produce permanent damage. Such is, for example, the loading of a railcar axel which is essentially a round beam subjected to bending. As the axle rotates, the stresses at the surface of the axle alternate between tensile and compressive. The damage accumulates until it develops into a crack which propagates and causes fracture (fast fracture or ductile fracture).

A comprehensive discussion regarding the different factors affecting the fatigue life of components can be found in Hertzberg (1996).

Usually, the fatigue life of machine components is a sum of a fatigue crack initiation life and life for fatigue crack propagation. *Fatigue life is associated with big scatter.* It has been reported (Ting and Lawrence, 1993) that in some alloys, for example, cast aluminium alloys, the dominant fatigue cracks (the cracks which caused fatigue failure) initiated from near-surface casting pores in polished specimens or from cast-surface texture discontinuities in as-cast specimens. For fatigue cracks emanating from casting pores, the nucleation life was almost non-existent (Ting and Lawrence, 1993).

A popular model related to the physics of fatigue failure initiated on a pre-existing crack leads to the Paris equation about the rate of crack propagation (Paris and Erdogan, 1963; Paris *et al.*, 1961):

$$\frac{da}{dN} = C\left(Y \Delta\sigma \sqrt{\pi a}\right)^m$$

where da/dN is the growth rate of the fatigue crack; $\Delta\sigma = \sigma_{max} - \sigma_{min}$ is the stress range of the loading cycle; C, m are constants depending on the material microstructure, environment, test temperature and load ratio $R = \sigma_{min}/\sigma_{max}$; Y is a geometry factor; a is the crack size and N is the number of cycles.

Consequently, the fatigue life is strongly impacted by the rate of fatigue crack growth. The crack growth rate da/dN is commonly estimated from the Paris power law (Paris and Erdogan, 1963; Paris *et al.*, 1961):

$$\frac{da}{dN} = C\Delta K^m \tag{12.11}$$

where a is the crack length; N is the number of load cycles and $\Delta K = K_{max} - K_{min}$ is the range of the stress intensity factor. $K_{max} = \sigma_{max} Y\sqrt{\pi a}$ and $K_{min} = \sigma_{min} Y\sqrt{\pi a}$ are the stress intensity factors which correspond to the maximum (σ_{max}) and minimum (σ_{min}) value of the uniform tensile stress perpendicular to the crack plane. Consequently, $\Delta K = \Delta\sigma Y\sqrt{\pi a}$, where $\Delta\sigma = \sigma_{max} - \sigma_{min}$ is the stress range. In some cases, $\sigma_{max} > 0$ and $\sigma_{min} < 0$. Because fatigue cracks do not propagate if the loading stress is compressive, in this case, in calculating the stress range, the minimum stress should be set to zero ($\sigma_{min} = 0$) and the stress range becomes $\Delta\sigma = \sigma_{max}$. Y is a dimensionless factor that depends on the geometry of the loaded crack.

Although the Paris law is empirical, it remains one of the most useful expressions for making conservative estimates of the life of components and structures. This equation is valid beyond a specific threshold value ΔK_{th} of the stress intensity factor which defines a fatigue crack which is capable of propagating in the material. For relatively short propagating cracks, it can be assumed that the geometry factor Y is independent of the crack length a and the differential equation

$$\frac{da}{dN} = C \times \left(Y \Delta\sigma \sqrt{\pi a}\right)^m$$

can be solved by separation of variables, where $\Delta\sigma = \sigma_{max} - \sigma_{min}$ is the range of the uniform tensile stress perpendicular to the crack plane.

Rearranging the Paris law results in

$$C\left(Y \Delta\sigma \sqrt{\pi}\right)^m dN = \frac{da}{a^{m/2}}$$

Integrating the above equation between the initial crack size a_0 and the final crack size a_f

$$C\left(Y \Delta\sigma \sqrt{\pi}\right)^m \int_0^{N_f} dN = \int_{a_0}^{a_f} \frac{da}{a^{m/2}} \tag{12.12}$$

yields the fatigue life in number of cycles N_f until failure

$$N_f = \frac{2}{(m-2)C\left(Y \Delta\sigma \sqrt{\pi}\right)^m}\left(\frac{1}{a_0^{(m-2)/2}} - \frac{1}{a_f^{(m-2)/2}}\right), \quad \text{for } m \neq 2 \tag{12.13}$$

and

$$N_f = \frac{1}{C\left(Y \Delta\sigma \sqrt{\pi}\right)^2} \ln\frac{a_f}{a_0}, \quad \text{for } m = 2 \tag{12.14}$$

If the shape factor Y is a function of the crack size a, the integration can be performed numerically. In the usual case where $m \neq 2$, the equation related to the number of load cycles to failure N_f is

$$N_f = A \times \left(\frac{1}{a_0^{(m-2)/2}} - \frac{1}{a_f^{(m-2)/2}}\right)$$

where A is a constant (see Equation 12.13).

Similar to the case of fast fracture, the analysis of the physics of failure model (12.13) reveals the effect of the parameters controlling the fatigue life and provides an insight into how to improve the fatigue life.

The sensitivity of the fatigue life to the initial or final crack size a_0 can be determined by differentiating expression (12.13) with respect to a_0. Indeed, $|dN_f/da_0|_{a_f=\text{const}} = A((m-2)/2)\left(1/a_0^{m/2}\right)$ and $|dN_f/da_f|_{a_0=\text{const}} = A((m-2)/2)\left(1/a_f^{m/2}\right)$. Since for $a_0 \ll a_f$, $\left(1/a_0^{m/2}\right) \gg \left(1/a_f^{m/2}\right)$, $|dN_f/da_0|_{a_f=\text{const}} \gg |dN_f/da_f|_{a_0=\text{const}}$.

In words, the fatigue life is much more sensitive to the initial crack size a_0 compared to the final crack size a_f. Most of the loading cycles are expended on the early stages of crack extension when the crack length is small. During the late stages of fatigue crack propagation, a relatively small number of cycles is sufficient to extend the crack to its final size triggering fracture.

As a result, fatigue life predictions depend strongly on the correct estimation of the *initial size* of the flaws in the material. Small relative errors in the estimated initial flaw size a_0 could lead to very large errors in the estimated fatigue life.

Suppose that an edge crack is loaded in tensile opening mode (mode I) by a pulsating force P as shown in Figure 12.3a. The final crack size at which fast fracture is initiated can be determined from the final crack size triggering fast fracture or from the final crack size triggering ductile fracture, whichever is smaller.

The final crack size triggering fast fracture is determined from the condition of fast fracture

$$K_{\mathrm{I}} = Y\sigma_{\max}\sqrt{\pi a_{\mathrm{fc}}} = K_{\mathrm{Ic}}$$

from which

$$a_{\mathrm{fc}} = \frac{1}{\pi}\left(\frac{K_{\mathrm{Ic}}}{Y\sigma_{\max}}\right)^2 \tag{12.15}$$

where $\sigma_{\max}$ is the maximum stress from the stress cycle.

The final crack size triggering ductile fracture can be determined from the condition of ductile fracture $\sigma_l = \sigma_{\mathrm{S}}$:

$$\sigma_l = \frac{P}{t\times(w - a_{\mathrm{fd}})} = \sigma_{\mathrm{S}}$$

where σ_l is the maximum tensile stress acting on the crack ligament with area $t\times(w-a)$ (Figure 12.3a) and σ_{S} is the yield stress of the material. From the condition $\sigma_l = \sigma_{\mathrm{S}}$, the expression

$$a_{\mathrm{fd}} = w - \frac{P}{t\sigma_{\mathrm{S}}} \tag{12.16}$$

is obtained for the final crack length a_{fd} causing plastic yielding.

The smaller of the two values $a_{\mathrm{f}} = \min\{a_{\mathrm{fc}}, a_{\mathrm{fd}}\}$ indicates the failure mode of the final fracture and is substituted as the upper limit a_{f} of the integral 12.12 related to the number of fatigue cycles N_{f} until failure.

During fatigue life predictions, the distribution of the initial lengths of the fatigue cracks is commonly assumed to be the size distribution of the surface (subsurface) discontinuities and pores. It is implicitly assumed that the probability of fatigue crack initiation on a particular defect is equal to the probability of its existence in the stressed volume. This assumption however is too conservative because it does not account for the circumstance that the probability of fatigue crack initiation on a particular defect depends not only on its size but also on its type. A more refined approach should include an assessment of the probability with which the defects in the material initiate fatigue cracks (Todinov, 2001c).

A powerful application of the fracture mechanics approach is in *deciding whether to remove components from service or leave them in service, based on evidence provided by a current inspection.*

If there is evidence from inspection that a flaw of given size is present in the stressed component, a decision is made whether the component can wait until the next inspection period or the component should be replaced immediately. In this way, components which would normally be removed from service after a certain operating period would be allowed to continue service if the inspection and the fracture mechanics calculations confirm safe service (no failure) until the next inspection. Allowing components to continue their service results in considerable savings to the world economy.

If the inspection has not indicated a presence of a flaw in the component, in the calculations, a flaw with size equal to the threshold flaw size of the non-destructive flaw detection method should be assumed. The life of the component is then estimated by also assuming that the flaw of threshold size is located in the region with the highest loading stresses and the flaw is with the most unfavourable orientation.

This approach will be illustrated by the next example.

Example

An NDT examination of a metal plate with width $w = 500\,\text{mm}$ and thickness $t = 6\,\text{mm}$ loaded in tension revealed a central through crack of size $2a = 12\,\text{mm}$ (Assume constant $Y \approx 1$), which developed from a corrosion pit in the plate (Figure 12.13). Every 3 hours, the plate is subjected to a repeated load F varying between 0 and 900 kN. The fracture toughness of the material is $66\,\text{MPa} \times m^{1/2}$, and the yield strength is 980 MPa. Based on experimental test results, the Paris exponent was estimated to be $m = 3.0$, and the Paris parameter C is $C = 2.7 \times 10^{-11}$. Should the plate be replaced now or it can wait for the inspection next year?

Solution

The crack length triggering a fully plastic yielding can be estimated from

$$\sigma_S = \frac{900 \times 10^3}{(0.5 - 2a_{fd}) \times 0.006}$$

where $2a_{fd}$ is the critical final crack length at which yielding starts:

$$2a_{fd} = 0.5 - \frac{900 \times 10^3}{980 \times 10^6 \times 0.006} = 0.347\,\text{m}$$

The crack half-length at fast fracture is estimated from

$$\sigma_{max} \times \sqrt{\pi a_{fc}} = K_{Ic}$$

$$\sigma_{max} = \frac{F_{max}}{0.5 \times 0.006} = \frac{900 \times 10^3}{0.5 \times 0.006} = 300\,\text{MPa}$$

$$a_{fc} = \frac{1}{\pi}\left(\frac{K_{Ic}}{\sigma_{max}}\right)^2 = \frac{1}{\pi}\left(\frac{66 \times 10^6}{300 \times 10^6}\right)^2 \approx 0.0154\,\text{m}$$

$a_f = \min\{a_{fd}, a_{fc}\} = a_{fc} = 15.4\,\text{mm}$; therefore, the fracture failure mode will be fast fracture and the final half-crack length is $a_f = 15.4\,\text{mm}$. The number of cycles until failure is estimated from integrating the Paris law $da/dN = C(\Delta K)^m$:

$$N = \int_{a_i}^{a_f} \frac{da}{C(\Delta K)^m}$$

$$\Delta K = \Delta\sigma\sqrt{\pi a},\ Y = 1$$

$$N = \frac{1}{C \times (\Delta\sigma \times \sqrt{\pi})^m} \int_{a_i}^{a_f} a^{-m/2}\, da = \frac{2}{C \times (\Delta\sigma \times \sqrt{\pi})^m (m-2)} \times \left[\frac{1}{a_i^{(m-2)/2}} - \frac{1}{a_f^{(m-2)/2}}\right]$$

$$m - 2 = 1,\ a_i = 0.006\,\text{m},\ a_f = 0.0154\,\text{m}$$

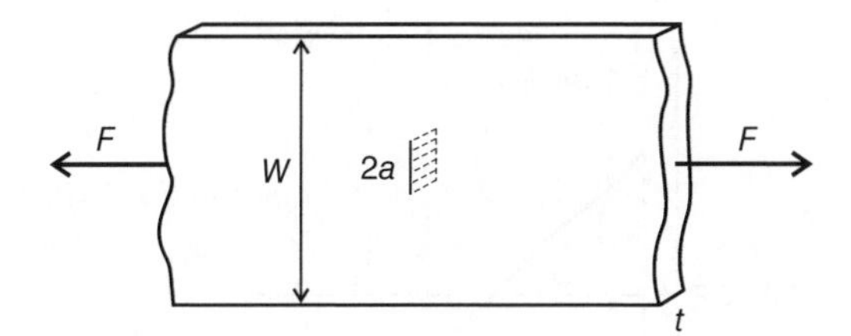

Figure 12.13 A plate with thickness t and width w, with a central through crack of size $2a$

$$C\times\left(\Delta\sigma\sqrt{\pi}\right)^{m}(m-2)=2.7\times10^{-11}\times\left(300\times\sqrt{\pi}\right)^{3}\times1=0.004$$

$$N=\frac{2}{0.004}\times\left(\frac{1}{0.006^{1/2}}-\frac{1}{0.0154^{1/2}}\right)\approx2425\,\text{cycles}$$

This number of cycles is equivalent to $2425\times3=7275$ hours of service until failure.

The time to failure of the plate in years will be $7275/24\approx303$ days.

The service life is smaller than the time for the new inspection; therefore, the plate should be replaced now.

Fatigue life calculations can also be used to determine the maximum permissible flaw size or stress range which guarantees a specified minimum operation period MFFOP free of fatigue failures (*a minimum failure-free operation period* (MFFOP)). In this case, the specified MFFOP is used to determine the maximum permitted stress range necessary to guarantee the specified MFFOP. Suppose that the minimum stress σ_{min} is zero. In this case, Equation 12.13 for the fatigue life should be solved with respect to the maximum stress σ_{max} in order to obtain the maximum permissible stress range.

The final crack length which triggers fast fracture is estimated from

$$Y\sigma_{max}\times\sqrt{\pi a_{fc}}=K_{Ic}$$

$$a_{fc}=\frac{1}{\pi}\left(\frac{K_{Ic}}{Y\sigma_{max}}\right)^{2}$$

which itself is a function of the maximum permitted stress σ_{max}. As a result, the fatigue life given by Equations 12.13 or 12.14 becomes a complex function of the maximum stress σ_{max}. The maximum value of the loading stress which still guarantees the specified MFFOP can be found easily by using a repeated bisection. The σ_{max} (in MPa) is varied between the limits 'low' and 'high'. For σ_{max} = low, the fatigue life is larger than the specified MFFOP (Figure 12.14). For σ_{max} = high, the fatigue life is smaller than the specified MFFOP. Because the MFFOP is a continuous function of σ_{max}, according to the intermediate value theorem, there exists σ_{max} for which the fatigue life is exactly equal to the specified MFFOP. The fatigue life is first calculated at the middle of the interval `mid = (low + high)/2`. If the calculated fatigue life is smaller than the specified MFFOP, the point `mid` becomes the new high limit. If the converse is true, the point `mid` becomes the new low limit. In this way, the optimal solution (the intersection of the fatigue life dependence with the horizontal line representing the MFFOP in Figure 12.14) has been bracketed in an interval with size half of the size of the initial interval [low, high]. Continuing the repeated bisection, the optimal solution will be bracketed within a very small interval defined by the selected precision.

Suppose that the procedure ***fatigue_life***`(value)` returns the fatigue life as a function of the maximum loading stress `value`. The next algorithm determines the maximum value of the loading stress which still guarantees the specified MFFOP. The constant `'eps'` contains the specified precision (in MPa) with which the optimal solution is obtained:

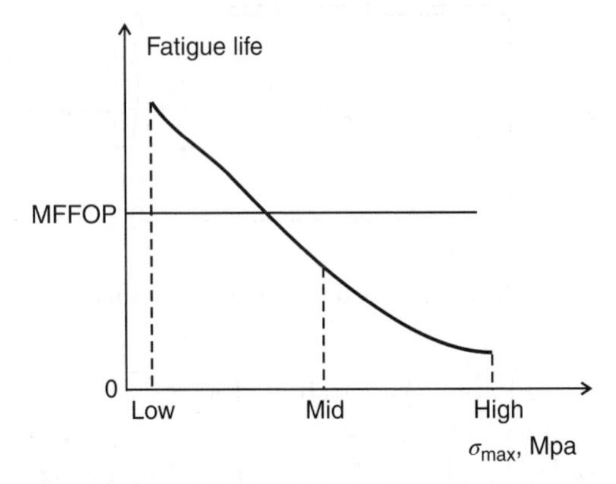

Figure 12.14 Optimisation method for determining the maximum loading stress which still guarantees a minimum failure-free operation period of specified length

Algorithm 12.2

```
left = low;
right = high;

while (right-left>eps) do
{
 mid = (left+right)/2;
   fl_mid = fatigue_life(mid);

     if (fl_mid<MFFOP) then right=mid;
        else left=mid;
}

return mid;
```

The application of this approach to guaranteeing a specified MFFOP has been illustrated by the next application.

Example
Every hour, a round bar made of a high-strength steel with fracture toughness 65 MPa × $m^{1/2}$ is loaded in tension with stress varying from $\sigma_{min} = 0$ to σ_{max}. The designed MFFOP of the bar is 5 years. The failure mechanism of the bar is a fatigue crack growth followed by a fast fracture. Based on experimental test results, the Paris exponent was estimated to be $m = 3.1$, and the Paris parameter C is $C = 2.1 \times 10^{-12}$. If the threshold of the non-destructive technique is 1 mm, what is the maximum permitted stress σ_{max}.

Solution
The worst case of an edge crack oriented perpendicular to the tensile stress is assumed ($Y = 1.12$). Running Algorithm 12.2 with the specified numerical data yielded $\sigma_{max} = 380.6$ MPa for the maximum loading stress still guaranteeing 5 years of failure-free operation. A loading stress σ for which $\sigma \leq \sigma_{max}$ is fulfilled guarantees that the fatigue life of the bar will be at least 5 years.

12.2.1 *Reducing the Risk of Fatigue Fracture*

The physics of failure model (12.11) gives not only an estimate of the time to failure of a cracked component but reveals also ways by which fatigue life can be increased.

12.2.1.1 Reducing the Size of the Flaws

The analysis shows that fatigue life is very sensitive to the initial crack size a_i and reducing the size of the flaws, which serve as places for fatigue crack initiation, increases significantly fatigue life. Most of the loading cycles are expended on the early stage of crack propagation when the crack is small. During the late stages of fatigue crack propagation, a relatively small number of cycles is sufficient to extend the crack until failure.

The flaw size can be decreased by using cleaner material, better material processing and better inspection for flaws.

12.2.1.2 Increasing the Final Fatigue Crack Length by Selecting Material with a Higher Fracture Toughness

The fatigue life is more sensitive to the initial size of the crack than the final crack size. Consequently, a reduction in the initial size of the flaws results in a significant improvement of the fatigue life, while the improvement in the fracture toughness which results in a smaller final size of the fatigue crack has a smaller impact.

12.2.1.3 Reducing the Stress Range by Appropriate Design

Decreasing the stress range has a big impact on the fatigue life. Reducing the stress range can be done by design modifications that transform large variations of the loads acting on the component into smaller stress amplitudes. An example of such design modification is given in Figure 12.15 (Todinov, 2007).

The two designs (Figure 12.15) are characterised by the same loading force F and different design angles α ($\alpha = 85°$ for design A and $\sigma' = 20°$ for design B). For a loading force with magnitude F, the force acting in the two struts '1' and '2' is $F/(2\cos\alpha)$. Its values, at different values of the loading force F, have been plotted in Figure 12.16.

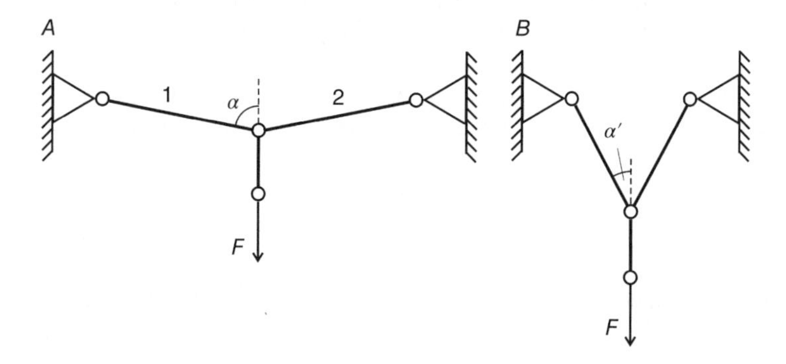

Figure 12.15 Design A can be made more fatigue resistant by reducing angle α to α' (B)

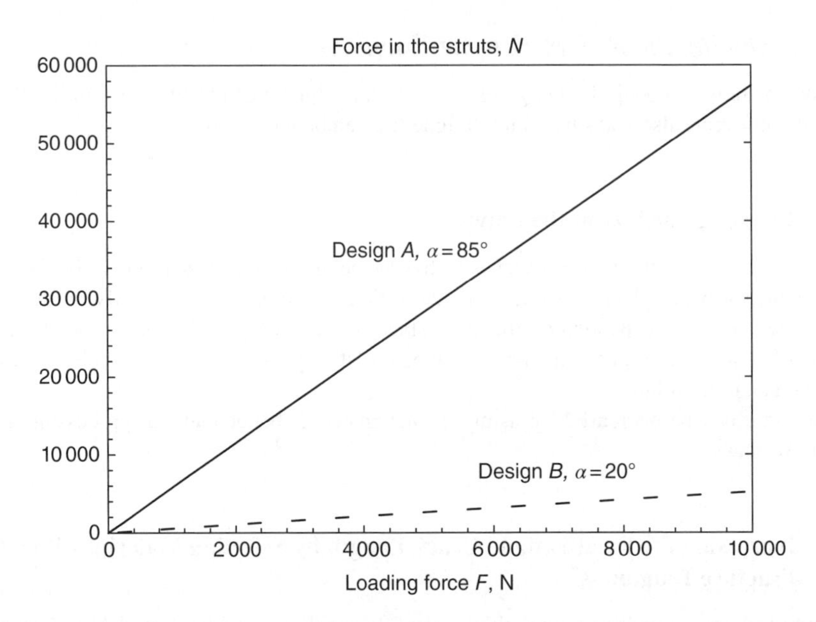

Figure 12.16 Variation of the force acting in struts '1' and '2' as a function of the variation of the loading force F

As can be verified, for design A, characterised by an angle $\alpha = 85°$, a variation of the loading force F within the range 0, 10 000 N results in a variation of the strut forces in the range 0, 57 369 N.

For design B, however, characterised by an angle $\alpha' = 20°$, the same variation of the loading force F results in more than ten times smaller range (0, 5321 N) of the strut forces. Given that the cross sections of the struts are not altered during the modification of angle α, the stress range characterising design B is also more than 10 times smaller.

Compared to design A, design B is characterised by a larger fatigue life. Because of the much smaller stress range, the fatigue crack growth rate for design B is much smaller compared to design A.

12.2.1.4 Reducing the Stress Range by Restricting the Springback of Elastic Components

Suppose that for two competing designs, the maximum stress σ_{max} from the fatigue loading (Figure 12.17), the material and the initial crack size a_i are the same. Because the maximum stress σ_{max} is the same for the competing designs, the final crack size a_f is also the same.

The only difference is the minimum stress $\sigma_{min,1}$ and $\sigma_{min,2}$ from the fatigue loading which yields different stress ranges $\Delta\sigma_1 = \sigma_{max} - \sigma_{min,1}$ and $\Delta\sigma_2 = \sigma_{max} - \sigma_{min,2}$ (Figure 12.17).

The different stress ranges characterising the designs will result in different fatigue life. Assume that $m = 3$. According to Equation 12.13, the ratio of the fatigue lives of the two designs is

$$\frac{N_{f1}}{N_{f2}} = \left(\frac{\Delta\sigma_2}{\Delta\sigma_1}\right)^m \tag{12.17}$$

Let the stress range characterising the second design be twice as small as the stress range characterising the first design ($\Delta\sigma_2 / \Delta\sigma_1 = 1/2$). From Equation 12.17, the fatigue life characterising the second design becomes

$$N_{f2} = N_{f1} \times 2^3 = 8N_{f1}$$

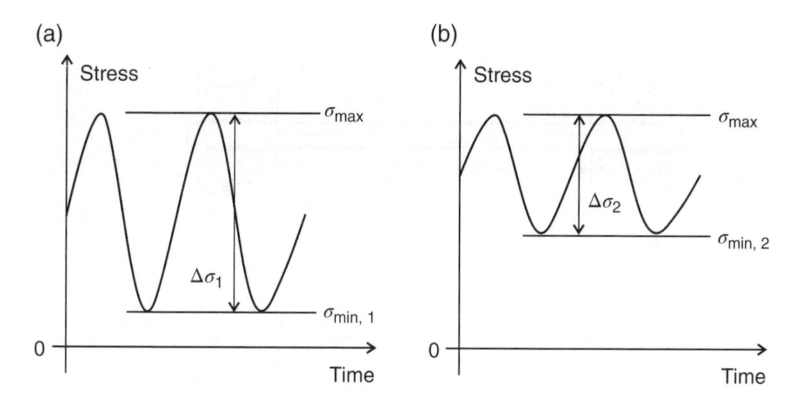

Figure 12.17 (a) The stress range from loading, decreased by (b) increasing the minimum loading stress

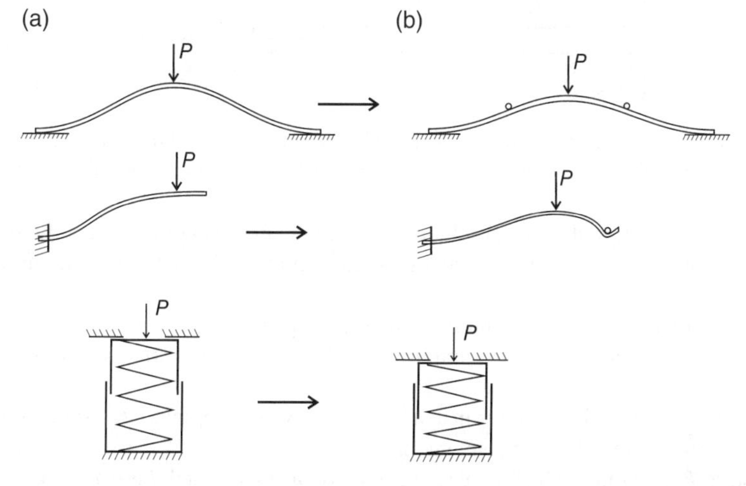

Figure 12.18 (a) Elastic components whose fatigue life has been increased by (b) increasing the minimum stress σ_{min} through mechanical restrictors of the springback

As a result of the reduction of the stress range by increasing the minimum stress, the fatigue crack propagation life has been increased eight times.

This analysis can be used for increasing the fatigue life of elastic components (Figure 12.18a) by increasing the minimum stress σ_{min} through restricting the springback (Figure 12.18b).

12.2.1.5 Reducing the Stress Range by Reducing the Magnitude of Thermal Stresses

The stress range $\Delta\sigma$ can be decreased and fatigue life increased by measures aimed at reducing the amplitude of the thermal stresses. Reducing the amplitude of thermal stresses can be achieved by:

- Using materials with low coefficient of thermal expansion.
- Using components in assemblies made of materials with similar coefficients of thermal expansion. For example, if a steel bolt works with bronze nut, because of the different coefficients of thermal expansions, thermal stresses of varying amplitude will develop as a result of the temperature variations.

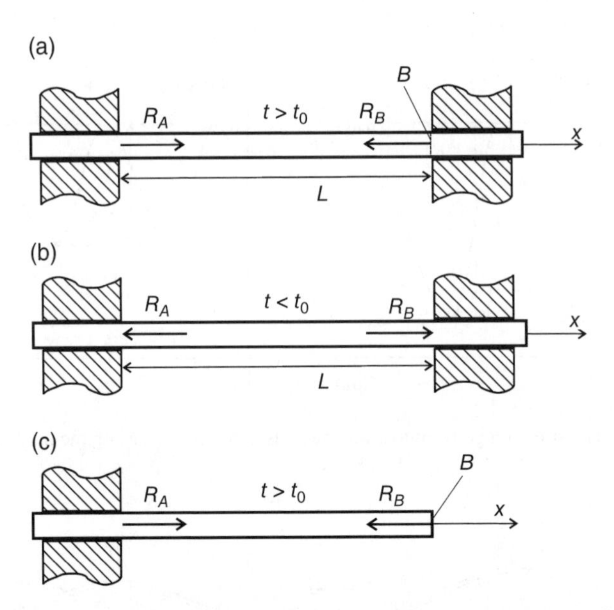

Figure 12.19 Reactions in the support of a statically indeterminate structure at (a) high temperature and (b) low temperature. (c) Determining the thermal stresses in indeterminate structures by eliminating one of the supports

- Using materials with high thermal conductivity. Using materials with high conductivity does not allow the formation of significant thermal gradients and the associated with them large thermal stress amplitudes.
- Using expansion offsets which accommodate thermal expansion.
- Using thermal insulation to reduce the amplitude of the temperature variation and with it the amplitude of the thermal stresses.
- Using statically determinate structures instead of statically indeterminate structures (statically determinate structures are free of thermal stresses).
- Avoiding start–stop regimes which induce cyclic thermal stresses and *thermal fatigue*.

The last point will be illustrated with the statically indeterminate section of the pipe in Figure 12.19. Suppose that the pipe section with length L and cross-sectional area F has been fixed at the walls at room temperature t_0. The standard approach to determining the thermal stresses in indeterminate structures is to eliminate one of the supports and replace it with the reaction force (Figure 12.19c).

With increasing the temperature beyond temperature t_0 during a start regime, the section B at the right end of the pipe undergoes two deformations – a temperature deformation and an elastic deformation. The temperature deformation is given by $\delta_T = \alpha_t L(t - t_0)$, where α_t is the linear coefficient of thermal expansion, and the elastic deformation is given by $\delta_e = R_B L/(EF)$, where E is the Young modulus, F is the area of the cross section and R_B is the axial force acting on the pipe. The resultant axial deformation of the pipe is zero because the supports are fixed; $\delta_T + \delta_e = 0$, which means that

$$\alpha_t L(t - t_0) + \frac{R_B L}{EF} = 0 \tag{12.18}$$

holds, from which

$$\sigma_x = \frac{R_B}{F} = -\alpha_t E(t - t_0) \tag{12.19}$$

Because $t > t_0$, $t - t_0 > 0$, with increasing temperature, the pipe experiences a compressive stress of magnitude $\alpha_t E(t - t_0)$. Conversely, with decreasing temperature below t_0, the pipe experiences tensile stress with magnitude $\alpha_t E(t_0 - t)$, because $t - t_0$ in Equation 12.19 will now be negative ($t < t_0$). For a steel pipe with $E = 1.96 \times 10^{11}$ MPa, $\alpha_t = 12 \times 10^{-6} K^{-1}$ and $t_0 = 20°C$, the increase of the temperature to $t = 120°C$ will generate thermal stress with magnitude 235.2 MPa. Frequent start–stop regimes inducing variation of the temperature between $t_0 = 20°C$ and $t = 120°C$ will generate cyclic thermal stresses with range (0, 235.2 MPa) which often leads to thermal fatigue.

12.2.1.6 Reducing the Stress Range by Introducing Compressive Residual Stresses at the Surface

Introducing compressive residual stresses at the surface delays the fatigue crack initiation by causing crack closure. Furthermore, compressive residual stresses also decrease the rate of crack propagation. The compressive residual stresses subtract from the loading stresses which results in *a smaller effective stress range*. This, according to the Paris law, results in a smaller fatigue crack growth rate and a longer fatigue life. According to the Paris model 12.13, the fatigue life is inversely proportional to the mth power of the stress range. This is in line with experimental observations that shot peening increases significantly the life of leaf suspension springs.

In order to compensate the tensile stresses from loading at the surface, and improve fatigue resistance, shot peening has been used as an important element of the manufacturing technology. Figure 12.20 shows the net stress distribution $\sigma_r(x)$ near the surface of a loaded, shot-peened helical spring, obtained from the superposition of the load-imposed principal tensile stress $\sigma_t(x)$ and the residual stress $\sigma_{rs}(x)$ from shot peening. Shot peening decreases the effective net stress range $\Delta\sigma_r^{eff}$ at the surface. During service, the compressive residual stress from shot peening delays the fatigue crack initiation and impedes the fatigue crack propagation.

Figure 12.21 depicts the residual stress variation measured in a shot-peened Si–Mn steel spring with spring wire diameter 12 mm.

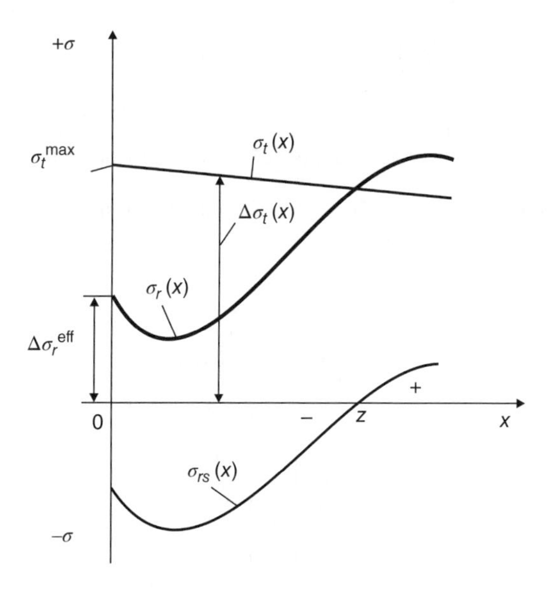

Figure 12.20 Net stress distribution near the surface of a loaded, shot-peened compression spring (Todinov, 2007)

The stress distribution with depth was produced by X-ray measurements followed by dissolving the surface layers (Todinov, 2000b).

Compressive residual stresses at the surface, decreasing the effective stress range during fatigue loading, can also be created by a special heat and thermochemical treatment such as case hardening, gas carburising and gas nitriding.

Residual stresses may also occur in the absence of thermal processing. During cold rolling of metals, for example, the surface fibres are stretched more than the inner material. After the cold rolling of a section, the requirement for compatibility of the deformations results in surface layers loaded in compression and core loaded in tension.

Cold expansion, discussed in Chapter 11 (Section 11.1.12), is also an efficient method for reducing the stress range and improving the fatigue life of components.

12.2.1.7 Reducing the Stress Range by Avoiding Excessive Bending

Excessive bending of flexible pipes in dynamic applications can also be associated with excessive stress and strain ranges. Bend restrictors and bend stiffeners (a bend stiffener is shown in Figure 12.22)

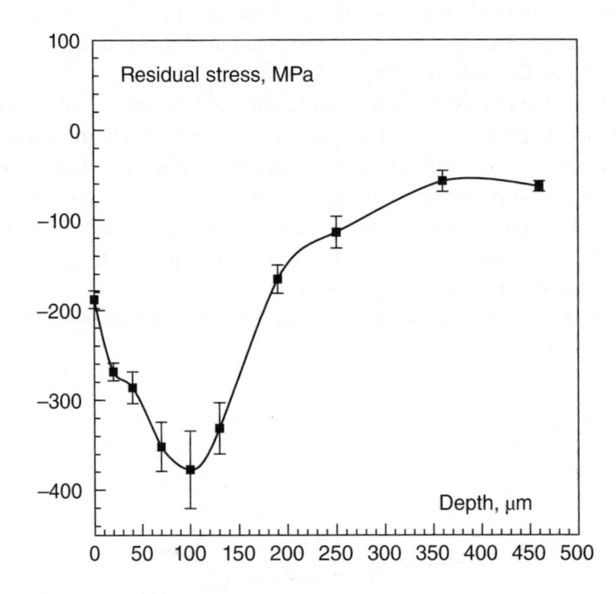

Figure 12.21 Residual stress variation in a shot-peened Si–Mn spring wire with diameter 12 mm (Todinov, 2007)

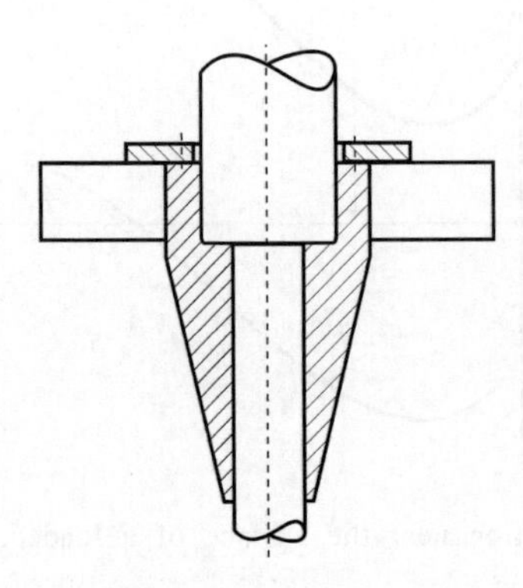

Figure 12.22 Bend stiffener

are common design measures which allow a certain degree of bending and restrict at the same time excessive bending. Bend restrictors consist of interlocking rings around the pipe which do not restrict decreasing the curvature until a particular critical value is reached. Bending beyond this critical value causes the rings to lock, and no further decrease of the curvature is possible.

12.2.1.8 Reducing the Stress Range by Avoiding Stress Concentrators

Sharp notches in components result in a high stress concentration which reduces the fatigue life and promotes early fatigue failures. Such are the sharp corners, keyways, holes, abrupt changes in cross sections, etc. Fatigue cracks on rotating shafts often originate on badly machined fillet radii which act as stress intensifiers. Because of this, they are reliability-critical elements and their appropriate design and manufacturing quality should be guaranteed. Reducing the stress magnitude in the vicinity of a fillet can be achieved by increasing its radius. In the vicinity of a notch, the stress is characterised by a sharp gradient. The smaller the curvature of the notch, the larger the stress magnitude and the lower the resistance to fatigue failures.

Fatigue crack initiation is also promoted at the grooves and the micro-crevices of *rough surfaces*. These can be removed if appropriate treatment (grinding, honing and polishing) is prescribed (Ohring, 1995).

12.2.1.9 Improving the Condition of the Surface and Eliminating Low-Strength Surfaces

The fatigue life of components depends strongly on the condition of the surface. Numerous observations confirmed that the fatigue failure usually starts from a surface imperfection. The major reason for this is the unfavourable combination of three circumstances:

1. The surface layers usually carry the largest stresses.
2. The surface layers are usually saturated with various discontinuities and defects.
3. The surface is exposed directly to the negative influence of the processing and working environment.

High-strength steels and alloys are particularly sensitive to surface defects. Some of the surface imperfections are a direct result from the manufacturing process. The surface roughness, for example, is a function of many parameters: the geometry of the cutting tool, the type of machined material, the homogeneity of its micro-structure, the cutting speed and feed, the rigidity of the fixtures, the vibration resistance of the cutting machine, the degree of wear of the cutting blade, the presence of lubricants and coolants, etc. The machined surface contains a large number of grooves of different depth and sharpness, causing local stress concentrations and reduced fatigue strength. The greater the material strength, the more detrimental is the effect of these stress concentrators. Surface roughness is decreased and fatigue life is improved if the machined materials have a homogeneous microstructure, characterised by a small grain size. Surface roughness is decreased by using sharp cutting blades, increasing the cutting speed, applying lubricants and coolants, eliminating vibrations by using damping devices and fixtures of high rigidity. The size of surface irregularities and their direction has a profound effect on fatigue. Consequently, surface roughness from machining can be reduced significantly and fatigue life further increased by *grinding*, *polishing*, *honing* and *superfinishing*.

Strain-hardening operations such as burnishing, rolling and *shot peening* increase fatigue life because the strain-hardened surface layers resist the formation and propagation of fatigue cracks. As a consequence, in strain-hardened components, the initiation of fatigue cracks occurs at higher stresses and after a greater number of loading cycles compared to components which have not been strain hardened. During burnishing, for example, the surface roughness is decreased, surface layers are strain hardened, and residual compressive stresses are generated. Burnishing also raises the fatigue limit at high temperatures. As a result, burnishing applied as a finishing operation to shafts, bars, pistons and cylinders ensures high reliability.

Even insignificant decarburisation of steels with martensitic structure causes a significant reduction of their fatigue strength. Decarburisation diminishes the fatigue resistance of steel components by:

1. Diminishing the local fatigue strength due to the decreased density of the surface layer
2. Increased grain size and diminished fracture toughness and yield strength
3. Creating low-cycle fatigue conditions for the surface

These factors promote early fatigue crack initiation and premature fatigue failure. Consequently, in order to delay the onset of fatigue failure, during the austenitisation of steel components, decarburisation and excessive grain growth should be avoided.

Eliminating low-strength surfaces can be achieved by:

- Eliminating soft decarburised surface after austenitisation by machining or carburizing in atmosphere with strictly controlled carbon potential.
- Eliminating surface discontinuities, folds and pores by machining and polishing.
- Eliminating coarse microstructure at the surface by a controlled heat treatment.
- Strengthening the surface layers by surface hardening, carburising, nitriding and deposition of hard coatings (Budinski, 1996). For example, TiC, TiN and Al_2O_3 coatings increase significantly the fatigue resistance. Early-life failures due to rapid wear can substantially be reduced by specifying appropriate lubricants. These create interfacial incompressible films that keep the surfaces from contacting.

12.2.1.10 Increasing the Fatigue Life of Automotive Suspension Springs

Delaying the onset of fatigue failure for hot-coiled automotive suspension springs, for example, requires (Todinov, 2007):

1. A spring steel with a small susceptibility to surface decarburisation (Gildersleeve, 1991).
2. A controlled austenitisation of the spring wire. The austenitisation time and temperature must be carefully selected in order to guarantee homogeneous microstructure and at the same time to prevent excessive grain growth.
3. Improved quenching of the hot-coiled spring, resulting in compressive residual stresses at the surface.
4. An optimal tempering to achieve a microstructure with optimal hardness, corresponding to a maximum fatigue resistance.
5. A carefully controlled shot peening (Todinov 2000b):
   - The shot peening must have a sufficient intensity in order to introduce residual stresses with a large magnitude.
   - The shot peening must be carefully conducted in order to guarantee a uniform compressive residual stresses at the surface of the spring coil.
6. Controlling the cleanliness of the spring steel:
   - The spring steel must have a small number density of oxide inclusions which serve as ready fatigue crack initiation sites.
   - The spring steel must have a small number density of sulphide inclusions, which promote anisotropy and reduce the toughness of the spring wire.
7. Painting to provide a durable film preventing corrosion damage.

These measures increase the number of cycles needed for fatigue crack initiation, slow down the rate of fatigue crack propagation and delay significantly the onset of fatigue failure.

12.3 Early-Life Failures

Early-life failures occur within a very short period from the start of the design life of the installed equipment. Because they usually occur during the payback period of the installed equipment and are also associated with substantial losses due to warranty payments, they have a strong negative impact on the financial revenue. Early-life failures result in loss of reputation and market share and are usually caused by:

- Poor design
- Defects in the material from processing
- Defects from manufacturing
- Inadequate material
- Poor inspection and quality control
- Misassembly, poor workmanship and mishandling
- Human errors

Most of the early-life failures are overstress failures which occur during the infant mortality region of the bathtub curve. They are often caused by inherent defects in the system, due to poor design, manufacturing and assembly. Human errors during design, manufacturing, installation and operation also account for a significant number of early-life failures. Some of the causes for these errors are listed in Section 11.1.11.

12.3.1 Influence of the Design on Early-Life Failures

Inadequate design is one of the most common reasons for early-life failures. A common design error is the underestimation of the actual load/stress magnitude the equipment is likely to experience in service. For example, underestimating the working pressure and temperature may cause an early-life failure of a seal and a release of harmful chemicals into the environment. An early-life failure of a critical component may also be caused by unanticipated eccentric loads due to weight of additional components or by high-amplitude thermal stresses during start-up regimes. Early-life failures are also promoted by failure to account for the extra loads during assembly. Installation loads are often the largest loads a component is ever likely to experience. Misalignment of components creates extra loads, susceptibility to vibrations, excessive centrifugal forces on rotating shafts and large stress amplitudes leading to early fatigue. Misassembly and mishandling often causes excessive yield and deformation or damaged protective coatings, which promotes rapid corrosion.

Operation outside the design specifications is also a major contributor to early-life failures. Early-life failures caused by inadequate operating procedures can be reduced significantly by implementing mistake proofing (*Poka-Yoke*) devices into the systems, which eliminate the possibility of violating the correct procedures, especially during assembly.

A special design (referred to also as *robust design*) can reduce significantly the variability of performance as a function of the inevitable variability of design parameters. Robust designs are characterised by a small sensitivity to variations of design parameters and are an important instrument in decreasing early-life failures of systems and assemblies. More information on robust designs is provided in Section 11.1.21.

Finally, designs can be strengthened and early-life failures reduced significantly by applying highly accelerated stress screens (HASS) which identify weak spots in the designs (Hobbs, 2000).

12.3.2 *Influence of the Variability of Critical Design Parameters on Early-Life Failures*

An important factor promoting early-life failures is the variability associated with critical design parameters (e.g. material properties and dimensions) which leads to deviations from the expected functions and performance. Material properties such as (i) yield stress, (ii) static fracture toughness, (iii) fatigue resistance, (iv) modulus of elasticity and elastic limit, (v) shear modulus, (vi) percentage elongation and (vii) density are often critical design parameters determining the expected functions and performance. Which of the material properties are relevant, depends on the required functions. Material properties depend on the materials processing and manufacturing and are characterised by a large variability. Often, defects and unwanted inhomogeneity are the source of variability. Residual stress magnitudes are also characterised by a large variation. Strength variability caused by production variability and variability of material properties is one of the major reasons for an increased interference of the strength distribution and the load distribution which promotes early-life failures. A heavy lower tail of the mechanical property distributions usually results in a heavy lower tail of the strength distribution. Low values of the material properties exert stronger influence on reliability than do high or intermediate values. Interestingly, deviations towards higher values can also be a cause for failures. For example, due to a deviation of the shear strength of a shear pin towards high values, the pin may fail to disconnect the driven shaft upon reaching a critical torque. The result is a failure of expensive equipment.

An important way of reducing the lower tail of the material properties distribution is the high-stress burn-in. The result is a substantial decrease of the strength variability and increased reliability on demand due to a smaller interference of the strength distribution and the load distribution. Most of the failures occurring early in life are quality failures due to the presence of substandard items which find their way into the final products because of deficient quality assurance procedures. Poor interfaces are a frequent cause for early-life failures. Interfaces are rarely manufactured to the same standards as the components involved. As a result, interfaces often appear as weak links in the chain, and their reliability limits the overall reliability of the assembly.

In this respect, inspection and quality control techniques are important means of weeding out substandard components and interfaces before they can initiate an early-life failure. Examples of inspection and quality control activities which help reduce early-life failures are:

- Checking for the integrity of protective coatings and assuring that the corrosion protection provided by the cathodic potential is adequate
- Using non-destructive inspection techniques such as *ultrasonic inspection technique* for testing components and welds for the presence of cracks and other flaws
- Inspection for water ingress in underwater installations
- Inspection for excessive elastic and plastic deformation

Defects like shrinkage pores, sand particles and entrained oxides from casting, micro-cracks from heat treatment and oxide inclusions from material processing are preferred sites for early fatigue crack initiation and often cause early fatigue failure. These flaws are also preferred sites for initiating fracture during an overstress loading. Segregation of impurities along the grain boundaries significantly reduces the local fracture toughness and promotes *intergranular brittle fracture*. Impurities like *sulphide stringers*, for example, cause a reduced corrosion resistance and anisotropy. Early-life failures caused by *lamellar tearing* beneath welds, *longitudinal splitting* of wire and increased susceptibility to *pitting corrosion* can all be attributed to anisotropy.

Production variabilities during manufacturing, not guaranteeing the specified tolerances or introducing flaws in the manufactured product, often lead to early-life failures. Depending on the supplier, the same component of the same material manufactured to the same specification is usually characterised by different properties. Between suppliers, variation exists even if the variations of property values characterising the individual suppliers are small (see Chapter 4). A possible way of reducing the 'between-suppliers variation' is to use only the supplier producing items with the smallest variation of properties.

Furthermore, due to the inherent variability of the manufacturing process, even items produced by the same manufacturer can be characterised by different properties. The 'within-supplier variation' can be reduced significantly by a better control of the manufacturing process, more precise tools, production and control equipment, specifications, instructions, inspection and quality control procedures. The manufacturing process, if not tightly controlled, can be the largest contributor to early-life failures. Because of the natural variation of critical design parameters, early-life failures are often due to unfavourable combinations of values (e.g. worst-case tolerance stacks) rather than due to particular production defects. Since variability is a source of unreliability (Carter, 1997), a particularly important factor reducing significantly early-life failures is the manufacturing process control. Process control based on computerised manufacturing processes reduces the variation of properties. Process control charts monitoring the variations of the output parameters, statistical quality control and statistical techniques are important tools for reducing the variation of properties (Montgomery *et al.*, 2001).

Another important way of decreasing early-life failures is adopting the six-sigma quality philosophy (Harry and Lawson, 1992) based on production with very small number of defective items (zero defect levels). Modern electronic systems, in particular, include a large number of components. For the sake of simplicity, suppose that a complex system is composed of N identical components, arranged logically in series. If the required system reliability is $R_s = R_0^N$, the reliability of a single component should be $R_0 = \left(R_s\right)^{1/N}$. Clearly, with increasing the number of components N, the reliability R_0 required from the separate components to guarantee the specified reliability R_s for the system approaches unity. In other words, in order to guarantee the required system reliability R_s, the number of defective components must be very small. Adopting a six-sigma process guarantees no more than two defective components out of a billion manufactured, and this is important strategy for eliminating early-life failures in complex systems.

13

Probability of Failure Initiated by Flaws

13.1 Distribution of the Minimum Fracture Stress and a Mathematical Formulation of the Weakest-Link Concept

Early-life failures are often the result of poor manufacturing and inadequate design. A substantial proportion of early-life failures is also due to the presence of flaws in the material.

Consider a bar made of brittle material containing random flaws, loaded in tension (Figure 13.1). Since the loading stress σ *is below* the minimum fracture stress σ_M of the homogeneous matrix, failure can only be initiated by a flaw residing in the stressed volume. A flaw that will initiate failure with certainty if it is present in the volume of the loaded bar will be referred to as *critical flaw* (Batdorf and Crose, 1974). For example, a critical flaw could be a flaw whose size exceeds a particular limit that depends on the loading stress. Assume a population of fracture-initiating flaws with finite number density λ, whose locations in the stressed volume of the bar follow a homogeneous Poisson process. The critical flaws whose number density at a loading stress σ will be denoted by $\lambda_c(\sigma)$, also follow a homogeneous Poisson process in the volume of the loaded bar (the filled circles in Figure 13.1). For a long time, the Weibull model (Weibull, 1951)

$$p_f(\sigma) = 1 - \exp\left[-V\left(\frac{\sigma - \sigma_l}{\sigma_0}\right)^m\right], \quad m > 0 \tag{13.1}$$

has been used to model the probability of failure $p_f(\sigma)$ of brittle components loaded in tension.

In Equation 13.1, σ is the loading tensile stress, σ_l is a location parameter or a threshold stress below which the probability of failure is zero, V is the stressed volume, and σ_0 and m are material constants. The utility of the Weibull distribution has been traditionally justified with its ability to fit well a wide range of data. The theoretical justification of the Weibull distribution is the extreme value theory (Gumbel, 1958; Trustrum and Jayatilaka, 1983). The Weibull model assumes no interaction between the flaws and no crack growth resistance. In other words, once a crack is initiated from a flaw, it leads to failure.

Reliability and Risk Models: Setting Reliability Requirements, Second Edition. Michael Todinov.

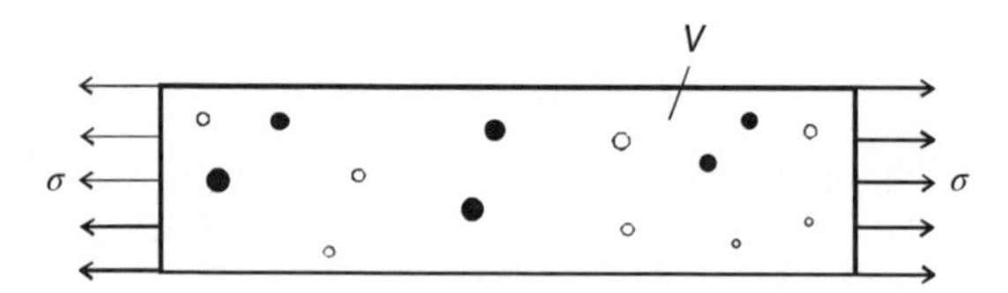

Figure 13.1 Stressed bar with volume V containing flaws with finite number density λ

In most publications related to the Weibull distribution, the utility of the Weibull distribution has also been justified with the belief that it is the mathematical formulation of the *weakest-link concept*. In other words, if a number of random flaws are present in a stressed volume, it is believed (e.g. Freudenthal, 1968) that the Weibull distribution is the distribution of the minimum failure stress characterising these flaws. This common belief has been challenged in Todinov (2009c, 2010) and proven false.

If the minimum failure stress $\sigma_{\text{min,f}}$ characterising the flaws in a stressed volume is greater than the loading stress σ ($\sigma_{\text{min}} > \sigma$), then no critical flaws are present in the stressed volume V. On the other hand, if no critical flaws reside in the stressed volume, the minimum failure stress $\sigma_{\text{min,f}}$ characterising the flaws in the stressed volume will certainly be greater than the loading stress σ.

The expected number of flaws in the stressed volume is λV, and if the probability that a flaw will be critical at a stress level σ is denoted by $F_c(\sigma)$, the expected number of critical flaws in the stressed volume V will be $\lambda V F_c(\sigma)$. Because the locations of the flaws follow a homogeneous Poisson process in the stressed volume V, for the probability that the minimum failure stress will be greater than the loading stress σ, $P(\sigma_{\text{min,f}} > \sigma) = \exp[-\lambda V F_c(\sigma)]$ holds. This is the probability that no critical flaw will be present in the stressed volume V. The probability distribution function of the minimum failure stress characterising the flaws in the stressed volume is therefore given by

$$P(\sigma_{\text{min,f}} \le \sigma) = 1 - \exp[-\lambda V F_c(\sigma)] \tag{13.2}$$

Suppose that there are M different types of flaws in the stressed volume V, each characterised by a number density λ_i and probability $F_{ci}(\sigma)$ that a flaw from the ith type will be critical. Equation 13.2 can then be easily generalised for M types of flaws:

$$P(\sigma_{\text{min,f}} \le \sigma) = 1 - \exp\left(1 - V\sum_{i=1}^{M} \lambda_i F_{ci}(\sigma)\right) \tag{13.3}$$

The comparison of Equation 13.2 with the strictly increasing Weibull distribution function (13.1) reveals that the dependence $F_c(\sigma)$ must necessarily be a strictly increasing function of the applied stress σ. The probability $F_c(\sigma)$ that a flaw will be critical however is not necessarily a strictly increasing function of the applied stress σ.

This point will be illustrated immediately by a thought experiment. Suppose that the brittle bar in Figure 13.2 contains a single type of surface flaws (e.g. random scratches). The number of scratches along the stressed length L follows a homogeneous Poisson process with lineal intensity λ (number of flaws/m).

The bar is subjected to a tensile loading in the range ($0 \le \sigma \le \sigma_2$) which is below the minimum fracture stress σ_M of the bar (with no flaws on it, Figure 13.3).

Suppose that the threshold stress level σ_1 is such that any flaw from the considered type (e.g. the scratches), will cause failure if present in the stressed bar. In other words, beyond the stress level σ_1, all flaws (scratches) are critical ($F_c(\sigma) = 1, \sigma > \sigma_1$; Figure 13.3). Note that the stress level σ_2 is not capable of initiating failure if no scratches are present. Hence, failure in the stress interval $0, \sigma_2$ can only be initiated by a scratch on the surface of the bar.

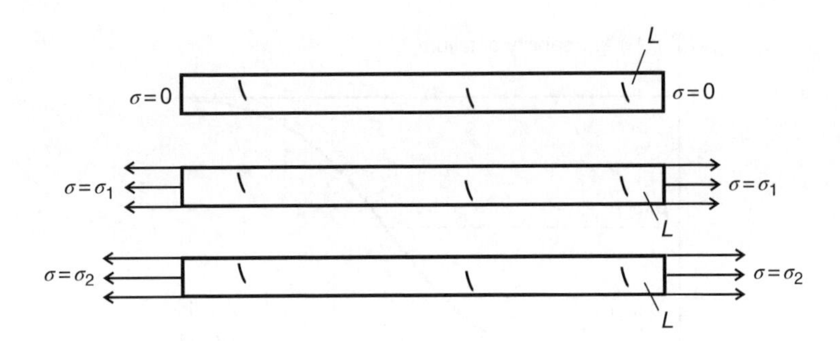

Figure 13.2 A brittle bar with random scratches, loaded in tension

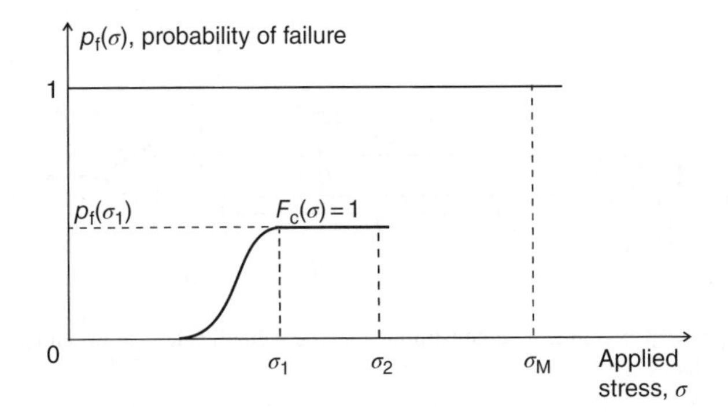

Figure 13.3 Variation of the probability of failure initiated by a flaw

Clearly, the probability of failure in the stress region (σ_1, σ_2) is equal to the probability of existence of a flaw (scratch) on the surface of the stressed bar. The probability of exactly x scratches in the stressed length L is given by the Poisson distribution

$$P(X = x) = \frac{(\lambda L)^x}{x!}\exp(-\lambda L) \tag{13.4}$$

Hence, the probability of no scratches in the stressed length L is given by

$$P(X = 0) = \frac{(\lambda L)^0}{0!}\exp(-\lambda L) = \exp(-\lambda L)$$

The probability of at least a single flaw in the stressed length L is equal to the probability of failure in the stress region (σ_1, σ_2) and is given by

$$p_f(\sigma) = 1 - \exp(-\lambda L) \tag{13.5}$$

Note that the probability of failure initiated by a flaw *is constant* in the stress region (σ_1, σ_2).

However, according to the Weibull distribution (13.1), within the stress range (σ_1, σ_2), as σ varies within the range σ_1, σ_2, the probability of failure *always increases* $p_f(\sigma_2) > p_f(\sigma_1)$; Figure 13.4.

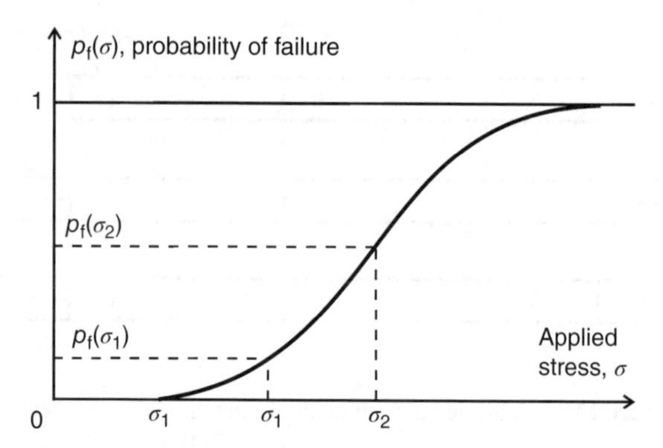

Figure 13.4 According to the Weibull model, the probability of failure is a strictly increasing function of the applied stress

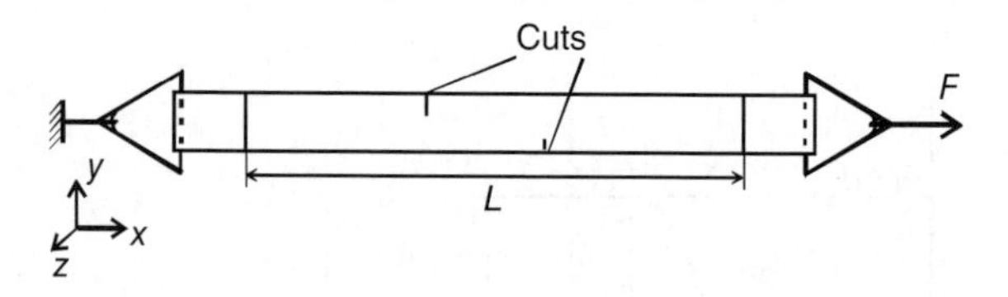

Figure 13.5 The paper strip experiment demonstrating that the Weibull distribution cannot model the probability of failure controlled by flaws

In fact, the probability of failure in the stress region (σ_1, σ_2) remains constant because *the expected number of random scratches on the stressed length has not been increased from loading in the stress region* σ_1, σ_2.

This example is applicable to a number of real cases involving *flaw populations where the flaws from a particular type become unstable at a smaller stress than the rest of the flaws*. In this case, there will be a stress region within which most of the flaws from a particular type will be critical, while the stress level will be insufficient to make the flaws from the other types critical. In this stress region, *increasing the loading stress will result in a kink on the probability of failure curve or even a plateau. In both cases, the strictly increasing Weibull distribution function will not be able to describe correctly the variation of the probability of failure with increasing load.*

To demonstrate experimentally that in general, the Weibull distribution cannot model correctly the probability of failure controlled by flaws, an easy-to-reproduce experiment has been proposed in Todinov (2010), involving 18 mm wide paper strips (Figure 13.5). On the loaded length of $L = 250$ mm, random cuts of 10 and 3 mm length were made, acting as flaws.

In order to make the axial loading more uniform, the paper strips were loaded through triangular frames, free to rotate along the perpendicular axes *x*, *y* and *z* (Figure 13.5). The locations of the cuts from each type follow a homogeneous Poisson process with number densities $\lambda_1 = 1\text{m}^{-1}$ and $\lambda_2 = 1\text{m}^{-1}$. The number of cuts from each size was generated by a specially designed generator of random numbers following a homogeneous Poisson process, according to the algorithm described in Chapter 8. Once the random number of cuts k from a particular type was obtained from sampling the Poisson distribution, the locations of the cuts x_i along the paper strip were found by distributing them uniformly along the length L according to the relationship $x_i = L \times u_i$, $i = 1, k$ where u_i is a uniformly distributed random number in the interval (0, 1). To minimise the interference of the stress fields around the cuts, adjacent cuts were alternatively placed on both sides of the strip, as shown in Figure 13.5.

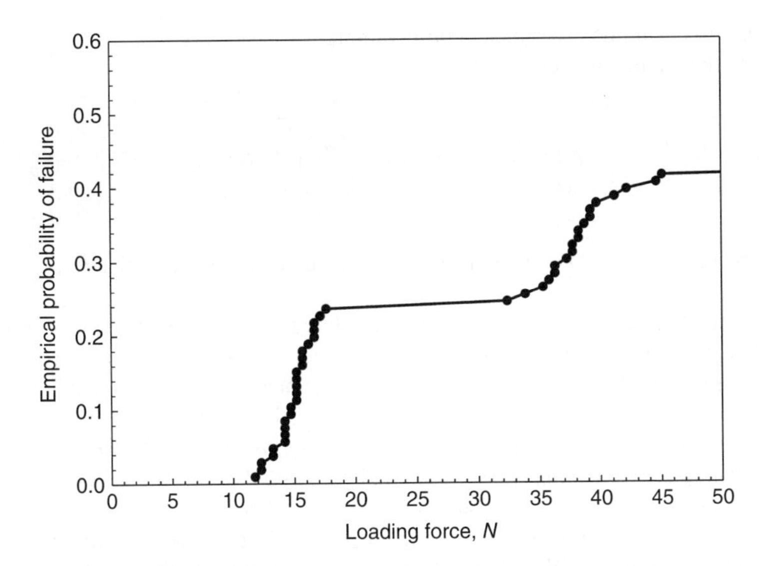

Figure 13.6 Empirical cumulative distribution of the probability of failure of paper strips with random cuts

The paper strip was then loaded gradually and the breaking load recorded. The results from $n=105$ experiments are shown in Figure 13.6. An empirical cumulative distribution of the probability of failure has been produced by ordering the breaking strength in ascending order and plotting the load versus the rank estimate $F_i = i/(n+1)$, where $i=1, 2, \ldots, 105$ is the index of the ordered measurements.

Only the lower part of the curve has been reproduced, corresponding to the loading range where at least one random cut was present on the strip.

The paper strip fails at the cut with the smallest strength. As can be seen from the experimental cumulative distribution of the probability of failure, the distribution of the minimum failure load does not follow the Weibull distribution. Note that the reason is certainly not that the strength distribution of the cuts is a two-point distribution, containing only two possible values for the strength. As can be seen from the graph in Figure 13.6, the cuts from each size cannot be made identical. Both the large cuts and the small cuts are characterised by their own unique strength distributions. The Weibull model fails to capture the variation of the probability of failure *because the Weibull model is characterised by a strictly increasing function of the probability of failure with the applied stress. A strictly increasing function cannot approximate a probability of failure which does not increase over a particular loading range*. The beginning of the loading range where the probability of failure is constant is marked by a load for which all big cuts are critical (will cause failure if present on the strip). The end of the loading range is marked by a load for which none of the small cuts are critical (Figure 13.6).

The empirical probability of failure obtained from the paper strip experiment is in agreement with the predictions from Equation 13.3. Indeed, according to Equation 13.3 in the loading region between 20 and 30 N where all large cuts are critical but none of the small cuts is critical, $F_{c1}(\sigma)=1$, $F_{c2}(\sigma)=0$ and substituting in Equation 13.3 where $M=2$ gives

$$P(\sigma_{\text{min,f}} \le \sigma) = 1 - \exp\left[-L\lambda_1 F_{c1}(\sigma)\right] = 1 - \exp(-0.25 \times 1 \times 1) \approx 0.22$$

In the loading region beyond 46 N where all cuts are critical, $F_{c1}(\sigma)=1$ and $F_{c2}(\sigma)=1$. Substituting in Equation 13.3 where $M=2$ gives

$$P(\sigma_{\text{min,f}} \le \sigma) = 1 - \exp\left\{-\left[L\lambda_1 F_{c1}(\sigma) + \lambda_2 F_{c2}(\sigma)\right]\right\} = 1 - \exp\left[-0.25 \times (1 \times 1 + 1 \times 1)\right] \approx 0.4$$

Both results agree with the experimental observations.

According to the Weibull model

$$p_{\mathrm{f}}(\sigma) = 1 - \exp\left[-L\left(\frac{\sigma - \sigma_l}{\sigma_0}\right)^m\right], \quad m > 0,$$

the probability of failure is a strictly increasing function of the applied stress $p_{\mathrm{f}}(\sigma_2) > p_{\mathrm{f}}(\sigma_1)$ (Figure 13.4).

The Weibull model yields incorrect probability of failure in the flat region of the empirical cumulative distribution. The probability of failure should remain the same: $p_{\mathrm{f}}(\sigma_2) = p_{\mathrm{f}}(\sigma_1)$ because the expected number of flaws has not been altered in the stress interval (σ_1, σ_2) belonging to the flat region of the dependence in Figure 13.6. The conclusion from the simple paper strip experiment is that the classical Weibull model, *with its strictly increasing function, is incapable of approximating a constant probability of failure over a loading region*. Such regions exist whenever the largest flaws are all critical, but the loading magnitude is insufficient to make the smaller flaws unstable. *If no new flaws are created in a particular loading region, there is no reason for the probability of failure to increase.*

Note that inventing separate Weibull functions to approximate the strength distributions of each flaw type (as it is commonly done by many researchers) does not resolve the fundamental problem with the Weibull model. The probability of failure within the flat regions of the probability distribution function cannot be approximated by any strictly increasing Weibull functions. A kink on the probability of failure curve also cannot be approximated by a single Weibull function.

Here are some of the reasons why the fundamental flaw in the Weibull model has evaded the attention of researchers for such a long time:

- Lack of a rigorous theoretical analysis of the Weibull model.
- Small data sets were fitted. Because of the flexibility of the Weibull model, a small sample appears to follow the Weibull distribution in almost any case.
- The confirmation bias of researchers (selecting data sets complying with the Weibull distribution and ignoring data sets contradicting it).
- Because of its popularity, a single Weibull distribution was frequently imposed even on data sets whose empirical cumulative distribution indicated that the data do not come from a single Weibull population.
- The Weibull distribution is the correct model in the important special case where fracture is controlled by the size of a single type of flaws (e.g. voids or inclusions) and the flaw size can be approximated well by an inverse power law distribution.
- For relatively small stress magnitudes, the lower tail of the probability that a flaw will be critical can often be approximated reasonably well by a power law dependence, and in this case, the Weibull distribution approximates well the probability of failure.

The outlined limitations associated with the traditional Weibull model means also that the software tools based on this model are not capable of predicting correctly the probability of failure initiated by flaws.

13.2 The Stress Hazard Density as an Alternative of the Weibull Distribution

Here, a general equation will be derived regarding the probability of failure of uniformly stressed component with volume V (Figure 13.1), irrespective of whether failure is initiated by flaws or not. This can be done by expressing the probability of failure using the concept *stress hazard density* $h(\sigma)$.

The stress hazard density stands for the quantity

$$h(\sigma)=\frac{f(\sigma)}{VR(\sigma)} \tag{13.6}$$

where $f(\sigma)=dF(\sigma)/d\sigma$ is the failure probability density function and $R(\sigma)=1-F(\sigma)$ is the probability of surviving a loading stress σ. The conditional probability of failure in the infinitesimally small stress interval $(\sigma, \sigma+d\sigma)$ given that the component has survived a loading stress σ is

$$P(\sigma<\sigma_{\mathrm{f}}<\sigma+d\sigma \mid \sigma_{\mathrm{f}}>\sigma)=\frac{f(\sigma)d\sigma}{R(\sigma)} \tag{13.7}$$

Using the concept stress hazard density, the conditional probability of failure in the infinitesimally small stress interval $(\sigma,\sigma+d\sigma)$ can be presented as $Vh(\sigma)d\sigma$.

Equation 13.6 can also be presented as

$$h(\sigma)=-\frac{R'(\sigma)}{VR(\sigma)} \tag{13.8}$$

which is a separable differential equation with initial condition $R(\sigma=0)=1$. Presenting Equation 13.8 as

$$-Vh(\sigma)d\sigma=\frac{dR(\sigma)}{R(\sigma)}$$

and integrating both sides from 0 to σ give

$$-V\int_0^{\sigma} h(v)dv=\ln R(\sigma)+C$$

From the initial condition $R(\sigma=0)=1$, we get $C=0$. Finally, the probability of failure of the component can be presented as

$$p_{\mathrm{f}}(\sigma)=1-\exp\left[-V\int_0^{\sigma} h(v)dv\right] \tag{13.9}$$

$H(\sigma)=\int_0^{\sigma} h(v)dv$ will be referred to as *cumulative stress hazard density* (Todinov, 2010). Hence, the probability of failure can be expressed as a function of the cumulative stress hazard density

$$p_{\mathrm{f}}(\sigma)=1-\exp[-VH(\sigma)] \tag{13.10}$$

As can be seen, Equation 13.9 is very general. During its derivation, the notions 'flaws', 'strength of flaws', 'critical flaws' or 'locally initiated failure by flaws' have not been used. As a result, Equation 13.9 is a general model, with wide validity.

The cumulative stress hazard density is a material property and reflects the properties of the matrix, the flaws, their location and orientation, etc., at different levels of the loading stress. For the same level of the loading stress, the cumulative stress hazard density is the same for different size of the uniformly stressed volume/gauge length. In the case of a fibre with gauge length L, the analogue of Equation 13.10 is the equation

$$p_{\mathrm{f}}(\sigma)=1-\exp\left[-LH(\sigma)\right] \tag{13.11}$$

13.3 General Equation Related to the Probability of Failure of a Stressed Component with Complex Shape

Suppose that a component with complex shape is loaded in an arbitrary fashion and contains non-interacting flaws. It is assumed that the flaw locations in the volume V follow a non-homogeneous Poisson process. The variation of the flaw number density in the volume of the component is described by the function $\lambda(x,y,z)$. It gives the flaw number density in the infinitesimal volume dv, at a location with coordinates x, y, z (Figure 13.7).

Suppose that a single flaw is characterised by a conditional individual probability F_{c} of initiating failure *given* that the flaw is present in the stressed component. The index c means that the individual probability of initiating failure has been conditioned on the existence of a flaw in the component. This probability is different from the probability p_{f} of failure of the component associated with a population of flaws. The probability p_{f} is related to the whole population of flaws and is not conditioned on the existence of flaws in the component. In other words, p_{f} is still meaningful even if no flaws are present in the component.

The probability p_{f} (unconditional) of failure associated with a population of flaws can be determined by subtracting from unity the probability p^0 of the complementary event: 'none of the flaws will initiate failure'. The probability $p^0_{(r)}$ of the compound event *exactly r flaws exist in the volume V of the component and none of them will initiate failure* is a product

$$p^0_{(r)}(V)=\exp\left(-\int_V \lambda(x,y,z)\,dv\right)\frac{\left(\int_V \lambda(x,y,z)\,dv\right)^r}{r!}\left[1-F_{\mathrm{c}}\right]^r. \tag{13.12}$$

of the probabilities of two statistically independent events: (i) 'exactly r flaws reside in the volume V', the probability of which is given by the non-homogeneous Poisson distribution

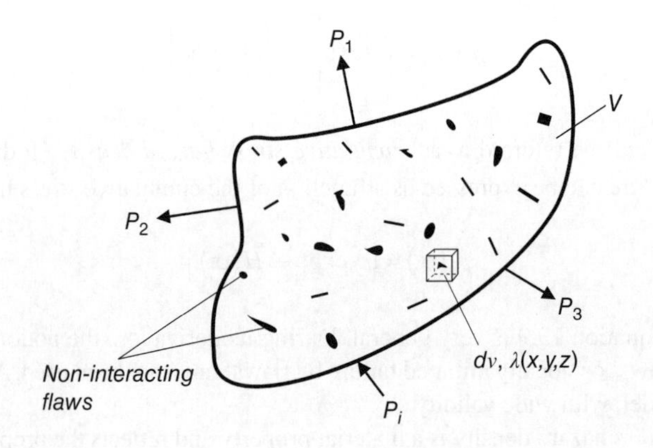

Figure 13.7 A component with complex shape, loaded with arbitrary forces P_i

$$P(X=r)=\exp\left(-\int_V \lambda(x,y,z)\,dv\right)\frac{\left(\int_V \lambda(x,y,z)\,dv\right)^r}{r!}$$

and (ii) 'none of the r flaws will initiate failure', the probability of which is $[1-F_c]^r$. The event *no failure will be initiated in the volume V* is the union of disjoint events characterised by probabilities $p^0_{(r)}$, (r=0,1, 2,...) and its probability p^0, according to the total probability theorem, is

$$p^0=\sum_{r=0}^{\infty} p^0_{(r)}=\exp\left(-\int_V \lambda(x,y,z)\,dv\right)\sum_{r=0}^{\infty}\frac{\left([1-F_c]\int_V \lambda(x,y,z)\,dv\right)^r}{r!} \tag{13.13}$$

Since

$$\sum_{r=0}^{\infty}\frac{\left([1-F_c]\int_V \lambda(x,y,z)\,dv\right)^r}{r!}=\exp\left([1-F_c]\int_V \lambda(x,y,z)\,dv\right),$$

Equation 13.13 can be simplified to

$$p^0=\exp\left(-F_c\int_V \lambda(x,y,z)\,dv\right)$$

The probability p_f of failure for the component with volume V becomes

$$p_f=1-\exp\left(-F_c\int_V \lambda(x,y,z)\,dv\right) \tag{13.14}$$

Equation 13.14 also holds for the two- and one-dimensional case if the volume V is replaced by the area S or the length L of the component. Correspondingly, the flaw number density will be a number of flaws per unit area or unit length.

Since $\bar{\lambda}=(1/V)\int_V \lambda(x,y,z)dv$ is the expected (average) number density of flaws in the volume V, Equation 13.14 can also be presented as

$$p_f=1-\exp\left(-\bar{\lambda}VF_c\right) \tag{13.15}$$

Equation 13.15 *is valid for a component with any shape, subjected to any type of loading* (Todinov 2005, 2006a).

A very important special case is obtained when the flaws follow a homogeneous Poisson process in the volume V of the specimen. In this case, the flaw locations are uniformly distributed in the bulk of the component. The defect number density is constant $\lambda(x,y,z)=\lambda=\text{const}$, and the probability of failure in Equation 13.15 becomes

$$p_f=1-\exp\left(-\lambda VF_c\right) \tag{13.16}$$

The parameter λ in Equation 13.16 is the number density of *all* flaws in the stressed volume V and is a measurable quantity.

In order to distinguish between a complex stress state and a uniaxial stress state, for a volume V subjected to a uniaxial stress σ, the probability F_c in Equation 13.16 will be denoted by $F_c(\sigma)$. As a result, for a uniform tensile stress of magnitude σ, Equation 13.16 becomes

$$p_f(\sigma) = 1 - \exp(-\lambda(\sigma) V F_c(\sigma)) \tag{13.17}$$

where $\lambda(\sigma)$ is the number density of all flaws at a loading stress σ. Equation 13.17 coincides with Equation 13.2, derived for the distribution of the minimum failure stress.

An upper bound of the probability of failure p_f can be produced if *weak flaws* ($F_c \approx 1$) are assumed. This is a very conservative assumption, suitable in cases where the upper bound of the probability of failure is required.

Equation 13.15 can be generalised for multiple type of flaws. Thus, if M types of flaws are present, the probability that no failure will be initiated is

$$p^0 = \exp(-\bar{\lambda}_1 V F_{1c}) \times \cdots \times \exp(-\bar{\lambda}_M V F_{Mc}) = \exp\left(-V \sum_{i=1}^{M} \bar{\lambda}_i F_{ic}\right)$$

where $\bar{\lambda}_i$ and F_{ic} are the average flaw number density and the conditional individual probability of initiating failure characterising the ith type of flaws. This equation expresses the probability that no failure will be initiated by the first, the second, … and the Mth type of flaws. The probability of failure then becomes

$$p_f = 1 - \exp\left(-V \sum_{i=1}^{M} \bar{\lambda}_i F_{ic}\right) \tag{13.18}$$

13.4 Link between the Stress Hazard Density and the Conditional Individual Probability of Initiating Failure

Comparing Equations 13.17 and 13.10 yields the important link

$$H(\sigma) = \lambda(\sigma) F_c(\sigma) \tag{13.19}$$

In the case of failure controlled by flaws, the cumulative hazard stress density in fact measures the *detrimental effect* of the flaws. Indeed, since $H(\sigma) = \bar{\lambda}(\sigma) F_c(\sigma)$, the detrimental effect (virulence) increases proportionally with increasing the number density of the flaws and their probability of initiating failure. For multiple types of flaws $H(\sigma) = \sum_{i=1}^{M} \bar{\lambda}_i(\sigma) F_{ci}(\sigma)$; the most detrimental type of flaws is the one with the largest $\lambda_i(\sigma) F_{ci}(\sigma)$. The *detrimental factor* $\lambda_i(\sigma) F_{ci}(\sigma)$ is an important parameter characterising the separate types of flaws (Todinov, 2000a). Consider, for example, two components with identical material and geometry. One of the components is characterised by flaws with a high number density λ_1 which initiate failure with small probability F_{c1}, and the other component is characterised by flaws with a low number density λ_2 which initiate failure with large probability F_{c2}. If both components are characterised by the same detrimental factors ($\lambda_1 F_{c1} = \lambda_2 F_{c2}$), the probabilities of failure initiated by flaws for both components will be the same. The most dangerous type of flaws is the one characterised by the largest detrimental factor $\bar{\lambda}_i F_{ic}$. Consequently, the efforts towards eliminating flaws from the material should concentrate on types of flaws with large detrimental factors.

The cumulative stress hazard density $H(\sigma)=\sum_{i=1}^{M}\bar{\lambda}_i(\sigma)F_{ci}(\sigma)$ for a small volume ΔV can be estimated from experimental measurements of the failure stress. In this way, there is no need to assume a power law stress dependence which is implied if the Weibull model is used. In other words, material is left 'to speak' for itself and not forced to obey the power law by fitting its properties with the Weibull function.

Suppose now that the graph of the cumulative hazard stress dependence $H(\sigma)=\bar{\lambda}_1(\sigma)F_{c1}(\sigma)+\bar{\lambda}_2(\sigma)F_{c2}(\sigma)+\cdots+\bar{\lambda}_M(\sigma)F_{cM}(\sigma)$ is known for a material with multiple types of flaws i ($i=1, 2, \ldots, M$). Suppose that the stress regions within which the flaws from the different types become critical are well separated. Since the number density of the flaws from any particular type does not vary with increasing stress (only the probability $F_{ci}(\sigma)$ varies), a plateau on the cumulative stress hazard dependence at a stress σ_x means that for all flaw types i ($i=1, 2, \ldots, k$), for which the flaws become critical at a stress below σ_x, the relationship

$$F_{c1}(\sigma)=F_{c2}(\sigma)=\cdots=F_{ck}(\sigma)=1$$

holds. The plateau on the curve giving the probability of failure then corresponds to the combined number density $\bar{\lambda}_1(\sigma)+\bar{\lambda}_2(\sigma)+\cdots+\bar{\lambda}_k(\sigma)$ of the flaws which all become critical below the stress $\sigma=\sigma_x$.

Accordingly, the first plateau on the cumulative stress hazard dependence corresponds to the number density of the largest (most virulent) critical flaws.

The cumulative stress hazard density is an important characteristic, which permits extrapolating the behaviour of the material under loading. It is a key to determining the probability of failure of components under load.

The cumulative hazard stress density can be built for an important range of materials (e.g. glass, ceramics and composites) whose failure is locally initiated by flaws.

13.5 Probability of Failure Initiated by Defects in Components with Complex Shape

For the important special case where the flaw number density is constant throughout the volume λ = const, the generated random locations are uniformly distributed in the volume V. For each random location, a random flaw size can be generated by sampling the size distribution of the flaws. Given the specified location and size of the flaw, a failure criterion is applied to check whether the flaw will be unstable (will initiate failure).

Equation 13.15 is very flexible and general because it permits the conditional individual probability F_c of initiating failure to be estimated using different methods. Indeed, the failure criterion is not restricted to fracture mechanics criteria only. It can also be based on other models related to the micromechanics of initiating failure. For the special case of brittle fracture and flaws whose shape can be approximated well by penny-shaped cracks, for example, a mixed-mode criterion can be adopted (Todinov and Same, 2014).

The conditional individual probability F_c of initiating failure characterising a single flaw is estimated by using the algorithm presented in Chapter 14. Finally, substituting the estimate F_c in Equation 13.15 yields the probability of failure of the stressed component, *irrespective of its geometry, type of loading and flaw number density*.

Equation 13.15 is valid *for an arbitrarily loaded component, with complex shape*. The power of the equation is in relating in a simple fashion the individual probability of failure F_c characterising a single flaw (with locations following the specified non-homogeneous flaw number density $\lambda(x, y, z)$) to the probability of failure p_f characterising the entire population of flaws.

Suppose that a direct Monte Carlo simulation was used to determine the probability of failure of the component. In this case, at each simulation trial, a large number of flaws need to be generated, and for each flaw, a check needs to be performed to determine whether there will be at least a single unstable flaw which initiates failure. In addition, there is no guarantee that small stress intensification zones will be sampled. If Equation 13.15 is used to determine the probability of failure of the component, only a single run through the finite elements composing the component would be necessary. The purpose is to collect information from all parts of the volume stressed in different ways, necessary to estimate the conditional individual probability F_c.

It is important to point out that F_c incorporates the influence of the particular loading (stress) state throughout the entire volume of the component. If the stress state in the loaded component is altered, F_c will be altered too.

Equation 13.15 avoids overly conservative estimates for the probability of failure, which results from equating the probability that a flaw will initiate failure in a stressed region with the probability that the flaw will reside in the region. The new concept 'conditional individual probability of initiating failure' characterising a single flaw, acknowledges the fact that not all flaws present in the material will initiate failure. In other words, flaws initiate failure with certain probability.

Important application areas of the derived equation are (i) determining the lower tail of the strength distribution for components containing flaws and (ii) assessing the vulnerability of designs to failure initiated by flaws.

13.6 Limiting the Vulnerability of Designs to Failure Caused by Flaws

Equation 13.15 can also be used for setting reliability requirements – setting the upper bound of the flaw number density λ which limits the probability of failure below a maximum acceptable level p_{fmax}.

By solving Equation 13.15 numerically with respect to $\bar{\lambda}$ (given a specified maximum acceptable probability of failure p_{fmax}), an upper bound $\bar{\lambda}_u$ of the average flaw number density can be determined:

$$\bar{\lambda}_u = -\frac{1}{VF_c}\ln\left(1 - p_{\mathrm{fmax}}\right) \tag{13.20}$$

This upper bound guarantees that whenever the average flaw number density $\bar{\lambda}$ satisfies $\bar{\lambda} \le \bar{\lambda}_u$, the probability of failure of the component will be smaller than p_{fmax}.

Figure 13.8 gives the dependence between the flaw number density upper bound λ_u and p_{fmax}, for different values of the stressed volume V, in case of very weak flaws ($F_c = 1$).

Consider now a component with volume V, which has been cut from material with flaw number density λ and subjected to a uniaxial stress σ. It is assumed that the flaws, whose locations follow a homogeneous Poisson process, are from a single type. Suppose that failure is controlled solely by the size of the flaws in the material and does not depend on their orientation and shape. The size distribution $G(d)$ of the flaws is the probability $G(d) = P(D \le d)$ that the size D of a flaw will not be greater than a specified value d. Let d_σ denote the critical flaw size for the stress level σ. In other words, a flaw with size greater than the critical size d_σ will initiate failure at a stress level σ.

Given the size distribution of the flaws, the maximum acceptable value V of the stressed volume that limits the probability of failure below a maximum acceptable level can be determined.

In case of failure controlled solely by the size of the flaws, $F_c(\sigma)$ in Equation 13.21 becomes $1 - G(d_\sigma)$ which is the probability that a flaw will initiate failure at the stress level σ. Substituting $F_c(\sigma) = 1 - G(d_\sigma)$ in Equation 13.21 gives

$$p_f = 1 - \exp\left\{-\lambda V\left[1 - G\left(d_\sigma\right)\right]\right\} \tag{13.21}$$

for the probability p_f of initiating failure at a stress level σ.

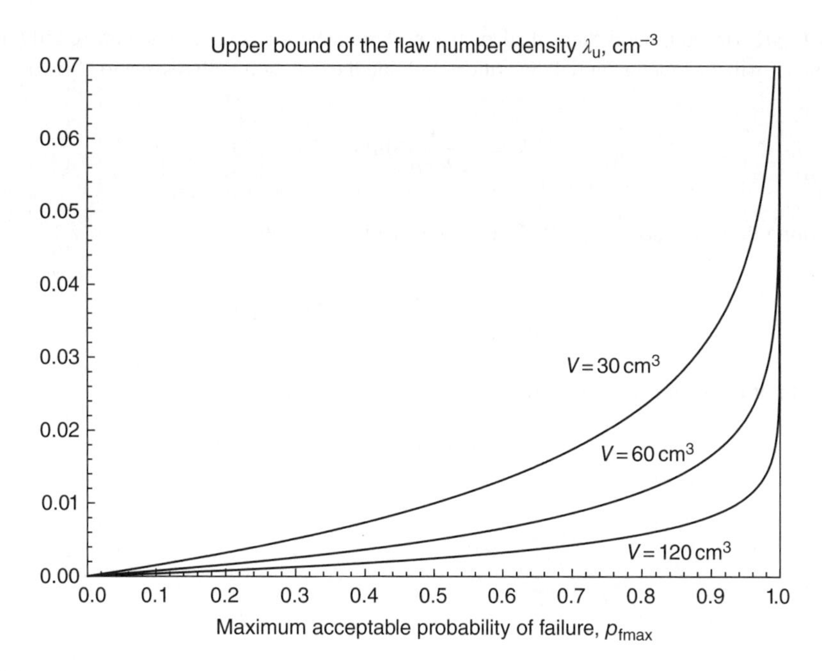

Figure 13.8 A flaw number density upper bound, as a function of the maximum acceptable probability of failure, for different values of the stressed volume

Equation 13.21 can be used for calculating the probabilities of failure from the lower tail of the strength distribution in case of failure controlled by the size of the flaws. Limiting the size of the stressed volume limits the probability of failure initiated by flaws, which is of significant importance to the design for reliability. By solving Equation 13.21 with respect to V (given a specified maximum acceptable probability of failure $p_{\sigma\max}$ at a stress level σ), an upper bound V^* for the stressed volume can be determined:

$$V^* = -\frac{1}{\lambda\left[1 - G(d_\sigma)\right]}\ln\left(1 - p_{\mathrm{fmax}}\right) \tag{13.22}$$

The upper bound V^* guarantees that if for the stressed volume, $V \leq V^*$ is satisfied, the probability of failure p_f will be smaller than the maximum acceptable level p_{fmax}.

Example

A piece of steel wire with diameter 2 mm may be subjected to tensile overloading. During the overloading, oxide inclusions with size greater than 200 μm will cause failure if present in the stressed section of the wire. In the steel, the number density of the oxide inclusions is 0.001 mm^{-3}. From quantitative microscopy studies, it is known that 20% of these inclusions have size greater than 200 μm.

What is the maximum length of the stressed section of the wire so that the probability of failure during overloading, caused by inclusions, does not exceed the maximum acceptable value of 0.05?

Solution

The probability of failure initiated by the inclusions is determined from Equation 13.17:

$$p_f = 1 - \exp\left[-\lambda V F_c(\sigma)\right]$$

where $\lambda = 0.001$ is the number density of all inclusions, V is the volume of the stressed wire and $F_c(\sigma) = 0.2$ is the probability that an inclusion will be critical. Solving the equation with respect to V yields

$$V = -\frac{1}{\lambda F_c(\sigma)} \ln(1 - p_f)$$

Substituting in the equation $p_f = 0.05$ and $\lambda F_c(\sigma) = 0.0002$ yields

$$V = -\frac{1}{0.0002} \ln(1 - 0.05) \approx 256.5\,\text{mm}^3$$

The length of the wire is determined from $[(\pi \times 2^2)/4] \times L = 256.5$

$$L = 81.6\,\text{mm}.$$

14

A Comparative Method for Improving the Reliability and Availability of Components and Systems

14.1 Advantages of the Comparative Method to Traditional Methods

Often, the absolute reliability of a product cannot be revealed. Here are some of the reasons:

- In many cases, reliability–critical data (failure frequencies, strength distribution of the flaws, fracture mechanism, repair times) are simply not available for the components.
- The physical processes and physical mechanisms underlying the failure modes, remain unknown or are associated with large uncertainties.
- The complex influence and uncertainty associated with the environment, the operational loads and the duty cycles.
- The variability associated with reliability–critical design parameters (e.g. the state of manufactured surfaces, components tolerances, unbalanced forces, internal environment, duty cycles, etc.).
- The non-robustness of the reliability prediction models.

Key reliability-controlling parameters are associated with uncertainty which does not allow to reveal the absolute reliability level. Major sources of uncertainty are associated with the natural variation of the material properties and the uncertainty associated with their measurement, the uncertainty in determining the times to failure, estimating load magnitudes, etc. Furthermore, even if this information were available, for common, widely used reliability models, even relatively small errors in the reliability parameters lead to large errors in the model predictions which renders such predictions of questionable value.

Here are some examples illustrating the problem. Reliability predictions during multiple loading are often based on the load–strength interference (Carter, 1986; Freudenthal, 1954) model which involves two basic random variables, 'load' and 'strength', characterised by distinct distributions. The reliability in this case is determined by the probability of a relative configuration in which the load is smaller than the strength.

Reliability and Risk Models: Setting Reliability Requirements, Second Edition. Michael Todinov.

Suppose that a load with cumulative distribution function $F(x)$ has been applied n times. The probability density function of the strength is given by $s(x)$. The reliability R (probability of no failure) during multiple loading is then given by the integral (Carter, 1986)

$$R = \int_{s_{\min}}^{s_{\max}} F^n(x)s(x)dx \tag{14.1}$$

where $s_{\min}$ and $s_{\max}$ are the lower and the upper limit of the strength. The application of expression (14.1) for reliability predictions however is associated with large errors if n is relatively large. For a large 'n', $F^n(x)$ is very sensitive to variations ΔF in the load distribution $F(x)$. Small inaccuracies in the parameters of $F(x)$ will cause large variations of $F^n(x)$. Indeed, if, for the sake of simplicity, the strength has been taken to be constant $s = c$ (the variance of $s(x)$ is zero), the integral (14.1) becomes $R = F^n(c)$ or simply $R = F^n$. An error in the parameters of the load distribution $F(x)$ for $x = c$ leads to an error in the reliability R. The differential of R is $dR = nF^{n-1}dF$ which can also be presented as

$$\frac{dR}{R} = n \times \frac{dF}{F}$$

Consequently, a small relative error $\Delta F/F$ will result in a large relative error $\Delta R/R$ of the estimated reliability R, magnified n times. In other words, the model (14.1) is not robust for large n and predicts the reliability R with large errors. Inevitable errors in the parameters of the load and strength distributions will cause very large errors in the calculated reliability value. Since load distributions are always associated with uncertainty in their parameters, using Equation 14.1 to make reliability predictions for a large number n of load applications is of a questionable value.

Very similar is the case where reliability of a system with a large number of components logically arranged in series is considered. With the continual increase in the complexity of the existing engineering equipment, such complex systems are now very common. The reliability of a system with components logically arranged in series, working independently from one another, is estimated from

$$R_{sys} = R_1 \times R_2 \times \cdots \times R_M \tag{14.2}$$

For the special case where the system is built on M identical components, each with reliability R, the reliability of the system becomes $R_{sys} = R^M$. Estimating the reliability R of the component however is always associated with uncertainty. Indeed, let us assume that the reliability of the component has been estimated from testing n components of the same type. At the beginning, the reliability is unknown and the initial distribution (prior) is uniform in the interval 0, 1:

$$f(R) = \begin{cases} 1, & \text{for } 0 \le R \le 1 \\ 0, & \text{otherwise} \end{cases} \tag{14.3}$$

If x denotes the number of components (out of n components) which survived a single test, the conditional probability $P(x|R)$ is given by

$$P(x \mid R) = R^x(1-R)^{n-x} \tag{14.4}$$

Note that $C_n^x = n!/[x!(n-x)!]$ is not present as a factor because the surviving sequence is given. The prior distribution $f(R)$ regarding the unknown reliability R can then be revised by using the Bayes' theorem (Ang and Tang, 1975):

$$f(R \mid x) = \frac{R^x(1-R)^{n-x}}{\int_0^1 R^x(1-R)^{n-x}dR}, \quad \text{for } 0 \le R \le 1 \tag{14.5}$$

and $f(R \mid x) = 0$, otherwise. The posterior distribution $f(R \mid x)$ is the beta probability distribution.

Increasing the number of tests n will reduce the uncertainty associated with the unknown reliability R but will never eliminate it. In other words, *uncertainty and errors associated with the reliability parameters in the reliability models are inherent and cannot be eliminated.* An error ΔR in the reliability of a single component could lead to a large error $\Delta R_{sys} / R_{sys} = n(\Delta R / R)$ in the predicted reliability for the system.

Predicting the probability of failure locally initiated by flaws is also associated with uncertainty related to the type of existing flaws initiating fracture, the size distributions of the flaws, the locations and the orientations of the flaws and the microstructure around the flaws (crystallographic orientation, chemical and structural inhomogeneity, local fracture toughness, etc.). Some of these random variables are not statistically independent. Capturing the uncertainty necessary for a correct prediction of the reliability of components is a formidable task.

An efficient way of resolving this predicament, when the focus is on improving the reliability of designs, is *not to attempt prediction of the absolute reliability.* Competing designs are simply compared on the basis of their reliability, which is calculated with the same predefined set of reliability parameters. *Despite that the absolute reliability level remains unknown, the relative reliability level can be increased irrespective of the absolute reliability level.*

14.2 A Comparative Method for Improving the Reliability of Components Whose Failure is Initiated by Flaws

In order to isolate and assess the impact of the design shape or type of loading on the probability of failure of brittle components, the same notional material properties, number density of flaws, fracture criterion and distribution of the flaw size can be assumed for the competing designs. Given these common assumed properties, the probabilities of overstress failure characterising the competing designs are compared, and the design characterised by the smallest probability of overstress failure is selected.

In order to estimate the probability of failure of a component with complex shape locally initiated by flaws, Equation 14.6 derived in Todinov (2005, 2006a) will be applied,

$$p_f = 1 - \exp(-\lambda V F_c) \tag{14.6}$$

where λ is the number density of the existing flaws and V is the volume of the component.

Equation 14.6 is part of a methodology proposed in (Todinov, 2006a) for determining the probability of failure of components with complex shape initiated by flaws, where F_c is the probability that a flaw will be critical (will cause failure) given that it resides with certainty in the component/structure. The methodology for determining the probability F_c however, was based on a Monte Carlo simulation. For large components and structures, characterised by small zones of stress intensification, the Monte Carlo simulation does not guarantee that representative statistical information will be collected from the small yet reliability–critical zones of stress intensification. Increasing the number of simulation trials n improves the chances of sampling the small stress concentration zones and decreases the error by a factor $1/\sqrt{n}$, but the price is increased computation time.

Consequently, a new algorithm was proposed in Todinov (2009a) which does not rely on Monte Carlo simulation to determine the probability F_c in Equation 14.6 that a flaw will be critical. The purpose is to guarantee that the computation will not be slowed down and, at the same time, representative statistical information will be collected from all stress concentration zones.

By using this algorithm, the resistance of design shapes to overstress failure initiated by flaws can be compared, and the design with the highest resistance selected. Essential part of the method is a block for reading the output data file from the ABAQUS software package (ABAQUS, 2007) for finite element analysis. For each finite element, the block extracts the principal stresses characterising the centroid of each finite element and its volume.

Weak spherical inclusions are assumed for the purpose of comparing the performance of competing designs. During loading, the weak spherical inclusions produce penny-shaped cracks perpendicular to the direction of the maximum principal tensile stress acting on the inclusions. The size of the crack is equal to the maximum diameter of the inclusion. For failure controlled by the size of the inclusions, the probability F_c can be determined by scanning all finite elements. Each finite element i is characterised by a volume ΔV_i and the principal stresses acting in it. For a particular inclusion, the remote stress is approximated with the stress tensor characterising the centre of the inclusion.

The critical flaw radius $a_{c,i}$ beyond which a fast fracture is triggered, can be calculated for each finite element i. In this way, information about the failure resistance is obtained from all parts of the stressed volume without missing even the smallest zone of stress concentration. It is assumed that brittle fracture is caused by the maximum principal tensile stress σ_t opening the penny-shaped cracks initiated by the inclusions. The penny-shaped cracks are oriented in the worst possible direction, perpendicular to the direction of the maximum tensile stress σ_t. Consequently, the fracture criterion

$$Y\sigma_t\sqrt{\pi a} = K_{Ic} \tag{14.7}$$

can be adopted. In the fracture criterion (14.7), K_{Ic} is the fracture toughness of the material, a is the radius of the flaw, and $Y = 2/\pi$ is the geometry factor. The failure criterion (14.7) follows the stress intensity approach (see Chapter 12). Fracture occurs if the stress intensity factor becomes equal to or greater than the fracture toughness of the material. From Equation 14.7, the critical size of the flaw which causes brittle fracture can be obtained:

$$a_c = \frac{\pi}{4}\left(\frac{K_{Ic}}{\sigma_t}\right)^2 \tag{14.8}$$

Given that a spherical flaw with a random location resides in the component, the conditional probability that the centre of the flaw will be in the ith finite element is $\Delta V_i/V$ where ΔV_i is the volume of the ith finite element and V is the total volume of the component ($V = \sum_i V_i$). The conditional probability that the flaw will reside in the ith finite element and will cause failure, given that the flaw is inside the component, is

$$p_{f,i} = \frac{\Delta V_i}{V} \times P\left(a > a_{c,i}\right) \tag{14.9}$$

where $P(a > a_c)$ is the probability that the flaw size a will be greater than the critical flaw size $a_{c,i}$ characterising the ith finite element. Since, $P(a > a_{c,i}) = 1 - F(a_{c,i})$ where $F(a)$ is the size distribution of the flaws, the conditional probability $p_{f,i}$ becomes

$$p_{f,i} = \frac{\Delta V_i}{V} \times \left[1 - F\left(a_{c,i}\right)\right] \tag{14.10}$$

Noticing that the events 'the centre of the flaw resides in the ith finite element' are mutually exclusive, the probability that a flaw will be critical, given that it resides in the component, is determined from

$$F_c = \frac{1}{V}\sum_{i=1}^{n} \Delta V_i \times \left[1 - F\left(a_{c,i}\right)\right] \tag{14.11}$$

The calculated value for F_c is substituted in Equation 14.6. As a result, the probability of failure of components with complex shape containing flaws with number density λ becomes

$$p_f = 1 - \exp\left\{-\lambda \sum_{1}^{n} \Delta V_i \times \left[1 - F\left(a_{c,i}\right)\right]\right\} \quad (14.12)$$

where n is the number of finite elements into which the volume V of the component has been divided. For equal finite element volumes $\Delta V_i = \Delta V$, considering that $n \times \Delta V = V$, equation (14.12) can be simplified further. The comparative method for assessing the resistance of components to failure initiated by flaws can therefore be summarised by the following algorithm:

Algorithm 14.1

- Assume common material properties, common spherical flaws, flaw number densities and flaw size distribution for the competing design alternatives.
- Determine the variation of stresses in the component by a standard finite elements package.
- By applying a failure criterion, for each finite element $i(i = 1,...,n)$, determine the probability p_{fi} that a flaw 'will initiate' failure given that its centre lies in the ith finite element.
- Estimate the probability that a flaw will be critical from

$$F_c = \sum_{i=1}^{n} \frac{\Delta V_i}{V} \times p_{fi}$$

- Estimate the probability of failure of the component from Equation 14.6.

Apart from being more precise, this algorithm is much more efficient, because it does not rely on Monte Carlo simulation. The probability of failure of the component is obtained after n steps, equal to the number of finite elements. In other words, the algorithm is of linear time complexity $O(n)$.

The described method can be applied to any material where failure is locally initiated by single, non-interacting flaws. Once a flaw becomes unstable, a crack is formed that propagates without a growth resistance until failure. Components with such a failure mechanism are usually manufactured from brittle materials like ceramics, high-strength steels, glasses, stone, concrete, etc. The described model is also valid for components from low carbon steels undergoing cleavage fracture at low temperature. Cleavage in steels usually propagates from cracked inclusions (Rosenfield, 1997). It usually involves a small amount of local plastic deformation to produce dislocation pile-ups and crack initiation from a particle which has cracked during the plastic deformation. The method is particularly suitable for optimising the shape of brittle components (e.g. the cross section of ceramic beams), in order to increase the resistance to failure locally initiated by flaws.

Here is a numerical example. Consider three design alternatives of a bracket shown in Figure 14.1. In order to compare the resistance to failure locally initiated by flaws of the competing design alternatives A, B and C, spherical flaws with the same number density of $\lambda_A = \lambda_B = \lambda_C = 0.1\,\text{cm}^{-3}$ has been assumed for all brackets. The same flaw size distribution

$$G(a) = 1 - \exp\left(-\frac{a}{a_{\text{mean}}}\right) \quad (14.13)$$

has also been assumed for all brackets, where a is the radius of the flaws and $a_{\text{mean}} = 1\,\text{mm}$ is the mean radius. Failure is assumed to occur if the stress intensity factor associated with a tensile crack tip opening mode of a penny-shaped crack emanating from a spherical flaw exceeds the critical stress intensity factor of the material:

$$K_I = \frac{\pi}{2}\sigma\sqrt{\pi a} \geq K_{Ic} \quad (14.14)$$

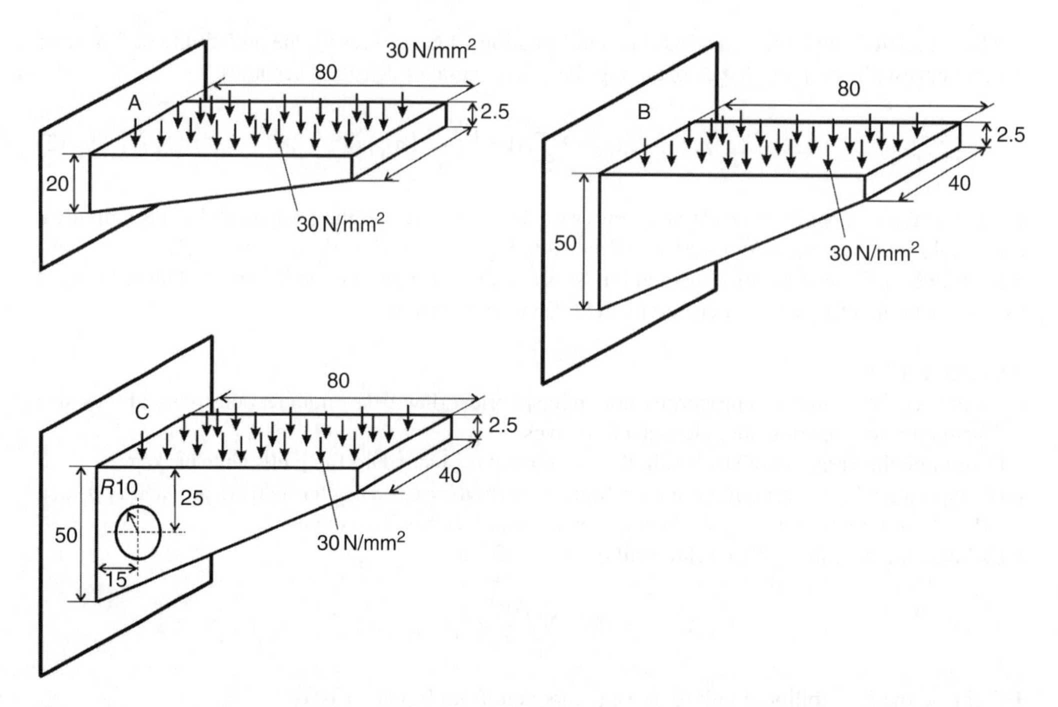

Figure 14.1 Comparing the resistance to failure locally initiated by flaws of three design alternatives A, B and C

Brittle material characterised by a fracture toughness $K_{Ic} = 6\,\text{MPa}\sqrt{m}$ has been assumed for the material of the brackets. It is assumed that no failure initiated by flaws can occur in a zone where the maximum principal stress is compressive. The volumes of the brackets are $V_A = 36\,\text{cm}^3$, $V_B = 84\,\text{cm}^3$ and $V_C = 71.4\,\text{cm}^3$, correspondingly.

The calculated probability of overstress failure for design A is

$$p_{f,A} = 1 - \exp(-\lambda_A V_A F_{cA})$$
$$= 1 - \exp(-0.1 \times 36 \times 0.33) \approx 0.70$$

for design B:

$$p_{f,B} = 1 - \exp(-\lambda_B V_B F_{cB})$$
$$= 1 - \exp(-0.1 \times 84 \times 0.043) \approx 0.30$$

and for design C:

$$p_{f,C} = 1 - \exp(-\lambda_C V_C F_{cC})$$
$$= 1 - \exp(-0.1 \times 71.4 \times 0.05) \approx 0.30$$

Compared to design A, designs B and C are more resistant to failure locally initiated by flaws, and design C is a lightweight design. Despite the hole in bracket C, the probability of failure initiated by flaws did not decrease noticeably because material has been removed from the zone where the maximum principal stress is mostly compressive.

As can be seen, no real reliability–critical data were necessary to compare the resistance of the brackets to fast fracture.

14.3 A Comparative Method for Improving System Reliability

Calculating the absolute reliability built in a product is often an extremely difficult task because in many cases reliability–critical data (e.g. failure frequencies) are simply not available for the system components. Often, the only available information is the ranking of the components in terms of their reliability, without being possible to attach any value to their failure frequencies. Such is the case where old and new components of the same type are used in the same system. Because of inevitable component wearout and deterioration, it is usually sensible to assume that the new components are more reliable than the old components.

Consider the functional diagrams of two systems built with pipes and valves of the same type: old valves A and new valves B (Figure 14.2). With respect to the function 'valve closure on demand', an old valve A is less reliable than a new valve B. The valves are working independently from one another and all of them are initially open. The question of interest is which system is more reliable with respect to stopping the fluid in the pipeline.

The reliability block diagrams of the systems, with respect to the function 'stopping the fluid in the pipeline', are given in the next figure (Figure 14.3).

System '1' is more reliable than system '2'. Let a denote the reliability of an old valve A and b denote the reliability of a new valve B, $(a < b)$. Then

$$R_1 = \left[1 - \left(1 - a^2\right)\left(1 - ab\right)\right]b = ab\left(a + b - a^2 b\right)$$

is the reliability of the first system and

$$R_2 = \left[1 - \left(1 - b^2\right)\left(1 - ab\right)\right]a = ab\left(a + b - ab^2\right)$$

is the reliability of the second system. Because $R_1 - R_2 = a^2 b^2 (b - a) > 0$, the first system is more reliable.

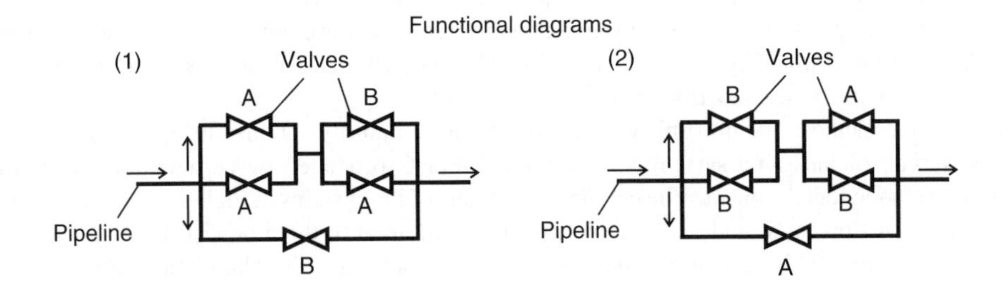

Figure 14.2 Functional diagrams of two different systems built with old valves A and new valves B

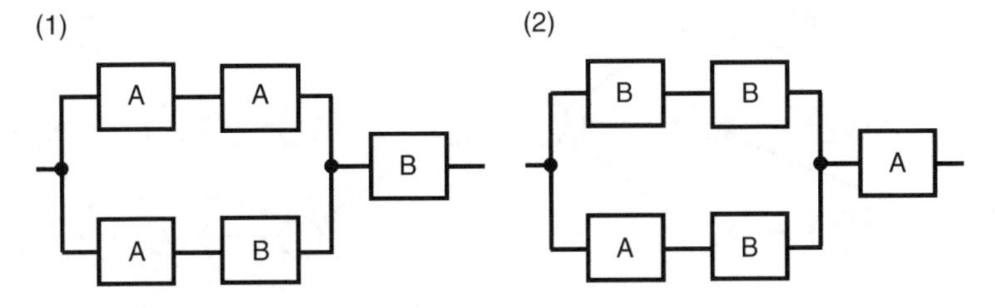

Figure 14.3 Reliability networks of the systems in Figure 14.2

14.4 A Comparative Method for Improving the Availability of Flow Networks

An important application of the proposed approach is in comparing quickly the performance of competing network topologies and selecting the topology with the best performance. The comparative method for assessing the performance of competing network topologies can be summarised by the following steps:

- Assume common flow capacities, failure frequencies and repair times for the corresponding components/edges of the compared networks.
- Determine the performance of the networks by using an appropriate software tool.
- Select the topology or variant with superior performance.

Production availability is an important indicator of the performance of repairable production systems (Ebeling, 1997). It is defined as the ratio of the total amount of production fluid delivered by the system for 1 year in the presence of failures of the pipeline sections to the total amount of production fluid which can be potentially delivered in the absence of failures (Todinov 2007; 2013a). Even a very small percentage decrease in the production availability (1–2%) translates into big financial losses over the entire period of operation. Maximising the production availability is already an essential part of the design of new production systems and telecommunication networks. The production availability of an oil and gas production network, for example, is its most important characteristic which determines the profitability of the installation. The availability of electrical distribution networks determines the quality of electricity supply to customers.

Consider, for example, the competing production networks in Figure 14.4a and b, where, for the sake of simplicity, all edges have the same flow rate capacity of 40 flow units per day, hazard rate of 4 expected failures per year and downtime for repair 10 days. Edges (3, 8) and (4, 9) from network 14.4a and edges (2, 8) and (4, 10) from network 14.4b are redundant. Without a supporting comparative model, it is very difficult to infer which network topology is superior. Applying the software tool for determining production availability developed in Todinov (2013a) yields production availability of $\psi_a = 70\%$ for the network in Figure 14.4a and $\psi_b = 75\%$ for the network topology in Figure 14.4b. As can be verified, despite the seemingly insignificant differences in the competing topologies, the impact on the production availability is significant. No real reliability data are necessary to deduce which system topology possesses superior production availability.

Comparative methods for maximising the production availability can also be applied for selecting the best variant of the same system from possible variants produced by permutations of its interchangeable components. Consider the competing variants of the systems in Figure 14.5 which includes three sources of flow s_1, s_2 and s_3, three old pipeline sections (O), three new sections (N) and three medium-age sections (M). Because of the inevitable deterioration which the sections undergo due to

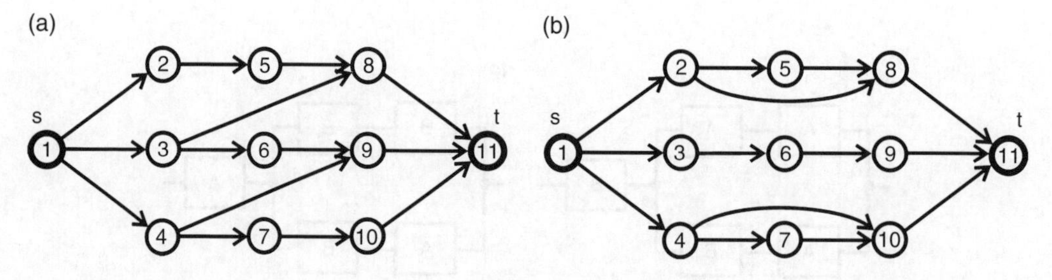

Figure 14.4 (a, b) Two competing networks with different types of redundancy topology

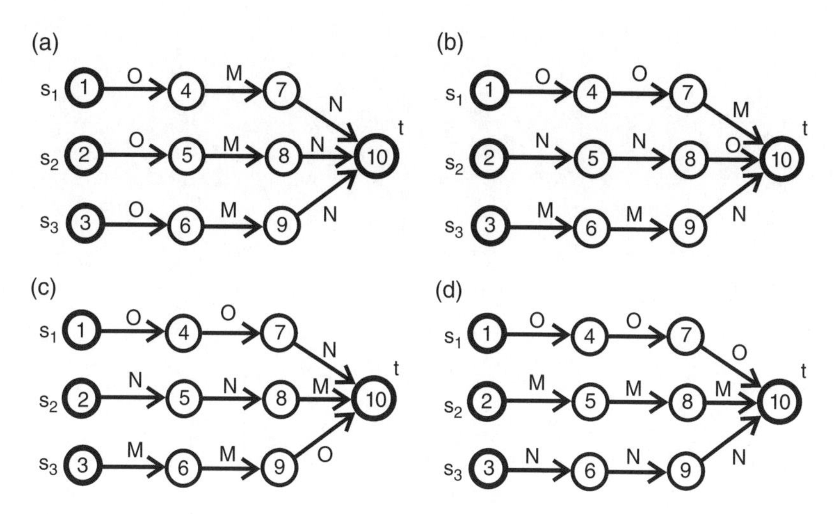

Figure 14.5 (a–d) Competing arrangements of the components in a gas production system consisting of three parallel branches and different state of deterioration of the pipeline sections

wearout, the new sections (N) are more reliable than the medium-age sections which in turn are more reliable than the old sections.

To compare the performance of the competing variants of the production system and decide which variant is most beneficial, some exemplary reliability data are assumed, consistent with the ranking of the reliabilities of the new sections, medium-age sections and old sections.

The capacity of each pipeline section was chosen to be 100 thousand cubic metres of fluid per day. A repair time of 20 days is assumed for each failed section (new, medium-age and old section). During the repair of a failed pipeline section, the corresponding parallel branch is not delivering any fluid. The failure frequency (expected number of failures per year) of the old sections is assumed to be 8 year^{-1}; the failure frequency of the new sections (N) is 0.1 year^{-1}, while the failure frequency of the medium-age sections (M) is 2 year^{-1}.

The production availability, for 1 year of operation, of the different variants in Figure 14.5 was assessed by using the discrete-event simulator for the production availability of repairable flow networks described in Todinov (2013a). The production availabilities characterising the different variants were as follows: system ‘a’, 62.3%; system ‘b’, 64.3%; system ‘c’, 64.6%; and system ‘d’, 68.4%. The largest production availability is demonstrated by system ‘d’.

The largest removed risk of lost production due to failures was achieved for variant ‘d’, which showed a significant (more than 6%) increase in production availability compared to the worst variant ‘a’.

The variant presented in Figure 14.5d is an example of *a well-ordered parallel-series system.* A well-ordered parallel-series arrangement is obtained if the available components are used to build the branch with the highest possible reliability/availability, the remaining components are used to build the next branch with the highest possible reliability/availability and so on until the entire parallel-series arrangement is built. A detailed discussion of well-ordered parallel-series systems has been presented in Chapter 19.

15

Reliability Governed by the Relative Locations of Random Variables in a Finite Domain

15.1 Reliability Dependent on the Relative Configurations of Random Variables

There exist numerous examples of reliability governed by the relative locations of a number of random variables, uniformly distributed in a finite domain. A commonly encountered problem is presented in Figure 15.1a.

During a finite time interval of length a, exactly n consumers connect to a supply system independently and randomly. Each connection is associated with a shock (increased demand) to the supply system which needs a minimum time interval s to recover and stabilise after a connection (demand). The supply system is overloaded if two or more successive demands follow within a critical time interval s (Figure 15.1a). The probability of overloading is equal to the probability that two or more random demands will cluster within the critical interval with length s.

Another common case is present when n users arrive randomly during a finite time interval of length a and use a particular piece of equipment for a fixed time s. The problem is to calculate the probability of a collision of demands, which occurs if two or more users arrive within a time interval s (Figure 15.1a).

A mechanical problem of a similar nature is presented in Figure 15.1b. A number of discs/wheels are attached to a common shaft, independently from one another. Each disc is associated with an eccentricity which creates a centrifugal force F_i, $i = 1, 2, 3$, on the shaft rotating at a high speed, as shown in Figure 15.1b. If the forces cluster within a critically small angle s, the shaft will suffer excessive deformation during rotation at a high speed. The problem is to calculate the probability of clustering the centrifugal forces within a critical angle s.

15.2 A Generic Equation Related to Reliability Dependent on the Relative Locations of a Fixed Number of Random Variables

Only configurations of uniformly distributed random variables in a common domain are considered where the safe/failure state depends only on the relative locations of the random variables and not on their absolute locations. The location of each random variable can be represented as a point in the

---

Reliability and Risk Models: Setting Reliability Requirements, Second Edition. Michael Todinov.

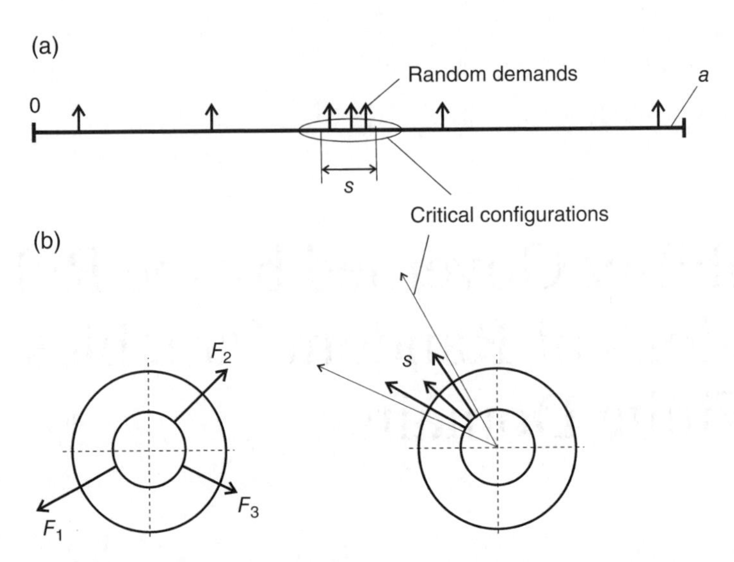

Figure 15.1 (a) Overloading a supply system due to clustering of a fixed number of random demands. (b) Safe and failure configurations of centrifugal forces

common domain, and each point in the common domain corresponds to a possible location of a random variable (see Figure 15.2a).

The random variables are not necessarily identical; only their distribution in the finite domain is uniform. Suppose that a domain with measure v has been defined as in Figure 15.2a, where n uniformly distributed random variables form a safe/failure configuration with probability p. We assume the probability that two or more random variables will reside simultaneously into a very small domain increment Δv is negligible. If p_i^* is the probability of a safe/failure configuration when the ith random variable is located at the boundary of the common domain, the link between the probability p and the probabilities p_i^* is given by

$$pv^n = C + \int v^{n-1} \sum_{i=1}^{n} p_i^*(v)\,dv \tag{15.1}$$

where C is an integration constant determined from the boundary conditions (Todinov, 2004a). In the case where all of the random variables are identical, all probabilities p_i^* are equal $\left(p_1^* = p_2^* = \cdots = p^*\right)$, and Equation 15.1 transforms into

$$pv^n = C + \int v^{n-1} n p^*(v)\,dv \tag{15.2}$$

The derivation of Equation 15.1 has been presented in Appendix 15.1.

In the case of distinct (non-identical) random variables, the probabilities p_i^* in Equation 15.1 are different in general, and determining the integration constant C is more complicated. Thus, in a finite time domain a, depending on which random variable (event) appears first, different boundary conditions may be present.

Example

Consider a problem present in many manufacturing processes where a number of machine centres demand particular resource (expensive measuring equipment, control equipment, production equipment or an operator), at a random time during the production process. Because the control equipment is expensive and unique, it is not feasible to equip each machine centre with a separate piece of equipment.

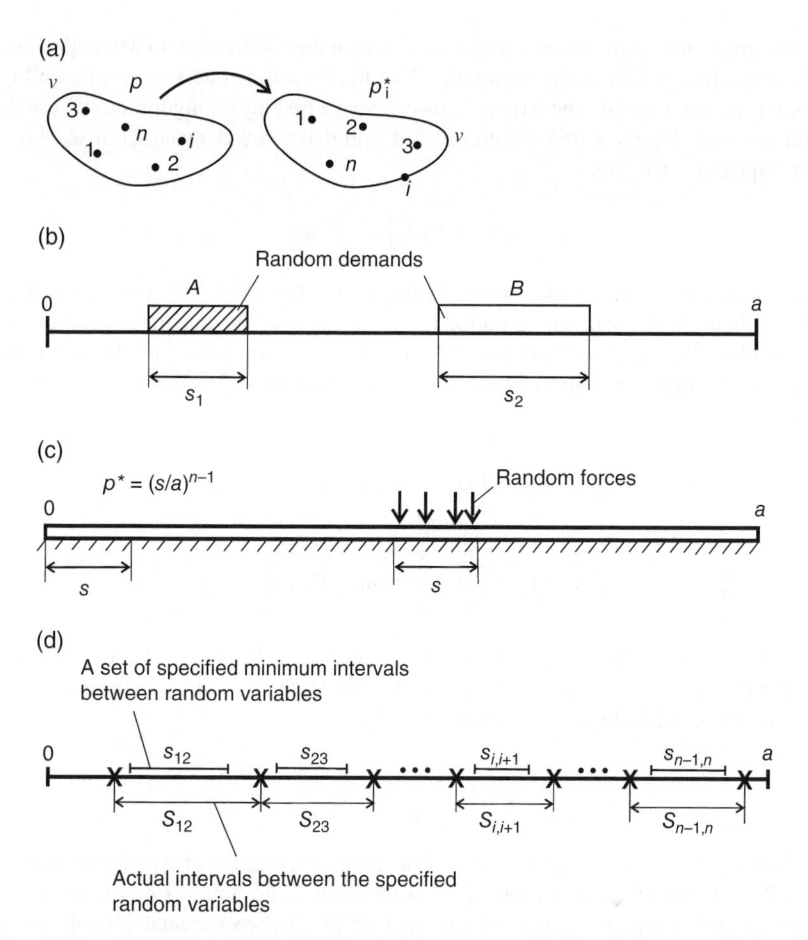

Figure 15.2 (a) The probability of a safe/failure configuration of uniformly distributed random variables in a common domain is given by the generic Equation 15.1. (b) Two uniformly distributed random demands of different length. (c) Total clustering of random forces acting on a structural member. (d) Specified distances between random variables, whose total number is known

The demand can be for a particular resource (e.g. water vapour), and there is only a single source available which is capable of supplying only one machine centre at a time. The demands for the resource may occur at random times during a shift with duration a.

Consider a simplified version of this problem where the services of the source are demanded by two machine centres A and B each placing a single demand randomly during the time interval a (Figure 15.2b). Suppose that machine centre A demands the resources for a duration s_1, while machine centre B demands the resource for a duration s_2. The probability p of no overlapped demands (smooth operation) can be presented as a sum of the probabilities p_1 and p_2 of smooth operation associated with the mutually exclusive and exhaustive events where either the demand from machine centre A or the demand from machine centre B arrives first. Because Equation 15.1 can also be applied to the case where one of the demands (e.g. the demand A) always appears first, the probability p_1 of smooth operation in this case can be calculated from

$$p_1 a^n = C_1 + \int a^{n-1} p_1^* \, da \tag{15.3}$$

where a is the length of the time interval and $n = 2$ is the number of the random demands. If the demand from machine centre A is fixed at the beginning '0' of the finite time interval, the probability of a safe configuration is $p_1^* = (a - s_1)/a$. The term p_2^* is missing because only configurations where the demand from machine centre A arrives first are considered. Similarly, if a demand from machine centre B arrives first, Equation 15.1 gives

$$p_2 a^n = C_2 + \int a^{n-1} p_2^* da \tag{15.4}$$

for the probability p_2 of a 'safe' configuration. If the demand B is fixed at the beginning '0' of the time interval, the probability of a safe configuration is $p_2^* = (a - s_2)/a$. The integration constants C_1 and C_2 and the probabilities p_1 and p_2 in Equations 15.3 and 15.4 are associated with the cases where either the demand A or the demand B arrives first. Integrating Equations 15.3 and 15.4 over the finite time interval a gives

$$p_1 a^2 = C_1 + \int a\left(\frac{a - s_1}{a}\right) da = C_1 + \frac{a^2}{2} - s_1 a \tag{15.5}$$

$$p_2 a^2 = C_2 + \int a\left(\frac{a - s_2}{a}\right) da = C_2 + \frac{a^2}{2} - s_2 a \tag{15.6}$$

From the boundary conditions: $p_1 = 0$ if $a = s_1$, and $p_2 = 0$ if $a = s_2$ – the integration constants $C_1 = s_1^2/2$ and $C_2 = s_2^2/2$ are determined. Adding the probabilities p_1 and p_2 as probabilities of mutually exclusive events and dividing by a^2 result in

$$p = 1 - \frac{s_1 + s_2}{a} + \frac{s_1^2 + s_2^2}{2a^2} \tag{15.7}$$

for the probability $p = p_1 + p_2$ of no collision of demands irrespective of which demand arrives first. Equation 15.7 has been verified by a Monte Carlo simulation. Thus, for $a = 1$, $s_1 = 0.1$ and $s_2 = 0.05$, the equation yields probability $p = 0.856$, which is close to the experimental probability $p = 0.8558$ obtained from the simulation.

Example

The application of Equation 15.1 can be illustrated by another simplified problem from determining the probability of overloading a structural component with length a from n loads randomly distributed during a time interval (0, a). Overloading of the structural member occurs if all n loads are concentrated (cluster) within a small length s anywhere within the length a (Figure 15.2c). In this case, the critical configuration of the random variables is a 'total clustering' within a distance s. Indeed, because $p_1^* = p_2^* = \cdots = p_n^* = p^*$ and $p^* = (s/a)^{n-1}$, if one of the loads is fixed at the beginning of the length a (point '0' in Figure 15.2c), according to Equation 15.2, the probability of total clustering is

$$pa^n = C + \int a^{n-1} n \left(\frac{s}{a}\right)^{n-1} da = C + ns^{n-1}a \tag{15.8}$$

For $a = s$, the probability of total clustering becomes $p = 1$. Substituting $a = s$ in Equation 15.8 yields $s^n = C + ns^n$, from which $C = -(n-1)s^n$. Finally, the probability of overloading the structural component becomes

$$p_f = n\left(\frac{s}{a}\right)^{n-1} - (n-1)\left(\frac{s}{a}\right)^n \tag{15.9}$$

15.3 A Given Number of Uniformly Distributed Random Variables in a Finite Interval (Conditional Case)

If the probabilities p_i^* in Equation 15.1 cannot be calculated easily, the problem can be solved by reducing its complexity and solving a series of problems involving a smaller number of random variables. This method will be illustrated by finding the probability that the actual distances $S_{12}, S_{23}, \ldots, S_{n-1,n}$ between the locations of a given number n of uniformly distributed random variables in a finite interval a will be greater than the corresponding specified minimum distances $s_{12}, s_{23}, \ldots, s_{n-1,n} \left(\sum_{i=1}^{n-1} s_{i,i+1} \le a\right)$, where $s_{i,i+1}$ is the specified minimum distance between adjacent random variables with indices i and $i+1$ (Figure 15.2d).

Suppose that p_n is the probability of existence of the specified minimum distances between the n random variables. Then, from Equation 15.2, it follows that

$$p_n a^n = C + \int a^{n-1} n \left(\frac{a - s_{12}}{a}\right)^{n-1} p_{n-1} da \tag{15.10}$$

In this way, the probability p_n that the actual distances $S_{12}, S_{23}, \ldots, S_{n-1,n}$ between the random variables will be greater than the specified minimum distances $s_{12}, s_{23}, \ldots, s_{n-1,n}$ has been expressed by the probability p_{n-1} related to $n-1$ variables. This is the probability that the actual distances between $n-1$ adjacent random variables over the shorter time interval $a - s_{12}$ will be greater than the specified minimum distances $s_{23}, \ldots, s_{n-1,n}$, if one of the random variables is 'fixed' at the beginning of the finite interval $a - s_{12}$. The complexity of the initial problem has been reduced. The complexity of the simpler problem can also be reduced in a similar fashion, and finally, a problem with a trivial solution will be obtained. Starting from the trivial solution, all necessary intermediate solutions can be produced as well as the solution of the initial complex problem.

Indeed, following Equation 15.10:

$$p_{12} a^2 = C + \int 2a p_{12}^* da = C + (a - s_{12})^2 \tag{15.11}$$

where $p_{12} = P(S_{12} \ge s_{12})$ is the probability of existence of a specified minimum distance s_{12} for two random variables only. Because $p_{12}^* = (a - s_{12})/a$ and because for $a = s_{12}$, $p_{12} = 0$ and $C = 0$, from Equation 15.11, it follows that $p_{12} = (1 - s_{12}/a)^2$. Similarly, for three random variables, the probability $p_{123} = P(S_{12} \ge s_{12} \cap S_{23} \ge s_{23})$ of existence of intervals greater than s_{12} and s_{23} is

$$p_{123} a^3 = C + \int 3a^2 \left(\frac{a - s_{12}}{a}\right)^2 \left(1 - \frac{s_{23}}{a - s_{12}}\right)^2 da = C + (a - s_{12} - s_{23})^3$$

from which $p_{123} = [1 - (s_{12} + s_{23})/a]^3$, because for $a = s_{12} + s_{23}$, $p_{123} = 0$ and $C = 0$. In a similar fashion,

$$p_{12\ldots n} = P\left(S_{12} \ge s_{12} \cap S_{23} \ge s_{23} \cap \cdots \cap S_{n-1,n} \ge s_{n-1,n}\right) = \left(1 - \frac{s_{12} + s_{23} + \cdots + s_{n-1,n}}{a}\right)^n \tag{15.12}$$

is obtained for the probability of existence of the specified minimum distances.

Next, an equation can be derived giving the probability p that the actual distances between n adjacent random variables will be at least $s_{12}, s_{23}, \ldots, s_{n-1,n}$ and the actual distance S_{01} of the first random variable location from the start of the interval a will be at least s_{01} $(S_{01} \ge s_{01})$. The probability p is a product of the probability $\left[(a - s_{01})/a\right]^n$ that all random variable locations will be at a distance larger than s_{01} from the beginning of the finite interval a and the probability that the distances between the random variables will be greater than the specified distances (see Eq. 15.12). As a result,

$$p \equiv P(S_{01} \geq s_{01} \cap S_{12} \geq s_{12} \cap S_{23} \geq s_{23} \cap \cdots \cap S_{n-1,n} \geq s_{n-1,n}) = \left(\frac{a - s_{01}}{a}\right)^n \left(1 - \frac{s_{12} + s_{23} + \cdots + s_{n-1,n}}{a - s_{01}}\right)^n$$

is obtained which, after simplifying, becomes

$$p \equiv P(S_{01} > s_{01} \cap S_{12} > s_{12} \cap S_{23} > s_{23} \cap \cdots \cap S_{n-1,n} > s_{n-1,n}) = \left(1 - \frac{s_{01} + s_{12} + s_{23} + \cdots + s_{n-1,n}}{a}\right)^n \tag{15.13}$$

where $\sum_{i=0}^{n-1} s_{i,i+1} < a$.

For the probability that some of the distances related to six random variable locations in an interval with length $a = 100$ units will be at least $s_{01} = 15$, $s_{12} = 7$ and $s_{45} = 20$ (Figure 15.2d), the Monte Carlo simulation yields the empirical probability of 0.038. This result is confirmed by the theoretical probability calculated from Equation 15.13: $p = [1 - (s_{01} + s_{12} + s_{45})/a]^6 = 0.038$. By setting $s_{01} = 0$ and $s_{12} = s_{23} = \cdots = s_{n-1,n} = s$ in Equation 15.13, the probability

$$p_f = 1 - \left[1 - \frac{(n-1)s}{a}\right]^n \tag{15.14}$$

of a cluster of two or more random variables within a distance s is obtained.

15.4 Probability of Clustering of a Fixed Number Uniformly Distributed Random Events

Now, let us return to the problem related to a given number of users arriving randomly within a finite time interval with length a and using the same piece of equipment for a fixed time s.

The graphs of Equation 15.3 in Figure 15.3 give the probability of existence of a cluster of two or more demands for three different fixed demand times $s = 0.1$, 0.05 and 0.01 hour. The length of the finite time interval was assumed to be $a = 3$ hours. Clearly, the probability of collision of two or more demands within the user time $s = 0.05$ hour (the middle curve) increases rapidly with increasing the number of users. For $n = 10$ users, the probability of collision is approximately 0.80, as shown in Figure 15.3. In this sense, the problem can be regarded as a continuous analogue of the *birthday problem* (DeGroot, 1989).

For a time interval of 3 hours, containing 20 users, and for user demand time $s = 0.05$ hour, the probability of collision is practically unity (Figure 15.4).

This probability decreases significantly if the number of users decreases, as is shown in Figure 15.4. If the finite time interval is increased, as shown in Figure 15.5, the probability of collision also decreases.

Decreasing the number of users while keeping the finite time interval constant, as shown in Figure 15.4, leads to a much faster decrease of the probability of collision compared to increasing the time interval a while keeping the number of users constant, as shown in Figure 15.5. A simple

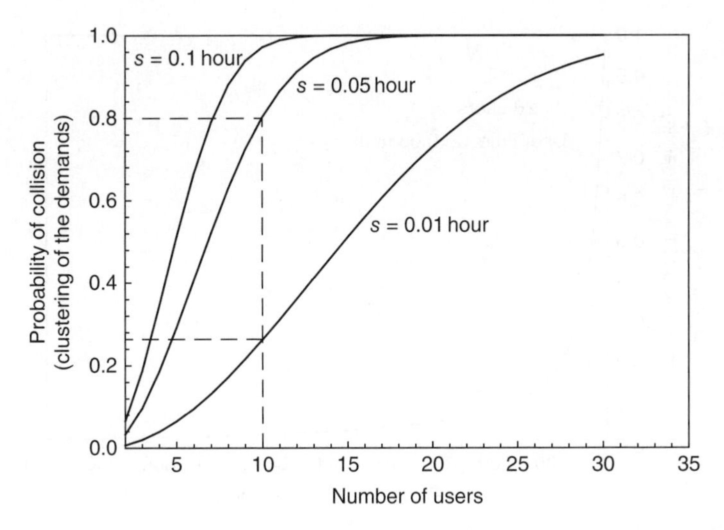

Figure 15.3 Probability of clustering uniformly distributed random demands from a given number of users

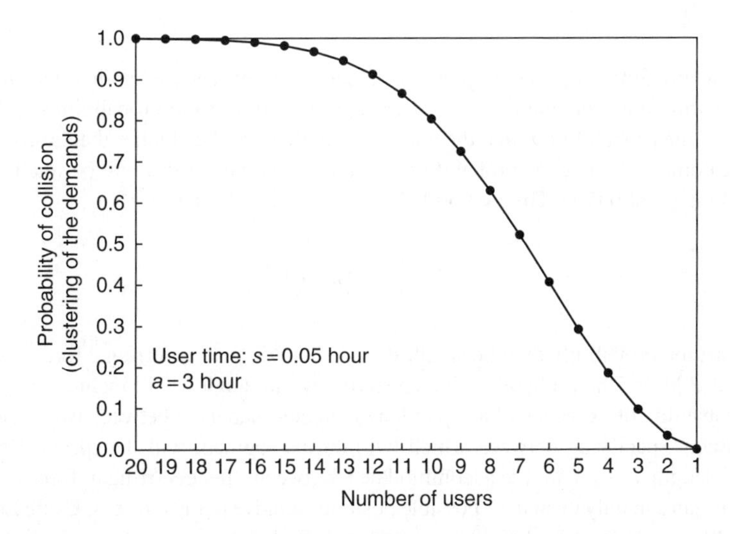

Figure 15.4 Probability of clustering of random demands from different number of users

calculation using Equation 15.14 shows that for $a = 1000$ hours, the probability of collision is approximately 2%. Even after increasing the time interval to $a = 10000$ hours, there still exists a 0.2% chance of collision.

According to Equation 15.13, the probability that before each random variable location there will be a distance greater than s is

$$p = \left(1 - \frac{n\,s}{a}\right)^n \tag{15.15}$$

Now, the solution of the problem related to the clustering of centrifugal forces due to eccentricity from a given number of discs on a rotating shaft (Figure 15.1b) follows from Equations 15.9 and 15.12.

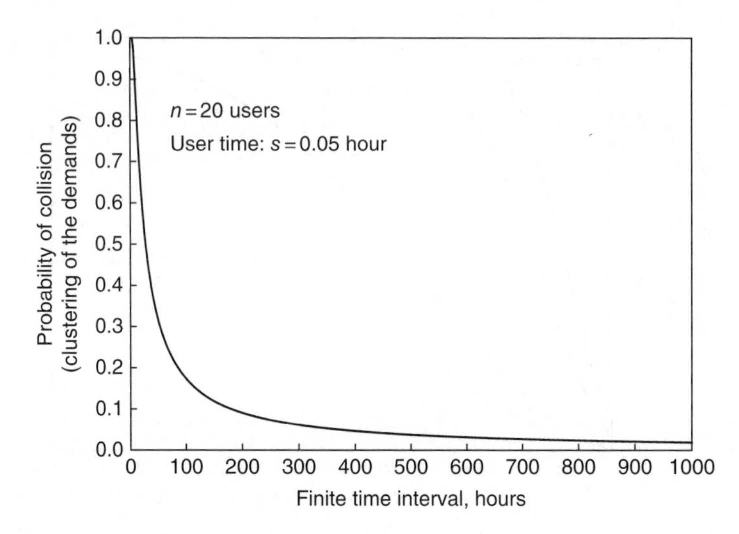

Figure 15.5 Variation of the probability of collision of random demands with increasing the length of the finite time interval

Let us present the full 360° angle as a segment with length $a = 2\pi$. The probability that all n centrifugal forces will be within a critical angle s ($s < a/2 = \pi$) is the sum of the probabilities of two mutually exclusive events: the probability p_1 that the clustering angle s will be smaller than s, given that it does not include the point 0 (2π) and the probability p_2 that the clustering angle will be smaller than s given that it includes the point 0 (2π). The first probability is given by Equation 15.9:

$$p_1 = n\left(\frac{s}{a}\right)^{n-1} - (n-1)\left(\frac{s}{a}\right)^{n}$$

and this is in fact the probability that the smallest angle bounding all centrifugal forces will be smaller than the critical angle s. The probability that clustering within angle s will include the point 0 (2π) is equal to the probability of existence of a gap of length greater than $a - s$ between two adjacent centrifugal forces (without specifying between which two) on the segment with length 2π. Because only a single gap of length $a - s$ can be accommodated between the centrifugal forces, $s < a/2$ and $a - s > a/2$, the gap can only be in $n - 1$ possible positions relative to the n forces. Using Equation 15.15 and the formula for a sum of probabilities of mutually exclusive events, the probability of a gap of length at least $a - s$ between two adjacent centrifugal forces (without specifying between which two) is

$$p_2 = (n-1)\times\left(1 - \frac{a-s}{a}\right)^{n} = (n-1)\times\left(\frac{s}{a}\right)^{n}$$

Finally, the probability that all n forces will cluster within the critical angle s becomes

$$p = p_1 + p_2 = n\left(\frac{s}{a}\right)^{n-1} = n\left(\frac{s}{2\pi}\right)^{n-1} \tag{15.16}$$

For the probability that all five centrifugal forces will be concentrated within a critical angle $s = 3\pi/5$ (Figure 15.1b), Equation 15.16 yields $p = 5 \times 0.3^4 = 0.04$. This result is in agreement with the empirical probability 0.04 obtained from a Monte Carlo simulation.

According to Equation 15.13, for a given number of random variables in a finite time interval, the probability that at least one of the specified minimum gaps will be violated is

$$p = P\left(S_{01} \le s_{01} \cup S_{12} \le s_{12} \cup S_{23} \le s_{23} \cup \cdots \cup S_{n-1,n} \le s_{n-1,n}\right)$$
$$= 1 - \left(1 - \frac{s_{01} + s_{12} + s_{23} + \cdots + s_{n-1,n}}{a}\right)^n$$

where $s_{01} \ge 0, s_{12} \ge 0, \ldots, s_{n-1,n} \ge 0$ (Todinov, 2004b). If in this equation $n-1$ specified minimum intervals are set to zero, the cumulative distribution $F(s)$ of the gap s between the locations of any two adjacent random variables is obtained:

$$P(S \le s) = F(s) = 1 - \left(1 - \frac{s}{a}\right)^n \tag{15.17}$$

Equation 15.1 is generic and gives the probability of a safe/failure configuration governed by the relative configuration of a given number of random variables uniformly distributed in a finite domain. Many intractable reliability problems can be solved easily using Equation 15.1 by reducing them to problems with trivial solutions.

Indeed, Equation 15.1 links the probability of a safe/failure configuration for arbitrary locations of the random variables in their domain with the probability of a safe/failure configuration, where one of the random variables is 'fixed' at the boundary of the domain. As a result, the initial complex problem is reduced to a simpler problem, which is in turn can be simplified until problems with trivial solutions are obtained.

The significance of Equation 15.1 stems also from the fact that potential problems are not restricted to one-dimensional problems only or to a simple function of the relative distances d_{ij} between the locations of random variables. The probability of a safe/failure configuration may also depend on more complicated functions $y = f(d_{ij})$ of the relative distances between locations uniformly distributed in a finite domain. As long as the function y depends only on the relative configuration of the random variables, not on their absolute locations, Equation 15.1 can be applied.

Equation 15.12 gives the probability of gaps of specified minimum lengths, between a given number of uniformly distributed random variables in a finite interval (conditional case).

Interestingly, according to Equation 15.12, the probability of existence of any specified set of free intervals between any selected pairs of adjacent random variables is the same, as long as the sum of the intervals is the same. It must be pointed out that Equations 15.12 and 15.13 can only be applied in cases where the number of random variables in the finite time interval is known and guaranteed to exist. In this context, Equation 15.13 appears also to be useful for making inferences in cases where only the number of random failures following a homogeneous Poisson process is known, but not the actual failure times. This is the case where an inspection at time a has identified a certain number of failures, whose actual times had not been recorded. Suppose that n random failures following a homogeneous Poisson process have been registered by an inspection at time a. Equation 15.13 can then be used to calculate:

(i) The probability $p = 1-(1-s/a)^n$ that the first failure has occurred before time s

(ii) The probability $p = (1-s/a)^n$ of existence of a continuous failure-free operation interval with length at least s until the first failure

(iii) The probability $p = [1-(s_0+s_n)/a]^n$ that the first failure had occurred after a time s_0 and the last failure had occurred before time s_n

15.5 Probability of Unsatisfied Demand in the Case of One Available Source and Many Consumers

Consider the important common case where a single source is satisfying demands from n consumers. It is assumed that a source can service only one consumer at a time.

Suppose that n consumers initiate their demands at random times $s_1, s_2, \ldots, s_n$, for durations $d_1, \ldots, d_n$, during a time interval (0, L) ($d_1 + d_2 + \cdots + d_n < L$) (Figure 15.6). Suppose that the times of the start of random demands are uniformly distributed along the length of the time interval (0, L). Let $A_1, A_2, \ldots, A_n$ denote the events 'the last demand has a duration $d_1, d_2, \ldots, d_n$', correspondingly. The probability of the event B that there will be no unsatisfied demand can be determined by the following probabilistic argument.

Initially, the conditional probability $P(B \mid A_n)$ will be determined – the probability that there will be no unsatisfied demand, given that the last demand has a duration d_n. Because every consumer has an equal chance to be the last consumer, the probabilities $p(A_i)$ of events A_i are all equal to $1/n$ ($p(A_i) = 1/n$).

According to Equation 15.12, the conditional probability that there will be no unsatisfied demand, given that the last demand has a length d_n, is

$$P(B \mid A_n) = \left[1 - \frac{(d_1 + \cdots + d_{n-1})}{L}\right]^n \tag{15.18}$$

The absence of unsatisfied demand however can occur in n different ways. The absence of unsatisfied demand can occur given that the demand of length d_n is the last demand, given that the demand of length d_{n-1} is the last demand and so on.

According to the total probability theorem,

$$P(B) = P(B \mid A_1)P(A_1) + \cdots + P(B \mid A_n)P(A_n) \tag{15.19}$$

The probabilities $P(B \mid A_i)(i = 1, \ldots, n-1)$ are determined in a similar fashion. As a result, the expression

$$P(B) = \frac{1}{n}\left[\left(1 - \frac{d_2 + d_3 + \cdots + d_n}{L}\right)^n + \left(1 - \frac{d_1 + d_3 + \cdots + d_n}{L}\right)^n + \cdots + \left(1 - \frac{d_1 + d_2 + \cdots + d_{n-1}}{L}\right)^n\right] \tag{15.20}$$

is obtained for the probability that there will be no unsatisfied random demand. If $D = \sum_1^n d_i$ denotes the sum of durations of all demands, the probability of unsatisfied demand $P(\bar{B}) = 1 - P(B)$ can be obtained as a probability of a complementary event:

$$P(\bar{B}) = 1 - \frac{1}{n}\left[\left(1 - \frac{D - d_1}{L}\right)^n + \left(1 - \frac{D - d_2}{L}\right)^n + \cdots + \left(1 - \frac{D - d_n}{L}\right)^n\right] \tag{15.21}$$

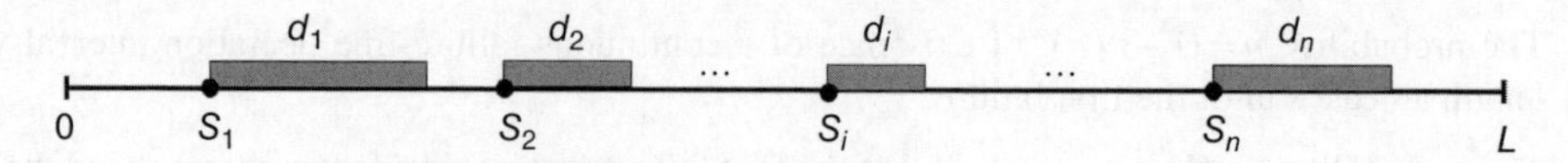

Figure 15.6 Random demands in a finite time interval (0, L)

This equation has been confirmed by the results from a simulation algorithm. Thus, for four consumers demanding a particular resource for $d_1 = 5$ minutes, $d_2 = 10$ minutes, $d_3 = 20$ minutes and $d_4 = 35$ minutes, respectively, during a time interval of 10 hours, the probability of unsatisfied demand calculated from Equation 15.21 is 0.3. This probability has been confirmed by the probability of 0.3 calculated from the simulation.

An important special case is obtained if the durations of all random demands are equal. Substituting $d_1 = d_2 = \cdots = d_n = d$ in Equation 15.21 then yields

$$P(\bar{B}) = 1 - \left(1 - \frac{D-d}{L}\right)^n \tag{15.22}$$

which is equivalent to Equation 15.14.

The analysis of Equation 15.22 reveals a useful result. Suppose that the total length of demand $D = n \times d$ is kept constant and only the number of customers n and the durations of their demands d are varied in such a way that $D = n \times d$ does not change. In other words, $d = D / n$.

Equation 15.22 then becomes

$$P(\bar{B}) = 1 - \left[1 - \left(1 - \frac{1}{n}\right) \times \left(\frac{D}{L}\right)\right]^n \tag{15.23}$$

With increasing n, $P(\bar{B})$ tends to unity. For $D = 50$ minutes total demand and a time interval $L = 600$ minutes, the dependence presented by Equation 15.23 has been given in Figure 15.7.

As can be seen from the graph in Figure 15.7, if the supplied resource is finite and sufficient for a total duration of supply of D hours (e.g. compressed gas in bottles, chemicals, etc.), a strategy involving splitting the resource among a fewer number of consumers is better than a strategy based on splitting the resource among a larger number of consumers. The first strategy is characterised by a significantly smaller probability of unsatisfied demand.

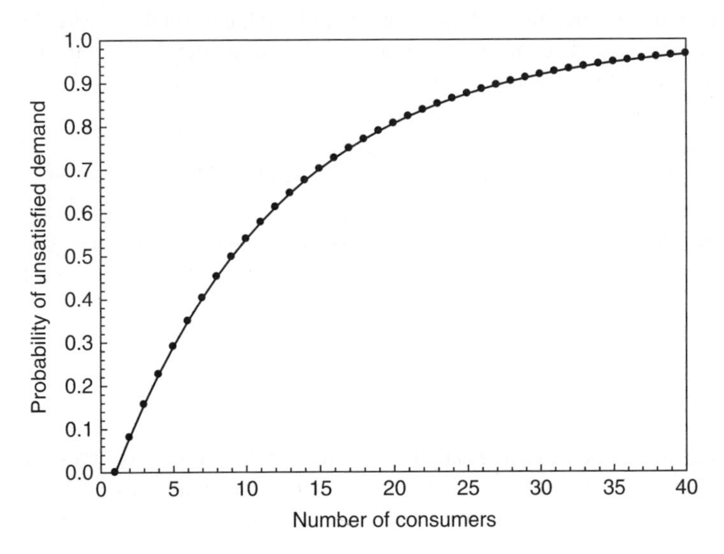

Figure 15.7 Probability of unsatisfied demand as a function of the number of consumers. The total duration of the demand from all consumers is the same – 50 minutes

15.6 Reliability Governed by the Relative Locations of Random Variables following a Homogeneous Poisson Process in a Finite Domain

An important, commonly encountered case is where the random variables follow a homogeneous Poisson process in a finite interval. The homogeneous Poisson process and the uniform distribution are closely related. A basic property of the homogeneous Poisson process, well documented in texts on probabilistic modelling (Gross and Harris, 1985; Ross, 2000), states: given n random variables following a homogeneous Poisson process in the finite time interval $0,a$, the coordinates of the random variables are distributed uniformly over the interval $0,a$. For example, when a calibration length a is cut from wire containing flaws following a homogeneous Poisson process and the number of flaws n is known (given) in the length a, the successive coordinates of the n flaws are jointly distributed along the length a as the order statistics in a sample of size n from the uniform distribution.

Assume that the random variables (not necessarily identical) follow a homogeneous Poisson process in the finite domain with measure ν (volume, area or length). Then, the probability of k random variables in the domain ν is given by the Poisson distribution $(\lambda\nu)^k e^{-\lambda\nu}/k!$, $k = 0,1,2,\ldots$, where $\lambda\nu$ is the mean number of random variables in the finite domain ν and λ is the number density of the random variables. The safe/failure states depend only on the relative configurations of the random variables in the common domain and not on their absolute locations. According to the total probability theorem, the probability of a safe/failure configuration is a sum of the probabilities of the mutually exclusive events involving all possible numbers of random variables in the finite domain ν. These mutually exclusive events are as follows: *k random variables reside in the finite domain with measure ν, and the variables form a safe/failure configuration where* $k = 0,1,2,\ldots$.

The probability $p(S)$ of a safe/failure configuration is then given by

$$p(S) = \sum_{k=0}^{\infty} \frac{(\lambda v)^k e^{-\lambda v}}{k!} p(S|k) \tag{15.24}$$

where $p(S\,|\,k)$ is the conditional probability of a safe/failure configuration, given that k random variables reside in the domain with measure ν. $p(S\,|\,k)$ can be determined considering that if the homogeneous Poisson process is conditioned on the number of random variables, the random variables will be uniformly distributed in the domain. In case of reliability dependent only on the relative configuration of the random variables, according to Equation 15.1,

$$p(S\,|\,k) = \left(\frac{1}{v^k}\right)\left(C_k + \int v^{k-1} \sum_{i=1}^{k} p_i^*(v)\,dv\right) \tag{15.25}$$

where the C_k are integration constants determined from the boundary conditions. If all of the random variables are identical, all probabilities p_i^* are equal ($p_1^* = p_2^* = \cdots = p^*$), and Equation 15.25 transforms into

$$p(S\,|\,k) = \left(\frac{1}{v^k}\right)\left(C_k + \int v^{k-1} k p^*(v)\,dv\right) \tag{15.26}$$

Equation 15.24 is generic and can be applied to calculate the probability of a safe/failure configuration governed by the relative configuration of random variables following a Poisson process in a finite domain. In this case, the number of random variables in the finite domain is a random variable.

Appendix 15.1

Suppose that the common domain v is incremented by a very small value Δv. Then, the random variables may all reside in v (event A_0) and form the safe/failure configuration with probability p. Alternatively, the ith random variable may reside in Δv and the rest of the random variables in v (this will be referred to as event A_i, $i = 1,n$) forming safe/failure configurations with probabilities p_i^*, depending on which random variable (i) is in the small domain increment Δv. As a result, events A_i, $i = 0,n$, form a set of mutually exclusive and exhaustive events partitioning the probability space: $P(A_i \cap A_j) = 0$, if $i \neq j$ and $\sum_{i=0}^{n} P(A_i) = 1$.

Let B denote the event *a safe/failure configuration of the random variables*. Because event B may occur with any of the events A_i, according to the total probability theorem, the probability of event B is

$$P(B) = P(B \mid A_0)P(A_0) + \sum_{i=1}^{n} P(B \mid A_i)P(A_i) \tag{15.A.1}$$

where $P(B \mid A_i)$ denotes the probability of B given A_i. Denoting $p = P(B \mid A_0)$, for a zero domain increment ($\Delta v = 0$), $P(B) = P(B \mid A_0) = p$, because in this case $P(A_0) = 1$ and $\sum_{i=1}^{n} P(B \mid A_i)P(A_i) = 0$.

Next, a small increment Δv of the domain v will cause a small increment Δp of the probability $P(B)$: $P(B) = p + \Delta p$. If a small domain increment Δv is present, $P(A_0) = [v/(v+\Delta v)]^n$ (this is the probability that all n random variables reside in v only). $P(A_i) = \Delta v / v$, $i = 1,2,\ldots,n$, are the probabilities that the ith random variable will reside in the small domain increment Δv, and $P(B \mid A_i) = p_i^*$ are the probabilities of a safe/failure configuration provided that the ith random variable resides in Δv. Substituting these values in Equation 15.A.1 gives

$$p + \Delta p = p\left(\frac{v}{v+\Delta v}\right)^n + \frac{\Delta v}{v}\sum_{i=1}^{n} p_i^* \tag{15.A.2}$$

For a small Δv, $[v/(v+\Delta v)]^n = (1+\Delta v/v)^{-n} \approx 1 - n\Delta v/v$, and Equation 15.A.2 becomes

$$\Delta p = -pn\frac{\Delta v}{v} + \frac{\Delta v}{v}\sum_{i=1}^{n} p_i^* \tag{15.A.3}$$

For an infinitesimal domain increment $\Delta v \to 0$, from Equation 15.A.3, the linear differential equation

$$\frac{dp}{dv} + \frac{pn}{v} = \frac{1}{v}\sum_{i=1}^{n} p_i^* \tag{15.A.4}$$

is obtained. The p_i^* are easier to calculate, because they correspond to calculating the probability of a safe/failure configuration when one of the random variables is fixed at the domain boundary. Generally, the p_i^* are functions of the measure v of the finite domain: $p_i^* = p_i^*(v)$. The integrating factor of Equation 15.A.4 is $\mu = \exp\left[\int (n/v)dv\right] = v^n$, and the general solution of differential Equation 15.A.4 is given by

$$pv^n = C + \int\left(v^{n-1}\sum_{i=1}^{n} p_i^*(v)\right)dv \tag{15.A.5}$$

16

Reliability and Risk Dependent on the Existence of Minimum Separation Intervals between the Locations of Random Variables on a Finite Interval

16.1 Applications Requiring Minimum Separation Intervals and Minimum Failure-Free Operating Periods

Reliability often depends on the existence of minimum separation intervals between or before the locations of random variables. Commonly encountered examples where reliability depends on the existence of minimum separation intervals between the locations of adjacent random variables are presented in Figure 16.1:

- A source servicing a number of randomly arriving demands, where the source can only service one request at a time. Unsatisfied demand occurs if two or more demands cluster within a critical time interval s (Figure 16.1a).
- Emergency calls for a nurse from critically ill patients in a hospital. The nurse can service only a single patient at a time. There will be no unsatisfied demands if the times between the emergency calls are greater than the maximum time s needed for servicing a call (Figure 16.1b).
- Stored spare equipment servicing the needs of customers arriving randomly during a specified time interval. After a demand from a customer, the warehouse needs a minimum time to replenish the dispatched equipment before the next demand can be serviced. In this case, the probability of unsatisfied demand equals the probability of clustering two or more customer arrivals within the critical period needed for making the equipment available for the next customer (Figure 16.1a).
- Supply systems which accumulate the supplied resource before it is dispatched for consumption (e.g. compressed gaseous substances). Suppose that after a demand for the resource, the system needs a minimum period of specified length to restore the amount of supplied resource to the level

Reliability and Risk Models: Setting Reliability Requirements, Second Edition. Michael Todinov.

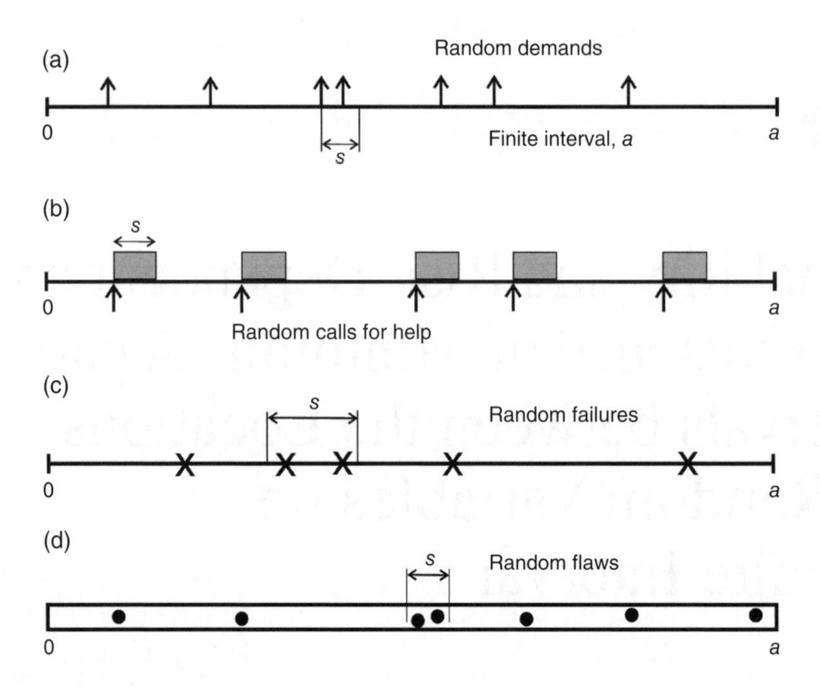

Figure 16.1 (a–d) Common examples where reliability depends on the existence of minimum separation intervals between the locations of random variables

existing before the demand. In this case, the probability of unsatisfied demand equals the probability of clustering two or more demands within the critical recovery period s (Figure 16.1a).

- Forces acting on a loaded component which fails if two or more forces cluster within a critical time interval s.
- Clustering of two or more random flaws over a small critical distance s (Figure 16.1d) dangerously decreases the load-carrying capacity of thin fibres and wires. As a result, a configuration where two or more flaws are closer than a critical distance cannot be tolerated during loading. Reliability in this case is governed by the probability of clustering of the random flaws.
- Failures associated with pollution to the environment (e.g. a leakage of chemicals) (Figure 16.1c). If such a failure is followed by another failure associated with leakage of chemicals, before a critical time interval has elapsed needed for recovery from pollution, irreparable damage to the environment could be done. For example, clustering of failures associated with a release of chemicals in the seawater could result in a dangerously high acidity which will destroy marine life.

Often, it is essential that before the first random failure and before each subsequent failure throughout the design life of a system, a minimum operating interval (Figure 16.2a) of length s exists, with high probability. Here are some examples:

- Specified rolling warranty periods are required before each failure followed by repair (Figure 16.2a) or before specified failures (Figure 16.2b) in a finite time interval. The violation of any of the specified rolling warranty periods is associated with a warranty payment.
- Systems for which a failure within a critical start-up period (of length s) is associated with severe consequences (Figure 16.2a).

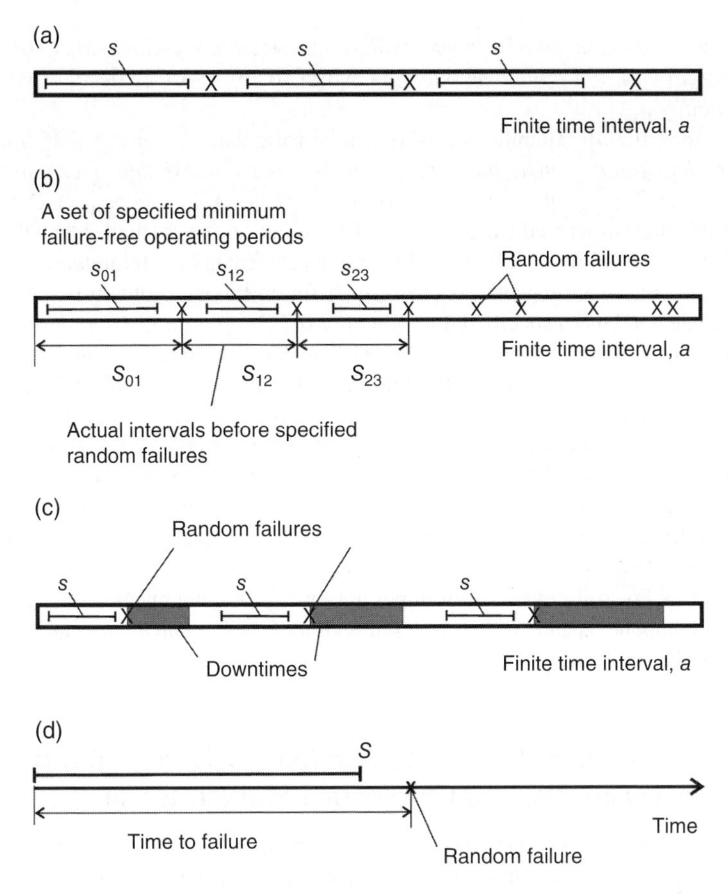

Figure 16.2 (a–d) Common examples where reliability depends on the existence of minimum separation intervals before the locations of random variables

- In cases where life of length at least s is expected immediately after each failure and replacement (Figure 16.2a) during a time interval a. Suppose that after a failure and replacement, the component is put on a mission of length s. Commonly, premature failure of the component during the mission with length s is associated with grave consequences and is highly undesirable.

  In this case, it is important to guarantee with a large probability failure-free operating period of length s *before each failure and replacement*. Guaranteeing with a large probability a failure-free operating period with length s before the first failure does not guarantee that a failure-free operating period with length s will exist with the same large probability, before each subsequent failure.

16.2 Minimum Separation Intervals and Rolling MFFOP Reliability Measures

All of the cases discussed in the previous section require a new reliability measure which can broadly be defined as *minimum separation intervals (MSI) between or before random variables in a finite time interval, whose existence is guaranteed with a minimum probability* $p_{\text{MSI}} = P(\text{All } S_{ij} \geq s_{ij})$ (Figure 16.2b) (Todinov, 2004c). The translation of the MSI reliability measure to random failures in a finite time interval is as follows: *specified minimum failure-free operating periods (MFFOPs)* s_{ij} *before or between random failures in a finite time interval* (Figure 16.2b) *whose existence is guaranteed with minimum probability* p_{MFFOP}.

Equivalently, the MFFOP reliability measure can be formulated as *specified MFFOPs* s_{ij} *before or between random failures in a finite time interval and a maximum acceptable probability* $p_{\text{fmax}} = P(\text{at least one } S_{ij} < s_{ij})$ *of violating at least one of them* ($p_{\text{fmax}} = 1 - p_{\text{MFFOP}}$). A violation of an MFFOP interval is present when the actual interval S_{ij} before failure is smaller than the corresponding specified MFFOP s_{ij} (Figure 16.2b). The MFFOP measure can be interpreted in a broader sense as minimum event-free operating period measure. If the events are random demands to a single source, and each demand has a specified duration, *an MFFOP guaranteed with large probability in fact guarantees with large probability that there will be no collision of random demands.*

Note that for a single specified MFFOP interval, the definition of the MFFOP reliability measure coincides with the definition of reliability associated with this interval (Figure 16.2d): 'the probability of surviving a specified minimum time interval of length s'. While for a non-constant hazard rate the classical reliability measure mean time to failure (MTTF) can be misleading (see Chapter 3), the MFFOP reliability measure is sound. Yet another reason for the importance of the MFFOP reliability measure is the possibility to link naturally reliability with the cost of failure (see Chapter 17).

The homogeneous Poisson process is an important model for component/system failures because the useful life of components and systems (the flat region of the bathtub curve) can be approximated well by a constant hazard rate.

16.3 General Equations Related to Random Variables following a Homogeneous Poisson Process in a Finite Interval

Random failures following a homogeneous Poisson process in a finite time interval with length a are considered. The number of failures in the finite time interval a is a random variable. It is assumed that after each failure, the component/system is brought by a replacement to as-new condition. Failure is understood to be a critical event leading to a system halt or degeneration of the required function below an acceptable level. In this sense, failure requires immediate intervention.

Specifying multiple MFFOPs is important in cases where before each random failure in a finite time interval, there should be a minimum period s of failure-free operation (free from intervention for unscheduled maintenance), guaranteed with large probability. If a rolling warranty period of length at least s is required before each failure during a finite time period with length a (Figure 16.2a), the MFFOP reliability measure consists of an MFFOP interval with length s and a minimum probability p_{MFFOP} with which this interval is guaranteed.

The maximum number of failure-free gaps of length s which can fit into the finite time interval with length a is $r = [a/s]$, where $[a/s]$ denotes the greatest integer part of the ratio a/s which does not exceed it. The probability that before k random failures in a finite interval with length a, there will be distances greater than a specified minimum distance s is given by Equation 15.15. According to Equation 15.18, the probability of existence of a minimum gap of length at least s before each random failure is

$$p(S) \equiv p_{\text{MFFOP}} = \sum_{k=0}^{r} \frac{(\lambda a)^k e^{-\lambda a}}{k!} \times \left(1 - \frac{ks}{a}\right)^k + \sum_{k=r+1}^{\infty} \frac{(\lambda a)^k e^{-\lambda a}}{k!} \times 0$$
$$= \sum_{k=0}^{r} \frac{(\lambda a)^k e^{-\lambda a}}{k!} \times \left(1 - \frac{ks}{a}\right)^k. \tag{16.1}$$

In Equation 16.1, $[(\lambda a)^k e^{-\lambda a}]/k!$ is the probability of exactly k failures in the finite time interval a. According to Equation 15.15, $p(S|k) = (1 - ks/a)^k$ is the conditional probability that given k random failures on the time interval with length a, before each failure, there will be a failure-free gap of length at least s.

Expanding the sum in Equation 16.1 results in

$$p_{\text{MFFOP}} = \exp(-\lambda a) \times \left(1 + \lambda(a - s) + \frac{\lambda^2 (a - 2s)^2}{2!} + \cdots + \frac{\lambda^r (a - rs)^r}{r!}\right) \tag{16.2}$$

for the probability p_{MFFOP} that before each random failure in the finite interval a, there will be a failure-free interval greater than s.

Consider now the important practical problem of requests arriving randomly in time to a source which can service only a single request at a time. Assume for the sake of simplicity that time s is needed to service each random request. The random requests could be for using unique equipment (e.g. X-ray equipment) or for a particular resource (e.g. water vapour, electrical power, compressed air, etc.). The list can be continued.

Now only minimum separation intervals (MSI) between adjacent random requests are considered, without specifying a MSI s_{01} from the beginning of the finite interval ($s_{01} = 0$). As a result, the equation

$$\begin{aligned} p(S) = p_{\text{MSI}} &= \sum_{k=0}^{r} \frac{(\lambda a)^k e^{-\lambda a}}{k!} \times \left(1 - \frac{(k-1)s}{a}\right)^k + \sum_{k=r+1}^{\infty} \frac{(\lambda a)^k e^{-\lambda a}}{k!} \times 0 \\ &= \sum_{k=0}^{r} \frac{(\lambda a)^k e^{-\lambda a}}{k!} \times \left(1 - \frac{(k-1)s}{a}\right)^k . \end{aligned} \tag{16.3}$$

is obtained from the general Equation 15.18, where $r = [a/s] + 1$.

In Equation 16.3, $[(\lambda a)^k e^{-\lambda a}]/k!$ is the probability of exactly k random requests in the finite time interval. According to Equation 15.12, $p(S|k) = [1 - (k-1)s/a]^k$ is the conditional probability that between any two adjacent random requests, there will be a gap of length at least s. Expanding the sum in Equation 16.3 results in

$$p_{\text{MSI}} = \exp(-\lambda a) \left\{1 + \lambda a + \frac{\lambda^2 (a - s)^2}{2!} + \cdots + \frac{\lambda^r [a - (r-1)s]^r}{r!}\right\} \tag{16.4}$$

for the probability p_{MSI} that the distance between any two adjacent random demands will be greater than the specified critical time s for satisfying a random demand. The probability p_c that two or more random demands will cluster within the critical time s needed for servicing a single demand is

$$p_c = 1 - \exp(-\lambda a) \left\{1 + \lambda a + \frac{\lambda^2 (a - s)^2}{2!} + \cdots + \frac{\lambda^r [a - (r-1)s]^r}{r!}\right\} \tag{16.5}$$

This is also the probability that there will be unsatisfied random demand.

Equations 16.2 and 16.4 can be used for setting reliability and risk requirements in cases where the random events follow a homogeneous Poisson process. For any specified MFFOP interval and a minimum probability p_{MFFOP} with which it must exist, solving the equations with respect to λ yields an upper bound (an envelope) for the number density of the random events. The number density envelope guarantees that if the actual number density of the events lies in it, the specified MFFOP will exist with probability at least equal to the specified minimum probability p_{MFFOP}.

It is important to point out that solving the exponential equation $p_{\text{MFFOP}} = \exp(-\lambda s)$ to specify the hazard rate guaranteeing an MFFOP of length at least s until the first failure does not mean that this period will exist before each subsequent random failure in the finite time interval.

Given a maximum acceptable probability of unsatisfied demand p_c, by solving Equation 16.5 numerically with respect to λ, an upper bound λ^* for the number density envelope of the random demands can be determined. This guarantees that whenever for the number density λ, $\lambda \le \lambda^*$ is fulfilled, the specified minimum separation interval (MSI) of length at least s will exist between random demands, with minimum probability $p = 1 - p_c$. In other words, the probability of unsatisfied demand will be smaller than p_c.

16.4 Application Examples

16.4.1 Setting Reliability Requirements to Guarantee a Specified MFFOP

Equation 16.2 has been used for setting MFFOP reliability requirements to guarantee a specified minimum failure-free operating interval of length at least $s = 30$ months before each random failure in a finite time interval of length $a = 100$ months. The value $p_{\text{fmax}} = 0.21$ has been specified as a maximum acceptable probability of violating at last one MFFOP interval. Equation 16.2 was solved numerically with respect to λ where $p_{\text{MFFOP}} = 1 - p_{\text{fmax}} = 0.79$. The numerical routine yielded a value $\lambda^* = 0.00614$ month^{-1} for the upper bound of the hazard rate which guarantees an MFFOP of length at least $s = 30$ months before each random failure, with probability equal to or greater than $p_{\text{MFFOP}} = 0.79$. This result has been verified by a Monte Carlo simulation. Given a hazard rate $\lambda^* = 0.00614$ month^{-1}, the Monte Carlo simulation yielded $p_{\text{MFFOP}} \approx 0.79$ for the probability that before each random failure, there will be a failure-free interval at least $s = 30$ months.

If the negative exponential distribution $p_{\text{MFFOP}} = \exp(-\lambda s)$ was used to calculate the probability of a failure-free interval of length at least $s = 30$ months before each random failure, it would have yielded the incorrect value:

$$\lambda^* = \left(-\frac{1}{s}\right)\ln\left(p_{\text{MFFOP}}\right) \approx 0.0079$$

The value $\lambda^* = 0.0079$ obtained from the negative exponential distribution guarantees a single MFFOP interval only, until the first failure. Larger discrepancies are obtained if the length of the specified MFFOP interval before each failure is reduced. Thus, the maximum hazard rate which guarantees with minimum probability $p_{\text{MFFOP}} = 0.79$ an MFFOP of length at least $s = 2$ months before each random failure is $\lambda^* = 0.031$ month^{-1}. The value $\lambda^* = 0.118$ month^{-1} obtained from solving the exponential equation $p_{\text{MFFOP}} = \exp(-\lambda s)$ guarantees with minimal probability $p_{\text{MFFOP}} = 0.79$ a single MFFOP interval only, until the first failure. In order to guarantee a failure-free interval of $s = 2$ months before each failure, the hazard rate needs to be decreased to the value $\lambda^* = 0.031$ month^{-1}. These examples show that if a specified MFFOP interval (a rolling warranty period) is required before each random failure following a homogeneous Poisson process in a finite interval, Equation 16.2 must be used to calculate the necessary hazard rate, not the negative exponential distribution.

16.4.2 Reliability Assurance That a Specified MFFOP Has Been Met

Suppose that a number of tests have been performed from which a constant hazard rate has been estimated in the way discussed in Chapter 3. On the basis of this estimate, a reliability assurance is required about the existence of an MFFOP before each random failure in a finite time interval with length a.

According to the discussion in Chapter 3, the MTTF estimated from data follows a χ^2-distribution. The probability $P(\theta_1 \le \theta \le \theta_2)$ that the true MTTF θ will lie between two specified bounds θ_1 and θ_2 is given by Equation 3.37.

Since the probability distribution of the MTTF can always be determined given the total operational time T and the number of failures k, assume for the sake of simplicity that $f(\theta)$ is the probability distribution of the true MTTF for a given number of failures and a total operational time T.

Given that the MTTF is in the interval $\theta, \theta + d\theta$, the probability that the actual failure-free operating intervals S_{ii+1} ($i=0, 1, \ldots$) before all random failures will be larger than the specified minimum failure-free operating interval s is

$$P(\text{all } S_{ii+1} > s \mid \theta) = e^{-a/\theta}\left(1 + \frac{(a-s)}{\theta} + \frac{(a-2s)^2}{\theta^2 2!} + \cdots + \frac{(a-rs)^r}{\theta^r r!}\right) \tag{16.6}$$

where $r = [a/s]$ is the greatest integer part of the ratio a/s which does not exceed it. The probability of the compound event that the MTTF will be in the interval $\theta,\ \theta + d\theta$ and the actual failure-free operating intervals before the random failures will be greater than the specified MFFOP with length s is equal to $P(\text{all } S_{ii+1} > s \mid \theta) f(\theta) d\theta$. According to the total probability theorem, the probability that all of the actual failure-free operating intervals before failures will be larger than the specified MFFOP with length s becomes

$$P(\text{all } S_{ii+1} > s) = \int_{\theta_{\min}}^{\theta_{\max}} P(\text{all } S_{ii+1} > s \mid \theta) f(\theta) d\theta$$

or, after the substitution,

$$P(\text{all } S_{ii+1} > s) = \int_{\theta_{\min}}^{\theta_{\max}} f(\theta) e^{-a/\theta}\left[1 + \frac{(a-s)}{\theta} + \frac{(a-2s)^2}{\theta^2 2!} + \cdots + \frac{(a-rs)^r}{\theta^r r!}\right] d\theta \tag{16.7}$$

where $\theta_{\min}$ and $\theta_{\max}$ are the lower and the upper bounds for θ, for which $f(\theta) \approx 0$ if $\theta < \theta_{\min}$ or $\theta > \theta_{\max}$. For a single specified MFFOP of length s before the first random failure only, the conditional probability of an MFFOP with length at least s is obtained on the basis of the negative exponential distribution:

$$P(S_{01} > s \mid \theta) = \exp\left(-\frac{s}{\theta}\right) \tag{16.8}$$

The probability $P(S_{01} > s)$ that the actual failure-free operating interval will be larger than the specified MFFOP with length s is

$$P(S_{01} > s) = \int_{\theta_{\min}}^{\theta_{\max}} f(\theta) \exp\left(-\frac{s}{\theta}\right) d\theta \tag{16.9}$$

Equations 16.7 and 16.9 can be used to provide reliability assurance that the specified MFFOP interval with length s has been met.

Increasing the number of tests alters the distribution $f(\theta)$ of the MTTF and also the probability $P(\text{all } S_{ii+1} > s)$. In order to minimise the number of tests needed to provide reliability assurance that the specified MFFOP is guaranteed with minimum probability p_{MFFOP}, the following steps can be repeated until the target probability p_{MFFOP} has been attained:

(i) Increment the number of tests by one
(ii) Update the distribution $f(\theta)$ of the true MTTF
(iii) Calculate the probability $P(\text{all } S_{ii+1} > s)$ from Equation 16.7 or 16.9

16.4.3 Specifying a Number Density Envelope to Guarantee Probability of Unsatisfied Random Demand below a Maximum Acceptable Level

In an illustrative example, the number density envelope of random demands will be determined which guarantees that the probability of unsatisfied demand will be below a specified level. A single source servicing random requests is available, and each random request requires a minimum time interval of 0.5 hour to be serviced. Demands follow a homogeneous Poisson process in a finite time interval of 100 hours, and if two or more demands follow within the critical service time interval of 0.5 hour, there will be unsatisfied demand. The maximum acceptable probability of unsatisfied demand has been specified to be $p_c = 0.1$. An upper bound $\lambda^* = 0.0467$ hour^{-1} of the number density of the demands was obtained by solving Equation 16.4 with respect to λ where $p_{MSI} = 1 - p_c = 0.9$. Whenever for the number density λ of demands $\lambda \le \lambda^* = 0.0467$ hour^{-1} is fulfilled, the probability of unsatisfied demand is smaller than 0.1. Monte Carlo simulations (one million trials) of a homogeneous Poisson process with density $\lambda^* = 0.0467$ yielded 0.1 for the probability of clustering of two or more demands within the critical interval of 0.5 hours, which confirms the result from solving Equation 16.4. Thus, for an expected number of five random demands in 100 hours, the probability of unsatisfied demand is substantial (≈0.1). Even for the mean number density of two random demands in 100 hours, the calculation from Equation 16.4 shows that there is still approximately 2% chance of unsatisfied demand.

Figure 16.3 gives the dependence of the probability of unsatisfied demand for a single source and time of $s = 1$ hour for servicing a single random demand. The operating time interval is $a = 100$ hours. The probability of unsatisfied demand has been plotted for different values of the number density of the demands.

For a mean number of 14 demands per 100 hours, there is already 80% probability of unsatisfied demand. Clearly, the probability of unsatisfied demand is substantial and should always be taken into consideration in risk assessments.

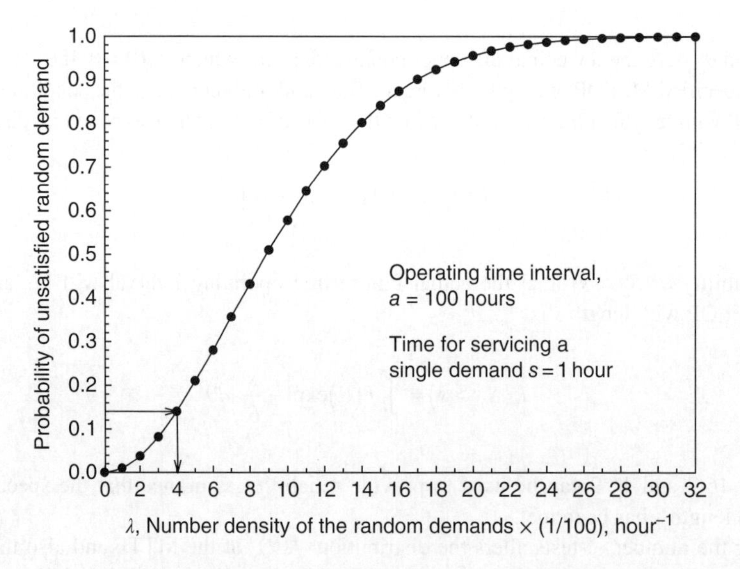

Figure 16.3 Probability of unsatisfied demand on a finite operational time interval of 100 hours. The random demands follow a homogeneous Poisson process, and each random demand requires 1 hour service time

Plots similar to the one in Figure 16.3 can be used for setting reliability requirements. For a specified maximum acceptable probability of unsatisfied demand, for example, 14%, the number density envelope of $\lambda^* = 0.04$ hour^{-1} can be determined (see the arrows in Figure 16.3). This envelope guarantees that whenever the number density of the demands does not exceed $\lambda^* = 0.04$ ($\lambda \le \lambda^*$), the probability of unsatisfied demand will not exceed the critical level of 14%.

Equation 16.5 can also be used to determine the probability of collision of demands from users of the same particular resource. Unlike the problem solved in Chapter 15, where the number of users was fixed, *here the number of users is a random variable following a Poisson distribution.*

This example demonstrates *the importance of setting reliability requirements not only to minimise the probability of unsatisfied demand below a maximum acceptable level but also to provide an optimal balance between risk and cost.*

This problem appears frequently in critical situations. For example, if the demands are emergency calls for a nurse arriving from critically ill patients in a hospital, the obtained hazard rate envelope can be used to determine the maximum number of such patients that could be looked after by a single nurse so that the probability of unattended call remains below a maximum acceptable level. The consequences of an unattended patient's call are grave (an unattended call from a critically ill patient could result in death or serious damage to health), and to keep the risk low, the tolerable probability of unsatisfied demand should be very low.

If the average number density of the calls characterising a single critically ill patient is λ_0, the total number density of the calls characterising all n critically ill patients is $\lambda = n \times \lambda_0$. Determining the maximum acceptable call rate λ^* which guarantees with a specified probability that there will be no patients' calls while the nurse is servicing another patient can be determined using a method similar to the method illustrated in Figure 16.3. Dividing the maximum acceptable call rate λ^* to λ_0 yields the maximum acceptable number n^* of patients that can be looked after by a single nurse:

$$n^* = \frac{\lambda^*}{\lambda_0} \tag{16.10}$$

Equations 16.2, 16.3, 16.4 and 16.5 are relevant to a wide class of reliability problems. Equation 16.4 can also be used to establish whether clustering of random events is a 'random fluctuation' or not. In addition, Equation 16.5 can also be used to determine the probability of clustering of two or more flaws in fibres or wires, within a critical distance s, given that the flaw number density is λ. It is assumed that the locations of the flaws follow a homogeneous Poisson process in the finite length a. Solving Equation 16.5 with respect to the flaw number density λ defines an upper bound for the flaw number density which guarantees with a specified minimum probability no clustering of flaws within a small critical distance. This is important in cases where the probability of failure during loading is strongly correlated with the probability of clustering of flaws. Solving Equation 16.4 with respect to the flaw number density λ in fact specifies requirements regarding the maximum acceptable flaw content in the material in order to reduce the probability of early-life failures caused by clustering of flaws.

The proposed models and algorithms form the core of a methodology for reliability analysis and setting reliability requirements based on minimum separation intervals (MSI) between random events in a finite time interval.

16.4.4 Insensitivity of the Probability of Unsatisfied Demand to the Variance of the Demand Time

Simulation experiments have been conducted involving constant (fixed) number of random demands on a specified time interval. Each consumer places exactly one demand randomly located in the specified time interval (Figure 16.4).

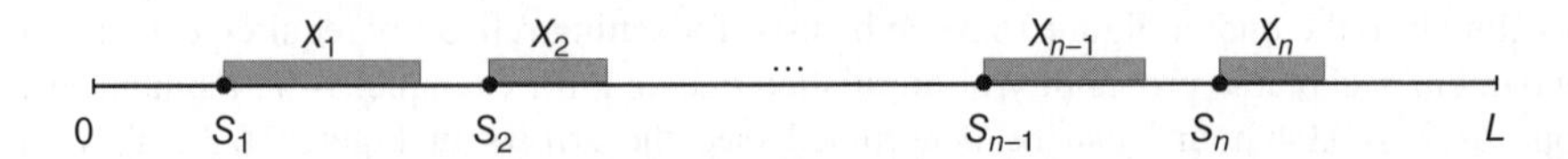

Figure 16.4 A fixed number (n) of random demands in a time interval (0, L). The duration of the demands is a random variable

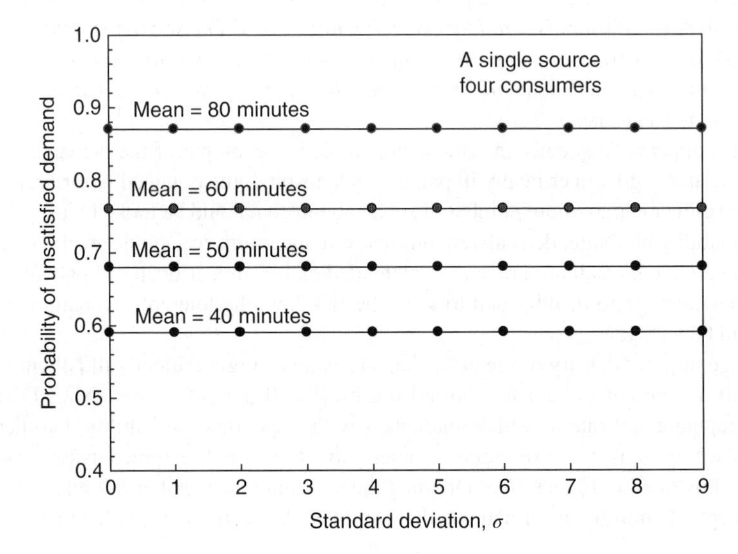

Figure 16.5 Dependence of the probability of unsatisfied demand on the standard deviation of the demand time for a single available source capable of servicing a single consumer at a time. The demand times follow a normal distribution with a specified mean, which is kept constant

Increasing the standard deviation of the duration of random demands reveals a rather unexpected trend (Figure 16.5). The simulation results clearly show that *the probability of unsatisfied demand practically does not vary with varying the standard deviation of the demand time.* In the experiments presented in Figure 16.5, the demand times follow a normal distribution with a specified mean (80, 60, 50 and 40 minutes) and a standard deviation varying in the interval 0–9 minutes. The results in Figure 16.5 feature a single source and four consumers.

To check whether these results are caused by the symmetry of the Gaussian distribution, the simulation experiments were repeated with an asymmetrical log-normal distribution of the demand times, with mean $\mu = 180$ minutes, a standard deviation varying in the range (0, 36 minutes) and duration of the operation time interval of 300 hours. All 15 consumers were characterised by the same mean demand time of 180 minutes (which was kept constant) and the same standard deviation. The common standard deviation characterizing the demand time of each consumer was varied in the interval 0,36. The trend was the same – the probability of unsatisfied demand practically did not vary with varying the standard variation of the consumers' demand time (Figure 16.6).

These results can be rationalised by referring to the analytical expression (Equation 15.12 from Chapter 15) regarding the probability that there will be no unsatisfied demand from n random consumers on a time interval (0, L), given by

$$p_n^0 = \left(1 - \frac{x_1 + x_2 + \cdots + x_{n-1}}{L}\right)^n$$

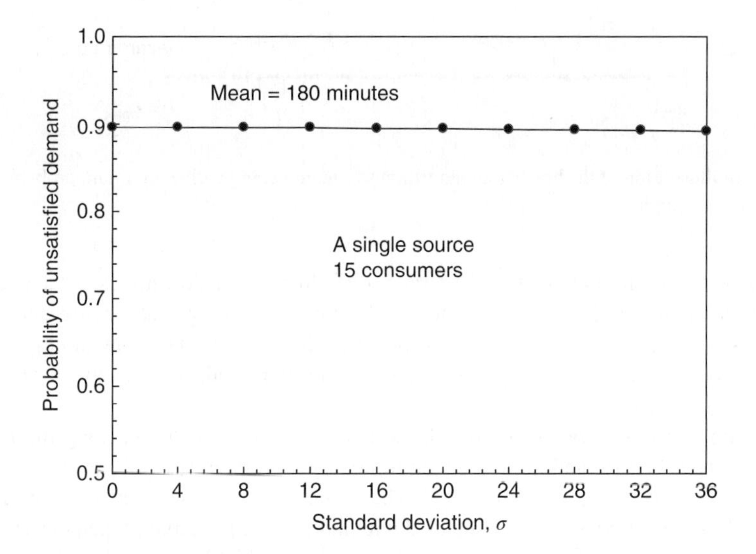

Figure 16.6 Dependence of the probability of unsatisfied demand on the standard deviation of the demand time for a duration of random demand following the log-normal distribution

where x_i $(i = 1, n-1)$ are the durations of the random demands from the consumers. The demand x_n from the last consumer has been discarded, because it cannot possibly contribute to the probability of unsatisfied demand.

If the durations $x_1, x_2, \ldots, x_{n-1}$ of the random demands are now realisations of a random variable X following a statistical distribution with mean μ and standard deviation σ, even for a relatively small number of consumers, the sum of the durations $\sum_{i=1}^{n-1} x_i$ in the above equation can be approximated reasonably well with $\sum_{i=1}^{n-1} x_i \approx (n-1)\mu$ and, as a result, the probability p_n^0 that there will be no unsatisfied demand will be insensitive to the variance of the random variable X standing for the durations of the demand times. The probability p_n^0 will depend only on the expected value $\mu = E(X)$ of the demand times. Because the probability of unsatisfied demand p_n is given by $p_n = 1 - p_n^0$, the probability of unsatisfied demand will be insensitive to the variance (standard deviation σ) of the random demand times X.

These results clearly show that reducing the variances of the demand times practically has no impact on the probability of unsatisfied demand. This rather unexpected result *provides the valuable opportunity to work with random demand times characterised by their means only and not requiring information related to the variance of the demand times.*

16.5 Setting Reliability Requirements to Guarantee a Rolling MFFOP Followed by a Downtime

Random failures are usually followed by downtimes which, in some cases, can be significant. For subsea oil and gas production, for example, the downtimes include the time for locating the failure, the time for mobilisation of resources and the time needed for intervention and repair/replacement. Downtimes can vary from several days to several months.

Suppose that the distribution of downtimes is known and a minimum probability p_{MFFOP} has been specified, with which an MFFOP of length s before each random failure is guaranteed. A hazard rate envelope can then be determined which guarantees that if the system hazard rate is in the envelope, the probability of existence of the minimum failure-free operating interval will be greater than the specified p_{MFFOP}.

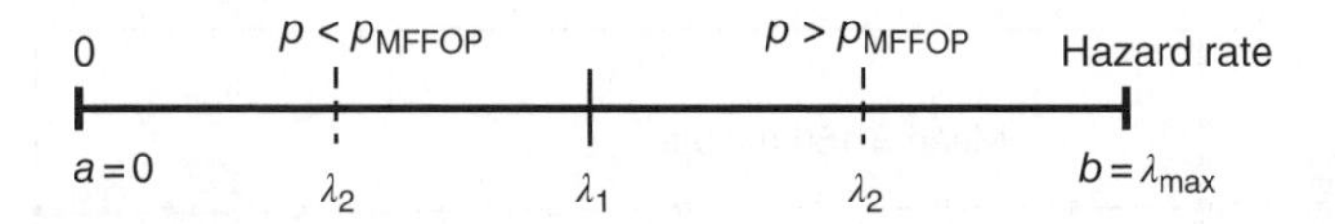

Figure 16.7 An illustration of the bracketing algorithm for guaranteeing with a minimum probability p_{MFFOP} an MFFOP of specified length

In the Monte Carlo simulation model of random failures with downtimes presented here, it is assumed that the random failures are characterised by a constant hazard rate; hence, failure times are produced by sampling from the negative exponential distribution. The downtimes are simulated by sampling from the distribution representing them (an empirical, uniform, negative exponential, log-normal distribution, etc.).

The algorithm consists of bracketing the hazard rate and subsequent checking the length of the failure-free distances before random failures until the target probability p_{MFFOP} is attained. The initial interval for the hazard rate is $a=0$, $b=\lambda_{max}$. The value $b=\lambda_{max}$ is selected to guarantee that the probability p of the specified MFFOP associated with hazard rate b is smaller than the specified target p_{MFFOP}. For a hazard rate $a=0$ the probability of the specified MFFOP is larger than the specified target p_{MFFOP}. Considering also that the probability of the specified MFFOP is a continuos function of the hazard rate, according to the *intermediate value theorem* (Ellis and Gulick, 1991), there exists a hazard rate from the closed interval [a,b] for which the probability of the specified MFFOP is exactly equal to the specified target p_{MFFOP}. The probability p of the specified MFFOP interval associated with hazard rate $\lambda_1 = (a+b)/2$ is calculated next. If the calculated probability p associated with hazard rate λ_1 is smaller than the specified target $p_{MFFOP}(p \le p_{MFFOP})$, the initial hazard rate interval (a, b) is truncated to the interval (a, λ_1) whose length is twice as small. A new probability p of existence of the specified MFFOP is then calculated for the hazard rate $\lambda_2 = (a+\lambda_1)/2$, which is in the middle of the truncated interval (Figure 16.7).

If the probability p associated with the value λ_1 is greater than the specified target $(p > p_{MFFOP})$, the initial interval for the hazard rate (a, b) is truncated to (λ_1, b). A new probability p is then calculated for the hazard rate $\lambda_2 = (\lambda_1 + b)/2$ in the middle of the truncated interval (Figure 16.7). Truncating intervals and calculating the new probability p for the hazard rate in the middle of each interval continue until the length of the last truncated interval becomes smaller than the required precision ε. Since at each step of the calculation, the current interval for the hazard rate is halved, the total number of calculations is $\ln_2[\lambda_{max}/\varepsilon]$. The algorithm in pseudocode is as follows:

Algorithm 16.1

```
function   Calc_Pmffop (λ)
{/* returns the probability with which the specified MFFOP of
    length s exists for a hazard rate λ and the specified
    distribution of downtimes */}

left = λmin;
right = λmax;
while (|right - left| > eps) do
{
  mid = (left + right)/2;
  p_mid = Calc_Pmffop(mid);
  if (p_mid < pMFFOP) then right = mid;
  else left = mid;
}
```

The variable eps specifies the desired precision. At the end of the calculations, the hazard rate which guarantees the specified MFFOP of length s with minimum probability p_{MFFOP} remains in the variable mid. The function ***Calc_Pmffop*** (λ) returns the probability with which the specified MFFOP with length s exists before each random failure, given the specified hazard rate λ and the distribution of downtimes. Its algorithm is given in the next section:

Algorithm 16.2
The algorithm of the routine ***Calc_Pmffop (λ)*** in pseudocode is as follows:

```
function Calc_Pmffop(λ)
{
  function  Lognormal_down_time()
           {/* Generates a log-normally distributed downtime with
               specified parameters */}

  function  Exponential_uptime()
           { /* Generates an exponentially distributed uptime with
                the specified hazard rate λ */}

violations = 0; /* Initialising the 'violations counter' (the
                   number of trials in which an MFFOP with length s
                   has been violated) */

for i = 1 to Number_of_trials do
  {
   t_cumul = 0; /* where the subsequent uptimes and downtimes are
                   accumulated  */

  repeat
   /* Generate an exponential uptime */
      exp_time = Exponential_uptime();
      t_cumul = t_cumul + exp_time;

      if (t_cumul > a)  then  break;
      else if (exp_time < s) then { violations = violations + 1;
      break; }

      /* Generate a lognormal downtime */
      Cur_downtime = Lognormal_down_time();
      t_cumul = t_cumul + Cur_downtime;
      if (t_cumul > a)  then  break;

  until 'break' is executed anywhere in the loop;

}
          /* Calculates the probability of existence of
             the MFFOP with length s before each failure */
Pmffop = 1 - violations / Num_of_trials;
return  Pmffop;
}
```

Central to the routine is the statement

```
else if (exp_time < s) then { violations = violations + 1; break;}
```

where a check is performed whether the currently generated time to the next failure is smaller than the MFFOP of length s. If a violation of the specified MFFOP is present, the variable counting the violations is incremented and the repeat-until loop is exited immediately, continuing with the next Monte Carlo simulation trial.

The probability of violating the specified MFFOP interval of length s is calculated by dividing the content of the violations counter to the total number of Monte Carlo trials. Subtracting this ratio from unity gives the probability `Pmffop` of existence of the specified MFFOP interval before each random failure.

Example

This example involves log-normally distributed downtimes. It is assumed that the natural logarithms of the downtimes (in days) follow a normal distribution with mean 3.5 and a standard deviation 0.5. For the length of the specified time interval and for the length of the specified MFFOP, $a=60$ months and $s=6$ months have been assumed, respectively. A minimum probability $p_{\mathrm{MFFOP}} = 0.90$ of existence of the MFFOP=6 months was also specified. Using Algorithm 16.1, the hazard rate envelope which guarantees this probability was determined to be $\lambda_s^* \approx 0.012$ months^{-1}. In other words, whenever the system hazard rate λ is smaller than $\lambda_s^* = 0.012$, the failure-free interval before each random failure will be greater than the specified MFFOP $s=6$ months (Figure 16.2a) with probability greater than or equal to $p_{\mathrm{MFFOP}} = 0.90$.

16.6 Setting Reliability Requirements to Guarantee an Availability Target

The current practice for setting reliability requirements for production systems is based solely on specifying a high availability target because it provides a direct link with cash flow. Given a particular distribution of downtimes and a specified minimum availability, reliability requirements can be set to guarantee that the availability will be greater than the specified target. Again, the algorithm is based on bracketing the hazard rate and subsequent calculation of the availability until the target availability is attained with the desired precision. It is assumed that failures follow a homogeneous Poisson process and after each failure, the system is brought by repair/replacement to 'as good as new' condition. The initial interval for the hazard rate is $a=0$, $b=\lambda_{\max}$. The right limit $b=\lambda_{\max}$ is selected to guarantee that the calculated availability for hazard rate b is smaller than the required availability target A_T. For a hazard rate $a=0$ the calculated availability is larger than the specified target A_T. Considering also that the availability is a continuos function of the hazard rate, according to the *intermediate value theorem* (Ellis and Gulick, 1991) there exists a hazard rate from the closed interval [a,b] for which the calculated availability is exactly equal to the specified target A_T. Next, the availability associated with the middle of the interval $\lambda_1 = (a+b)/2$ is calculated. If the availability A associated with the value λ_1 is smaller than the specified target $A_T (A < A_T)$, the hazard rate interval is truncated to (a, λ_1) and a new availability value is calculated at $\lambda_2 = (a + \lambda_1)/2$ (Figure 16.8).

If the availability associated with value λ_1 is greater than the specified target $(A > A_T)$, the hazard rate interval is truncated to (λ_1, b) and the new availability value is calculated at $\lambda_2 = (\lambda_1 + b)/2$ (Figure 16.8). The calculations, which are very similar to the calculations described in the previous section, continue until the final truncated interval containing the last calculated approximation λ_n of the hazard rate becomes smaller than the desired precision ε. If $N = [\lambda_{\max}/\varepsilon]$ denotes the integer part of the ratio $\lambda_{\max}/\varepsilon$, after $\ln_2 N$ calculations, the desired precision will be attained. This algorithm is very efficient because, for example, even for $[\lambda_{\max}/\varepsilon] = 2^{100}$, the solution is attained after only 100 calculations. The algorithm in pseudocode is as follows:

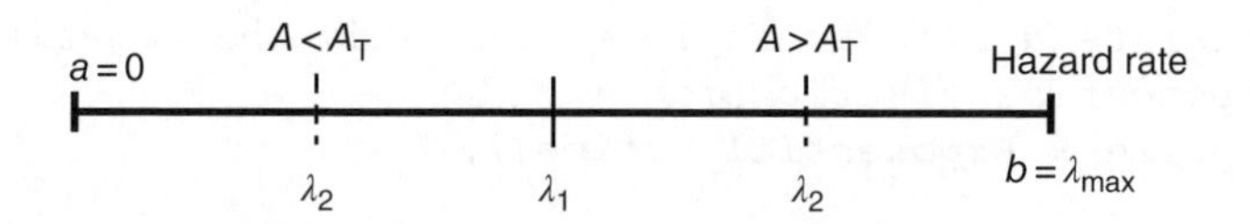

Figure 16.8 An illustration of the algorithm for guaranteeing a minimum availability A_T

Algorithm 16.3

```
function  Simulate_availability (λ)
{/* returns the availability associated with hazard rate λ */}

left  = λmin;
right  = λmax;

while (|right - left| > eps) do
 {
  mid = (left + right)/2; a_mid = Simulate_availability(mid);
  if (a_mid < AT) then right = mid;
  else left = mid;
 }
```

The variable eps contains the desired precision. At the end of the calculations, the hazard rate which guarantees the specified availability target A_T remains in the variable 'mid'. The function ***Simulate_availability*** (λ), whose algorithm is presented next, returns the average availability during the finite time interval for the specified downtime distribution and hazard rate λ. The failure times are produced by sampling from the exponential distribution, while the downtimes are produced by sampling the distribution representing the downtimes (empirical, uniform, log-normal, etc.).

Here is the computer simulation algorithm in pseudocode for determining the mean availability on a finite time interval:

Algorithm 16.4

```
Availability [Number_of_trials];   /* Array containing the
                                      availability values calculated
                                      from each Monte Carlo trial */

function  Lognormal_down_time()
          {/* Generates a log-normally distributed downtime with
              specified parameters */}

function  Exponential_uptime()
          {/* Generates an exponentially distributed time to
              failure with a specified hazard rate λ */}

function Simulate_availability(λ)
{
 for i = 1 to Number_of_trials do
  {
     a_remaining = a; Total_uptime = 0;

   repeat
```

```
      /* Generate a time to failure following the negative
         exponential distribution */
      Cur_uptime = Exponential_uptime();

      If (Cur_uptime > a_remaining) then
                  {Total_uptime = Total_uptime + a_remaining; break;}

      Total_uptime = Total_uptime + Cur_uptime;
      a_remaining = a_remaining - Cur_uptime;

      /* Generate a lognormal downtime */
        Cur_downtime = Lognormal_down_time ();

      If (Cur_downtime > a_remaining) then break;

      a_remaining = a_remaining - Cur_downtime;
   until 'break' is executed anywhere in the loop;

  Availability [i] = Total_uptime / a;
  Sum = Sum + Availability[i];
  }

Mean_availability = Sum / Number_of_trials;
return Mean_availability;
}
```

The algorithm consists of generating alternatively an *uptime* (operational time) from sampling the negative exponential distribution, followed by a *downtime* (repair time) obtained from sampling the log-normal distribution describing the distribution of the repair time. This process continues until the finite time interval with length a is exceeded. In the variable 'Total_uptime', the total time during which the system is operational is accumulated. In the indexed variable Availability[i], the availability characterising the ith Monte Carlo simulation trial is obtained by dividing the total uptime to the length of the finite time interval a. The sorted values stored in the Availability array can subsequently be used to plot the empirical distribution of the availability. Averaging the values stored in the Availability array gives the mean availability associated with the finite time interval a.

Example

This numerical example is also based on log-normally distributed downtimes. Similar to the previous example, it is assumed that the natural logarithms of the downtimes (in days) are normally distributed with mean 3.5 and a standard deviation 0.5. The length of the specified time interval was assumed to be $a=60$ months. An average availability $A_T = 95\%$ has been specified. By using the described algorithm implemented in C++, a hazard rate $\lambda^*_{A(0.95)} \approx 0.044$ months^{-1}, which guarantees the availability target $A_T = 95\%$, was determined. In other words, whenever the system hazard rate λ is smaller than $\lambda^*_{A(0.95)}$, the average system availability A is greater than the target value $A_T = 95\%$. Surprisingly, the probability of premature failure before time $a=60$ months, for a constant hazard rate $\lambda^*_{A(0.95)} \approx 0.044$ month^{-1}, is $p_f = 1-\exp(-0.044 \times 60) \approx 0.93$. This result shows that even for a relatively high availability target, it is highly likely that there will be a premature failure. Consequently, setting a high availability target does not exclude a substantial probability of premature failure. In subsea deep-water production, for example, the cost of intervention to fix failures is high; therefore, the risk (the expected losses) associated with failure is also high even for a large specified availability.

16.7 Closed-Form Expression for the Expected Fraction of the Time of Unsatisfied Demand

This section treats the important practical problem where n consumers demand a particular resource at random times, for specified durations, during a specified time interval $(0, L)$. Each source of the supplied resource can only service a single consumer at a time.

This problem is present in many manufacturing processes where a number of machine centres demand expensive measuring equipment, control equipment, production equipment or an operator, at a random time during the production process. Because the control equipment is expensive and unique, it is not feasible to equip each machine centre with a separate piece of equipment. The demands for the resource may occur at random times during a shift. After the duration of the demand, the control equipment is released and made available for future demands.

If m sources (pieces of control equipment) are available, a simultaneous demand from not more than m consumers can be satisfied, but not a simultaneous demand from $m+1$ or more consumers. As a result, the problem of unsatisfied demand can be reduced to a problem of geometrical probability where a segment of specified length L is covered by randomly located smaller segments with different lengths. The probability of unsatisfied demand if m sources are present can then be estimated by the probability of an overlap by more than m segments. The expected time of unsatisfied demand, given that m sources are available, is numerically equal to the expected fraction of area covered simultaneously by more than m segments.

For the sake of simplicity, consider a case where the duration of the demand from the ith consumer is equal to d_i, during the operation period with length L. The ratio of the duration of the demand and the time interval 'L' will be denoted by $\psi_i = d_i / L$.

Before determining the expected time fraction of unsatisfied demand, the following theorem related to a coverage of space with volume V by n 3-D interpenetrating objects with volumes $v_i (i = 1,\ldots,n)$, randomly placed in the volume V, will be stated and proved. The volume fractions of the separate objects will be denoted by $\psi_i = v_i / V$. The coverage of a point from the volume V is a 'coverage of order k if exactly k objects cover the point. The following theorem then holds.

Theorem 16.1

The expected covered fraction of order k (k=0, 1, …, n) from the volume V, by n interpenetrating objects with volume fractions ψ_i, is given by the $k+1$st term of the expansion $\prod_{i=1}^{n}[(1-\psi_i)+\psi_i]$.

Proof

The volume fraction covered by exactly m objects can be determined from the probability that a randomly selected point in the volume V will sample simultaneously m overlapping (interpenetrating) random objects. The probability that a randomly selected point in the volume V will sample simultaneously m overlapping objects is equal to the probability that a fixed point from the volume V will be covered exactly m times by randomly placed objects in the volume V.

Let p_0 denote the probability that the fixed point will not be covered, p_1 denote the probability that the fixed point will be covered by exactly one random object, …, and p_n denote the probability that the fixed point will be covered by all n random objects.

Because the locations of the random objects are statistically independent events, the probability of the event A_0 that a fixed point in the volume will not be covered by any of the random objects is given by

$$P(A_0) = \prod_{i=1}^{n}(1-\psi_i), \tag{16.11}$$

which is the probability that the fixed point will not be covered by the first, the second, …, the nth object.

The probability of the event A_1 that exactly one random object will cover the fixed point is a sum of the probabilities of the following mutually exclusive events: the first object covers the fixed point and the rest of the random objects do not, the second object covers the fixed point and the rest of the random objects do not and so on. As a result, the probability $P(A_1)$ that the fixed point will be covered by exactly one random object is given by

$$P(A_1)=\sum_{i=1}^{n}\left[\psi_i \prod_{\substack{k=1\\ k\neq i}}^{n}(1-\psi_k)\right] \tag{16.12}$$

The probability that exactly two random objects will cover the fixed point is a sum of the probabilities of the following mutually exclusive events: the first and the second random object cover the fixed point and the rest of the objects do not, the first and the third random object cover the fixed point and the rest of the random objects do not and so on, until all possible combination of two objects out of n are exhausted. As a result, the probability that the fixed point will be covered by exactly two random objects is given by

$$P(A_2)=\sum_{i1,i2}\left[(\psi_{i1}\,\psi_{i2})\prod_{\substack{k=1\\ k\neq i1;\ k\neq i2}}^{n}(1-\psi_k)\right] \tag{16.13}$$

where $\sum_{i1,i2}$ denotes the sum over all possible combinations of two indices $i1$ and $i2$ out of n. The number of these combinations is $\binom{n}{2}=\frac{n!}{2!(n-2)!}=\frac{n(n-1)}{2}$.

Continuing this reasoning through the cases 3, 4, …, n, the probability $P(A_m)$ that the fixed point will be covered by exactly m random objects is given by

$$P(A_m)=\sum_{i1,\ldots,im}\left[(\psi_{i1}\psi_{i2}\cdots\psi_{im})\prod_{\substack{k=1\\ k\neq i1;\ldots k\neq im}}^{n}(1-\psi_k)\right] \tag{16.14}$$

where $\sum_{i1,\ldots,im}$ denotes the sum over all distinct combinations of m indices $i1, i2, \ldots, im$ out of n. The number of these combinations is $\binom{n}{m}=\frac{n!}{m!(n-m)!}$.

The fixed point can either remain uncovered or covered by exactly one, two, …, n objects, and there are no other alternatives. Therefore, the events $A_0, A_1, \ldots, A_n$ constitute a set of mutually exclusive and exhaustive events. According to the third axiom of the probability theory, their probabilities add up to one:

$$\sum_{i=0}^{n}P(A_i)=1 \tag{16.15}$$

$$\prod_{i=1}^{n}(1-\psi_i)+\cdots+\sum_{i1,\ldots,im}\left[\psi_{i1}\psi_{i2}\cdots\psi_{im}\prod_{\substack{k=1\\ k\neq i1;\ldots;k\neq im}}^{n}(1-\psi_k)\right]+\cdots+\prod_{i=1}^{n}\psi_i=1 \tag{16.16}$$

Equation 16.16 can also be presented as an expansion of the expression $\prod_{i=1}^{n}[(1-\psi_i)+\psi_i]$. The theorem has been proved. ■

Because the proof does not make a reference to the shape of the random objects, the theorem is valid for interpenetrating random objects of any shape. It is also valid in the two-dimensional (2-D) and one-dimensional (1-D) case of area or a segment covered by 2-D or 1-D objects, correspondingly.

An immediate corollary of the theorem is related to objects with the same volume fraction $\psi = v / V$. In this case, the expected fractions from the volume V covered by the separate random objects with volumes v are given by the terms of the binomial expansion of $\left[(1-\psi)+\psi\right]^n$:

$$\left[(1-\psi)+\psi\right]^n = \sum_{i=0}^{n} \binom{n}{i} \psi^i (1-\psi)^{n-i} = 1 \tag{16.17}$$

The expected fraction of the volume covered by exactly m random objects is given by

$$P(A_m) = \binom{n}{m} \psi^m (1-\psi)^{n-m} \tag{16.18}$$

Now consider a case where n consumers demand a particular resource, during an operation period with length L. The durations of the demands from the consumers are d_i ($i=1, \ldots, n$). The ratios of the durations of the demands from the separate consumers are given by $\psi_i = d_i / L$. The maximum number of consumers whose demand can be simultaneously satisfied by the sources is m.

Theorem 16.2

The expected fraction of time of unsatisfied demand from m sources and n consumers is given by the expression

$$1 - \left\{ \prod_{i=1}^{n}(1-\psi_i) + \sum_{i=1}^{n}\left[\psi_i \prod_{\substack{k=1 \\ k\neq i}}^{n}(1-\psi_k)\right] + \cdots + \sum_{i1,\ldots,im}\left[\psi_{i1}\psi_{i2}\cdots\psi_{im} \prod_{\substack{k=1 \\ k\neq i1;\ldots;k\neq im}}^{n} (1-\psi_k)\right]\right\} \tag{16.19}$$

Proof

Unsatisfied demand for m sources and n consumers ($n>m$) is present in the case where more than m consumers require the supplied resource. Let p_0 denote the probability that a fixed point in the interval (0, L) will not sample any demand, p_1 denote the probability that the fixed point will sample exactly one random demand, …, and p_m denote the probability that the fixed point will sample exactly m random demands.

The probability $p_{\geq m+1}$ that the fixed point will sample more than m random demands, randomly placed in the time interval (0, L), is then given by

$$p_{\geq m+1} = 1-(p_0 + p_1 + \cdots + p_m) \tag{16.20}$$

According to Theorem 16.1, the sum of the probabilities $p_0 + p_1 + \cdots + p_m$ is given by

$$\prod_{i=1}^{n}(1-\psi_i) + \cdots + \sum_{i1,\ldots,im}\left[\psi_{i1}\psi_{i2}\cdots\psi_{im} \prod_{\substack{k=1 \\ k\neq i1;\ldots;k\neq im}}^{n} (1-\psi_k)\right].$$

Hence, the theorem has been proved. ■

If all random demands are characterised by the same duration d, $\psi = d/L$, the time fraction of unsatisfied demand is given by

$$p_{\geq m+1} = 1 - \sum_{i=0}^{m} \binom{n}{i} \psi^i (1-\psi)^{n-i} \tag{16.21}$$

Note that the sum in the right-hand side of Equation 16.21 is part of the binomial expansion of the expression $[(1-\psi)+\psi]^n$, which is identically equal to unity:

$$[(1-\psi)+\psi]^n = \sum_{i=0}^{m} \binom{n}{i} \psi^i (1-\psi)^{n-i} = 1 \tag{16.22}$$

All equations in this section have been verified by a Monte Carlo simulation, by measuring and accumulating directly the multiple intersections.

17

Reliability Analysis and Setting Reliability Requirements Based on the Cost of Failure

17.1 The Need for a Cost-of-Failure-Based Approach

Critical failures in many industries (e.g. in the nuclear or deep-water oil and gas industry) can have disastrous environmental and health consequences. Such failures entail loss of production for very long periods of time and extremely high costs of the intervention for repair. Consequently, for industries characterised by a high cost of failure, setting quantitative reliability requirements must be driven by the cost of failure. The author strongly disagrees with some risk experts who advocated that setting reliability requirements is unnecessary. Many technical failures, with disastrous consequences to the environment, could have been prevented easily by using the cost-of-failure-based approach to set correct reliability requirements for critical components.

Higher reliability does not necessarily mean low cost of failure. To demonstrate that selecting the more reliable system does not necessarily mean selecting the system with the smaller losses from failures, consider two very simple systems consisting of two components, logically arranged in series (Figure 17.1).

Both systems consist of an electronic control module (EC) and mechanical device (*M*). The components are logically arranged in series, and each system fails whenever the electronic control module fails or the mechanical device fails.

For the first system (Figure 17.1a), suppose that the hazard rate of the electronic control module EC1 is $\lambda_{EC1} = 4$ year^{-1} and its replacement after failure costs $C_{EC1} = \$300$, while the hazard rate of the mechanical device M1 is $\lambda_{M1} = 1$ year^{-1} and its failure is associated with $C_{M1} = \$200\,000$ cost for replacement. Suppose now that for an alternative system consisting of the same type of electronic control module and mechanical device (Figure 17.1b), the losses associated with failure of the separate components are the same, but the hazard rates are different. The electronic control module EC2 is now characterised by a hazard rate $\lambda_{EC2} = 1$ year^{-1} and the mechanical device by a hazard rate $\lambda_{M2} = 2$ year^{-1}. Clearly, the second system (b) is more reliable than the first system (a) because it is characterised by a hazard rate $\lambda_2 = \lambda_{EC2} + \lambda_{M2} = 3$ year^{-1}, whereas the first system is characterised by a hazard rate $\lambda_1 = \lambda_{EC1} + \lambda_{M1} = 5$ year^{-1}. Because the first system fails whenever either component EC1 or component M1 fails, the expected (average) losses from failures for the system, during 1 year of operation, are

$$\bar{L}_1 = \lambda_{EC1} C_{EC1} + \lambda_{M1} C_{M1} = 201200 \tag{17.1}$$

Reliability and Risk Models: Setting Reliability Requirements, Second Edition. Michael Todinov.

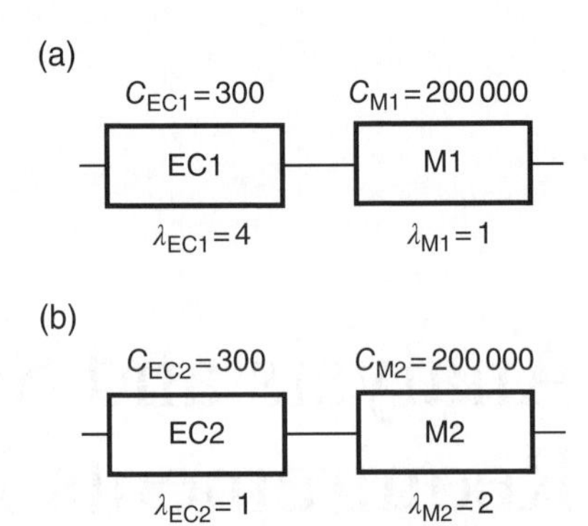

Figure 17.1 Systems composed of two components, demonstrating that the more reliable system (b) is associated with larger losses from failures compared to the less reliable system (a)

For the second system, the expected losses from failures during 1 year of operation are

$$\bar{L}_2 = \lambda_{EC2} C_{EC2} + \lambda_{M2} C_{M2} = 400300 \tag{17.2}$$

As can be verified, the more reliable system (the second system) is associated with the larger losses from failures.

This simple example shows that a selection of a system solely based on its reliability can be misleading. In case of component failures associated with similar cost, a system with larger reliability does mean a system with smaller losses from failures. In the common case of component failures associated with very different costs however, the system with the largest reliability is not necessarily the system with the smallest losses from failures.

17.2 Risk of Failure

The purpose of risk analysis is to provide support in making correct management decisions. By evaluating the risk associated with a set of decision alternatives, the risk analysis helps to identify the alternative which maximises the expected utility for the stakeholders while complying with a set of specified criteria and constraints. According to a classical definition (Henley and Kumamoto, 1981; Vose, 2000), the risk of failure is a product of the probability of failure and the cost given failure:

$$K = p_f C \tag{17.3}$$

where p_f is the probability of failure and C is the cost given failure. To an operator of production equipment, for example, the cost given failure C may include several components: cost of lost production, cost of cleaning up polluted environment, medical costs, insurance costs, legal costs, costs of mobilisation of emergency resources, loss of business due to loss of reputation and low customer confidence, etc. The cost of failure to the manufacturer of production equipment may include warranty payment if the equipment fails before the agreed warranty time, loss of sales, penalty payments, compensation and legal costs. Most of the losses from technical failures can be classified in several major categories:

Loss of life or damage to health

Losses associated with damage to the environment and the community infrastructure

Financial losses including warranty costs, loss of production, loss of capital assets, cost of intervention and repair, compensation payments, penalty payments and legal costs

Loss of reputation including loss of market share, loss of customers, loss of contracts, impact on share value, loss of confidence in the business, etc.

Depending on the category, the losses can be expressed in monetary units, number of fatalities, lost time, volume of lost production, volume of pollutants released into the environment, number of lost customers, amount of lost sales, etc. Often, losses from failures are expressed in monetary units and are frequently referred to as cost of failure.

The theoretical justification of Equation 17.3 can be made on the basis of the following thought experiment. Suppose that a non-repairable equipment is put in operation for a length of time a which can be the warranty period for the equipment. If the equipment fails before the specified warranty time a, its failure is associated with a constant loss C due to warranty replacement. Suppose that the experiment involves N identical pieces of equipment, of which N_f fail before time a. Since only a failure before time a is associated with losses, the total loss generated by failures among N pieces of equipment is $N_f \times C$. The average (expected) loss K is then $K = (N_f \times C)/N$. This expression can be rearranged as

$$K = \frac{(N_f \times C)}{N} = \left(\frac{N_f}{N}\right) \times C$$

If the number of pieces of equipment N is sufficiently large, $p_f \approx N_f/N$ gives an estimate of the probability of failure of the equipment before time a and the expression (17.3) is obtained.

Usually, a relatively small number N gives already a sufficiently accurate estimate of the true probability of failure. As a result, Equation 17.3 describes the average (expected) warranty loss per single piece of equipment. This is one of the reasons why, in the engineering context, the risk is often treated as *expected loss from failure*.

The traditional approach to the reliability analysis of engineering systems however does not allow setting reliability requirements which limit the risk of failure below a maximum acceptable level because it does not take into consideration the cost of failure. Incorporating the cost of failures is vital in bridging the gap between the traditional engineering approach to risk and the society's perception of risk.

Let us consider a typical problem facing the mechanical engineer–designer.

Exercise

Suppose that a failure-free service for a time interval of length at least a is required from an electrical connector. A premature failure of the connector (before time a) entails a loss of expensive processing unit, and the warranty costs C are significant. Because the cost of failure C is significant, the designer wants to limit the expected losses from warranty payments within the warranty budget of K_{max} per electrical connector. What should be the MTTF characterising the designed connectors so that the expected losses from premature failure remain within the warranty budget?

In the next sections, a new generation of models will be presented which form the basis of the cost-of-failure-based setting of reliability requirements.

Two main concepts related to the expected losses will be used: (i) *a risk of premature failure* before a specified time and (ii) *expected losses from multiple failures* on a specified time interval. The concept 'risk of premature failure' applies to both non-repairable and repairable systems, while the concept 'expected losses from failures' applies only to repairable systems.

17.3 Setting Reliability Requirements Based on a Constant Cost of Failure

Assume that $K_{max} = p_{fmax}C$ in Equation 17.3 is the maximum acceptable risk of failure, where p_{fmax} is the maximum acceptable probability of failure. This equation can also be presented as

$$p_{fmax} = \frac{K_{max}}{C} \tag{17.4}$$

Limiting the risk of failure K below K_{max} is then equivalent to limiting the probability of failure p_f below the maximum acceptable level p_{fmax} ($p_f \leq p_{f,max}$). This leads to the cost-of-failure concept for setting reliability requirements limiting the risk of premature failure (Todinov, 2003, 2004b):

$$p_f \leq \frac{K_{max}}{C} \tag{17.5}$$

As a result, whenever $p_f \leq p_{fmax} = K_{max} / C$, the relationship $K \leq K_{max}$ is fulfilled.

The maximum acceptable risk of premature failure K_{max} can conveniently be assumed to be the maximum budget for warranty payments per piece of equipment.

An important application of relationship (17.5) can be obtained immediately for a system characterised by a constant hazard rate λ. Such is, for example, the system including components arranged logically in series and characterised by constant hazard rates $\lambda_1, \lambda_2, \ldots, \lambda_n$. Because the system fails whenever any of the components fail, its hazard rate λ is $\lambda = \lambda_1 + \lambda_2 + \cdots + \lambda_n$. For a maximum acceptable risk K_{max} of premature failure (related to a finite time interval with length a) and a constant hazard rate λ, inequality (17.5) becomes $p_f(\lambda) \equiv 1 - \exp(-\lambda a) \leq K_{max} / C$, from which

$$\lambda^* = -\left(\frac{1}{a}\right)\ln\left(1 - \frac{K_{max}}{C}\right) \tag{17.6}$$

is obtained for the system hazard rate envelope λ^* which limits the risk of failure below K_{max} (Todinov, 2003, 2004b). In other words, whenever the system hazard rate λ lies within the as-determined envelope ($\lambda \leq \lambda^*$), the risk of failure (expected loss per unit equipment) remains below the maximum acceptable level K_{max} ($K \leq K_{max}$). Equation 17.6 provides a solution to the exercise at the end of the previous section. An electrical connector with MTTF greater than or equal to $\text{MTTF}^* = 1/\lambda^*$ guarantees that the expected loss per connector from warranty payment remains below K_{max}.

Substituting, for example, $a = 2$ years and $K_{max} = \$10$, $C = \$100$ yields a hazard rate envelope $\lambda^* = 0.052$ year^{-1}. An electrical connector with a hazard rate smaller than 0.052 year^{-1} limits the risk of failure (expected loss) below \$10 per connector.

It is important to point out that even in the absence of a specified maximum acceptable risk K_{max}, reliability requirements limiting the risk of failure can still be set by using the principle of the risk-based design (see Chapter 11). Suppose that a system consists for M components characterised by different costs of failure $C_1, \ldots, C_M$. Without loss of generality, assume that the first components is characterised by the largest risk of failure $K_1 = p_1C_1$. Without loss of generality, suppose that the component with the second largest risk of failure is the second component: $K_2 = p_2C_2$. If the risk of failure of the first component is much larger than the risk of failure of the second component ($K_1 >> K_2$), there is no point in investing resources for reducing the risks associated with the component failures before the risk of failure K_1 of the first component has been reduced to at least the second largest risk K_2. Suppose that no control over the cost of failure C_1 is available but controls over the probability of failure p_1 do exist (e.g. by implementing preventive risk reduction measures, see Chapter 11). As a result, for the new probability of failure p_1', which reduces the risk from K_1 to K_2, a setting

$$p_1'C_1 = p_2C_2 \tag{17.7}$$

can be made.

From Equation 17.7, the equation

$$p_1' = p_2\frac{C_2}{C_1} \tag{17.8}$$

can be obtained which captures the essence of the risk-based design principle: *The probability of failure of the component should be inversely proportional to its cost of failure.*

Because $p_1C_1 > p_2C_2$, $p_1 > p_2C_2 / C_1$ is always fulfilled, and because $p_1' < p_1 \le 1$, it is always guaranteed that $p_1' < 1$.

It is important to emphasise that to maintain the same risk level, even identical type of components, whose failure is associated with very different costs, C_1 and C_2 must have different probabilities of failure. Indeed, from Equation 17.7, it follows that the ratio of the probabilities of premature failure is $p_1' / p_2 = C_2 / C_1$.

For components with constant hazard rates, Equation 17.8 becomes

$$1-\exp(-\lambda_1'a) = [1-\exp(-\lambda_2 a)]\times\frac{C_2}{C_1}$$

where λ_1' is the hazard rate of the component associated with higher risk of failure and λ_2 is the hazard rate of the component associated with the lower risk of failure.

After some algebra, the above equation is reduced to

$$\lambda_1' = -\frac{1}{a}\ln\left\{1-[1-\exp(-\lambda_2 a)]\times\frac{C_2}{C_1}\right\} \tag{17.9}$$

which expresses the hazard rate of the first component, necessary to bring the risk of failure of the component down to the next largest risk of component failure.

Contrary to the classical approach which always starts the reliability improvement with the component with the smallest reliability in the system, the risk-based approach may actually start with the most reliable component in the system if it is associated with the largest risk of failure. This defines a principal difference between the classical approach to reliability analysis and setting reliability requirements and the cost-of-failure-based approach.

The cost-of-failure-based approach to setting reliability requirements will be illustrated by the next example.

Example

A system combines six clevis joints supporting various structural components. The first clevis joint, with a hazard rate 0.1 year^{-1}, is operated for a period of $a = 1$ year and its failure causes a loss of equipment worth \$1 million. Each of the remaining clevis joints has a hazard rate 0.2 year^{-1} and is operated for a period of $a = 1$ year. Failure of any of these clevis joints causes a loss of equipment worth \$20 000.

Reduce the risk associated with this system.

Solution

The risk associated with the first clevis joint is $K_1 = 0.1 \times 1000\,000 = \$100\,000$, while the risk associated with each of the remaining clevis joints is $K_2 = 0.2 \times 20\,000 = \$4000$.

Contrary to the conventional wisdom, to reduce the risk in this case, the reliability improvement should start not with the least reliable component but with the most reliable component in the system because its failure is associated with the largest risk. This is the first clevis joint. To reduce the risk of failure of the first clevis joint to a level comparable with the next largest risk, from Equation 17.9, the new hazard rate λ_1 should become

$$\lambda_1' = -\frac{1}{a}\ln\left\{1 - \left[1 - \exp(-\lambda_2 a)\right] \times \frac{C_2}{C_1}\right\}$$

where $\lambda_2 = 0.2$ year^{-1}, $a = 1$ year^{-1}, $C_2 = \$20\,000$ and $C_1 = \$1000\,000$.

Substituting these values in the above equation yields

$$\lambda_1' = -\frac{1}{1}\ln\left\{1 - \left[1 - \exp(-0.2 \times 1)\right] \times \frac{20\,000}{1000\,000}\right\} \approx 0.004 \text{ year}^{-1}$$

As a result, the reliability of the most reliable clevis joint should be increased further by reducing its hazard rate from $\lambda_1 = 0.1$ year^{-1} to $\lambda_1' = 0.004$ year^{-1}.

17.4 Drawbacks of the Expected Loss as a Measure of the Potential Loss from Failure

The risk Equation 17.4 only estimates the average value of the potential loss from failure. A decision criterion based on the expected loss would prefer the design solution characterised by the smallest expected potential losses. What is often of primary importance however *is not* the expected (average) loss, but the deviation from the expected loss (the unexpected loss). This is, for example, the case where a company estimates the probability that its potential loss will exceed a particular critical value after which the company will essentially be insolvent. Despite that the expected loss gives the long-term average of the loss, *there is no guarantee that loss will revert quickly to such average* (Bessis, 2002). This is particularly true for short time intervals where the variation of the number of failures is significant.

Let us consider a real-life example where a selection needs to be made between two competing identical systems which differ only by the time to repair. A critical failure of the first system is associated with a time for repair which follows a normal distribution. As a consequence, the lost production due to the critical failure also follows a normal distribution. Suppose that this distribution is characterised by mean $\bar{C}_1$ and variance σ_1^2. The second system is associated with a constant time for repair and constant cost of lost production $\bar{C}_2 > \bar{C}_1$. The two systems are characterised by the same probability of failure $p_{1f} = p_{2f} = p_f$. Equation 17.3 yields $K_1 = p_f \bar{C}_1$ for the risk of failure characterising the first system and $K_2 = p_f \bar{C}_2$ for the risk of failure characterising the second system. Clearly, $K_1 < K_2$ because $\bar{C}_1 < \bar{C}_2$. However, as can be verified from Figure 17.2, the probability that the loss given failure will exceed a critical maximum acceptable value x_{max} is zero for the system characterised by the larger risk and non-zero for the system characterised by the smaller risk.

In other words, *smaller expected loss does not necessarily mean smaller probability that the loss will exceed a particular limit.*

If the expected value of the loss given failure was selected as a utility function of the consequences from failure, the first system would be selected by a decision criterion based on minimising the expected loss.

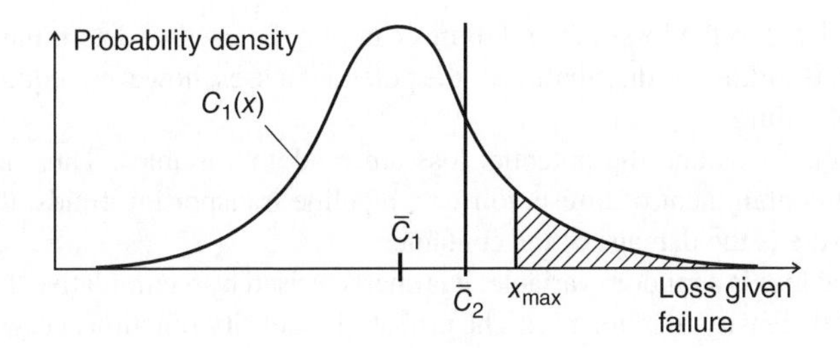

Figure 17.2 Distributions of the loss given failure for two systems

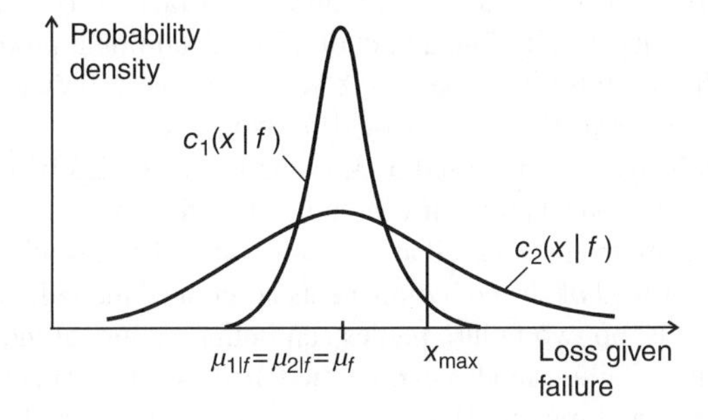

Figure 17.3 The variance of the loss given failure is strongly correlated with the probability that the loss will exceed a specified quantity

Suppose that x_{max} is the maximum amount of reserves available for covering the loss from critical failure. No recovery can be made from a loss exceeding the amount of x_{max}, and production cannot be resumed. With respect to whether a recovery from a critical failure can be made, the first system is associated with risk, while the second system is not.

In order to make a correct selection of a system minimising the risk of exceeding a maximum acceptable limit, the utility function should reflect whether the loss exceeds the critical limit x_{max} or not.

Increasing the variance of the loss given failure increases the risk that the loss will exceed a specified maximum tolerable level. This is illustrated in Figure 17.3 by the probability density distributions of the loss given failure $c_1(x \mid f)$ and $c_2(x \mid f)$ of two systems characterised by different variances $(\sigma_1^2 < \sigma_2^2)$ of the loss.

A new measure of the loss from failure which avoids the limitations of the traditional risk measure (Eq. 17.3) is the *cumulative distribution of the potential loss.*

17.5 Potential Loss, Conditional Loss and Risk of Failure

The concepts *potential loss* and *conditional loss* apply to both non-repairable and repairable systems. The quantity *loss given failure* is a conditional quantity because it is defined *given that failure has occurred.* This is in sharp contrast with the *potential loss* which is *unconditional quantity* and is defined *before failure occurs.* While the conditional distribution of the loss given failure can be used to determine the probability that given failure, the loss will be larger than a specified limit, the distribution of the potential loss combines the probability that there will be failure and the probability that the loss associated with it will be larger than a specified limit. In other words, the measure 'potential loss' incorporates the uncertainty associated with the exposure to losses and the uncertainty associated with the consequences given that exposure.

Historical data related to the losses from failures can only be used to determine the distribution of the conditional loss. Building the distribution of the potential losses, however, requires also an estimate of the probability of failure.

Both the conditional loss and the potential loss are random variables. Thus, in the failure event leading to a loss of containment of a reservoir or a pipeline transporting fluids, the conditional loss depends on how severe is the damage of the container.

Since the potential loss is a random variable, it is characterised by a cumulative distribution function $C(x)$ and a probability density function $c(x)$. The probability density function $c(x)$ gives the probability $c(x)\,dx$ (before the failure occurs) that the potential loss X will be in the infinitesimal interval $(x, x+dx)$: $P(x < X \le x+dx) = c(x)dx$.

Accordingly, the conditional loss (the loss *given* failure) is characterised by a cumulative distribution function $C(x|f)$ and probability density function $c(x|f)$. The conditional probability density function $c(x|f)$ gives the probability $c(x|f)dx$ that the loss X will be in the infinitesimal interval $(x, x+dx)$, given that the failure has occurred: $P(x < X \le x+dx|f) = c(x|f)dx$.

Let S be a non-repairable system composed of M components, logically arranged in series, which fails whenever any of the components fails. It is assumed that the components' failures are mutually exclusive; that is, no two components can fail at the same time. The reasoning below and the derived equations are also valid if instead of a set of components, a set of M mutually exclusive system failure modes are considered; that is, no two failure modes can initiate failure at the same time. Essentially, the set of mutually exclusive failure modes can be modelled as 'components' logically arranged in series. Because the system is non-repairable, the losses are associated with the first and only failure of the system. The reasoning in the following text, however, is also valid for a repairable system if the focus is on the loss from the first failure only.

The cumulative distribution function $C(x) \equiv P(X \le x)$ of the potential loss gives the probability that the potential loss X will not be greater than a specified value x. A loss is present only if failure is present. Consequently, the unconditional probability $C(x) \equiv P(X \le x)$ that the potential loss X will not be greater than a specified value x is equal to the sum of the probabilities of two mutually exclusive events: (i) failure will not occur and the loss will be not be greater than x, and (ii) failure will occur and the loss will not be greater than x. The probability of the first compound event is $(1-p_f)\times H(x)$, where p_f is the probability of failure and $H(x)$ is the conditional probability that the loss will not be greater than x given that no failure has occurred. This conditional probability can be presented by the Heaviside unit step function (Abramowitz and Stegun, 1972) $H(x) = \left\{\begin{smallmatrix} 1, & x\ge 0 \\ 0, & x<0 \end{smallmatrix}\right.$. The probability of the second compound event is $p_f C(x \mid f)$ where $C(x \mid f)$ is the conditional probability that *given* failure, the loss will not be greater than x. Consequently, the probability $C(x)$ that the potential loss X will not be greater than x is given by the distribution mixture (Todinov, 2006b):

$$C(x) \equiv P(X \le x) = (1-p_f)\times H(x) + p_f \times C(x|f) \tag{17.10}$$

The difference between a potential and conditional loss is well illustrated by their distributions in Figure 17.4. A characteristic feature of the cumulative distribution of the potential loss is the concentration of probability mass with magnitude $1-p_f$ at the origin (Figure 17.4b) because with probability $1-p_f$, failure will not occur and the potential loss will be zero.

If a level α for the probability of obtaining as extreme or more extreme loss is specified, a maximum potential loss x_α can be determined which corresponds to the specified level; α is the probability that the potential loss will exceed the maximum specified loss x_α ($\alpha = P(X > x_\alpha)$), Figure 17.4b. Then, the maximum potential losses $x_{\alpha,i}$ characterising different design solutions can be compared.

The maximum potential loss at a preset level is the limit, whose probability of exceeding is not greater than the preset level.

The maximum potential loss at a preset level determines the risk-based capital required to absorb the loss associated with failure. If x_α is the available resource of capital, the preset level α is the

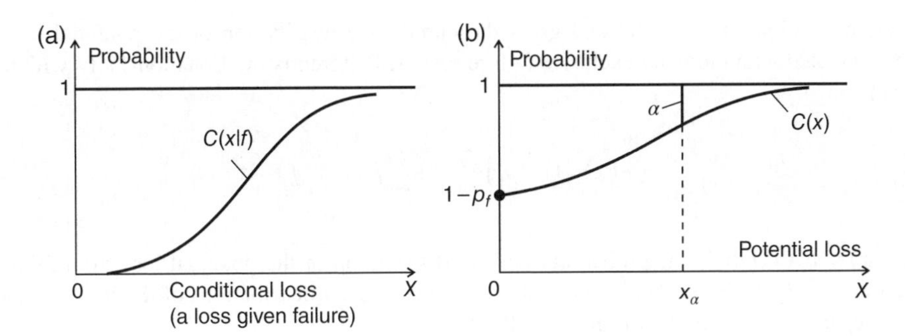

Figure 17.4 (a) A conditional loss (a loss given failure). (b) A potential loss and maximum potential loss x_α at a preset level α

probability that the actual loss will exceed it, thereby triggering insolvency. The expected loss cannot be used to define the necessary resource of capital because the actual loss varies randomly around it.

Let $C_k(x)$ be the conditional cumulative distribution of the loss (the loss *given* failure) characterising the kth individual component/failure mode, and $p_{k|f}$ be the conditional probability that *given* failure, the kth component/failure mode has initiated it first $\left(\sum_{k=1}^{M} p_{k|f} = 1\right)$. The conditional probability distribution $C(x \mid f) \equiv P(X \le x \mid f)$ that the loss X *given* that failure has occurred will not be greater than a specified value x can be presented by the union of the following mutually exclusive and exhaustive events: (i) given that failure has occurred, it is the first failure mode that has initiated it, and the loss X is not greater than x (the probability of which is $p_{1|f} C_1(x \mid f)$); (ii) given that failure has occurred, it is the second failure mode that has initiated it, and the loss X is not greater than x (the probability of which is $p_{2|f} C_2(x \mid f)$) and so on. The final event is: given that failure has occurred, it is the Mth failure mode that has initiated it, and the loss X is not greater than x (the probability of which is $p_{M|f} C_M(x \mid f)$). The probability of a union of mutually exclusive events equals the sum of the probabilities of the separate events. As a result, the conditional distribution of the loss given that failure has occurred is the mixture distribution:

$$C(x \mid f) = \sum_{k=1}^{M} p_{k|f} C_k(x \mid f) \tag{17.11}$$

The conditional probability distribution $C(x \mid f)$ is a mixture of the conditional distributions $C_k(x \mid f)$ characterising the individual failure modes, scaled by the conditional probabilities $p_{k|f}$ that the failure has been initiated by the kth failure mode, *given* that failure has occurred $\left(\sum_{i=1}^{M} p_{k|f} = 1\right)$. Finally, Equation 17.10 regarding the cumulative distribution of the potential loss becomes

$$C(x) = (1 - p_f) H(x) + p_f \sum_{k=1}^{M} p_{k|f} C_k(x \mid f) \tag{17.12}$$

The product of the probability of failure p_f and the probability $p_{k|f}$ that given failure, it is the kth failure mode that has initiated it, is simply equal to the probability p_k that the kth component/failure mode will initiate failure first ($p_f \times p_{k|f} = p_k$). Considering this and also the relationship $\sum_{i=1}^{M} p_k = p_f$, the above equation can also be presented as

$$C(x) = (1 - p_f) H(x) + \sum_{k=1}^{M} p_k C_k(x \mid f) \tag{17.13}$$

Equation 17.13 is fundamental and gives the cumulative distribution of the potential loss from failure associated with mutually exclusive failure modes. Differentiating Equation 17.13 with respect to x results in

$$c(x) = (1 - p_{\mathrm{f}})\delta(x) + \sum_{k=1}^{M} p_k c_k(x \mid f) \tag{17.14}$$

where $c(x) \equiv dC(x)/dx$ is the probability density distribution of the potential loss from failure and $c_k(x \mid f) \equiv dC_k(x \mid f)/dx$ are the conditional probability density distributions of the losses associated with the separate failure modes (Todinov, 2006b).

In Equation 17.14, $\delta(x)$ is the Dirac's delta function which is the derivative of the Heaviside function $dH(x)/dx$ (Abramowitz and Stegun, 1972). The expected value of the potential loss from failure $\bar{C}$ is obtained by multiplying Equation 17.14 by x and integrating it ($\int x\delta(x)dx = 0$):

$$\bar{C} = \int xc(x)dx = \sum_{k=1}^{M} p_k \int xc_k(x \mid f)dx = \sum_{k=1}^{M} p_k \bar{C}_{k\mathrm{lf}} \tag{17.15}$$

where $\bar{C}_{k\mathrm{lf}} = \int xc_k(x \mid f)dx$ is the expected value of the loss given that failure has occurred, characterising the kth failure mode.

For a single failure mode, Equation 17.15 transforms into

$$\bar{C} = p_{\mathrm{f}} \bar{C}_{\mathrm{f}} \tag{17.16}$$

which is equivalent to the risk Equation 17.3. Clearly, *the risk of failure K in* Equation 17.3 *can be defined as the expected value of the potential loss.*

Equation 17.13 can be used for determining the probability that the potential loss will exceed a specified critical value x. This probability is

$$P(X > x) = 1 - C(x) = 1 - (1 - p_{\mathrm{f}})H(x) - \sum_{k=1}^{M} p_k C_k(x \mid f)$$

which, for $x > 0$, becomes

$$P(X > x) = \sum_{k=1}^{M} p_k \left[1 - C_k(x \mid f)\right] \tag{17.17}$$

The probability that the potential loss will exceed a specified quantity is always smaller than the probability that the conditional loss will exceed the specified quantity.

17.6 Risk Associated with Multiple Failure Modes

Suppose now that the times to failure characterising M statistically independent failure modes are given by the cumulative distribution functions $F_k(t)$, $k = 1,2,\ldots,M$ with corresponding probability density functions $f_k(t) = dF_k(t)/dt$. The probability that the first failure mode will initiate failure can then be determined using the following probabilistic argument.

Consider the probability that the first component/failure mode will initiate failure in the infinitesimal time interval $t, t + dt$. This probability can be expressed as a product $f_1(t) \times [1 - F_2(t)] \times \cdots \times [1 - F_M(t)]dt$ of the probabilities of the following statistically independent events: (i) the first failure mode will initiate failure in the time interval $t, t+dt$, the probability of which is $f_1(t)dt$, and (ii) the other failure modes will not initiate failure before time t, the probability of which is given by $[1-F_2(t)]\times\cdots\times[1-F_M(t)]$. According to the total probability theorem, the total probability that the first component/failure mode will initiate failure in the time interval (0, a) is

$$p_1 = \int_0^a f_1(t)\left[1-F_2(t)\right]\ldots\left[1-F_M(t)\right]dt \tag{17.18}$$

Similarly, for the kth failure mode, this probability is

$$p_k = \int_0^a f_k(t)\left[1-F_1(t)\right]\ldots\left[1-F_{k-1}(t)\right]\left[1-F_{k+1}(t)\right]\ldots\left[1-F_M(t)\right]dt \tag{17.19}$$

Substituting these probabilities in Equation 17.17 yields the probability that the potential loss from multiple failure modes with known distributions of the time to failure will exceed a critical value x.

17.6.1 An Important Special Case

For failure modes characterised by constant hazard rates λ_k, $f_k(t) = \lambda_k \exp(-\lambda_k t)$ and $F_k(t) = 1-\exp(-\lambda_k t)$. Substituting these in Equation 17.19 and integrating yields

$$p_k = \int_0^a \lambda_k \exp\left[-(\lambda_1+\cdots+\lambda_M)t\right]dt = \left\{1-\exp\left[-(\lambda_1+\cdots+\lambda_M)a\right]\right\}\frac{\lambda_k}{\lambda_1+\cdots+\lambda_M} \tag{17.20}$$

Thus, for a system with failure modes characterised by constant hazard rates $\lambda_1, \lambda_2, \ldots, \lambda_M$, where M is the number of failure modes, the probability that the potential loss will exceed a specified value x in a specified time interval (0, a), is given by Equation 17.17 where the probabilities p_k, $k = 1, 2, \ldots, M$ are given by Equation 17.20. For components/failure modes characterised by constant hazard rates λ_k, the probability of failure before time a is $p_{\mathrm{f}} = 1-\exp\left[-(\lambda_1+\lambda_2+\cdots+\lambda_M)a\right]$, and from $p_k = p_{\mathrm{f}} \times p_{k|\mathrm{f}}$, the relationship

$$p_{k|\mathrm{f}} = \frac{\lambda_k}{\lambda_1+\cdots+\lambda_M}$$

is obtained.

Following Equation 17.15, regarding the expected loss given failure, the expected value of the potential loss (the risk) becomes (Todinov, 2004d)

$$K = \left\{1-\exp\left[-(\lambda_1+\cdots+\lambda_M)a\right]\right\}\times\sum_{k=1}^{M}\frac{\lambda_k}{\lambda_1+\cdots+\lambda_M}\bar{C}_{k|\mathrm{f}} \tag{17.21}$$

where the sum

$$\bar{C}_{\mathrm{f}} = \sum_{k=1}^{M}\frac{\lambda_k}{\lambda_1+\cdots+\lambda_M}\bar{C}_{k|\mathrm{f}} \tag{17.22}$$

can be interpreted as the expected conditional loss given that failure has occurred before time a.

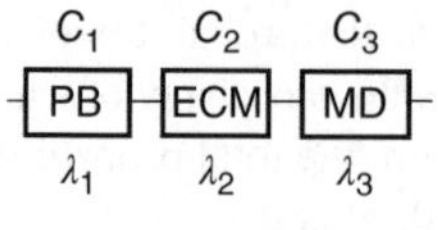

C_1 C_2 C_3
PB ECM MD
λ_1 λ_2 λ_3

Figure 17.5 A device composed of a power block (PB), electronic control module (ECM) and mechanical device (MD), logically arranged in series

Considering that for a non-repairable system, the probability of failure before time a is $1-\exp\left(-\int_0^a \lambda(t)dt\right)$, where $\lambda(t)$ is the hazard rate of the system, the expected value of the potential loss (the risk) from failure before time a becomes

$$\bar{C} \equiv K = \left[1-\exp\left(-\int_0^a \lambda(t)\,dt\right)\right] \times \sum_{i=1}^{M} p_{klf}\bar{C}_{klf} \tag{17.23}$$

Example

Suppose that a system contains a number of critical components. If failure of any of these critical components leads to a system failure, the risk of premature failure can be modelled by Equation 17.21. The equation is the basis of a new technology for a cost-of-failure reliability analysis and setting reliability requirements. The use of Equation 17.21 can be illustrated by a numerical example involving a device (Figure 17.5) composed of (i) a power block (PB) characterised by a hazard rate $\lambda_1 = 0.0043$ months^{-1} and cost of replacement $C_1 = 350$, (ii) an electronic control module (ECM) characterised by a hazard rate $\lambda_2 = 0.0014$ months^{-1} and cost of replacement $C_2 = 850$, and (iii) a mechanical device (MD) characterised by a hazard rate $\lambda_3 = 0.016$ months^{-1} and cost of replacement $C_3 = 2300$. The logical arrangement in series (Figure 17.5) means that failure of any block causes a system failure. For a warranty period of $a = 12$ months, substituting the numerical values in Equation 17.21 yields

$$K = \{1-\exp[-(\lambda_1+\lambda_2+\lambda_3)a]\} \times \sum_{i=1}^{3} \frac{\lambda_i}{\lambda_1+\lambda_2+\lambda_3} C_i \approx 417$$

for the risk (expected loss per device) of warranty payments due to premature failure.

In order to limit the risk of premature failure below a maximum acceptable level K_{max}, the hazard rates of the components must satisfy

$$K = \{1-\exp[-(\lambda_1+\cdots+\lambda_M)a]\} \times \left(\frac{\lambda_1}{\lambda_1+\cdots+\lambda_M} C_1 + \cdots + \frac{\lambda_M}{\lambda_1+\cdots+\lambda_M} C_M\right) \le K_{max} \tag{17.24}$$

Unlike the case of a single component/failure mode for which the equation $K = K_{max}$ has a unique solution given by Equation 17.6, for more than one component/failure mode, the equality $K = K_{max}$ in Equation 17.24 is satisfied for an infinite number of values $\lambda_1, \ldots, \lambda_M$ for the hazard rates. To prevent excessive costs associated with improving the reliability of the components, the hazard rates λ_i can be selected such that equality is attained in expression 17.24.

17.7 Expected Potential Loss Associated with Repairable Systems Whose Component Failures Follow a Homogeneous Poisson Process

Consider the case where the individual components are logically arranged in series and characterised by constant hazard rates λ_i. Since after each failure of a component, it is replaced by an identical component, the system failures are a superposition of the components' failures. The sequential failures of component i is a homogeneous Poisson process with density numerically equal to the hazard rate λ_i of the component. Since a superposition of several homogeneous Poisson processes with densities λ_i is a homogeneous Poisson process with density

$$\lambda = \sum_{i=1}^{M} \lambda_i, \tag{17.25}$$

Equation 17.25 also holds for the rate of occurrence of failures λ of a system with M components logically arranged in series. Combining Equation 17.22 related to the expected losses given failure and Equation 17.25 results in

$$\bar{L} = \lambda a \bar{C} = \lambda a \sum_{k=1}^{M} \frac{\lambda_k}{\lambda_1 + \cdots + \lambda_M} \bar{C}_k \tag{17.26}$$

for the expected losses from failures $\bar{L}$ in the finite time interval (0, a), $\left(\lambda = \sum_{k=1}^{M} \lambda_k\right)$. Considering (17.25), Equation 17.26 becomes

$$\bar{L} = \sum_{k=1}^{M} \lambda_k a \bar{C}_k \tag{17.27}$$

If L_{max} are the maximum acceptable expected losses and the losses given failure $\bar{C}_k$ associated with each component are the same ($\bar{C}_k = \bar{C}$), Equation 17.27 gives

$$\lambda^* = \frac{L_{max}}{\bar{C}a} \tag{17.28}$$

for the upper bound of the rate of occurrence of failures which guarantees that if for the system's rate of occurrence of failures $\lambda \le \lambda^*$ is fulfilled, the expected losses from failures $\bar{L}$ will be within the maximum acceptable level L_{max}. This is, in effect, setting reliability requirements to limit the expected losses from failures of a repairable system below a maximum acceptable limit. The acceptable limit L_{max} is often the budget allocated for unscheduled maintenance. In cases where the expected losses $\bar{C}_k$ associated with the different failure modes/components are different, the hazard rates λ_i which satisfy the inequality

$$\sum_{i=1}^{M} \lambda_i \bar{C}_i \le \frac{L_{max}}{a} \tag{17.29}$$

limit the expected losses from failures below the maximum acceptable level L_{max}.

Example

Setting reliability requirements from the type given by Equation 17.28 can be illustrated by the following example. Suppose that for a particular unit, the losses from system failures caused by component failures are approximately the same. This is common in cases where upon a component failure the whole unit has to be replaced. In this case, despite variations in the costs of the separate components belonging to the unit, the losses from failures are dominated by the cost of replacement of the whole unit. Suppose that the cost of replacement of the unit is approximately \$100 000. If the maximum budget covering unscheduled maintenance within 15 years is \$600 000, the hazard rate λ of the unit cannot exceed the value 0.4 years^{-1}.

$$\lambda \le \lambda_L = \frac{L_{max}}{\bar{C}a} = \frac{600\,000}{15 \times 100\,000} = 0.4$$

This is, in effect, setting reliability requirements to limit the expected losses from failures below a specified limit.

If the actual losses from failures are of importance, and not the expected losses, the approach to setting reliability requirements is different. For a system whose failures follow a homogeneous Poisson process with density λ and the cost given failure is constant $\bar{C}$ (irrespective of which component has failed), the probability that the losses from failures X will be smaller than a maximum acceptable value $L_{\max}$ is

$$P(X \le L_{\max}) = \sum_{y=0}^{r} \frac{(\lambda a)^y}{y!} \exp(-\lambda a) \tag{17.30}$$

where $r = [L_{\max} / \bar{C}]$ and the right-hand side of Equation 17.30 is the cumulative Poisson distribution. $L_{\max}$ could be, for example, the maximum tolerable losses from failures (e.g. the budget allocated for unscheduled maintenance).

Equation 17.30 can be used to verify that the potential losses from failures will not exceed a critical limit and will be illustrated by a simple example.

Example

Suppose that the available amount of repair resources for the first 6 months ($a = 6$ months) of operating a system is \$2000. Each system failure requires \$1000 of resources for intervention and repair. Suppose also that the system failures follow a homogeneous Poisson process with intensity $\lambda = 0.08$ months^{-1}. The probability that within the first 6 months of operation the potential losses X will exceed the critical limit of \$2000 can be calculated by subtracting from unity the probability of the complementary event: 'the potential losses within the first 6 months will not exceed \$2000':

$$P(X > L_{\max} = 2000) = 1 - P(X \le L_{\max} = 2000)$$

Since $r = [L_{\max} / \bar{C}] = 2$, applying Equation 17.30 results in

$$P(X \le L_{\max} = 2000) = \sum_{y=0}^{2} \frac{(\lambda a)^y}{y!} \exp(-\lambda a) = 0.987$$

for the probability that the potential losses will not exceed the critical value. Hence, the probability that the potential losses will exceed the critical value of 2000 units is

$$P(X > L_{\max} = 2000) = 1 - P(X \le L_{\max} = 2000) = 0.013$$

If an upper bound of the system failure density is required, which guarantees with confidence $q\%$ that the potential losses in the interval (0, a) will not exceed the maximum acceptable limit $L_{\max}$, the equation

$$\frac{q}{100} = \sum_{y=0}^{r} \frac{(\lambda a)^y}{y!} \exp(-\lambda a) \tag{17.31}$$

where $r = \left[L_{\max} / \bar{C}\right]$ must be solved numerically with respect to λ.

The solution λ_L gives the upper bound of the rate of occurrence of failures which guarantees that if for the system rate of occurrence of failures $\lambda \le \lambda_L$ is fulfilled, the potential losses L will be within the available resources $L_{\max}$ $(L \le L_{\max})$ allocated for unscheduled maintenance.

17.8 A Counterexample Related to Repairable Systems

The fact that a larger reliability of a repairable system does not necessarily mean smaller expected losses from failures can be demonstrated on the simple system in Figure 17.5 composed of a power block (PB), an electronic control module (ECM) and a mechanical device (MD). Two systems of this type are compared; their components' hazard rates and losses from failure are listed in Table 17.1.

The reliability of the first system for $t = 2$ years is

$$R(t) = \exp[-(\lambda_1 + \lambda_2 + \lambda_3)t] = 0.24$$

while the expected losses from failures during $t=2$ years of operation are given by Equation 17.27:

$$\bar{L}_1 = \sum_{i=1}^{3} t\lambda_i C_i = \$98\,500$$

Correspondingly, the reliability of the second system associated with a time interval of $t=2$ years is

$$R(t) = \exp[-(\lambda_1' + \lambda_2' + \lambda_3')t] = 0.40$$

while the expected losses from failures during $t=2$ years of operation are

$$\bar{L} = \sum_{i=1}^{3} t\lambda_i' C_i \approx \$120\,400$$

As can be verified, although the second system has a superior reliability, it is also associated with larger expected losses. This counterexample shows that for a system which consists of components associated with different losses from failure, a larger system reliability does not necessarily mean smaller losses from failures.

Now, let us assume that the losses given failure characterising all components in the system are the same: $C_1 = C_2 = C_3 = C$. In this case, the expected losses become

$$\bar{L} = \sum_{i=1}^{3} \lambda_i tC = Ct\sum_{i=1}^{3} \lambda_i$$

for the first system and

$$\bar{L}' = \sum_{i=1}^{3} \lambda_i' tC = Ct\sum_{i=1}^{3} \lambda_i'$$

Table 17.1 Reliability parameters of two repairable systems with components arranged in series (Figure 17.5)

	Hazard rate (year^{-1})		
Component	First system	Second system	Losses from failures
PB	$\lambda_1 = 0.2$	$\lambda_1' = 0.1$	$C_1 = C_1' = \$2500$
ECM	$\lambda_2 = 0.3$	$\lambda_2' = 0.1$	$C_2 = C_2' = \$1500$
MD	$\lambda_3 = 0.21$	$\lambda_3' = 0.26$	$C_3 = C_3' = \$230\,000$

for the second system. Clearly, in this case, the smaller the system hazard rate $\lambda = \sum_{i=1}^{3} \lambda_i$, the larger the reliability of the system and the smaller the expected losses. This example shows that for a system which consists of components associated with the same cost of system failure, a higher system reliability always means smaller losses from failures. Further detailed discussion related to cost-of-failure-based reliability analysis and cost-of-failure-based setting of reliability requirements can be found in Todinov (2007).

17.9 Guaranteeing Multiple Reliability Requirements for Systems with Components Logically Arranged in Series

The hazard rate envelope which guarantees more than a single reliability requirement can be found as an intersection of the hazard rate envelopes guaranteeing the separate reliability requirements. Suppose that a hazard rate envelope is required to guarantee expected losses from failures of a repairable system below the maximum acceptable limit K_{max}. The hazard rate envelope λ_R^* guaranteeing that the expected losses from failures will be below the limit L_{max} is obtained from Equation 17.28. If a minimum availability target and an MFFOP have been specified (i.e. availability at least A_T and an MFFOP of length at least s before the first failure), corresponding hazard rate envelopes λ_A^* and λ_S^* can be determined. These hazard rate envelopes guarantee that whenever $\lambda \le \lambda_A^*$ is fulfilled for the rate of occurrence of system failures, the availability will be greater than the specified minimum target value A_T, and whenever $\lambda \le \lambda_S^*$ is fulfilled, an MFFOP with length at least s will exist before the first failure, with a specified minimum probability p_{MFFOP}.

The intersection of all hazard rate envelopes is a hazard rate envelope $\lambda^* = \min\{\lambda_R^*, \lambda_A^*, \lambda_S^*\}$ which guarantees that all reliability requirements will be met if for the system hazard rate, the relationship $\lambda \le \lambda^*$ is fulfilled (Figure 17.6a).

A significant drawback of the current practice for setting quantitative reliability requirements in certain industries, is that they are based solely on specifying a minimum availability target. As it was shown in Chapter 16, specifying a high availability target does not necessarily guarantee a low risk of premature failure. It must also be pointed out that an availability target does not necessarily account for the cost of failure. Although availability does account for the cost of lost production which is proportional to the downtime, it does not account for the cost of resources for cleaning up polluted environment, the cost of repair and the cost of intervention for repair. Setting reliability requirements should be made by finding the intersection of the hazard rate envelopes which limit the losses from

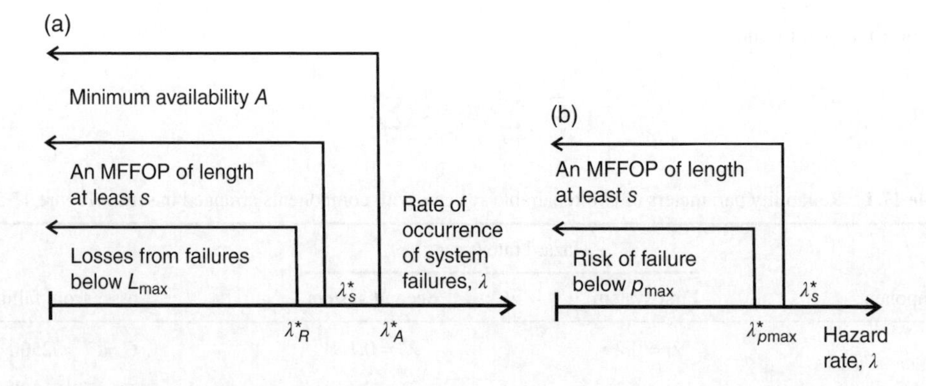

Figure 17.6 Setting reliability requirements as an intersection of hazard rate envelopes (a) for a repairable system and (b) for a non-repairable system

failure below a specified value, guarantee an MFFOP and production availability larger than specified target values.

Figure 17.6b, illustrates the process of setting reliability requirement for a non-repairable system. In this case, the reliability requirements are determined from the intersection of the hazard rate which delivers a specified MFFOP of a particular length and risk below a maximum acceptable level.

Equations 17.6 and 17.9, for example, can be applied to determine the maximum hazard rate that guarantees that the risk of failure within a time interval 0, *a* does not exceed a particular level. It must be pointed out that the cost-of-failure-driven reliability requirements do not necessarily entail unrealistic hazard rates, which are difficult to achieve by suppliers. Indeed, according to Equation 17.6, the hazard rate envelope λ^* is small when the maximum acceptable risk K_{max} is small and can be very large if the maximum acceptable risk K_{max} is close to the cost of failure C.

Ignoring the cost of failure and not setting reliability requirements linked to the cost of failure has lead and still leads to failures with grave consequences to the environment. Many high-consequence failures in deep-water oil and gas production, for example, could have been avoided if cost-of-failure-based reliability requirements had been used in the design of components associated with high risk of failure.

A set of quantitative reliability requirements in deep-water oil and gas production, for example, should include several basic components: (i) a specified maximum acceptable level of the expected potential loss due to failures (asking for minimum risk), (ii) a specified minimum availability target (asking for availability) and (iii) a requirement for a minimum failure-free operating period before the first failure (asking for reliability). The intersection of the hazard rate envelopes guaranteeing the individual requirements is a hazard rate envelope which guarantees all of the requirements.

The input data for setting basic reliability requirements at a system level which limit the risk of premature failure and deliver a required minimum availability target are the downtimes due to a critical failure, the average cost of the lost production per unit downtime and the cost of intervention for repair (includes the cost of locating the failure, the cost of mobilisation of resources, the cost of intervention and the cost of repair/replacement). Another component of the input data is the required minimum availability. Detailed discussions related to determining the risk of failure for a time-dependent cost of failure and complex systems with complex topology can be found in Todinov (2007).

18

Potential Loss, Potential Profit and Risk

18.1 Deficiencies of the Maximum Expected Profit Criterion in Selecting a Risky Prospect

The optimal choice from a number of risky prospects, each containing a set of *risk–reward activities*, is part of an important class of risk decisions made in business, economics, technology, medicine, etc. Currently, the *maximum expected profit criterion* is used for making an optimal choice among risky prospects (Denardo, 2002; Moore, 1983). According to this criterion, a rational decision maker compares the expected profits from a number of risky prospects and selects the prospect with the largest expected profit. An expected profit from a risky prospect is obtained by adding the monetary outcomes characterising the prospect multiplied by their probabilities. In this chapter, we demonstrate that *a choice based on maximising the expected profit is deeply flawed if a small number of risk–reward bets are present in the risky prospects.*

In the past, the expected profit from an infinite number of statistically independent repeated bets led to the *Petersburg paradox*. Its avoidance was one of the reasons for proposing the expected utility theory by Bernoulli (1738), later developed by von Neumann and Morgenstern (1944). The effort towards understanding statistically independent repeated bets did not stop with the work of D. Bernoulli. Statistically independent repeated bets, for example, have been at the focus of a paper from Samuelson (1963). In this paper, the author brings the following argument, through a story in which he offered to his colleague a good bet (with a positive expected value): 50–50 chance of winning 200 or losing 100. The colleague refused by saying that he would feel the 100 loss more than the 200 gain. He said that he would agree to participate if he was offered to make 100 such bets. Samuelson criticised the reasoning of his colleague and went on to propose and prove a 'theorem' which stated that if a single good bet is unacceptable, then any finite sequence of such bets is unacceptable too. Samuelson claimed that increasing the number of unacceptable bets does not reduce the risk of a net loss and termed accepting a sequence of individually unacceptable bets 'a fallacy of large numbers'. Samuelson's 'theorem' has been reproduced in several related papers (Ross, 1999). This 'theorem' spawned several related papers where researchers have extended Samuelson's condition to assure that they would not allow the 'fallacy of large numbers'. In this chapter, it is shown that contrary to Samuelson's theory, increasing the number of unacceptable bets does reduce the risk of a net loss, and this is demonstrated by using Samuelson's own example.

Reliability and Risk Models: Setting Reliability Requirements, Second Edition. Michael Todinov.

A frequently pointed out weakness of the expected profit criterion is that it assumes too much knowledge necessary to make a decision. The information regarding the likelihood of an event and the consequences associated with the event is rarely available. In Todinov (2013c), it was shown that for a limited number of risk–reward activities in the risky prospects, *the maximum expected profit criterion could lead to accepting a decision associated with a large risk of a net loss*. This is true even with a full knowledge related to the likelihood of an event and its consequences and without the existence of a subjective bias while making a decision.

The case considered in the developments to follow is where the results from the different outcomes can be adequately measured in monetary terms and the analysis is confined to linear utility functions. For the sake of simplicity, the inadequacy of the maximum expected profit criterion is demonstrated for the simplest case, involving *statistically independent risk–reward bets* in the risky prospects.

The maximum expected profit criterion does not account for the significant impact of the actual number of risk–reward events/bets in a risky prospect. The choice under risk by using the maximum expected profit criterion is made as if each compared risky prospect contains a very large number of risk–reward events/bets. The critical dependence of the choice on the number of risk–reward bets in the risky prospects has not been discussed in studies related to ranking risky alternatives (Hador and Russel, 1969; Hansch and Leuy, 1969; Pflug, 2000; Roberts, 1979; Tobin, 1958). This is also true for more recent models related to ranking risky alternatives (Nielsen and Jaffray, 2006; Richardson and Outlaw, 2008; Rockafellar and Uryasev, 2002; Starmer, 2000). Even in a recent and probably the most comprehensive treatise of the theory of betting (Epstein, 2009), no discussion has been provided on the impact of the limited number of risk–reward bets on the choice of a risky prospect. The number of risk–reward events/bets in a risky prospect, however, has a crucial impact on the choice of the risky prospect and cannot be ignored.

18.2 Risk of a Net Loss and Expected Potential Reward Associated with a Limited Number of Statistically Independent Risk–Reward Bets in a Risky Prospect

Companies and entrepreneurs often make decisions under risk. Investing in an activity whose outcome is uncertain is a commonly made decision. Such is, for example, the drilling for oil and the advertising campaign for particular products on particular markets. The commonly used method for selection among risky prospects is the maximum expected profit criterion. According to the maximum expected profit criterion, the activity characterised by the largest expected profit is selected. Often, the opportunities for sequential bets are limited. A common cause for the bankruptcy of small companies is their inability to sustain financially the losses from several unsuccessful investments, despite that each investment may be characterised by a positive expected gain. Drilling for oil, for example, or running a large advertising campaign may be associated with a large expected profit/pay-off and a large probability of success, but if unsuccessful, they can also sink significant financial resources and bankrupt a small company. In contrast, large companies can sustain losses from a number of unsuccessful oil drillings or advertising campaigns and still be profitable in the long run. In short, small companies, because of their limited resources, are more likely to be affected by the large risk associated with a small number of sequential bets. This is an example where the blind adherence to the maximum expected profit criterion has been and has remained a source of heavy losses.

The inadequacy of the maximum expected profit criterion as a basis for making a choice between risky prospects containing a limited number of risk–reward events/bets activities will be demonstrated in the simplest case, where the risk–reward bets in the risky prospects are statistically independent.

Risk–reward events/bets can materialise as benefit or loss. An investment in a particular enterprise is a typical example of a risk–reward bet. A successful investment is associated with returns (benefits), while an unsuccessful investment is associated with losses. Usually, for risk–reward bets, the larger the magnitude of the potential loss, the larger is the magnitude of the potential benefit.

An example of existing framework for dealing with opportunity and failure events is the double probability–impact matrix for opportunities and threats proposed by Hillson (2002). Accordingly, expected potential reward R can be defined as a product of the probability p_s that a risk–reward event will materialise as 'success' and the benefit $\bar{B}$ given success:

$$R = p_s \bar{B} \tag{18.1}$$

In our opinion, it is not beneficial to treat the potential benefits as risk. Reserving the term risk for the potential loss only provides a better analysis structure for risk reduction and profit increase. Separating potential benefit from potential loss focuses attention on eliminating hazards and creating opportunity events as a way of increasing the potential profit.

Suppose that $0 \le p_s \le 1$ is the probability that the risk–reward event/bet will be a 'success' and will bring benefits characterised by the conditional cumulative distribution $B_s(x \mid s)$ (*given* that the risk–reward event has materialised as success). Correspondingly, $0 \le p_f = 1 - p_s$ is the probability that the risk–reward event/bet will generate a loss, associated with a conditional cumulative distribution function $C_f(x|f)$ (given that the risk–reward event has materialised as a loss).

The expected values of the benefit and the loss given that the risk–reward event has materialised are denoted by $\bar{B}_s$ and $\bar{C}_f$, respectively.

The expected profit $\bar{G}$ from a risk–reward event/bet is given by

$$\bar{G} = p_s \times \bar{B}_s + p_f \times \bar{C}_f \tag{18.2}$$

where p_s is the probability of a beneficial outcome with magnitude $\bar{B}_s$ of the expected benefits and $p_f = 1 - p_s$ is the probability of a loss with expected magnitude $\bar{C}_f$ (the loss $\bar{C}_f$ is taken with a negative sign).

An example of such a risk–reward event/bet has already been given with drilling for oil at a particular spot. Suppose that the geological analysis suggests that the probability of recovering oil by drilling at a particular spot is p_s. If oil is recovered, the benefit (after covering the cost of drilling) will be $\bar{B}_s$. With probability $p_f = 1 - p_s$ however, oil will not be recovered and a loss of magnitude C_f (the cost of drilling) will be incurred. Drilling an oil well is essentially a risk–reward bet.

A risky prospect may contain a number of risk–reward bets. The expected profit G_A from a risky prospect A containing M risk–reward bets is given by

$$\bar{G}_A = \sum_{i=1}^{M} \bar{G}_i \tag{18.3}$$

where $\bar{G}_i$ is the expected profit characterising the ith risk–reward bet in the risky prospect. Risk–reward bets with a positive expected potential profit ($\bar{G} > 0$) will be referred to as *risk–reward opportunities* or *opportunity bets*, while risk–reward bets with a negative potential expected profit ($\bar{G} < 0$) will be referred to as *risk–reward gambles*. Note that the concept 'opportunity bet' is the same as the concept 'good bet' used in Samuelson (1963).

Consider initially risk–reward opportunities only. The decision to be made is whether to invest in a risk–reward opportunity or not and which risk–reward opportunity should be preferred. The potential profit G from an investment is a random variable following a Bernoulli distribution with parameter p_s. For constant values of the benefit given success $\bar{B}_s$ and the loss given failure $\bar{C}_f$, the probability distribution of the potential profit G is given by $P\left(G = \bar{B}_s\right) = p_s$ and $P\left(G = \bar{C}_f\right) = p_f$. The probability distribution of the potential profit G can be considered to be a distribution mixture including two distributions with means

$\mu_1 = \bar{B}_s$ and $\mu_2 = \bar{C}_f$ and variances $V_1 = 0$ and $V_2 = 0$, sampled with probabilities $p = p_s$ and $1 - p = p_f$. According to Equation 4.7, for the variance Var(G) of the potential profit, we get

$$\text{Var}(G) = E(G^2) - [E(G)]^2 = p_s p_f (\bar{B}_s - \bar{C}_f)^2 \tag{18.4}$$

For a large number of risk–reward bets, the expected profit is approximated well by Equation 18.2. In short, for risky prospects each containing a large number of statistically independent risk–reward bets, choosing the prospect characterised by the maximum expected profit is a sound decision-making criterion. For a risky prospect containing a limited number of risk–reward bets however, despite the existence of a large expected profit, the risk of a net loss can be significant.

The next example involves two risky prospects containing a different number of risk–reward bets. The first risky prospect contains a single risk–reward bet with parameters $p_s = 0.3$, $\bar{B}_s = 300$, $p_f = 0.7$ and $\bar{C}_f = -90$, and its expected profit is $E(G) = 0.3 \times 300 - 0.7 \times 90 = 27$. The probability of a net loss from this risky prospect is 70%. The risk of a net loss is $-0.7 \times 90 = -63$.

The second risky prospect contains three risk–reward bets with the same probability of success and failure but with three times smaller magnitudes for the benefit given success and the loss given failure: $p_s = 0.3$, $\bar{B}_s = 300/3 = 100$, $p_f = 0.7$ and $\bar{C}_f = -90/3 = -30$. The expected profit from the risky prospect containing the three split bets is $E(G_{123}) = 3 \times (0.3 \times 100 - 0.7 \times 30) = 27$. Because a net loss from the second risky prospect can be generated only if a loss is generated from every single bet, the probability of a net loss from the second risky prospect is $p_{f,123} = 0.7^3 \approx 0.34$. The risk of a net loss is $-0.34 \times 90 = -30.6$.

Clearly, the maximum expected profit criterion cannot distinguish between these two risky prospects, characterised by the same expected profit. This is despite that the probability of a net loss and the risk of a net loss from the second risky prospect are much smaller compared to the probability of a net loss and the risk of a net loss from the first risky prospect. As a result, the second risky prospect is to be preferred to the first risky prospect. The serious problem from applying the maximum expected profit criterion becomes apparent if the three opportunity bets from the second risky prospect are characterised by a smaller benefit given success $\bar{B}_s = 300/3 - 1 = 99$. In this case, the expected profit characterising the second risky prospect will be smaller

$$E(G_{123}) = 3 \times (0.3 \times 99 - 0.7 \times 30) = 26.1$$

than the expected profit from the first risky prospect. The probability of a net loss $p_{f,123} = 0.7^3 \approx 0.34$ for the second risky prospect remains the same. Adherence to the maximum expected profit criterion will favour the selection of the first risky prospect, characterised by the larger expected profit of 27, the significantly larger probability of a net loss of 70% and more than twice magnitude −63 of the risk!

The significant hidden risk of a net loss associated with one of the risky prospects has not been revealed by the maximum expected profit criterion.

18.3 Probability and Risk of a Net Loss Associated with a Small Number of Opportunity Bets

The last example shows how the risk associated with a risk–reward bet can be reduced significantly if the risk–reward bet is split into several risk–reward bets characterised by the same probability of success and failure as the original bet but with proportionally smaller benefit and loss. Indeed, consider a risky prospect containing a single risk–reward bet, characterised by a probability of success p_s, benefit given success $\bar{B}_s$, probability of failure p_f and loss given failure $\bar{C}_f$. This risk–reward bet can be split into m risk–reward sub-bets, each characterised with the same probability of success and failure p_s and p_f and with m times smaller expected benefit and loss $\bar{B}_s / m$ and $\bar{C}_f / m$. The expected profit from the risk–reward bets:

$$E\left(G_{1,\ldots,m}\right) = m \times \left(\frac{p_s \bar{B}_s}{m} + \frac{p_f \bar{C}_f}{m}\right) = E(G) \tag{18.5}$$

is equal to the expected profit from the original bet. Considering Equation 18.4, the variance

$$V\left(G_{1,\ldots,m}\right) = \sum_{i=1}^{m} V_i = \sum_{i=1}^{m} p_s p_f \left(\frac{\bar{B}_s}{m} - \frac{\bar{C}_f}{m}\right)^2 = \frac{1}{m} p_s p_f \left(\bar{B}_s - \bar{C}_f\right)^2 \tag{18.6}$$

of the profit from the risky prospect with m risk–reward bets is m times smaller than the variance $p_s p_f \left(\bar{B}_s - \bar{C}_f\right)^2$ of the profit characterising the initial risk–reward bet.

In summary, the maximum expected profit criterion is a fundamentally flawed criterion, because it does not reflect the impact of the number of risk–reward activities in the risky prospects. *The number of the risk–reward activities should be a key consideration in selecting a risky prospect* (Todinov, 2013c).

Suppose that a finite number n of risk–reward opportunities of the same type are present. A series of n statistically independent risk–reward bets, characterised by the same probability of success in each trial, is a binomial experiment, where the number of successful outcomes is modelled by the binomial distribution.

The risk of a net loss from n risk–reward bets can be derived by the following probabilistic argument.

Let x denote the number of bets which materialise as 'benefit' among n bets, $x = 0, 1, \ldots, n$. Correspondingly, $n - x$ will be the number of bets which materialise as losses. The probability of a net loss equals the probability that the sum of the benefits from x benefit-generating bets will be smaller than the sum of the losses from $n - x$ loss-generating bets.

Let $\bar{B}_s$ be the expected value of the benefit given a successful bet and $\bar{C}_f$ be the expected value of the loss given a loss-generating bet. The probability that the sum of the benefits from x benefit-generating bets will be smaller than the sum of the losses from $n - k$ loss-generating bets is equal to the probability that the number of benefit-generating bets x does not exceed k, the largest integer satisfying the inequality

$$k \times \bar{B}_s < (n - k) \times \left|\bar{C}_f\right| \tag{18.7}$$

Condition (18.7) is equivalent to $k = n\left|\bar{C}_f\right| / \left(\bar{B}_s + \left|\bar{C}_f\right|\right) - 1$, if $n\left|\bar{C}_f\right| / \left(\bar{B}_s + \left|\bar{C}_f\right|\right)$ is an integer number. Otherwise, condition (18.7) is equivalent to $k = \left[n\left|\bar{C}_f\right| / \left(\bar{B}_s + \left|\bar{C}_f\right|\right)\right]$ where $\left[n\left|\bar{C}_f\right| / \left(\bar{B}_s + \left|\bar{C}_f\right|\right)\right]$ is the largest integer not exceeding the ratio $n\left|\bar{C}_f\right| / \left(\bar{B}_s + \left|\bar{C}_f\right|\right)$. The probability that there will be a net loss then becomes

$$P(\text{net loss}) = \sum_{x=0}^{k} \frac{n!}{x!(n-x)!} p_s^x (1 - p_s)^{n-x} \tag{18.8}$$

The risk of a net loss is equal to the expected value of the potential loss. This can be determined by adding the probability of zero benefit-generating bets times the losses $n\left|\bar{C}_f\right|$ from n loss-generating bets plus the probability of a single benefit-generating bet and $n - 1$ loss-generating bets times the net loss $(n-1)\bar{C}_f + \bar{B}_s$ and so on, plus the probability of k benefit-generating bets and $n-k$ loss-generating bets times the net loss $(n-k)\bar{C}_f + k\bar{B}_s$ from $n-k$ loss-generating bets and k benefit-generating bets. As a result, the risk of a net loss K becomes

$$K = \sum_{x=0}^{k} \frac{n!}{x!(n-x)!} p_s^x (1-p_s)^{n-x} \times \left[(n-x)\bar{C}_f + x\bar{B}_s\right], \tag{18.9}$$

The expected potential reward can be determined by adding the probability of $k+1$ benefit-generating bets times the net benefit $(k+1)\bar{B}_s + (n-k-1)\bar{C}_f$ from $k+1$ benefit-generating bets and $(n-k-1)$ loss-generating bets plus and so on, plus the probability of n benefit-generating bets times the net benefit $n\bar{B}_s$ from them. As a result, the expected potential reward from n risk–reward bets becomes

$$R = \sum_{x=k+1}^{n} \frac{n!}{x!(n-x)!} p_s^x (1-p_s)^{n-x} \times \left[(n-x)\bar{C}_f + x\bar{B}_s\right] \tag{18.10}$$

Equations 18.8, 18.9 and 18.10 (Todinov, 2013c) have also been verified by a Monte Carlo simulation. Thus, for $\bar{B}_s = 290$, $\bar{C}_f = -100$, $p_s = 0.3$, $p_f = 0.7$ and 12 opportunity bets, Equation 18.8 gives 0.49 for the probability of a net loss, Equation 18.9 gives $K = -152$ for the risk of a net loss, and Equation 18.10 gives $R = 356$ for the expected potential reward. These results have been confirmed by the empirical values ($P(\text{net loss}) = 0.49$, $K = -152$ and $R = 356$) obtained on the basis of one million simulation trials.

The expected profit from n bets can be obtained by using a standard result from the theory of probability stating that the expected value of a sum of random variables is the sum of the expected values of the random variables. Since the expected profit from a single bet is $\bar{G} = p_s \times \bar{B}_s + p_f \times \bar{C}_f$, the expected profit from a sequence of n bets is

$$E(X_1 + \cdots + X_n) = np_s\bar{B}_s + n(1-p_s)\bar{C}_f \tag{18.11}$$

In other words, with increasing the number of opportunity bets, the expected profit increases proportionally to the number of bets in the sequence. For statistically independent random variables, the variance of their sum is equal to the sum of the variances of the random variables (DeGroot, 1989):

$$V\left(\sum_{i=1}^{n} X_i\right) = \sum_{i=1}^{n} V(X_i) \tag{18.12}$$

Because the variance of the profit from a single bet is given by Equation 18.6, the variance of the profit from n statistically independent bets is given by

$$V(G) = np_s(1-p_s)\left(\bar{B}_s - \bar{C}_f\right)^2 \tag{18.13}$$

In other words, the variance of the profit from a sequence of a number of statistically independent bets increases proportionally to the number of bets in the sequence.

In the case of a very large number n of bets, the conditions for the validity of the central limit theorem are fulfilled, and a Gaussian distribution with mean

$$\mu = n\left[p_s\bar{B}_s + (1-p_s)\bar{C}_f\right] \tag{18.14}$$

and standard deviation

$$\sigma = \left(\bar{B}_s - \bar{C}_f\right)\sqrt{np_s(1-p_s)} \tag{18.15}$$

can be used for approximating the distribution of the potential profit. The probability of a net loss will be

$$P(\text{net loss}) = \Phi\left(\frac{0-\mu}{\sigma}\right) \tag{18.16}$$

where $\Phi(\bullet)$ is the cumulative distribution of the standard normal distribution with mean '0' and standard deviation '1'. Correspondingly, the probability of a net profit is

$$P(\text{net profit}) = 1 - \Phi\left(\frac{0-\mu}{\sigma}\right) \tag{18.17}$$

From the distribution, the probability that the net loss will exceed any specified quantity can be determined. It is necessary to point out, however, that approximations (18.16) and (18.17) are valid *only for a large number of bets in the risky prospect* and do not hold for a limited number of bets. In the case of a limited number of bets in the risky prospect, Equations 18.8, 18.9 and 18.10 should be used.

18.4 Samuelson's Sequence of Good Bets Revisited

Following Samuelson's paper (Samuelson, 1963), the same proposed 'good bet' will be used: probability $p_s = 0.5$ of winning 200 ($\bar{B}_s = 200$) and probability $p_f = 0.5$ of losing 100 ($\bar{B}_s = 200$). Table 18.1 lists the results from the calculations using Equations 18.8, 18.9 and 18.15. These calculations have also been confirmed by a Monte Carlo simulation.

As can be verified from the table, despite that with increasing the number of good bets the variance of the net profit increases, the risk of a net loss has decreased significantly. In other words, despite that selecting an individual bet is not acceptable because of the high probability of a loss, selecting repeated bets is beneficial, because a longer sequence of repeated bets is characterised by an increased expected profit, a small probability of a net loss and a small risk of a net loss. Contrary to the view expressed by Samuelson, which has also been adopted in several related papers (e.g. Ross, 1999), repeating the unacceptable bet reduced significantly the risk of a net loss.

With increasing the number of opportunity bets, the variance of the profit increases significantly (see Table 18.1). This has led some researchers to conclude incorrectly that because of the increased variance of the profit, the risk will also increase. The careful analysis shows that the variance of the profit can

Table 18.1 Expected profit and risk of a net loss with increasing the number of good bets

Number of good bets	Expected profit	Standard deviation	Risk of a net loss	Probability of a net loss
1	50	150	−50	0.5
10	500	474.3	−37.1	0.17
20	1000	670.8	−20	0.057
30	1500	821.6	−9.9	0.049
50	2500	1060.7	−3.1	0.0077
80	4000	1341.6	−0.5	0.0012
90	4500	1423	−0.26	0.0010
100	5000	1500	−0.15	0.00044
130	6500	1710.3	−0.025	0.00007
150	7500	1837	−0.007	0.00003

increase with a simultaneous decrease of the risk of a net loss. In this case, the commonly accepted rule that a larger variance means a larger risk does not hold. In this case, a larger variance actually coexists with a smaller risk, and the variance of the profit cannot serve as a risk measure.

In summary, *the popular view started with the Samuelson's paper* (1963) *that if a single bet is unacceptable, then a sequence of such bets is also unacceptable is incorrect.*

18.5 Variation of the Risk of a Net Loss Associated with a Small Number of Opportunity Bets

For a large number n of opportunity bets, the expected value of the net profit is approximated well by $\bar{G} \approx p_s \bar{B}_s - p_f |\bar{C}_f|$. According to the definition of an opportunity bet, $p_s \bar{B}_s - p_f |\bar{C}_f| > 0$, and consequently, with increasing the number of opportunity bets, the probability of a negative net profit (net loss) approaches zero.

It seems that with increasing the number of opportunity bets, the probability of a net loss always decreases. There is a common belief that increasing the number of opportunity bets is always beneficial because this increases the exposure to successful outcomes. Interestingly, this conventional belief is not necessarily correct. Here, we show that multiple opportunity bets (characterised with a positive expected gain) can in fact be associated with a larger risk of a net loss than a single opportunity bet.

This is shown in Figure 18.1 (Todinov, 2013c), where the parameters characterising the risk–reward bets are $p_s = 0.3$, $p_f = 0.7$ and $\bar{C}_f = -100$. Different values $E(G)$ of the expected profit were obtained by varying the expected benefit $\bar{B}_s$. Thus, for $\bar{B}_s = 299$, $E(G) = 19.7$; for $\bar{B}_s = 493$, $E(G) = 77.9$; and for $\bar{B}_s = 247$, $E(G) = 4.1$. These correspond to the three curves in Figure 18.1. Increasing the number of opportunity bets is associated with a decrease of the absolute value of the risk of a net loss. The decrease, however, does not occur monotonically as indicated by all three graphs. The analysis shows that the third curve, corresponding to $E(G) = 4.1$, also tends to zero absolute value of the risk of a net loss but after thousands of opportunity bets.

While for a large value of the expected gain, the absolute value of the risk of a net loss is quickly reduced to zero, for small values of the expected gain, the risk of a net loss may actually increase with

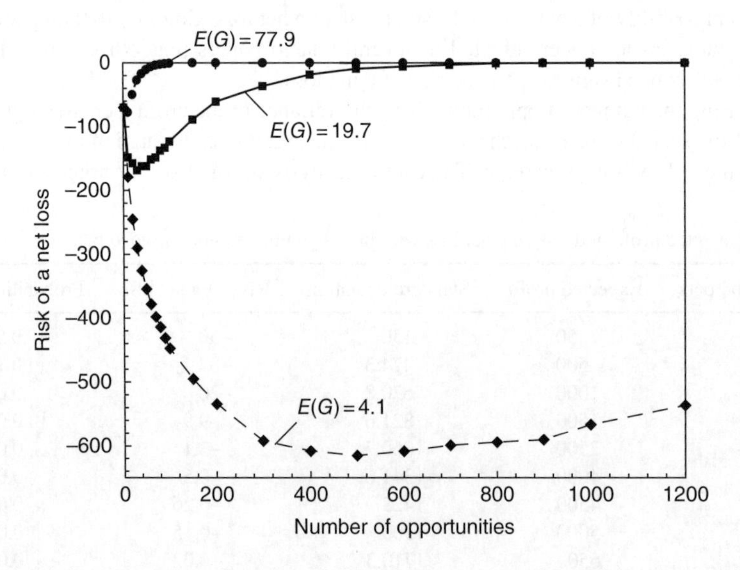

Figure 18.1 Increasing the number of opportunity bets may not necessarily result in a monotonic reduction of the risk of a net loss.

increasing the number of opportunity bets as indicated by the graph $E(G)=4.1$ and $E(G)=19.7$ in Figure 18.1. As a result, increasing the number of opportunity bets characterised by a small expected gain may have the opposite effect on the risk of a net loss! A simulation is necessary in each particular case to reveal the hidden risk.

18.6 Distribution of the Potential Profit from a Limited Number of Risk–Reward Activities

For a risky prospect containing multiple risk–reward events, if no analytical solution exists, building the distribution of the potential profit and evaluating the risk of a net loss and the expected potential reward can be made by a Monte Carlo simulation, whose algorithm in pseudocode is described in Algorithm 18.1 (Todinov, 2013c).

In Figure 18.2, the cumulative distribution of the potential profit has been built for risk–reward events following a homogeneous Poisson process with density 0.8 year^{-1} in the interval (0, 2) years. The benefit given success follows a uniform distribution in the interval (0, 320), $B_s(x\,|\,s)=U(0,320)$, and the loss given failure follows a uniform distribution in the interval (−50, 0), $C_f(x\,|\,f)=U(-50,0)$. The empirical risk of a net loss is −23.9, the expected potential reward is 47.9, and the expected potential profit is 19.3. It is interesting to note the jump of the net profit dependence at zero. This is caused by all outcomes for which no risk–reward events occurred in the specified finite time interval. The expected profit in this case is zero.

Figure 18.3 gives the distribution of the potential profit from five opportunity bets with parameters $p_s=0.4$, $\bar{B}_s=450$, $p_f=0.6$ and $\bar{C}_f=-180$. The empirical risk of a net loss is −140; the empirical expected potential reward is 499.6.

Figure 18.4 gives the distribution of the potential profit if the number of opportunity bets is increased to 30. The risk of a net loss has decreased to −75.3, while the expected potential benefit has increased to 2237.2.

In each particular case, a simulation is required to reveal the risk of a net loss. Simulation is also necessary for risk–reward bets characterised by different probabilities of success or by a complex distribution of the loss given failure or the benefit given success.

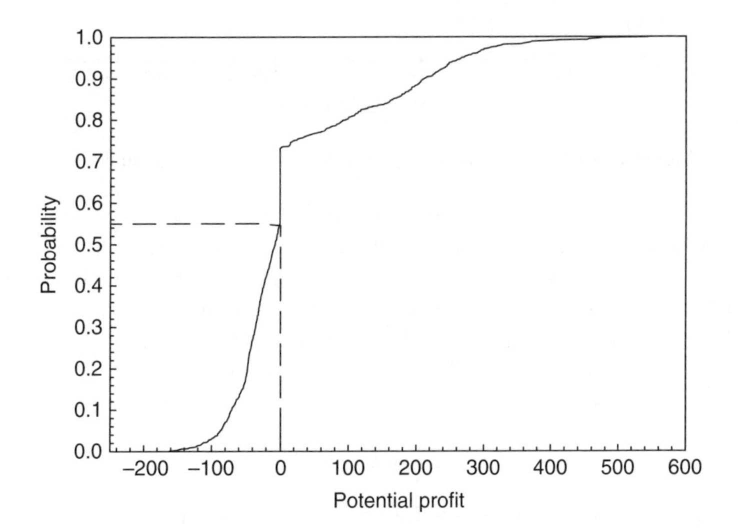

Figure 18.2 Distribution of the potential profit from risk–reward events following a homogeneous Poisson process in a specified time interval ($a=2$ years)

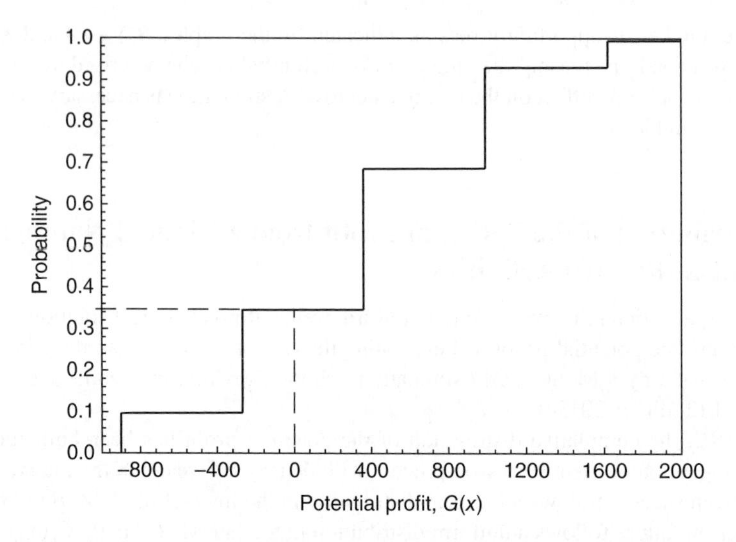

Figure 18.3 Distribution of the potential profit from five opportunity bets

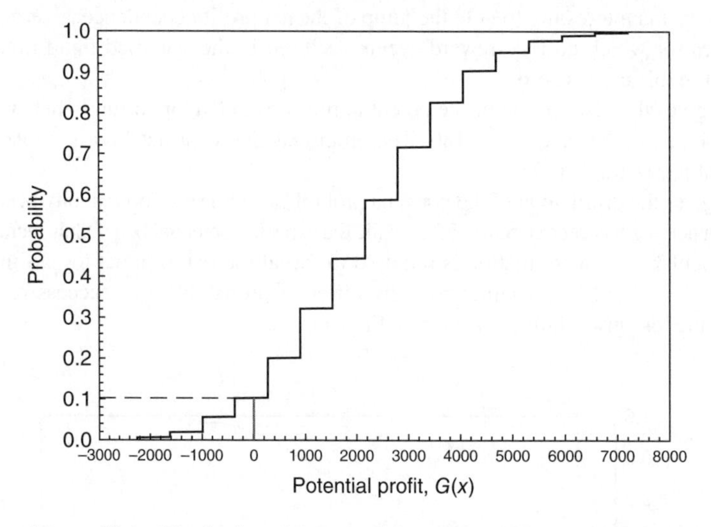

Figure 18.4 Distribution of the potential profit from 30 opportunity bets

The results from the simulation have been confirmed by an alternative method for calculating the expected value of the total profit based on analytical reasoning. The expected value of the total profit can be calculated as a sum of the expected value of the total benefit from risk–reward bets which materialise as success and the expected value of the total loss from risk–reward bets which materialise as a loss.

The total benefit T_B from N_s risk–reward bets which materialise as 'success' is $T_B = N_s\bar{B}_s$. Correspondingly, the total loss T_L from N_f bets which materialise as 'loss' is $T_L = N_f\bar{C}_f$. The total profit T_P is therefore given by

$$T_P = N_s\bar{B}_s + N_f\bar{C}_f \tag{18.18}$$

Taking expected values from both sides of Equation 18.18 results in

$$\bar{T}_P = \bar{N}_s\bar{B}_s + \bar{N}_f\bar{C}_f \tag{18.19}$$

for the expected value of the total profit. The expected number of the risk–reward bets $\bar{N}_s$ which materialise as success is

$$\bar{N}_s = \rho \times a \times p_s \tag{18.20}$$

where ρ is the density of the homogeneous Poisson process modelling the occurrence of the risk–reward bets, a is the length of the time interval, and p_s is the probability that a given risk–reward bet will materialise as 'success'. Similarly, the expected number of the risk–reward bets which materialise as a loss is

$$\bar{N}_s = \rho \times a \times p_f \tag{18.21}$$

Substituting Equations 18.20 and 18.21 in Equation 18.19 results in

$$\bar{T}_P = \rho a \left(p_s \bar{B}_s + p_f \bar{C}_f \right) \tag{18.22}$$

In the first example, $B_s(x \mid s) = U(0{,}320)$ and $C_f(x \mid f) = U(-50{,}0)$. Therefore, $\bar{B}_s = 160$ and $\bar{C}_f = -25$. Substituting these values and also the values $\rho = 0.8$, $a = 2$ years, $p_s = 0.2$ and $p_f = 0.8$ in Equation 18.22 results in $\bar{T}_P = 19.2$ which confirms the empirical result of 19.3 obtained by the simulation algorithm.

Algorithm 18.1

```
function Benefit_given_success(k);  //Samples the conditional
  distribution of the benefits given success of the k-th risk-
  reward bet and returns a random value
function Loss_given_failure(k);  //Samples the conditional
  distribution of the loss given failure of the k-th risk-reward
  bet and returns a random value

ps[Number_of_bets];          //Contains the probabilities of success
                               characterizing the
                               risk-reward bets in the risky prospect;
Net_revenue = 0;             //where the net revenue will be accumulated
Sum_net_Benefit = 0;         //where only the net benefit will be
                               accumulated
Sum_net_Loss = 0;            //where only the net loss will be
                               accumulated

  for i=1 to Number_of_trials do
{
  for j=1 to Number_of_bets do
   {
    t = generate_random_number(); // Generates a random number
                        uniformly distributed in the interval (0,1)
    if (t < ps[j]) then
          Net_revenue = Net_Revenue + Benefit_given_success(j);
           else Net_revenue = Net_Revenue + Loss_given_failure(j);
   }
```

```
    distr_pot_profit[i] = Net_revenue;
    if (Net_revenue > 0) then
            Sum_net_Benefit = Sum_net_Benefit+Net_revenue;
            else Sum_net_Loss = Sum_net_Loss+Net_Revenue;
 }
  Risk_of_net_loss = Sum_net_Loss / Number_of_trials;
  Potential_expected_reward = Sum_net_Benefit / Number_of_trials;

  Sort the array distr_pot_profit[] in ascending order.
```

For a number of risk–reward bets 'Number_of_bets' in a risky prospect, each characterised by different probabilities of success and failure, determining the risk of a net loss consists of the following steps. A simulation loop with control variable *i* is entered first, within which a nested loop with control variable *j* is entered, scanning through all risk–reward bets in the risky prospect. In the nested loop, a random variable *t* following Bernoulli distribution is simulated by generating a uniformly distributed random number in the interval (0, 1) and comparing it with the probability of success ps[*j*] of the scanned risk–reward bet. This random variable simulates success or failure outcome from the separate risk–reward bets. If success is simulated ($t \le ps[j]$), the distribution of the benefits given success is sampled by calling the function ***Benefits_given_success(j)***; if failure is simulated ($t > ps[j]$), the distribution of the loss given failure is sampled by calling the function ***Loss_given_failure(j)***. For all bets/events in the risky prospect, the sampled quantity is accumulated in the variable 'Net_revenue' which contains the net revenue (profit) from the risk–reward bets in the current simulation trial. The magnitude of the profit characterising the current simulation trial is also stored in the array 'distr_pot_profit[]'. At the end of the simulation, the elements of the array 'distr_pot_profit[]' are sorted in ascending order by using the *Quicksort algorithm* (Cormen *et al.*, 2001). The empirical cumulative distribution of the potential profit is built by plotting the sorted values of the cumulative array versus the probability rank estimates, $i = 1, 2, \ldots,$ Number_of_trials.

During each simulation trial, after obtaining the net revenue (profit) in the variable 'Net_revenue', its sign is checked. If the sign is negative, the net revenue is accumulated in the variable 'Sum_net_Loss'. Correspondingly, if the sign of the net profit in 'Net_revenue' is positive, the net benefit is accumulated in the variable 'Sum_net_Benefit'. The risk of a net loss and the potential expected reward are obtained at the end of the simulation trials by dividing the variables 'Sum_net_Loss' and 'Sum_net_Benefit' to the number of simulation trials.

19

Optimal Allocation of Limited Resources among Discrete Risk Reduction Options

19.1 Statement of the Problem

The problem of optimal allocation of limited safety resources to attain a maximum risk reduction is an important problem, which appears frequently in the budget planning of companies and enterprises and during the design of complex safety-critical systems. For a company or an enterprise, for example, it is important to determine how to allocate its budget in order to mitigate a number of sources of risk. In the railway industry in particular, the central question is how to allocate a fixed safety budget among a number of available risk reduction options, to prevent the maximum possible number of fatalities and injuries resulting from railway accidents.

Huge amount of invested resources are often wasted because the resource allocation is usually far from optimal and does not guarantee efficient risk reduction.

A fixed budget constraint is always present in cases where the total cost of the available risk-reduction options is greater than the amount of available resources. A common example of a fixed budget constraint is the reduction of the risk of infectious diseases (Richter *et al.*, 1999).

Dynamic programming techniques (Bellman, 1957; Dasgupta *et al.*, 2008; Horowitz and Sahni, 1997) can be used with success for solving the problem related to optimal allocation of safety resources to achieve a maximum risk reduction. Dynamic programming has been around for a long time, yet very few attempts (Pigman *et al.*, 1974; Reniers and Sorensen, 2013; Todinov and Weli, 2013) have been made to use it for optimal allocation of risk reduction resources.

The risk is often reduced by well-defined discrete options: purchasing new, more reliable and safer equipment, investing in personnel training, investing in improved security and control, investing in new systems, etc. Each risk reduction option can either be accepted (included) in the optimal set of options or not. For each specified risk reduction option, it is usually known from statistical data and experience how much risk reduction effect is achieved from implementing it. For example, in the railway industry, the risk reduction effect is commonly measured by estimating the number of prevented fatalities and injuries from implementing a particular option (Weli and Todinov, 2013).

In the case of discrete risk reduction options, *cost–benefit analysis* has been adopted by many industries and, in particular, by the railway industry as a tool for optimal allocation of safety resources. In the railway industry, for example, the safety budget allocation starts with assigning risk reduction

Reliability and Risk Models: Setting Reliability Requirements, Second Edition. Michael Todinov.

options to the different risk contributors or risk scenarios resulting in a major railway accident. Each risk reduction option is assessed in terms of the benefit it brings and its cost of implementation. The risk reduction options are ranked according to their benefit–cost ratio. By starting with the risk reduction option with the largest benefit–cost ratio, the options are sequentially included in the optimal set, and a check is performed whether the aggregated cost of the selected risk reduction options has exceeded the allocated budget. The risk reduction options whose aggregated cost is within the allocated budget are included in the optimal set. Consequently, the algorithm of the cost–benefit approach can be described by the following basic steps (Weli and Todinov, 2013):

1. Rank the risk reduction options in descending order, according to their benefit–cost ratio.
2. Choose the feasible risk reduction option with the highest benefit–cost ratio.
3. Update the total cumulative cost of all selected risk reduction options.
4. Repeat steps 2 and 3 until no other feasible risk reduction option can be included in the optimal set.

A feasible risk reduction option means an option that has not been selected and whose cost can be covered by the remaining budget. Each option can be selected only once.

For n risk reduction options, the ranking in descending order, according to the benefit–cost ratio, can be done in $O(n\log_2 n)$ time. As a result, the selection of risk reduction options by following the cost–benefit method can always be made in $O(n\times\log_2 n)$ time.

In the case of discrete and independent alternatives, the (0-1) knapsack dynamic programming approach has been used for a long time for optimal allocation of resources and, in particular, as a resource allocation method among competing projects.

Central to the optimal budget allocation problem considered in Todinov and Weli (2013) was the concept 'amount of removed risk' characterising the individual risk reduction options and measuring the benefit from their application. The removed risk can be expressed in monetary terms – the cost of prevented accidents, fatalities, injuries and delays.

All available risk reduction options $i=1, 2, \ldots, n$ are initially placed in a set Ω. The individual risk reduction options i $(i=1, 2, \ldots, n)$ are characterised by the amount of removed risk r_i and by the cost of their implementation c_i. Risk reduction options can be selected only once; hence, each risk reduction option can either be accepted or rejected.

By following the classical (0-1) knapsack budget allocation approach, to achieve a maximum risk reduction, the task should be reduced to determining the optimal subset $P\subseteq\Omega$ of risk reduction options associated with the maximum possible removed risk $\sum_{k\in P} r_k$. The imposed constraint is the specified limited budget – the maximal total cost must not exceed the available risk reduction budget B:

$$\textbf{Maximize}: \sum_{i=1}^{n} x_i \times r_i \tag{19.1}$$

subject to the constraint

$$\sum_{i=1}^{n} x_i \times c_i \le B \tag{19.2}$$

where $x_i \in \{0,1\}$ are decision variables; $x_i = 1$ if the risk reduction option is accepted and $x_i = 0$, otherwise.

A recent study, however, revealed an unexpected result (Todinov, 2014b): *in many cases, the exact (0-1) knapsack approach could actually lead to solutions inferior to the solutions produced by the cost–benefit approach.*

In addition, recent studies also revealed that in order to apply the knapsack (0-1) method, *the magnitude of the removed system risk should depend only on the sum of the removed risks from the selected individual options and should not depend on the particular selected subset of risk reduction options.*

After initial analysis demonstrating these weaknesses, relevant conditions will be proposed and a new model will be presented which avoids these weaknesses.

19.2 Weaknesses of the Standard (0-1) Knapsack Dynamic Programming Approach

19.2.1 A Counterexample

From the analyses published in the literature so far, it seems that the standard (0-1) knapsack dynamic programming approach is a real alternative to the cost–benefit approach. This perception however is rather deceptive as the next counterexample reveals (Todinov, 2014b). Suppose that the benefits and the costs of four risk reduction options are according to Table 19.1. The available safety budget is 30 million.

All risk reduction options are characterised by a benefit–cost ratio greater than one. The standard (0-1) knapsack algorithm selects risk reduction options *C* and *D*, which, within the fixed budget of 30 million, yield the largest risk reduction (54 million). Clearly, this is a flawed solution because if risk reduction options *A* and *B* had been selected, the risk reduction would be marginally smaller (53.9 million), but 13 million unnecessary expenses (43% of the budget) would have been saved. In fact, the standard (0-1) knapsack algorithm 'chooses' to spend 13 million towards a risk reduction of only 0.1 million, which effectively has a benefit–cost ratio $0.1/13 = 0.0077$. This is an indication of an extremely wasteful use of valuable resources!

It needs to be pointed out that in this counterexample, the cost–benefit approach selects correctly the risk reduction options *A* and *B* and avoids the problem associated with the standard (0-1) knapsack approach.

Suppose that $TR_{(0-1)}$ and $TC_{(0-1)}$ denote the total removed risk and the total cost of the risk reduction options characterising the (0-1) standard knapsack dynamic programming solution, respectively. Similarly, TR_{CB} and TC_{CB} denote the total removed risk and the total cost of the risk reduction options characterising the cost–benefit solution. The comparative risk reduction effectiveness ratio ρ can then be calculated from

$$\rho = \frac{TR_{(0-1)} - TR_{CB}}{\left|TC_{(0-1)} - TC_{CB}\right|} \tag{19.3}$$

Table 19.1 Four risk reduction options each characterised with cost of implementation and magnitude of the removed risk. The total safety budget is 30 million

Risk reduction option	Removed risk (in millions $)	Cost of implementation (in millions $)	Benefit–cost ratio
A	33	10	3.3
B	20.9	7	2.98
C	26	14	1.86
D	28	16	1.75

This ratio measures the effectiveness of the extra budget used by the standard (0-1) knapsack algorithm in reducing risk, compared to the cost–benefit solution.

If the risk reduction ratio ρ is too small, the standard (0-1) dynamic programming solution achieves only a marginal risk reduction, at a very large cost, and should be discarded in favour of the cost–benefit analysis solution. If the risk reduction ratio ρ indicates that a substantial risk reduction has been achieved, the (0-1) knapsack solution results in a cost-effective risk reduction and should be accepted as an alternative of the cost–benefit solution.

From Table 19.1, $TR_{(0-1)} = 26 + 28 = 54$, $TR_{CB} = 33 + 20.9 = 53.9$, $TC_{(0-1)} = 16 + 14 = 30$ and $TC_{CB} = 10 + 7 = 17$. The comparative ratio given by Equation 19.3 then becomes

$$\rho = \frac{54 - 53.9}{|30 - 17|} = 0.0077 \tag{19.4}$$

which is only 0.7%. This ratio indicates a very inefficient use of safety resources, and the standard (0-1) knapsack solution is worse than the cost–benefit solution.

The main reason for this problem is that the standard (0-1) knapsack approach has actually been devised to maximise the total value derived from items filling space with no intrinsic value. The budget, however, does have intrinsic value, and its utilisation is also important, as well as the maximisation of the risk reduction. It needs to be pointed out that if the budget had no intrinsic value or if spending the budget was analogous to filling empty space with no value, the (0-1) dynamic programming approach would always yield an optimal solution.

19.2.2 The New Formulation of the Optimal Safety Budget Allocation Problem

The counterexample from Table 19.1 exposes the dangers behind the standard (0-1) knapsack dynamic programming solution for optimising the allocation of risk reduction resources as a substitute of the cost–benefit approach. Despite that the standard (0-1) knapsack dynamic programming algorithm always yields the exact solution in maximising the risk reduction within a fixed budget, it could still generate 'solutions' wasting valuable resources with very small return in extra benefit.

The counterexample from Table 19.1 shows that there is clearly a need for an approach incorporating the value of the remaining safety budget. Consequently, the requirement the total removed risk $\sum_{k \in P} r_k$ to be maximal should be abandoned, because it leads to a wasteful use of safety resources.

This problem can be resolved by introducing weights ($\alpha, (1-\alpha)$; $0 \le \alpha \le 1$) assigned to both the amount of removed risk and the remaining budget, to reflect the value associated with the remaining budget. For risk reduction options, each characterised by benefit–cost ratio greater than unity, what needs to be maximised is not the total amount of removed risk $\sum_{k \in P} r_k$ but the weighted sum of the total removed risk $\alpha \sum_{k \in P} r_k$ and the weighted remaining budget $(1-\alpha)[B - \sum_{i=1}^{n} c_i \times x_i]$. This formulation prevents using up most of the remaining budget for a marginal risk reduction.

In the light of the presented discussion, the appropriate model of the optimal budget allocation among discrete independent risk reduction options is given next (Todinov, 2014b):

Considering the constraint

$$\sum_{i=1}^{n} x_i \times c_i \le B \tag{19.5}$$

Maximise the sum

$$X = \alpha \sum_{i=1}^{n} x_i \times r_i + (1-\alpha)\left(B - \sum_{i=1}^{n} c_i \times x_i\right) \tag{19.6}$$

where $x_i \in \{0,1\}$ are decision variables; $x_i = 1$, if the risk reduction option is selected and $x_i = 0$, otherwise. Because the available budget B is a constant, maximising the sum X in Equation 19.6 is equivalent to maximising $X = \alpha\sum_{i=1}^{n} x_i \times r_i - (1-\alpha)\sum_{i=1}^{n} c_i \times x_i$. The two summations can be combined, and as a result, what should be maximised is the expression

$$X = \sum_{i=1}^{n} x_i \times \left[\alpha r_i - (1-\alpha)c_i\right] \tag{19.7}$$

The weights can be conveniently altered to reflect correctly the value of unit removed risk and the value of unit remaining budget. Usually, both the removed risk and the available budget are measured in the same monetary units, and $\alpha = 0.5$ is the weighting factor reflecting that the value of unit risk is the same as the value of unit budget.

Thus, for ($\alpha = 0.5$), Equation 19.7 becomes

$$X = 0.5\sum_{i=1}^{n} x_i \times (r_i - c_i) \tag{19.8}$$

which is equivalent to maximising

$$X = \sum_{i=1}^{n} x_i \times (r_i - c_i) \tag{19.9}$$

19.2.3 Dependence of the Removed System Risk on the Appropriate Selection of Combinations of Risk Reduction Options

The application of the (0-1) knapsack allocation method assumes that (i) the risk removed by any risk reduction option does not depend on the presence/absence of other options and (ii) the application of any risk reduction option does not require the application of other options.

The next example shows that even for identical risk reduction options, removing individually the same amount of system risk, for which the previous conditions are fulfilled, the total amount of removed system risk depends on the selection of the particular combination of risk reduction options.

Consider the system in Figure 19.1 which transports cooling liquid from three sources s_1, s_2 and s_3 to the chemical reactor t.

The cooling system consists of identical pipeline sections (the arrows in Figure 19.1). Each pipeline section is coupled with a pump for transporting the cooling fluid through the section. Suppose that the pipeline sections and the pumps are old and prone to failure due to corrosion, fatigue, wear, deteriorated seals, etc. The cooling system fulfils its mission, if at least one cooling line delivers cooling fluid to the chemical reactor. Suppose for the sake of simplicity that all pipeline sections are in the same state of deterioration and each section is characterised with the same reliability 0.4, associated with 1 year of operation. Because of the deteriorated sections, the cooling system will benefit from risk reduction options consisting of purchasing and replacing deteriorated pipeline sections with new sections. Consequently, the replacement of any of the nine pipeline sections is a possible risk reduction option. Now, suppose that the available budget is sufficient for implementing exactly three options (for purchasing and replacing exactly three pipeline sections). Each new pipeline section is characterised by a reliability 0.9 for 1 year of operation.

Because of the symmetry of the system in Figure 19.1a, the replacement of any pipeline section is associated with the removal of the same amount of system risk. The pipeline sections work

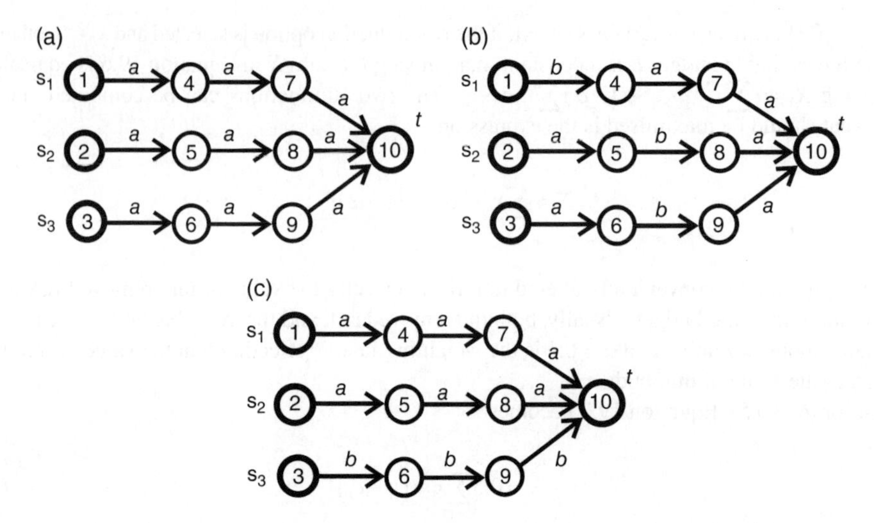

Figure 19.1 (a–c) A safety-critical cooling system consisting of three parallel branches

independently from one another, and because all of them are identical, it seems that any three risk reduction options can be selected (any three pipeline sections can be replaced with new ones; Figure 19.1b).

This impression however is incorrect. The removed risk of system failure is highest if the available budget is spent preferentially on replacing pipeline sections forming an entire cooling branch (Figure 19.1c), as opposed to replacing randomly selected sections inside the system (Figure 19.1b) (Todinov, 2014b).

Indeed, the reliability of the parallel-series arrangement in Figure 19.1b is

$$R_b = 1 - \left(1 - 0.4^2 \times 0.9\right)^3 = 0.373 \tag{19.10}$$

while the reliability of the parallel-series arrangement in Figure 19.1c is significantly higher:

$$R_c = 1 - \left(1 - 0.4^3\right)^2 \times \left(1 - 0.9^3\right) = 0.76 \tag{19.11}$$

The variant presented in Figure 19.1c is an example of *a well-ordered parallel-series system*. A well-ordered parallel-series arrangement is obtained if the available components are used to build the branch with the highest possible reliability/availability, the remaining components are used to build the next branch with the highest possible reliability/availability and so on, until the entire parallel-series arrangement is built.

Another example of a well-ordered parallel-series system is the system in Figure 19.2, where in the parallel branches there are three pre-existing components with reliabilities $r_{01} = 0.78$, $r_{02} = 0.75$ and $r_{03} = 0.80$. These components are always attached to the corresponding branches. There are also empty sockets which can accommodate type-A components and type-B components. The type-*A* components are of two varieties: there is one old component with reliability 0.57 and one medium-age component with reliability 0.67. The type-*B* components are also of two varieties: three old components with reliability 0.7 and two medium-age components with reliability 0.8. The system with the highest possible

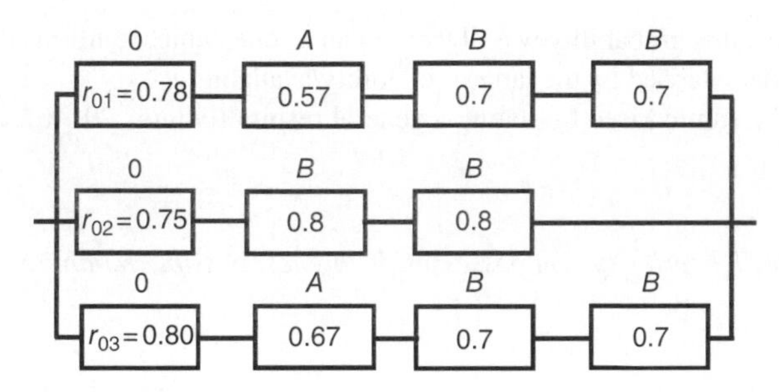

Figure 19.2 A parallel-series logical arrangement with three pre-existing components with specified reliabilities and eight interchangeable components

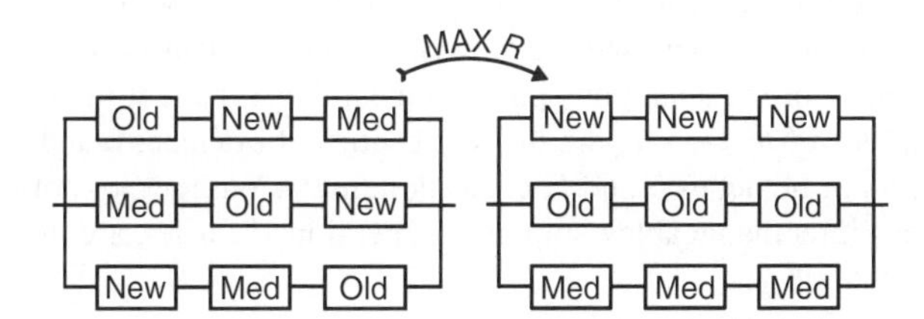

Figure 19.3 Minimising the risk of failure of a parallel-series system

reliability (removed risk of system failure) is the system shown in Figure 19.2. In this system, the reliability of the second branch cannot be improved by interchanging components with other branches. The reliability of the third branch cannot be improved by interchanging components with the less reliable first branch.

Parallel-series arrangements are very common. Consider a safety-critical system for detecting the release of toxic gas, based on n detectors working in parallel. Upon a toxic gas release, the system detects the release if at least one of the detectors working in parallel detects the toxic gas release. This system is a parallel-series system if the parts building the separate detectors are logically arranged in series.

Consider an example where there are three types of components with different age – new, medium and old components (Figure 19.3). The maximum reliability is achieved if all new components are arranged in a single branch, the medium-age components are arranged in another branch and all old-age components are grouped in a separate branch (Figure 19.3).

These results, for a number of well-ordered parallel-series systems, have been verified by a computer simulation. The computer simulation consisted of specifying the reliabilities of the interchangeable components in the branches and calculating the reliability/availability of the well-ordered system. The second phase of the procedure is a 'random scrambling' of the interchangeable components in the branches, by generating random indices of interchangeable components from different branches and swapping their reliability values. The swapping guarantees that any resultant system includes exactly the same set of interchangeable components on a specified branch as the initial system. After the 'random scrambling', the reliability/availability of the scrambled system was calculated and compared with the reliability/availability of the well-ordered system. If the reliability/availability of the well-ordered system was greater than or equal to the reliability/availability of the scrambled system, the content of a counter was increased. At the end, the probability that the well-ordered system has reliability/availability not smaller than the reliability/availability of the scrambled system was calculated. In all of the conducted simulations, this probability was always equal to one, which confirms that the well-ordered systems are indeed characterised by the largest reliability/availability.

These results can be summarised by stating a general result (Todinov, 2014b):

Theorem

The well-ordered parallel-series system possesses the highest possible reliability.

Proof

This theorem will be proved by contradiction and the extreme principle. Suppose that there is a system which is not well ordered and which possesses the highest possible reliability. Without loss of generality, suppose that the branches in this system have been rearranged in such a way that for any two branches i, j for which $i < j$, the branch with index 'i' is not less reliable than branch 'j' ($R_i \geq R_j$). If the system is not a well-ordered system, then there will be two branches a and b with reliabilities $R_a \geq R_b$, where there will be at least one component in branch b with larger reliability than the reliability of the analogous interchangeable component in branch a. Suppose that $R_a = a_1 a_2 \times \cdots \times a_{na}$ and $R_b = b_1 b_2 \times \cdots \times b_{nb}$ are the reliabilities of branches a and b and na, nb are the number of components in branches a and b, correspondingly. Without loss of generality, suppose that the two analogous interchangeable components mentioned earlier, are the last components in the branches a and b ($a_{na} < b_{nb}$).

The reliability of the initial system can be presented as

$$R_{\text{sys1}} = 1 - \left(1 - a_1 a_2 \times \cdots \times a_{na}\right)\left(1 - b_1 b_2 \times \cdots \times b_{nb}\right) \times \left(1 - R_{\text{rest}}\right) \quad (19.12)$$

where R_{rest} is the reliability of the rest of the parallel-series arrangement.

After swapping components a_{na} and b_{nb}, the reliability of the resultant system becomes

$$R_{\text{sys2}} = 1 - \left(1 - a_1 a_2 \times \cdots \times a_{na-1} b_{nb}\right)\left(1 - b_1 b_2 \times \cdots \times b_{nb-1} a_{na}\right) \times \left(1 - R_{\text{rest}}\right) \quad (19.13)$$

Subtracting (19.13) from (19.12) yields

$$R_{\text{sys1}} - R_{\text{sys2}} = \left(a_{na} - b_{nb}\right)\left(a_1 a_2 \times \cdots \times a_{na-1} - b_1 b_2 \times \cdots \times b_{nb-1}\right) \times \left(1 - R_{\text{rest}}\right) \quad (19.14)$$

Because $R_a = a_1 a_2 \times \cdots \times a_{na} \geq R_b = b_1 b_2 \times \cdots \times b_{nb}$ by the way the branches have been arranged in descending order according to their reliability ($R_a \geq R_b$), and because $a_{na} < b_{nb}$ by assumption, the inequality

$$a_1 a_2 \times \cdots \times a_{na-1} > b_1 b_2 \times \cdots \times b_{nb-1} \quad (19.15)$$

holds, which means that in Equation 19.14, $a_1 a_2 \times \cdots \times a_{na-1} - b_1 b_2 \times \cdots \times b_{nb-1} > 0$.

Since $1 - R_{\text{rest}} > 0$, the right-hand side of Equation 19.14 is negative, which means that the resultant system (after the swap of components) has a higher reliability. This contradicts the assumption that the initial system (before the swap) was the system with the highest possible reliability. Therefore, the reliability of a system which is not well ordered can be improved by swapping components between parallel branches until a well-ordered system is finally obtained. A well-ordered system is unique, and there can be no two different well-ordered systems. Because a parallel-series system can either be a well-ordered or not well-ordered system, the well-ordered system has a higher reliability compared to any other arrangement. The theorem has been proved.

This result also provides the valuable opportunity to improve the reliability/availability of common engineering systems with parallel-series logical arrangement of their components *without the knowledge of their reliabilities and without any investment.* Unlike all traditional approaches, which invariably require resources to achieve reliability improvement and system risk reduction, a system risk reduction can also be achieved by an appropriate permutation of the available interchangeable components in the parallel branches.

Components of similar level of deterioration (reliability levels) should be placed in the same parallel branch.

The example in Figure 19.1 clearly shows that the amount of removed system risk depends on the selected set of options despite that the individual options are identical and remove the same amount of system risk.

Going back to the limitations of the classical (0-1) dynamic programming approach, even if the two sets of risk reduction options yield the same sum of the individually removed risks, one of the sets removed larger amount of system risk. The classical (0-1) knapsack dynamic programming approach does not account for this situation.

This section establishes an important requirement for the application of the (0-1) dynamic programming method: *The total amount of removed system risk should depend only on the sum of the removed system risks from the individual options and should not depend on the selection of the risk reduction options or their number.*

Any two sets of risk reduction options with the same combined removed risk should result in the same amount of removed system risk. In what follows, only risk reduction options possessing this property will be considered. Many risk reduction options preventing fatalities in the railway industry, for example, possess this property. These will be considered in detail in the next section.

19.2.4 A Dynamic Algorithm for Solving the Optimal Safety Budget Allocation Problem

Considering the magnitude of the implementation costs for the risk reduction options in the industry and the magnitude of removed risks, it can be assumed that the costs and the amount of removed risk can always be expressed as integer numbers. These express the removed risk and the cost of implementation of the options in thousands, tens of thousands or hundreds of thousands of dollars. It is also assumed that the available budget can also be specified by an integer number.

As a result, the problem of optimal allocation of a risk reduction budget is reduced to a combinatorial optimisation problem, involving integers only.

Dynamic programming will be used for solving the problem formulated by the inequality 19.5 and Equation 19.9. The advantage of the dynamic programming consists of the fact that it finds solutions to sub-problems increasing in size, stores them in the memory and describes the solution of each sub-problem in terms of already solved and previously stored solutions of smaller sub-problems. As a result, sub-problems are solved only once, which makes the dynamic programming significantly more efficient than a brute-force method based on the enumeration of all possible subsets in the set of available risk reduction options S. The number of possible subsets in the set S is 2^n, and the computational time of a brute-force method based on scanning all possible subsets increases dramatically with increasing the number n of risk reduction options.

The description of the algorithm in pseudocode is presented next.

Algorithm 19.1: Building the dynamic risk reduction table

```
  Initialising array x[][] with zeroes in the first row and in the
  first column.
   for i=1 to n do
   for j=1 to B do
 {
   cur_budget = j;
   if(c[i]>cur_budget) then {
                                x[i][j]=x[i-1][j];
                                trac[i][j]=0;
                               }
                           else
                           {
         rem = cur_budget-c[i];
         tmp = rr[i] - c[i]+x[i-1][rem];

         if(x[i-1][cur_budget] > tmp) then {
         x[i][j] = x[i-1][j];
         trac[i][j]=0;
                                           }
          else {
               x[i][j]=tmp;
               trac[i][j]=1;
               }
          }
 }
```

The algorithm works as follows. The solutions of the sub-problems are kept in the array x[][]. The information necessary to restore the optimal solution is kept in the array trac[][]. The size of the x[][] array is (n+1) × *B* elements. The first row of the array x[][] corresponds to zero number of selected options in the optimal set *P*; the first column of array x[][] corresponds to zero budget.

The sub-problems are defined by the size of the current budget which varies from 1 to *B* units. The cost of the *i*th risk reduction option is compared with the value of the current budget, and if it is greater than the current budget, the *i*th risk reduction option is not included in the optimal set, which is reflected by the zero value in the trac array (trac[i][j]=0). In the case where the current budget is greater than the cost of the *i*th risk reduction option, a decision is taken whether to include the *i*th risk reduction option or not.

Initially, the statement 'rem=cur_budget - c[i];' determines the amount of remaining budget if the *i*th risk reduction option is included. The sub-problem marked by x[i-1][rem] however has already been solved, and its solution has been recorded in the x[][] array. The entry x[i-1][rem] gives the sum *X* in Equation 19.9 for available risk reduction options from 1 to i-1. Consequently, the solution of the sub-problem does not need to be determined again; it can simply be read out from the x[][] array. The amount of risk removed by the *i*th risk reduction option is rr[i] and the cost of the *i*th risk reduction option is c[i]. Consequently, the sum *X*, for budget cur_budget=j, if the *i*th risk reduction option is included, is given by 'tmp = rr[i]-c[i]+x[i-1][rem];'. If the *i*th option is not included in the optimal set *P,* the sum *X* of the total amount of removed risk and remaining budget, within the budget cur_budget, is given by x[i-1][cur_budget], (cur_budget=j). Consequently, the decision whether to include the *i*th risk reduction option in the optimal set or not depends on the outcome of the comparison made in the statement 'if(x[i-1][cur_budget] > tmp)' where tmp = rr[i]-c[i]+x[i-1][rem].

If 'x[i-1][cur_budget] > tmp', not including the *i*th risk reduction option yields a greater sum *X* and the entry 'trac[i][j]=0' in the track[][] array is set to zero, which indicates that the *i*th risk reduction option has not been included in the optimum set of options *P*. The maximum of the sum *X* is equal to the maximum sum *X* within the current budget 'j', for i-1 total number of available options. This maximum however has already been computed and is in the array x[][]; this is the entry x[i-1][j].

If 'x[i-1][cur_budget] < tmp', including the *i*th risk reduction option yields a greater sum *X*, and the entry in the trac array is set to one (trac[i][j]=1;), which indicates that the *i*th risk reduction option has been included in the optimal set *P*. The maximum sum *X* is equal to x[i][j] = rr[i]-c[i]+x[i-1][rem].

The optimal set of options is restored by the next algorithm in pseudocode.

Algorithm 19.2: Restoring the optimal set of risk reduction options from the dynamic tables

```
//Initialise all entries of the 'solution[]' array with zeroes.
cur_bud=B;
cur_opt=n;

tmp=trac[cur_opt][cur_bud];

while (cur_opt > =1 ) do
                    {
     if (trac[cur_opt][cur_bud]=1)  then  {
                                            solution[cur_opt] = 1;
                                            cur_bud=cur_bud
                                            - c[cur_opt];
                                            cur_opt=cur_opt - 1;
                                            }
            else  cur_opt=cur_opt-1;
     }
```

The algorithm starts with the entry trac[n][B] of the track[][] array, which corresponds to a full budget *B* and all *n* available risk reduction options. If the *n*th option has been included in the optimal set, this will be indicated by a non-zero entry in the trac array (trac[n][B]=1). In this case, the solution array marks the *n*th option as 'included' in the optimal set P, by the statement 'solution[n]=1'. The current budget is then reduced by the statement 'cur_bud = cur_bud - c[cur_opt]' with the cost of the current (*n*th) option. The current option to be considered should now be the n-1st option. This is ensured by the statement 'cur_opt = cur_opt - 1'.

If the *n*th option has not been included in the optimal set, this will be indicated by a zero entry in the trac array (trac[n][B]=0). In this case, the current budget is not altered because the *n*th risk reduction option has not been implemented.

The process of considering the options in reverse order continues until the first option is reached. At this point, the entries of the solution array will contain '1' for the options, which have been included in the optimal set P.

The running time of Algorithm 19.1 for building the dynamic tables is determined by the two nested loops: `for i=1 to n do {for j=1 to B do}`, which contain a set of operations that are executed in constant time. The maximum number of steps after which Algorithm 19.1 will terminate is $n \times B$. The maximum number of steps for Algorithm 19.2 is n because after each iteration of the while-do loop, the number of options is reduced by 1. As a result, after at most n steps, Algorithm 19.2 will terminate. The total number of steps is therefore $n \times B + n = n \times (B+1)$. The worst-case running time of the algorithm for optimal allocation of a safety budget is therefore $O(n \times (B+1))$.

This algorithm, applied to the counterexample from Table 19.1, yields the correct solution. Options A and B are selected as optimal options and not options C and D. The proposed model produced a superior solution compared to the standard (0-1) dynamic programming algorithm.

The proposed model also yields a solution superior to a cost–benefit solution. This point will be illustrated by the next example from the railway industry. A similar example has been considered in (Weli and Todinov, 2013). Table 19.2 lists five risk reduction measures (A, B, C, D and E) associated with different amount of removed risk and different costs.

Suppose that a total budget $B = 2.6$ million has been allocated for the reduction of platform train accidents with passengers. This is a major risk which is located in the high-risk region of the risk matrix. The first risk reduction option A requires the train driver to operate a CCTV monitoring of the platform. The train will not be started if there are passengers stuck at the door, fallen onto the track or fallen between train and platform. Option B requires introducing stop plungers – wall-mounted alarm devices at specified locations/intervals within the platform area which can be operated by platform staff or passengers. Trains in the platform area will be brought to a halt by operating any of these plungers. Option C includes equipping the train doors with sensors to reduce the possibility of trapping and dragging passengers. Option D consists of gap fillers between train and platform to reduce accidents where passengers fall between train and platform while boarding the train. Option E includes a system preventing opening the train doors on the wrong side of the platform.

The five key risk reduction options, A, B, C, D and E, have been evaluated, and the corresponding magnitudes of removed risk and costs are according to Table 19.2.

Following the cost–benefit approach, the risk reduction measures C and A, with the largest benefit–cost ratio will be selected. The combined cost of the selected risk reduction measures is 2.3 million – well within the fixed budget of 2.6 million. The removed risk is 4.7 million.

The proposed algorithm yields an optimal set including risk reduction options B, C and D with a combined cost exactly 2.6 million (equal to the available budget) and removed risk equal to 5.1 million.

The risk reduction ratio

$$\rho = \frac{5.1 - 4.7}{|2.6 - 2.3|} = 1.33 \quad (19.16)$$

equals 133%, which indicates that the solution produces a substantial return on the extra budget resources. The proposed model in Section 19.2.2 yields a solution superior to the cost–benefit solution.

Table 19.2 Risk reduction measures with the associated costs and magnitudes of the removed risk. The total budget is 2.6 million

Risk reduction option	Removed risk (in millions \$)	Cost of implementation (in millions \$)	Benefit–cost ratio
A	2.4	1.2	2.0
B	1.3	0.7	1.857
C	2.3	1.1	2.09
D	1.5	0.8	1.875
E	1.6	0.9	1.777

Table 19.3 Risk reduction options with the associated costs and magnitude of the removed risk. The available budget is $B = 170$ thousand

Risk reduction option	Removed risk (in thousand $)	Cost of implementation (in thousand $)	Benefit–cost ratio
A	442	41	10.78
B	525	50	10.5
C	511	49	10.4
D	593	59	10.05
E	546	55	9.927
F	564	57	9.89
G	617	60	10.28

Table 19.3 lists seven risk reduction options (*A*, *B*, *C*, *D*, *E*, *F* and *G*) with removed risks and costs, according to the table. The total budget is $B = 170$ thousand. Following the cost–benefit approach, risk reduction options *A*, *B* and *C*, associated with the largest benefit–cost ratio, will be selected. The combined cost of these risk reduction options is $\mathrm{TC}_{\mathrm{CB}} = 140$ thousand, well within the fixed budget of 170 thousand. The removed risk is $\mathrm{TR}_{\mathrm{CB}} = 1478$ thousand.

Applying the algorithm discussed earlier, yields an optimal set including risk reduction options *B*, *D* and *G*. The combined cost of these options is $\mathrm{TC} = 169$ thousand (within the fixed budget of 170 thousand) with a total removed risk $\mathrm{TR} = 1735$ thousand. The comparative ratio is

$$\rho = \frac{1735 - 1478}{|169 - 140|} = 8.86 \tag{19.17}$$

As can be verified, despite that the comparative ratio is smaller than the benefit–cost ratio of each risk reduction option, the extra risk reduction is substantial (257 thousand) which provides a very good return on the invested extra budget of 29 thousand. Clearly, the solution from the proposed algorithm should be preferred to the cost–benefit solution.

Computationally, the proposed in Section 19.2.3 dynamic algorithm is very efficient. This is illustrated with the example from Table 19.4, listing 24 different risk reduction options and available budget $B = £6\,404\,180$.

The options selected in the optimal set by the proposed algorithm are shown in the last column of the table.

19.3 Validation of the Model by a Recursive Backtracking

For a small number of risk reduction options (up to 12), the proposed algorithm has been validated by using a recursive backtracking algorithm by which all possible combinations of risk reduction options are generated, evaluated and compared, after which the best combination, associated with the largest sum $X = \sum_{i=1}^{n} x_i (r_i - c_i)$, is selected. Recursive backtracking has been used for a long time to solve combinatorial problems (Wirth, 1976) and guarantees that (i) all possible potential combinations of risk reduction options are generated and (ii) no possible combination of risk reduction measures has been missed.

The algorithm in pseudocode for generating and evaluating the possible valid combinations of risk reduction options is as follows:

Table 19.4 Risk reduction options with the associated costs and magnitude of the removed risk. The total budget is £6 404 180

Risk reduction option	Removed risk, $	Cost of implement, $	Benefit–cost ratio	Selection indicator
1	825 594	382 745	2.1570	1
2	1 677 009	799 601	2.0973	1
3	1 676 628	909 247	1.8440	0
4	1 523 970	729 069	2.0903	1
5	943 972	467 902	2.0175	1
6	97 426	44 328	2.1978	1
7	69 666	34 610	2.0129	0
8	1 296 457	698 150	1.8570	0
9	1 679 693	823 460	2.0398	0
10	1 902 996	903 959	2.1052	1
11	1 844 992	853 665	2.1613	1
12	1 049 289	551 830	1.9015	0
13	1 252 836	610 856	2.0510	1
14	1 319 836	670 702	1.9678	0
15	953 277	488 960	1.9496	0
16	2 067 538	951 111	2.1738	1
17	675 367	323 046	2.0906	0
18	853 655	446 298	1.9127	0
19	1 826 027	931 161	1.9610	0
20	65 731	31 385	2.0943	0
21	901 489	496 951	1.8140	0
22	577 243	264 724	2.1805	1
23	466 257	224 916	2.0730	1
24	369 261	169 684	2.1762	1

```
procedure evaluate_solution(num_opt)
{
 X=0;
 for j=1 to num_opt do    {
                          tmp=current_sol[j];
                          S=S+rrem[tmp]-cost[tmp];
                      }
 if (global_max < X) then {
                           global_max = X;
                           save the current optimal solution;
                      }
}
procedure extend_solution(k, rem_B)
{
 flag=0;
 k=k+1;
 for (each risk reduction option i) do
  {
  if (opt_assigned[i]=0 and cost[i]<rem_B) then
             {
              flag=1;
              current_sol[k]=i;
```

```
                opt_assigned[i]=1;
                rem_B=rem_B-cost[i];
                extend_solution(k,rem_B);    //try to extend the
                                               solution
                opt_assigned[i]=0;           // undo the option to
                                                allow further
                                                exploration
                rem_B=rem_B+cost[i];         // restore the remaining
                                                budget
               }
   }
   if (flag=0) then evaluate_solution(k-1);
}

for each risk-reduction option i do opt_assigned[i]=0;
extend_solution(0,B).
```

The recursive backtracking procedure ***extend_solution***`(k, rem_B)` has two parameters – 'k', the number of selected options in the partial solution, and 'rem_B', the size of the remaining budget. Initially, the backtracking procedure is called with parameters 0, B. For each risk reduction option i (from 1 to m), a check is performed whether the risk reduction option has not been assigned. If the option i has been assigned, this will be indicated by the value '1' in the array 'opt_assigned[]'. Initially, all entries of this array are set to '0'. The removed risks by the individual risk reduction options are kept in the array 'rrem[]', while the costs of the individual risk reduction options are kept in the array cost[].

If the risk reduction option has not been assigned and if it fits in the remaining budget 'rem_B', a flag is set to one after which the risk reduction option is assigned and the remaining budget is reduced by the cost of the risk reduction option. Subsequently, an attempt is made to extend the partial solution by calling the procedure ***extend_solution***`()` recursively. If there is no unassigned risk reduction option or if the available unassigned risk reduction options do not fit in the remaining budget, the flag remains equal to zero, and the attained partial solution is evaluated.

The evaluation of the partial solution is reduced to calculating the sum $X = \sum_{i=1}^{\text{num\_opt}} (r_i - c_i)$ of the assigned number of options 'num_opt' and comparing it with the current global maximum 'global_max'. If X is greater than the current global maximum, then X replaces the current global maximum, and the currently selected risk reduction options are saved.

After a return from a recursive call, it is very important to undo the risk option selection in order to allow the exploration of the other branches of the recursion tree. This is done by the two statements:

```
opt_assigned[i]=0;
rem_B=rem_B+cost[i];
```

which unmark the ith risk reduction option as 'unassigned' and increase the remaining budget by the cost of the option. This permits visiting all leaves of the recursion tree, to each of which corresponds a valid distinct permutation of risk reduction options fitting in the limited budget B.

A validation test has been conducted including 12 risk reduction options, with removed risks and costs according to Table 19.5. The optimal selection produced by the dynamic algorithm is given in the last column of Table 19.5. The execution of the recursive backtracking algorithm selected options 2, 3, 4, 6, 9, 10 and 11 as optimal options, with total cost 1577. This result matched exactly the result from the proposed in Section 19.2.3 algorithm.

Table 19.5 A validation test example with 12 risk reduction options, the associated and the magnitude of the removed risk. The total budget is £1600

Risk reduction option	Removed risk	Cost of option, £	Benefit–cost ratio	Selection indicator
1	245	182	1.35	0
2	311	166	1.87	1
3	412	240	1.72	1
4	567	378	1.5	1
5	188	112	1.68	0
6	443	277	1.6	1
7	116	79	1.47	0
8	89	45	1.98	0
9	398	217	1.83	1
10	178	98	1.82	1
11	477	201	2.37	1
12	289	245	1.18	0

A number of additional tests have also been conducted, with different number of risk reduction options. Invariably, the results from the recursive backtracking procedure matched exactly the results from the proposed dynamic programming algorithm. All tests have been done on small sets (up to 12) of risk reduction options because the running time of the recursive backtracking procedure increases exponentially with increasing the size of the tested set. In contrast, the worst-case running time of the (0-1) dynamic programming algorithm for optimal allocation of a safety budget is $O(n \times B)$, where n is the number of available options and B is the size of the budget as an integer number. Expressing the available budget B as an integer number (e.g. to the nearest thousand) and following this for the removed risk and the cost of implementation, make the (0-1) dynamic programming algorithm very fast, which is indicated by the results for the set of options in Table 19.4. Despite the large budget and the presence of risk reduction options with very different costs, the solution was reached by the (0-1) knapsack algorithm after 1.75 s, on a computer with a processor Intel (R) Core (TM) 2 Duo CPU T9900 @ 3.06 GHz.

Appendix A

A.1 Random Events

Sample space Ω

The union of all outcomes in an experiment.

Example

If the experiment is a toss of two independently rolled dice, the sample space has 36 equally likely outcomes (elements), each of probability 1/36:

$$
\begin{array}{cccc}
\multicolumn{4}{l}{\Omega:} \\
1,1 & 1,2 & \dots & 1,6 \\
2,1 & 2,2 & \dots & 2,6 \\
\dots & \dots & \dots & \dots \\
6,1 & 6,2 & \dots & 6,6
\end{array}
$$

The sample space from the toss of three coins has $2 \times 2 \times 2 = 8$ equally likely outcomes (H(eads) or T(ails)):

Ω: HHH, HHT, HTH, THH, HTT, THT, TTH, TTT

The sample space of the states of a system containing three components A, B and C each of which can be in 'working' (e.g. A) or 'failed' state (e.g. $\bar{A}$):

$$\Omega: ABC, AB\bar{C}, A\bar{B}C, A\bar{B}\bar{C}, \bar{A}BC, \bar{A}B\bar{C}, \bar{A}\bar{B}C, \bar{A}\bar{B}\bar{C}$$

Reliability and Risk Models: Setting Reliability Requirements, Second Edition. Michael Todinov.

Generally, a system containing n components each characterised by two distinct states contains a total of 2^n different states.

Event
A subset of the sample space Ω (of all outcomes) (the event A in the *Venn diagram*, $A \subseteq \Omega$).

Venn Diagram
Pictorial representation of subsets in the sample space.

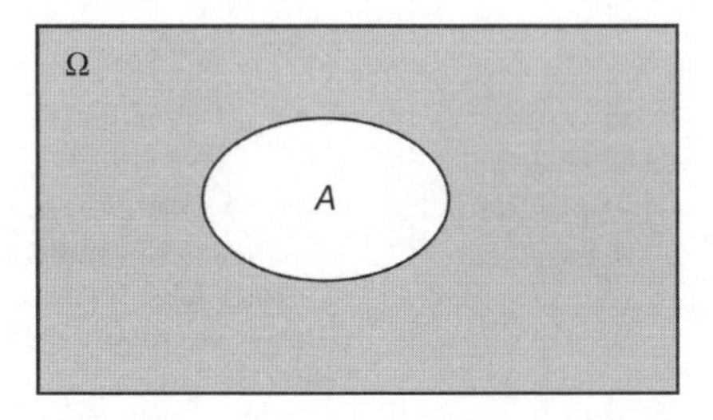

$A \subseteq \Omega$ (A is a subset of the sample space)

Certain Event
The sample space Ω. Contains all possible outcomes.

Impossible Event (Null Event)
$\varnothing$ Does not contain any outcomes. An empty set.

An Elementary Event
Consists of a single element (outcome).

Disjoint (Mutually Exclusive) Events
Cannot occur simultaneously (e.g. the events denoting the two possible states of a system or component: *working* or *failed*).

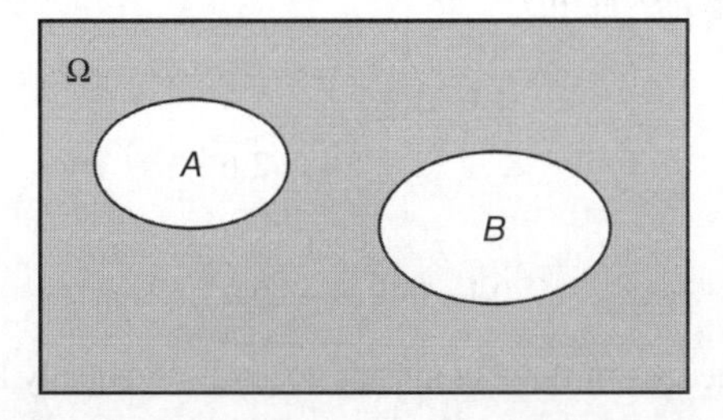

Complementary Events
Whenever one does not occur, the other does. The complement of event A is denoted by $\bar{A}$. $\bar{A}$ includes all outcomes (elementary events) x which do not belong to A:

$$\bar{A} = \{x \mid x \notin A\}$$

Note: The notation $\bar{A} = \{x \mid P\}$ means all outcomes x with property P; $\{x \mid x \notin A\}$ means all x with the property $x \notin A$.

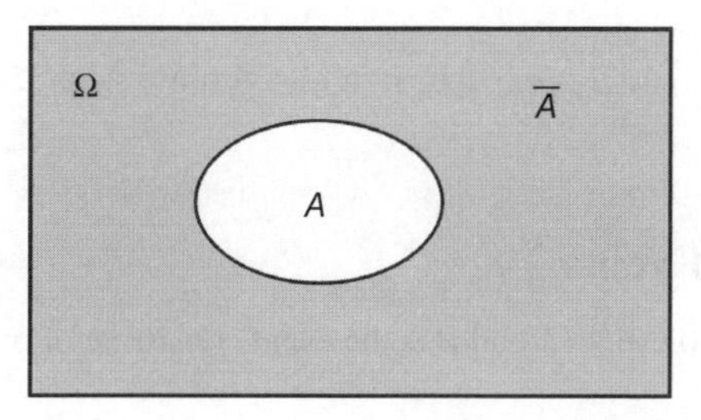

The Null event and the certain event are complementary events: $\overline{\varnothing} = \Omega$; $\overline{\Omega} = \varnothing$.

Two events are equivalent and we write $A = B$ if A and B have the same outcomes x. $A = B$ is fulfilled whenever for $x \in A$ then $x \in B$ and whenever $x \in B$ then $x \in A$.

Suppose that A and B are events. If every outcome of B is an outcome of A, we say that the outcomes of B are a subset of the outcomes of A. In other words, whenever there is a realisation of the event B, there is automatically a realisation of the event A.

$$B \subseteq A \text{ (} B \text{ is a subset of } A\text{)}$$
$$x \in B \rightarrow x \in A \text{ (from } x \in B\text{, it follows ('}\rightarrow\text{') that } x \in A\text{)}$$

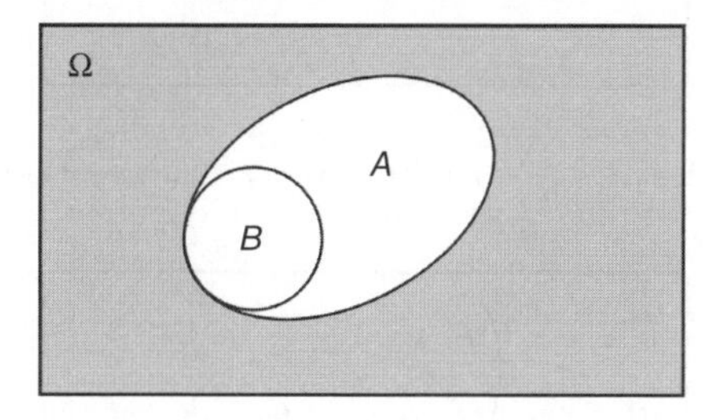

A.2 Union of Events

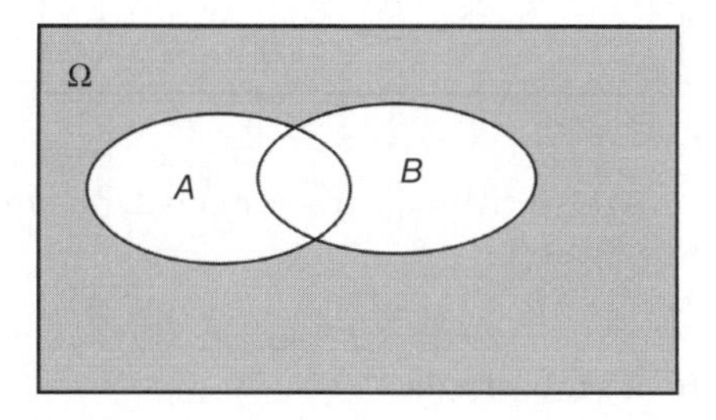

The union $A \cup B$ of two events A and B is the event consisting of outcomes x belonging to A or B or both.

$$A \cup B = \{x \mid x \in A \text{ or } x \in B\}$$

The union $\bigcup_{i=1}^{n} A_i = A_1 \cup A_2 \cup \cdots \cup A_n$ of n events is the event which contains all outcomes x belonging to *at least one* of the events A_i.

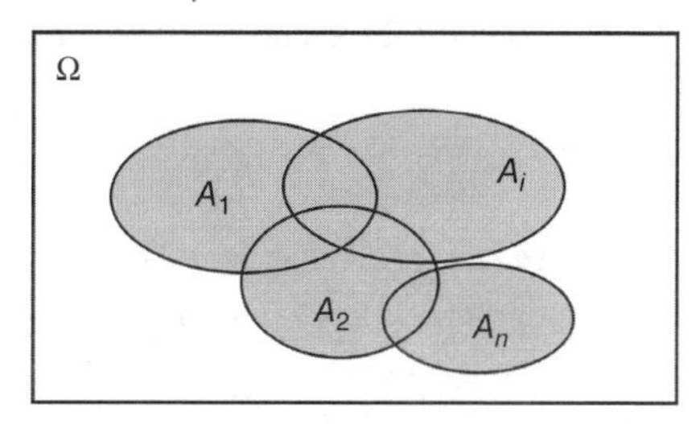

$$\bigcup_{i=1}^{n} A_i = A_1 \cup A_2 \cup \cdots \cup A_n = \{x \mid x \in A_1 \text{ or } x \in A_2 \text{ or} \ldots x \in A_n\}$$

A.3 Intersection of Events

The intersection $A \cap B$ of two events A and B is the event consisting of outcomes x *common to both* A and B:

$$A \cap B = \{x \mid x \in A \text{ and } x \in B\}$$

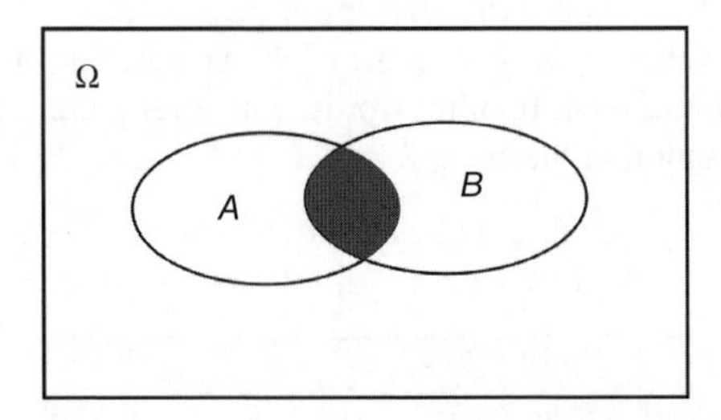

The intersection $\bigcap_{i=1}^{n} A_i = A_1 \cap A_2 \cap \cdots \cap A_n$ of n events is the event consisting of outcomes x *common to all events* A_i.

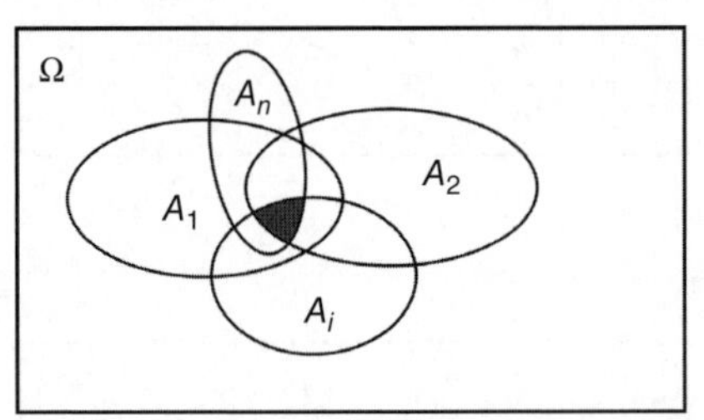

$$\bigcap_{i=1}^{n} A_i = A_1 \cap A_2 \cap \cdots \cap A_n = \{x \mid x \in A_1 \cap x \in A_2 \cap \cdots \cap x \in A_n\}$$

Example

Sample space: the outcomes from a toss of a die

Event A: *the result is a number greater than 3*; $A = \{4,5,6)$

Event B: *the result is an odd number*; $B = \{1,3,5)$

$$A \cup B = \{1,3,4,5,6\}$$
$$A \cap B = \{5\}.$$

Let Ω be the sample space and let A, B and C be events (subsets) in Ω.

The following laws hold:

Associative Laws

$$(A \cup B) \cup C = A \cup (B \cup C)$$
$$(A \cap B) \cap C = A \cap (B \cap C)$$

Commutative Laws

$$A \cup B = B \cup A$$
$$A \cap B = B \cap A$$

Distributive Laws

$$A \cap (B \cup C) = (A \cap B) \cup (A \cap C)$$
$$A \cup (B \cap C) = (A \cup B) \cap (A \cup C)$$

Identity Laws

$$A \cup \varnothing = A$$
$$A \cap \Omega = A$$

Complement Laws

$$A \cup \overline{A} = \Omega$$
$$A \cap \overline{A} = \varnothing$$

Idempotent Laws

$$A \cup A = A$$
$$A \cap A = A$$

Bound Laws

$$A \cup \Omega = \Omega$$
$$A \cap \varnothing = \varnothing$$

Absorption Laws

$$A \cup (A \cap B) = A$$
$$A \cap (A \cup B) = A$$

Involution Law

$$\overline{\overline{A}} = A$$

0/1 Laws

$$\overline{\varnothing} = \Omega$$
$$\overline{\Omega} = \varnothing$$

De Morgan's Laws for Sets

$$\overline{(A \cup B)} = \overline{A} \cap \overline{B}$$
$$\overline{(A \cap B)} = \overline{A} \cup \overline{B}$$

Principle of Duality

If any statement involving '∪', '∩' and '$\bar{\ }$' is true for all sets, then the dual statement obtained by replacing '∪' by '∩', '∩' by '∪', ∅ by Ω and Ω by ∅ is also true for all sets.

Partition of the Sample Space Ω

A collection of events $\{A_i\}$ is said to be a *partition* of the sample space if every element k of Ω belongs to exactly one event A_k. In other words, a partition of Ω divides Ω into non-overlapping subsets. A_i are pairwise disjoint and their union is the sample space Ω.

$$A_i \cap A_j = \varnothing, \quad \text{if}\, i \neq j$$
$$\bigcup_i A_i = \Omega$$

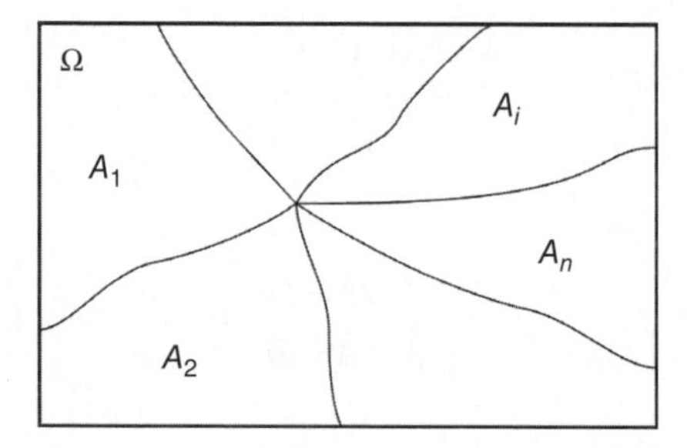

A.4 Probability

Classical Approach to Defining Probability

The ratio of the number of favourable outcomes to the total number of outcomes.

$$P(A) = \frac{\text{Number of outcomes leading to event } A}{\text{Total number of possible outcomes}}$$

This approach can only be used when symmetry is present, that is, if all outcomes are equally likely (the outcomes are interchangeable).

Example

What is the probability $P(A)$ of the event A that the sum of the faces of two independently thrown dice will be equal to 5?

Since there exist only four favourable (successful) outcomes leading to event

A: (1+4, 2+3, 3+2, 4+1) in the total sample space of 6×6=36 possible symmetrical (equally likely) outcomes, the probability of event A is

$$P(A) = \frac{4}{36} = \frac{1}{9}$$

Empirical Definition of Probability

Suppose that an experiment is performed in which a large number of components are tested under the same conditions. Thus, if N is the number of components tested and n is the number of failures, the probability of failure (event A) can be defined formally as

$$P(A) = \lim_{N \to \infty} \frac{n}{N}$$

According to the empirical definition, probability is defined as *a limit of the ratio of occurrences from a large number of trials.*

Usually, a relatively small number of trials N gives a sufficiently accurate estimate of the true probability $P(A) \approx n/N$.

Axiomatic Approach

Probability can also be defined using Kolmogorov's axioms:

Axiom 1: $0 \le P(A) \le 1$.
Axiom 2: $P(\Omega) = 1$.
Axiom 3: This axiom is related to an infinite number of *mutually exclusive events*:

$$A_1, A_2, \ldots \left(A_i \cap A_j = \varnothing, \text{ when } i \neq j\right)$$
$$P\left(A_1 \cup A_2 \cup \cdots\right) = P\left(A_1\right) + P\left(A_2\right) + \cdots$$

The probability of the certain event is unity (Axiom 2): $P(\Omega) = 1$.

The probability of the null event is zero: $P(\varnothing) = 0$. According to the first axiom, the probability of events is measured on a scale from 0 to 1, with '0' being impossibility and '1' being certainty.

A.5 Probability of a Union and Intersection of Mutually Exclusive Events

From the third Kolmogorov's axiom, it follows that

if A and B are *mutually exclusive (disjoint)* events ($P(A \cap B) = \varnothing$), then

$$P\left(A \cup B\right) = P\left(A\right) + P\left(B\right)$$

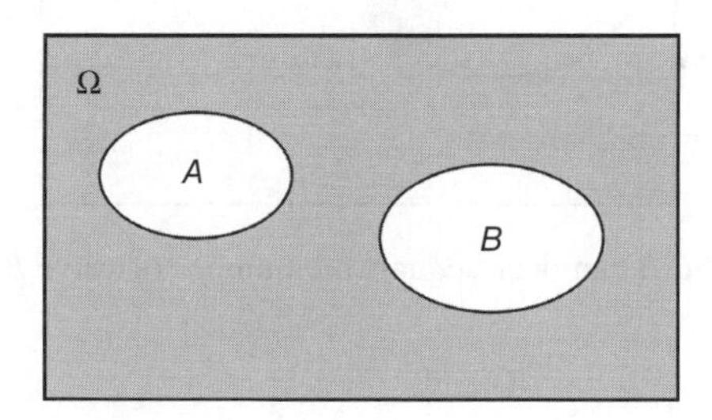

Example

Event A: 'The number of points from rolling a die is 5'.
Event B: 'The number of points from rolling a die is 4'.

The probability of obtaining 5 or 4 points is

$$P\left(A \cup B\right) = P\left(A\right) + P\left(B\right) = \frac{1}{6} + \frac{1}{6} = \frac{1}{3}$$

Probability of Complementary Events

$$P\left(A\right) = 1 - P\left(\bar{A}\right)$$

Indeed,

$$A \cup \bar{A} = \Omega,\ P(A) + P(\bar{A}) = P(\Omega) = 1,\quad \therefore P(A) = 1 - P(\bar{A})$$

An Important Application

The probability that a measuring device will fail during operation is p. If three devices are present, the probability that at least one device (one or two or three) will fail can be found using probability of complementary events. The probability of the event *at least one of the devices will fail* is equal to one minus the probability of the event *none of the devices will fail* because the two events are complementary:

$$P(\text{at least one device will fail}) = 1 - P(\text{none of the devices will fail})$$

Example

For an arrangement of five sensors, the probability that at least three sensors will work can be expressed as follows:

$$P(\text{at least three will work}) = 1 - P(\text{two or fewer will work}).$$

A.6 Conditional Probability

Probability of Intersection of Events

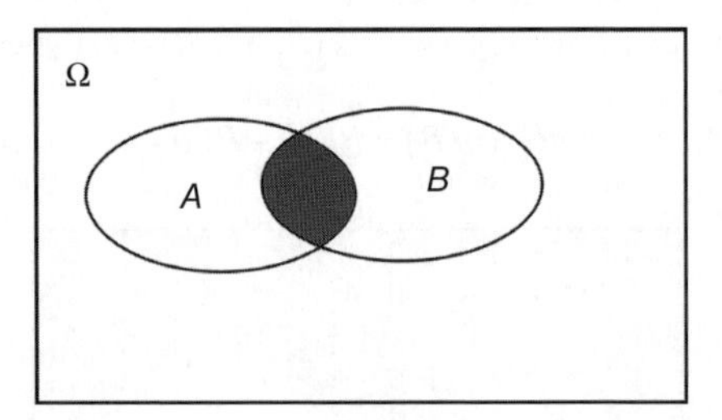

The number of ways A and B can occur equals the number of ways B can occur given that A has occurred:

$$N_{A \cap B} = N_{B|A}$$

The last expression can also be presented as

$$N_{A \cap B} = \frac{N_A N_{B|A}}{N_A}$$

Dividing the two sides by the total number of trials N results in

$$\frac{N_{A \cap B}}{N} = \frac{N_A \times N_{B|A}}{N_A N} = \left(\frac{N_A}{N}\right)\left(\frac{N_{B|A}}{N_A}\right)$$

Since according to the classical definition of probability

$$P(A \cap B) = \frac{N_{A \cap B}}{N},\ P(A) = \frac{N_A}{N} \text{ and } P(B \mid A) = \frac{N_{B|A}}{N_A},$$

finally, $P(A \cap B) = P(A)P(B \mid A)$, where $P(B \mid A)$ is the probability of B given that A has occurred (*conditional probability*).

Alternatively, the number of ways A and B can occur equals the number of ways A can occur given that B has occurred. Therefore, the corresponding probability expression becomes

$$P(A \cap B) = P(B)P(A \mid B)$$

where $P(A \mid B)$ is the probability of A given that B has occurred (*conditional probability*).

Example

A fault has occurred in one of the three sections of a communication line. The probability of finding the fault after inspecting any of the three sections is 1/3. The sections are searched sequentially, starting with the first section. Let event A be *the fault will not be found after searching the first section* and event B stand for *the fault will not be found after searching the second section.*

The probability of event B depends on the outcome of event A:

$$P(B \mid A) = \frac{1}{2}$$

$$P(B \mid \bar{A}) = 1$$

As a result,

$$P(A \cap B) = P(A)P(B \mid A)$$

Since $P(A) = 2/3$ and $P(B \mid A) = 1/2$,

$$P(A \cap B) = P(A)P(B \mid A) = \frac{2}{3} \times \frac{1}{2} = \frac{1}{3}$$

The same result could have been obtained by figuring out that the probability that the fault will not be found by searching the first and the second section is equal to the probability that the fault will be located in the third section which is equal to 1/3.

Probability of Intersection of Three Events

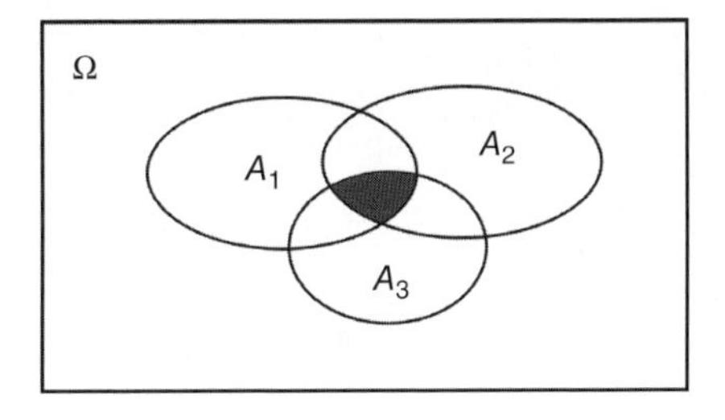

$$P(A_1 \cap A_2 \cap A_3) = P(A_1)P(A_2 \mid A_1)P(A_3 \mid A_1 A_2).$$

Example

A component is put into service. The probability that the component will be defective is 0.03. If the component is defective, with probability 0.4, the defect promotes an increased corrosion rate for the component. The increased corrosion rate causes 13% of the defective components to fail shortly after being put into service, as opposed to non-defective components which do not fail in such a short time.

Find the probability that a particular component will fail shortly after being put into service due to intensive corrosion promoted by the defect.

Let A denote the event *the component is defective*, B denote the event *the defect has promoted increased corrosion rate* and C denote the event *the component will fail shortly after being put into service*.

Then

$$P(A \cap B \cap C) = P(A)P(B \mid A)P(C \mid AB)$$

Since $P(A) = 0.03$, $P(B \mid A) = 0.4$ and $P(C \mid AB) = 0.13$,

$$P(A \cap B \cap C) = 0.03 \times 0.4 \times 0.13 \approx 0.00156$$

Probability of Intersection of n Events

The formula related to probability of intersection of three events can be generalised for n events. The probability of intersection of n events A_i, $i = 1,\ldots,n$ is

$$P(A_1 \cap A_2 \cap \cdots \cap A_n) = P(A_1)P(A_2 \mid A_1)P(A_3 \mid A_1A_2)\ldots P(A_n \mid A_1A_2\ldots A_{n-1})$$

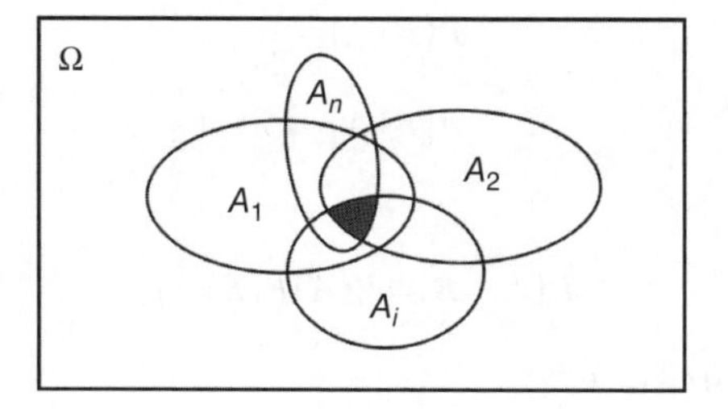

The probabilities $P(A \mid B)$ of the event A *given that B has occurred* and of the event B *given that A has occurred* can be determined as follows:

$$P(A \mid B) = \frac{\text{Number of ways } A \text{ and } B \text{ can occur}}{\text{Number of ways } B \text{ can occur}}$$

$$P(A \mid B) = \frac{P(A \cap B)}{P(B)}$$

$$P(B \mid A) = \frac{P(A \cap B)}{P(A)}$$

From the last two equations, it follows

$$P(A \mid B)P(B) = P(B \mid A)P(A).$$

Example

It has been observed that 3% of the components arriving on an assembly line are both defective and from supplier X. If 30% of the components come from supplier X, find the probability that a purchased component will be defective given that it comes from supplier X.

Let A denote the event *the component comes from supplier X* and B denote the event *the component is defective*.

Then $P(B \mid A) = \dfrac{P(A \cap B)}{P(A)} = \dfrac{0.03}{0.3} = 0.1$.

Thus, 10% of the components from supplier X are likely to be defective.

A.7 Probability of a Union of Non-disjoint Events

Non-disjoint events: $A \cap B \neq \varnothing$

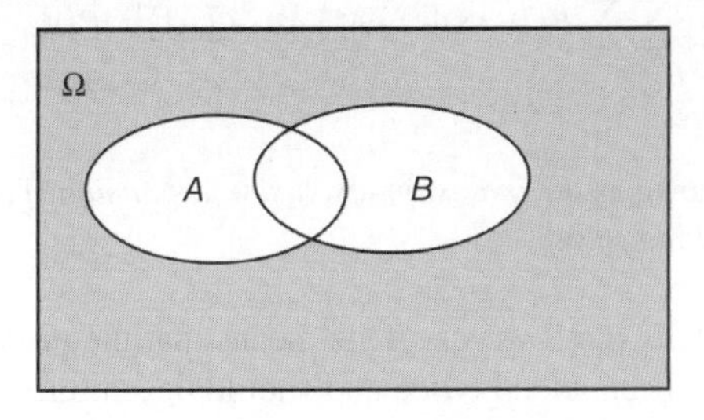

$$P(A \cup B) = P(A) + P(B) - P(A \cap B)$$

Indeed,

$$\begin{aligned} A &= (A \cap \bar{B}) \cup (A \cap B) \\ B &= (\bar{A} \cap B) \cup (A \cap B) \\ A \cup B &= (A \cap \bar{B}) \cup (A \cap B) \cup (\bar{A} \cap B) \\ P(A) + P(B) &= P(A \cap \bar{B}) + P(\bar{A} \cap B) + 2P(A \cap B) \\ P(A \cup B) &= P(A \cap \bar{B}) + P(\bar{A} \cap B) + P(A \cap B) \\ \therefore P(A \cup B) &= P(A) + P(B) - P(A \cap B) \end{aligned}$$

Similarly, the probability of a union of three non-disjoint events can be calculated:

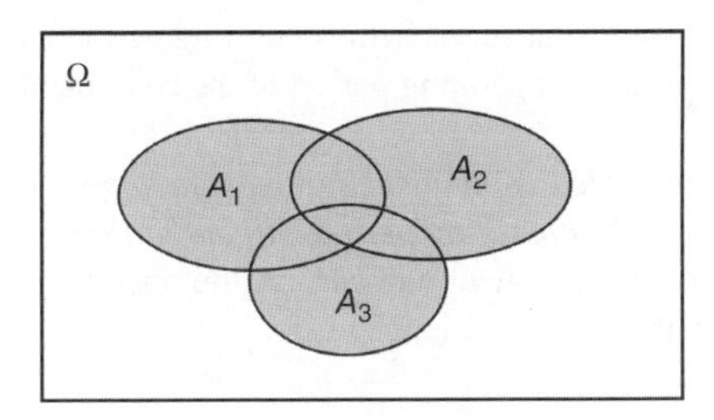

$$\begin{aligned} P(A_1 \cup A_2 \cup A_3) = {} & P(A_1) + P(A_2) + P(A_3) - P(A_1 \cap A_2) \\ & - P(A_2 \cap A_3) - P(A_1 \cap A_3) + P(A_1 \cap A_2 \cap A_3) \end{aligned}$$

The expression regarding the probability of a union of non-disjoint events can easily be generalised for n events:

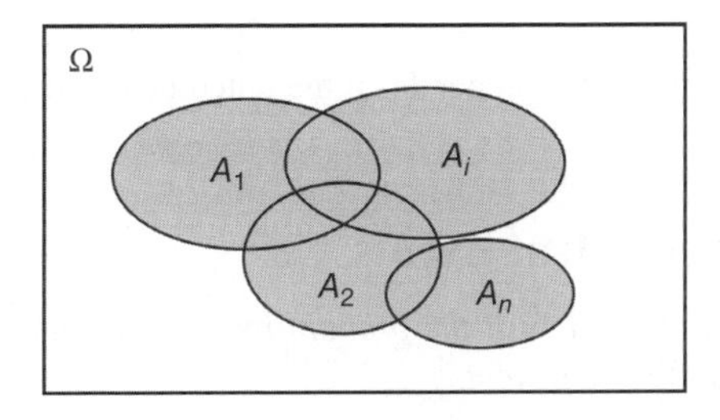

$$P\left(\bigcup_{i=1}^{n} A_i\right) = \sum_{i=1}^{n} P(A_i) - \sum_{i<j}\sum P\left(A_i \cap A_j\right)$$
$$+\sum_{i<j<k}\sum\sum P\left(A_i \cap A_j \cap A_k\right) - \cdots + (-1)^{n+1} P\left(A_1 \cap A_2 \cap \cdots \cap A_n\right)$$

This expression is also known as *the inclusion–exclusion expansion*. The rule for the expansion can be summarised by the following steps:

1. *Add the probabilities of all single events*. This means that the probability of the intersections of any pair of events has been added twice and should be subtracted.
2. *Subtract the probabilities of all intersections of two events from the previous result*. Since the contribution of the intersection of any three events has been added three times through the single events and subsequently has been subtracted three times from the twofold intersections, the probabilities of all threefold intersections must be added.
3. For higher-order intersections, *terms with odd number of events are added*, while *terms with even number of events are subtracted* from the sum.

A.8 Statistically Dependent Events

Two events are statistically dependent if the outcome of one of the events affects the probability of occurrence of the other event.

Example

An electronic component is cooled by a fan. If the cooling fan does not fail during the operating period of the component, the cooled component survives the operating period without failure with probability 0.99. If the cooling fan fails during the operating period of the component, the cooled component fails immediately.

Denote with A the event *the cooled component survives the operating period without failure* and with B the event *the fan survives the operating period without failure*.

The probability that the cooled component will survive the operating period, given that the fan has survived the operating period, is

$$P\left(A \mid B\right) = 0.99$$

The probability that the cooled component will survive the operating period given that the fan has failed is

$$P\left(A \mid \overline{B}\right) = 0$$

As a result, the probability of event A depends on the outcome of event B.

A.9 Statistically Independent Events

Two events are statistically independent if the probability of occurrence of one of the events is not influenced by the outcome of the other event.

Example
In an experiment of flipping two different coins, the events *head on the first coin* and *tail on the second coin* are statistically independent.

Example
For two components coming from separate suppliers, event A: *the first component is non-defective* and event B: *the second component is non-defective* are statistically independent.

In this case,

$$P(A \mid B) = P(A)$$

and

$$P(B \mid A) = P(B)$$

For statistically independent events, the probability that they will *both* occur simultaneously is

$$P(A \cap B) = P(A)P(B)$$

Indeed, this follows immediately from $P(A \cap B) = P(A)P(B \mid A)$ and $P(B \mid A) = P(B)$. If there are n independent events, the probability of all of them occurring simultaneously is

$$P(A_1 \cap A_2 \cap \cdots \cap A_n) = \prod_{i=1}^{n} P(A_i)$$

A.10 Probability of a Union of Independent Events

For two statistically independent events A and B, the probability that at least one will occur is

$$P(A \cup B) = P(A) + P(B) - P(A)P(B)$$

Indeed, the above follows from

$$P(A \cup B) = P(A) + P(B) - P(A \cap B)$$

and $P(A \cap B) = P(A)P(B)$, which is valid for statistically independent events.

A.11 Boolean Variables and Boolean Algebra

Boolean variables are *indicator variables* for the events. Boolean variables represent the two states of an event – occurrence and non-occurrence – and this defines the link between events and Boolean variables. Thus, a Boolean indicator variable can only take values '0' or '1' and is defined in the following way:

$$a = \begin{cases} 0 & \text{if event } A \text{ does not occur} \\ 1 & \text{if event } A \text{ occurs} \end{cases}$$

Boolean variables are often used in risk analysis to represent two states of a system or component (let A be the event *the system works*). Then $a = 1$ corresponds to the event A, *the system works*, and

$a = 0$ corresponds to the event $\bar{A}$, *the system is in a failed state*. For example, the state of components arranged in series or parallel can be presented with Boolean variables:

The series arrangement fails if and only if at least one component fails:

$$s = c_1 c_2 \ldots c_N \ (s = 0 \text{ if any } c_k = 0)$$

The system logically arranged in series works if and only if all of the components work:

$$s = c_1 c_2 \ldots c_N \ (s = 1 \text{ if all } c_k = 1)$$

The system logically arranged in parallel fails if and only if all of the components fail:

$$s = c_1 + c_2 + \cdots + c_N \ (s = 0 \text{ if all } c_i = 0)$$

The system logically arranged in parallel works if and only if at least one of the components works:

$$s = c_1 + c_2 + \cdots + c_N \ (s = 1 \text{ if at least one } c_k = 1)$$

Boolean algebra defines the operations with Boolean variables and is used in risk analysis to translate the state of a system into Boolean expressions. Boolean expressions represent the structure function of the system. It can be shown that the expected value of the structure function is the reliability of the system (Barlow and Proschan, 1975). Boolean expressions are also a useful tool for representing the structure of *fault trees* and *reliability block diagrams* (Hoyland and Rausand, 1994). The basic gate types used in a fault tree (AND, OR and NOT gate) correspond one-to-one to the basic Boolean operations.

Boolean Operations

Boolean algebra consists of Boolean variables and logical operations defined over them. The operations AND, OR and NOT in Boolean algebra are analogous to the operations intersection, union and complementation in set theory.

The Boolean operations will be illustrated by the event *fluid supply* in case where the fluid is delivered by two pumps *A* and *B* working independently. Values $a = 1$ ($b = 1$) correspond to the events *pump A (B) in working state*, whereas values $a = 0$ ($b = 0$) correspond to the events *pump A (B) in failed state*.

1. Logical AND, '.'

Truth table:

a	b	ab
1	1	1
1	0	0
0	1	0
0	0	0

Example

$ab = 1$; Full capacity fluid supply is present (both pumps *A* and *B* work).

A schematic presentation of an AND gate is given below. It is a basic building block of fault trees. The output event of AND gates occurs if *all* input events occur simultaneously.

2. Logical OR, '+'

a	b	$a+b$
1	1	1
1	0	1
0	1	1
0	0	0

Example

$a+b=1$: Fluid supply is present (at least one pump A or B is working).

A schematic presentation of an OR gate is given below. It is a basic building block of fault trees. The output event of an OR gate occurs if *any* of the input events occurs or both events occur.

3. Logical NOT, '-'

Associated with the complementary event of A. A value $a=1$ corresponds to the event *the pump works*; $\bar{a}=0$ corresponds to the event *the pump does not work*:

a	$\bar{a}$
1	0
0	1

Laws of Boolean Algebra

In the Boolean expressions, OR is symbolised by '+', but note that the sign '+' does not imply algebraic addition. The sign for AND is omitted.

1. ***Associative Laws***

$$(a+b)+c=a+(b+c)$$
$$(ab)c=a(bc)$$

2. ***Commutative Laws***

$$a+b=b+a$$
$$ab=ba$$

3. ***Distributive Laws***

$$a(b+c)=ab+ac$$
$$a+bc=(a+b)(a+c)$$

(Note that the last rule is not analogous to the distributive law in ordinary algebra.)

4. ***Identity Laws***

$$a+0=a$$
$$a.1=a$$

5. ***Complementary Laws***

$$a + \bar{a} = 1$$
$$a\bar{a} = 0$$

6. ***Idempotent Laws***

$$a + a = a$$
$$aa = a$$

7. ***Bound Laws***

$$a + 1 = 1$$
$$a.0 = 0$$

8. ***Absorption Laws***

$$a + ab = a$$
$$a(a + b) = a$$

9. ***Involution Law***

$$\bar{\bar{a}} = a$$

10. ***0/1 Laws***

$$\bar{0} = 1$$
$$\bar{1} = 0$$

11. ***De Morgan's Laws***

$$\overline{(a + b)} = \bar{a}\bar{b}$$
$$\overline{(ab)} = \bar{a} + \bar{b}$$

The following relationships help remove redundancies in Boolean expressions:

$$a + ab = a$$
$$a(a + b) = a$$
$$a + \bar{a}b = a + b$$

All laws can be proved using truth tables.
Proof of the distributive law $a + bc = (a + b)(a + c)$:

a	*b*	*c*	*a+bc*	*(a+b)(a+c)*
1	1	1	1	1
1	1	0	1	1
1	0	1	1	1
1	0	0	1	1
0	1	1	1	1
0	1	0	0	0
0	0	1	0	0
0	0	0	0	0

which shows that the two Boolean expressions $a + bc$ and $(a + b)(a + c)$ are logically equivalent.

Proof of the De Morgan's law $\overline{(ab)} = \bar{a} + \bar{b}$:

a	$\bar{a}$	b	$\bar{b}$	ab	$\overline{ab}$	$\bar{a}+\bar{b}$
1	0	1	0	1	0	0
1	0	0	1	0	1	1
0	1	1	0	0	1	1
0	1	0	1	0	1	1

Clearly, some of the Boolean operations resemble the operations in ordinary (numerical) algebra. However, in ordinary algebra, there is no unary operation equivalent to complementation, and idempotency laws do not hold. Furthermore, although the union is distributive over intersection $A \cup (B \cap C) = (A \cup B) \cap (A \cup C)$, addition in ordinary algebra is not distributive over multiplication $a + bc \neq (a+b)(a+c)$.

Proof of the relationship $a + \bar{a}b = a + b$:

$$a + \bar{a}b = (a + \bar{a})(a + b) = a + b$$

Simplifying Boolean Expressions

Reducing Boolean Expressions to a Sum-of-Products Form

Boolean algebra rules are mainly used to rewrite complex Boolean expressions to their simplest sum-of-products form.

Example

$$(a\bar{b} + c)(\bar{a}b + c)(ab + \bar{c}) = (ac\bar{b} + \bar{a}cb + c)(ab + \bar{c}) = abc$$

Example

$$\begin{aligned}
&(a\bar{b} + c)(\bar{a}b + c)(ab + \bar{c}) + \overline{abc}(ab + bc + ac) \\
&= abc + (\bar{a} + \bar{b} + \bar{c})(ab + bc + ac) = abc + \bar{a}bc + a\bar{b}c + ab\bar{c} \\
&= bc(a + \bar{a}) + a\bar{b}c + ab\bar{c} = bc + a\bar{b}c + ab\bar{c} \\
&= c(b + a\bar{b}) + ab\bar{c} = c(b + a) + ab\bar{c} \\
&= cb + ca + ab\bar{c} = cb + a(c + b\bar{c}) = cb + a(c + b) \\
&= ab + bc + ac
\end{aligned}$$

Appendix B

B.1 Random Variables: Basic Properties

Random variables can be discrete and continuous. A discrete random variable X takes only discrete values $X = x_1, x_2, \ldots, x_n$ with probabilities $f(x_1), f(x_2), \ldots, f(x_n)$ and no other value:

X	x_1	x_2	...	x_n
$P(X=x)$	$f(x_1)$	$f(x_2)$	...	$f(x_n)$

where $f(x) \equiv P(X = x)$ is the probability (mass) function of the random variable

$$\sum_{i=1}^{n} f(x_i) = 1$$

Example

The probability distribution of the score X from throwing a perfect die:

x	1	2	3	4	5	6
$P(X = x)$	1/6	1/6	1/6	1/6	1/6	1/6

Distribution (Cumulative Distribution) Function of Discrete Random Variables

$$P(X \le x) \equiv F(x) = \sum_{x_i \le x} f(x_i)$$

Expected Value (Mean)

$$E(X) = \mu = \sum_{i=1}^{n} x_i f(x_i).$$

Reliability and Risk Models: Setting Reliability Requirements, Second Edition. Michael Todinov.

Variance

$$V(X) = E\left([X-\mu]^2\right) = \sum_{i=1}^{n}(x_i-\mu)^2 f(x_i)$$

Examples (discrete random variables):

- The state of the system (working or failed) at a particular time
- The number of failures of a repairable component (system) in a finite time interval
- The number of failed structural elements during a test
- The number of defects in a structural element after manufacturing

B.2 Boolean Random Variables

An important class of discrete random variables are the *Boolean random variables*. The random variable X takes on only two values: '$X=1$' (true) and '$X=0$' (false) with specified probabilities. For example, X can take on: '1' (occurrence); '0' (non-occurrence) or represent the state of a component or a system: '1' (working) and '0' (failed).

The state of a system can be defined by the discrete distribution of the Boolean variable X where 'p' is the probability that the system will be working.

State	Working	Failed
X	1	0
$f(x)$	p	$1-p$

B.3 Continuous Random Variables

Continuous random variables take on continuous values from a specified interval. The probability that the continuous random variable X will take on values from the interval $[a, b]$ is

$$P(a \le X \le b) = \int_a^b f(x)dx$$

where $f(x)$ is the probability density function (p.d.f.) of X. $f(x) \ge 0$ for all x and $\int_{-\infty}^{\infty} f(x)dx = 1$.

Examples

- The value of a design parameter (e.g. length, yield strength)
- The operating load acting on a structural component
- Time to failure, time between failures
- The magnitude of the residual stress at the surface of a component

B.4 Probability Density Function

The distribution of the random variable X is specified by its *probability density function* (p.d.f.) $f(x)$. The p.d.f. describes how the probability of obtaining particular values of the random variable is spread over the range of all possible values. It is important to understand that the function $f(x)$ is not itself probability but a probability density (probability per unit value). Thus, the probability of a value in the infinitesimal interval x, $x+dx$ is $f(x)\,dx$. The probability of obtaining a value in any specified interval $x_1 \le X \le x_2$ is

$$P(x_1 \le X \le x_2) = \int_{x_1}^{x_2} f(x)dx.$$

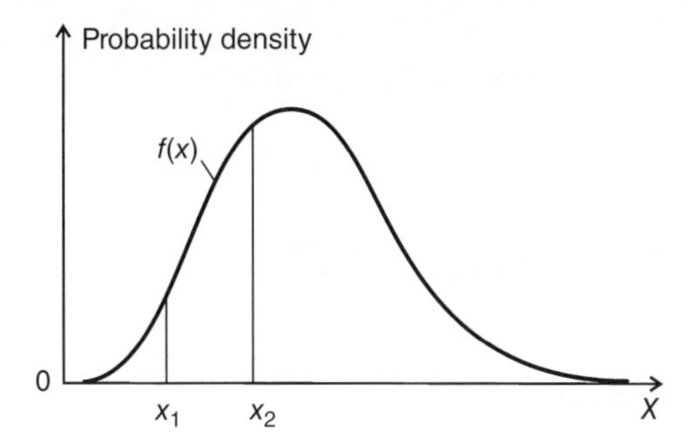

Basic Property of the p.d.f.
The total area beneath the p.d.f. $f(x)$ must always equal 1:

$$\int_{-\infty}^{\infty} f(x)dx = 1$$

The integral represents the fact that $f(x)$ is a probability distribution, that is, the probabilities of all outcomes must add up to unity (the random variable X will certainly accept some value from its domain).

B.5 Cumulative Distribution Function

Let $F(x)$ denote the (cumulative) distribution function (c.d.f.) of the random variable X. $F(x)$ gives the probability $P(X \le x)$ that the random variable X will be smaller than or equal to a specified value x:

$$F(x) \equiv P(X \le x) = F(x) = \int_{0}^{x} f(v)dv$$

where $f(x)$ is the p.d.f. Accordingly, $F(x)$ gives the area beneath the p.d.f. $f(x)$ until value x. The cumulative distribution function is related to the p.d.f. by $f(x) = dF(x)/dx$; $F(x)$ is a monotone, non-decreasing function of x. The probability that the random variable will accept values from the interval (x_1, x_2) is

$$P(x_1 < X \le x_2) = \int_{x_1}^{x_2} f(v)dv = F(x_2) - F(x_1)$$

B.6 Joint Distribution of Continuous Random Variables

Joint probability density function
$f(x, y)$ is a joint probability density function of two random variables X and Y if

$$f(x,y) \ge 0, \text{ for } -\infty < x < \infty \text{ and } -\infty < y < \infty$$

and

$$\int_{-\infty}^{\infty}\int_{-\infty}^{\infty} f(x,y)\,dx\,dy = 1$$

Joint Distribution Function

$$F(x,y) = P(X \le x \text{ and } Y \le y)$$
$$F(x,y) = \int_{-\infty}^{y}\int_{-\infty}^{x} f(r,s)\,dr\,ds$$
$$f(x,y) = \frac{\partial^2 F(x,y)}{\partial x \partial y}$$

Marginal probability density functions

$$f_1(x) = \int_{-\infty}^{\infty} f(x,y)\,dy,\ f_2(y) = \int_{-\infty}^{\infty} f(x,y)\,dx$$

Marginal Distribution Functions

$$F_1(x) = P(X \le x) = \lim_{y\to\infty} F(x,y)$$
$$F_2(y) = P(Y \le y) = \lim_{x\to\infty} F(x,y)$$

B.7 Correlated Random Variables

Suppose that X_1 and X_2 are random variables with expected (mean) values $E(X_1) = \mu_1$, $E(X_2) = \mu_2$ and variances $V(X_1) = \sigma_1^2$ and $V(X_2) = \sigma_2^2$.

The *covariance* of X_1 and X_2 is defined as

$$\text{Cov}(X_1,X_2) = E\left[(X_1 - \mu_1)(X_2 - \mu_2)\right] = E(X_1X_2) - E(X_1)E(X_2)$$

The *correlation coefficient* $\rho_{1,2}$ characterising X_1 and X_2 defined as

$$\rho_{1,2} = \frac{\text{Cov}(X_1,X_2)}{\sigma_1\sigma_2};\quad 0 \le \rho_{1,2} \le 1$$

measures the degree of linear association between X_1 and X_2. The random variables are not correlated if $\rho_{12} = 0$.

Properties of the Covariance

$$\text{Cov}(X_1,X_2) = \text{Cov}(X_2,X_1)$$
$$\text{Cov}(X,X) = V(X)$$
$$V\left(\sum_{i=1}^{n} X_i\right) = \sum_{i=1}^{n} V(X_i) + 2\sum_{i<j}\sum \text{Cov}(X_i,X_j)$$

Covariance Matrix

$$C = \begin{pmatrix} V(X_1) & \mathrm{Cov}(X_1,X_2) & \cdots & \mathrm{Cov}(X_1,X_n) \\ \cdots & \cdots & \cdots & \cdots \\ \cdots & \cdots & \cdots & \cdots \\ \mathrm{Cov}(X_n,X_1) & \mathrm{Cov}(X_n,X_2) & \cdots & V(X_n) \end{pmatrix}$$

If the random variables are statistically independent, $\mathrm{Cov}(X_i,X_j)=0$ for all $i \neq j$. Any pair of statistically independent random variables X_i, X_j are not correlated ($\mathrm{Cov}(X_i,X_j)=0$), but the converse is not necessarily true. The random variables may not be correlated and still be statistically dependent. For statistically independent random variables,

$$V\left(\sum_{i=1}^{n} X_i\right) = \sum_{i=1}^{n} V(X_i)$$

Examples

- Correlated random variables related to steels: X_1 (fatigue strength), X_2 (yield strength) and X_3 (hardness)
- Strongly correlated random variables: X_1 (fatigue strength) and X_2 (residual stress at the surface)
- Non-correlated random variables: X_1 (service stress) and X_2 (fracture toughness of the material)

B.8 Statistically Independent Random Variables

1. Random variables X and Y are statistically independent if the values accepted by X do not depend on the values accepted by Y and vice versa.
   Let us define events A and B as follows: $A: a \leq X \leq b$ and $B: c \leq Y \leq d$.
   Random variables X and Y are statistically independent if and only if A and B are independent events ($P(A \cap B) = P(A)\,P(B)$) for any choice of real numbers a, b, c and d.
2. Two random variables X and Y are statistically independent if and only if the factorisation

$$P(X \leq x \text{ and } Y \leq y) = P(X \leq x) \times P(Y \leq y)$$

   or $F(x,y) = F_1(x) \times F_2(y)$ is satisfied for all values x and y, where $F(x,y)$ is the joint distribution function (d.f.) of X and Y. $F_1(x)$ is the marginal d.f. of X and $F_2(y)$ is the marginal d.f. of Y.
3. Two random variables X and Y are statistically independent if and only if the factorisation $f(x,y) = f_1(x) f_2(y)$ is possible, where $f(x,y)$ is the joint p.d.f. of X and Y, $f_1(x)$ is the marginal p.d.f. of X and $f_2(y)$ is the marginal p.d.f. of Y.
4. Two random variables X and Y are statistically independent if and only if $f(x,y) = g(x)h(y)$, where $g(x)$ and $h(y)$ are non-negative functions of x and y; $f(x,y) > 0$ in $a \leq x \leq b$ and $c \leq y \leq d$ and $f(x,y) = 0$, elsewhere.
   If the random variables X and Y are statistically independent, then any two functions $h(X)$ and $g(Y)$ are also statistically independent.

Examples of Statistically Independent Random Variables

- X_1 (service load) and X_2 (strength), in cases where the service load does not cause any strength degradation
- X_1 (material property) and X_2 (dimensions)

Question
Is the strength of a load-carrying metal wire statistically independent from its length?

B.9 Properties of the Expectations and Variances of Random Variables

Suppose that X is a random variable with p.d.f. $f(x)$. The expected value of a function $g(X)$ of the random variable is

$$E[g(X)] = \int_{-\infty}^{\infty} g(x) f(x) dx$$

Expected Value of a Linear Function

$$E[aX + b] = aE(X) + b$$

Expected Value of a Sum of Random Variables

$$E[X_1 + X_2 + \cdots + X_n] = E(X_1) + E(X_2) + \cdots + E(X_n)$$

This formula is valid irrespective of whether the random variables are statistically independent or not.

Example
The mean load from several collinear loads is a sum of the means of the separate loads irrespective of whether the loads are statistically independent or not.

Alternative formula for the variance of a random variable:

$$V(X) = E(X^2) - [E(X)]^2$$

Variance of a random variable added to a constant:

$$V(a + X) = V(X)$$

Variance of a random variable multiplied by a constant:

$$V(aX) = a^2 V(X)$$

Expected value of a product of statistically independent random variables:

$$E(XY) = E(X)E(Y)$$

Important Property
The variance of a sum of statistically independent random variables is equal to the sum of the variances of the random variables:

$$V(X_1 + X_2 + \cdots + X_n) = V(X_1) + V(X_2) + \cdots + V(X_n)$$

Example

Calculate the standard deviation σ_d of the design parameter d from the machine part in the figure if the standard deviations of the lengths d_A, d_B and d_C of parts A, B and C are σ_A, σ_B and σ_C, correspondingly. The edges of parts A and B are aligned with the edges of part C.

Solution

$$\begin{aligned} d &= d_C - d_A - d_B \\ \sigma_d^2 &\equiv V(d) = V(d_C) + V(-1 \times d_A) + V(-1 \times d_B) \\ &= V(d_C) + (-1)^2 V(d_A) + (-1)^2 V(d_B) \\ &= V(d_C) + V(d_A) + V(d_B) \\ &= \sigma_C^2 + \sigma_A^2 + \sigma_B^2 \\ \sigma_d &= \sqrt{\sigma_A^2 + \sigma_B^2 + \sigma_C^2} \end{aligned}$$

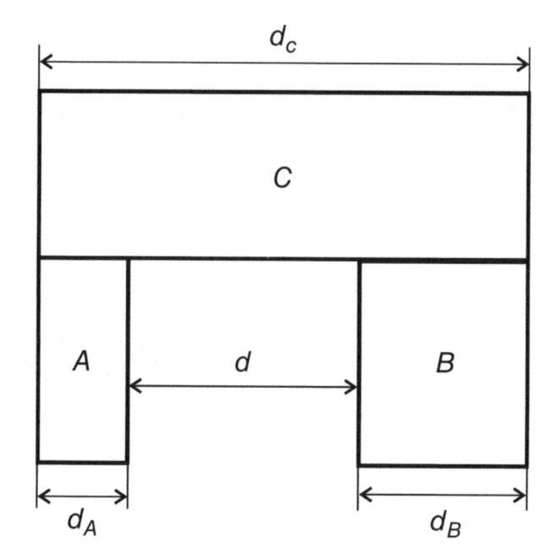

B.10 Important Theoretical Results Regarding the Sample Mean

The mean of a sum of identically distributed and statistically independent random variables will, with probability one, converge to the mean of the distribution characterising the random variables when the number of random variables approaches infinity.

Let $X_1, X_2, \ldots, X_n$ be n statistically independent and identically distributed random variables, where $E(X_i) = \mu$ is the mean of the common distribution. Then, with probability one,

$$\frac{X_1 + X_2 + \cdots + X_n}{n} \to \mu$$

as $n \to \infty$.

Expected Value and Variance of a Sample Mean

Random samples $x_1, x_2, \ldots, x_n$ of size n are taken from a statistical distribution with mean μ and variance σ^2: $E(x_i) = \mu$; $V(x_i) = \sigma^2$.

If $\bar{x} = (1/n)\sum_{i=1}^{n} x_i$ is the sample mean, then

$$E(\bar{x}) = \frac{1}{n}\sum_{i=1}^{n} E(x_i) = \frac{n\mu}{n} = \mu$$

and

$$V(\bar{x}) = \frac{1}{n^2}\sum_{i=1}^{n} V(x_i) = \frac{n\sigma^2}{n^2} = \frac{\sigma^2}{n}$$

The variance σ^2/n of the sample mean of n measurements is always smaller than the variance σ^2 of a single measurement.

Appendix C: Cumulative Distribution Function of the Standard Normal Distribution

	0.00	**0.01**	**0.02**	**0.03**	**0.04**	**0.05**	**0.06**	**0.07**	**0.08**	**0.09**
0.0	0.5000	0.5040	0.5080	0.5120	0.5160	0.5199	0.5239	0.5279	0.5319	0.5359
0.1	0.5398	0.5438	0.5478	0.5517	0.5557	0.5596	0.5636	0.5675	0.5714	0.5753
0.2	0.5793	0.5832	0.5871	0.5910	0.5948	0.5987	0.6026	0.6064	0.6103	0.6141
0.3	0.6179	0.6217	0.6255	0.6293	0.6331	0.6368	0.6406	0.6443	0.6480	0.6517
0.4	0.6554	0.6591	0.6628	0.6664	0.6700	0.6736	0.6772	0.6808	0.6844	0.6879
0.5	0.6915	0.6950	0.6985	0.7019	0.7054	0.7088	0.7123	0.7157	0.7190	0.7224
0.6	0.7257	0.7291	0.7324	0.7357	0.7389	0.7422	0.7454	0.7486	0.7517	0.7549
0.7	0.7580	0.7611	0.7642	0.7673	0.7703	0.7734	0.7764	0.7793	0.7823	0.7852
0.8	0.7881	0.7910	0.7939	0.7967	0.7995	0.8023	0.8051	0.8078	0.8106	0.8133
0.9	0.8159	0.8186	0.8212	0.8238	0.8264	0.8289	0.8315	0.8340	0.8365	0.8389
1.0	0.8413	0.8438	0.8461	0.8485	0.8508	0.8531	0.8554	0.8577	0.8599	0.8621
1.1	0.8643	0.8665	0.8686	0.8708	0.8729	0.8749	0.8770	0.8790	0.8810	0.8830
1.2	0.8849	0.8869	0.8888	0.8906	0.8925	0.8943	0.8962	0.8980	0.8997	0.9015
1.3	0.9032	0.9049	0.9066	0.9082	0.9099	0.9115	0.9131	0.9147	0.9162	0.9177
1.4	0.9192	0.9207	0.9222	0.9236	0.9251	0.9265	0.9279	0.9292	0.9306	0.9319
1.5	0.9332	0.9345	0.9357	0.9370	0.9382	0.9394	0.9406	0.9418	0.9429	0.9441
1.6	0.9452	0.9463	0.9474	0.9484	0.9495	0.9505	0.9515	0.9525	0.9535	0.9545
1.7	0.9554	0.9564	0.9573	0.9582	0.9591	0.9599	0.9608	0.9616	0.9625	0.9633
1.8	0.9641	0.9648	0.9656	0.9664	0.9671	0.9678	0.9686	0.9693	0.9699	0.9706
1.9	0.9713	0.9719	0.9726	0.9732	0.9738	0.9744	0.9750	0.9756	0.9761	0.9767
2.0	0.9772	0.9778	0.9783	0.9788	0.9793	0.9798	0.9803	0.9808	0.9812	0.9817
2.1	0.9821	0.9826	0.9830	0.9834	0.9838	0.9842	0.9846	0.9850	0.9854	0.9857
2.2	0.9861	0.9864	0.9868	0.9871	0.9875	0.9878	0.9881	0.9894	0.9887	0.9890
2.3	0.9893	0.9896	0.9898	0.9901	0.9904	0.9906	0.9909	0.9911	0.9913	0.9916
2.4	0.9918	0.9920	0.9922	0.9924	0.9927	0.9929	0.9930	0.9932	0.9934	0.9936

(*Continued*)

Reliability and Risk Models: Setting Reliability Requirements, Second Edition. Michael Todinov.

	0.00	0.01	0.02	0.03	0.04	0.05	0.06	0.07	0.08	0.09
2.5	0.9938	0.9940	0.9941	0.9943	0.9945	0.9946	0.9948	0.9949	0.9951	0.9952
2.6	0.9953	0.9955	0.9956	0.9957	0.9959	0.9960	0.9961	0.9962	0.9963	0.9964
2.7	0.9965	0.9966	0.9967	0.9968	0.9969	0.9970	0.9971	0.9972	0.9973	0.9974
2.8	0.9974	0.9975	0.9976	0.9977	0.9977	0.9978	0.9979	0.9979	0.9980	0.9981
2.9	0.9981	0.9982	0.9982	0.9983	0.9984	0.9984	0.9985	0.9985	0.9986	0.9986
3.0	0.9986	0.9987	0.9987	0.9988	0.9988	0.9989	0.9989	0.9989	0.9990	0.9990
3.1	0.9990	0.9991	0.9991	0.9991	0.9992	0.9992	0.9992	0.9992	0.9993	0.9993
3.2	0.9993	0.9993	0.9994	0.9994	0.9994	0.9994	0.9994	0.9995	0.9995	0.9995
3.3	0.9995	0.9995	0.9995	0.9996	0.9996	0.9996	0.9996	0.9996	0.9996	0.9996
3.4	0.9997	0.9997	0.9997	0.9997	0.9997	0.9997	0.9997	0.9997	0.9997	0.9998
3.5	0.9998	0.9998	0.9998	0.9998	0.9998	0.9998	0.9998	0.9998	0.9998	0.9998
3.6	0.9998	0.9998	0.9998	0.9999	0.9999	0.9999	0.9999	0.9999	0.9999	0.9999

Appendix D: χ^2-Distribution

n / α	**0.9995**	**0.999**	**0.995**	**0.990**	**0.975**	**0.95**	**0.90**	**0.80**	**0.70**	**0.60**
1	$0.0^{6}393$	$0.0^{5}157$	$0.0^{4}393$	$0.0^{3}157$	$0.0^{3}982$	$0.0^{2}393$	0.0158	0.0642	0.148	0.275
2	$0.0^{2}100$	$0.0^{2}200$	0.0100	0.0201	0.0506	0.103	0.211	0.446	0.713	0.02
3	0.0153	0.0243	0.0717	0.115	0.216	0.352	0.584	1.00	1.42	1.87
4	0.0639	0.0908	0.207	0.297	0.484	0.711	1.06	1.65	2.19	2.75
5	0.158	0.210	0.412	0.554	0.831	1.15	1.61	2.34	3.00	3.66
6	0.299	0.381	0.676	0.872	1.24	1.64	2.20	3.07	3.83	4.57
7	0.485	0.598	0.989	1.24	1.69	2.17	2.83	3.82	4.67	5.49
8	0.710	0.857	1.34	1.65	2.18	2.73	3.49	4.59	5.53	6.42
9	0.972	1.15	1.73	2.09	2.70	3.33	4.17	5.38	6.39	7.36
10	1.26	1.48	2.16	2.56	3.25	3.94	4.87	6.18	7.27	8.30
11	1.59	1.83	2.60	3.05	3.82	4.57	5.58	6.99	8.15	9.24
12	1.93	2.21	3.07	3.57	4.40	5.23	6.30	7.81	9.03	10.2
13	2.31	2.62	3.57	4.11	5.01	5.89	7.04	8.63	9.93	11.1
14	2.70	3.04	4.07	4.66	5.63	6.57	7.79	9.47	10.8	12.1
15	3.11	3.48	4.60	5.23	6.26	7.26	8.55	10.3	11.7	13.0
16	3.54	3.94	5.14	5.81	6.91	7.96	9.31	11.2	12.6	14.0
17	3.98	4.42	5.70	6.41	7.56	8.67	10.1	12.0	13.5	14.9
18	4.44	4.90	6.26	7.01	8.23	9.39	10.9	12.9	14.4	15.9
19	4.91	5.41	6.84	7.63	8.91	10.0	11.7	13.7	15.4	16.9
20	5.40	5.92	7.43	8.26	9.59	10.9	12.4	14.6	16.3	17.8
21	5.90	6.45	8.03	8.90	10.3	11.6	13.2	15.4	17.2	18.8
22	6.40	6.98	8.64	9.54	11.0	12.3	14.0	16.3	18.1	19.7
23	6.92	7.53	9.26	10.2	11.7	13.1	14.8	17.2	19.0	20.7
24	7.45	8.08	9.98	10.9	12.4	13.8	15.7	18.1	19.9	21.7
25	7.99	8.65	10.5	11.5	13.1	14.6	16.5	18.9	20.9	22.6
26	8.54	9.22	11.2	12.2	13.8	15.4	17.3	19.8	21.8	23.6

(*Continued*)

Reliability and Risk Models: Setting Reliability Requirements, Second Edition. Michael Todinov.

n / *α*	0.9995	0.999	0.995	0.990	0.975	0.95	0.90	0.80	0.70	0.60
27	9.09	9.80	11.8	12.9	14.6	16.2	18.1	20.7	22.7	24.5
28	9.66	10.4	12.5	13.6	15.3	16.9	18.9	21.6	23.6	25.5
29	10.2	11.0	13.1	14.3	16.0	17.7	19.8	22.5	24.6	26.5
30	10.8	11.6	13.8	15.0	16.8	18.5	20.6	23.4	25.5	27.4
31	11.4	12.2	14.5	15.7	17.5	19.3	21.4	24.3	26.4	28.4
32	12.0	12.8	15.1	16.4	18.3	20.1	22.3	25.1	27.4	29.4
33	12.6	13.4	15.8	17.1	19.0	20.9	23.1	26.0	28.3	30.3
34	13.2	14.1	16.5	17.8	19.8	21.7	24.0	26.9	29.2	31.3
35	13.8	14.7	17.2	18.5	20.6	22.5	24.8	27.8	30.2	32.3
36	14.4	15.3	17.9	19.2	21.3	23.3	25.6	28.7	31.1	33.3
37	15.0	16.0	18.6	20.0	22.1	24.1	26.5	29.6	32.1	34.2
38	15.6	16.6	19.3	20.7	22.9	24.9	27.3	30.5	33.0	35.2
39	16.3	17.3	20.0	21.4	23.7	25.7	28.2	31.4	33.9	36.2
40	16.9	17.9	20.7	22.2	24.4	26.5	29.1	32.3	34.9	37.1
41	17.5	18.6	21.4	22.9	25.2	27.3	29.9	33.3	35.8	38.1
42	18.2	19.2	22.1	23.7	26.0	28.1	30.8	34.2	36.8	39.1
43	18.8	19.9	22.9	24.4	26.8	29.0	31.6	35.1	37.7	40.0
44	19.5	20.6	23.6	25.1	27.6	29.8	32.5	36.0	38.6	41.0
45	20.1	21.3	24.3	25.9	28.4	30.6	33.4	36.9	39.6	42.0
46	20.8	21.9	25.0	26.7	29.1	31.4	34.2	37.8	40.5	43.0
47	21.5	22.6	25.8	27.4	30.0	32.3	35.1	38.7	41.5	43.9
48	22.1	23.3	26.5	28.2	30.8	33.1	35.9	39.6	42.4	44.9
49	22.8	24.0	27.2	28.9	31.6	33.9	36.8	40.5	43.4	45.9
50	23.5	24.7	28.0	29.7	32.4	34.8	37.7	41.4	44.3	46.9
51	24.1	25.4	28.7	30.5	33.2	35.6	38.6	42.4	45.3	47.8
52	24.8	26.1	29.5	31.2	34.0	36.4	39.4	43.3	46.2	48.8
53	25.5	26.8	30.2	32.0	34.8	37.3	40.3	44.2	47.2	49.8
54	26.2	27.5	31.0	32.8	35.6	38.1	41.2	45.1	48.1	50.8
55	26.9	28.2	31.7	33.6	36.4	39.0	42.1	46.0	49.1	51.7
56	27.6	28.9	32.5	34.3	37.2	39.8	42.9	47.0	50.0	52.7
57	28.2	29.6	33.2	35.1	38.0	40.6	43.8	47.9	51.0	53.7
58	28.9	30.3	34.0	35.9	38.8	41.5	44.7	48.8	51.9	54.7
59	29.6	31.0	34.8	36.7	39.7	42.3	45.6	49.7	52.9	55.6
60	30.3	31.7	35.5	37.5	40.5	43.2	46.5	50.6	53.8	56.6
61	31.0	32.5	36.3	38.3	41.3	44.0	47.3	51.6	54.8	57.6
62	31.7	33.2	37.1	39.1	42.1	44.9	48.2	52.5	55.7	58.6
63	32.5	33.9	37.8	39.9	43.0	45.7	49.1	53.5	56.7	59.6
64	33.2	34.6	38.6	40.6	43.8	46.6	50.0	54.3	57.6	60.5
65	33.9	35.4	39.4	41.4	44.6	47.4	50.9	55.3	58.6	61.5
66	34.6	36.1	40.2	42.2	45.4	48.3	51.8	56.2	59.5	62.5
67	35.3	36.8	40.9	43.0	46.3	49.2	52.7	57.1	60.5	63.5
68	36.0	37.6	41.7	43.8	47.1	50.0	53.5	58.0	61.4	64.4
69	36.7	38.3	42.5	44.6	47.9	50.9	54.4	59.0	62.4	65.4
70	37.5	39.0	43.3	54.4	48.8	51.7	55.3	59.9	63.3	66.4
71	38.2	39.8	44.1	46.2	49.6	52.6	56.2	60.8	64.3	67.4
72	38.9	40.5	44.8	47.1	50.4	53.5	57.1	61.8	65.3	68.4
73	39.6	41.3	45.6	47.9	51.3	54.3	58.0	62.7	66.2	69.3
74	40.4	42.0	40.4	48.7	52.1	55.2	58.9	63.6	67.2	70.3
75	41.1	42.8	47.2	49.5	52.9	56.1	59.8	64.5	68.1	71.3
76	41.8	43.5	48.0	50.3	53.8	56.9	60.7	65.5	69.1	72.3
77	42.6	44.3	48.8	51.1	54.6	57.8	61.6	66.4	70.0	73.2

n / α	0.9995	0.999	0.995	0.990	0.975	0.95	0.90	0.80	0.70	0.60
78	43.3	45.0	48.8	51.9	55.5	58.7	62.5	67.3	71.0	74.2
79	44.1	45.8	50.4	52.7	56.3	59.5	63.4	68.3	72.0	75.2
80	44.8	46.5	51.2	53.5	57.2	60.4	64.3	69.2	72.9	76.2
81	45.5	47.3	52.0	54.4	58.0	61.3	65.2	70.1	73.9	77.2
82	46.3	48.0	52.8	55.2	58.8	62.1	66.1	71.1	74.8	78.1
83	47.0	48.8	53.6	56.0	59.7	63.0	67.0	72.0	75.8	79.1
84	47.8	49.6	54.4	56.8	60.5	63.9	67.9	72.9	76.8	80.1
85	48.5	50.3	55.2	57.6	61.4	64.7	68.8	73.9	77.7	81.1
86	49.3	51.1	56.0	58.5	62.2	65.6	69.7	74.8	78.7	82.1
87	50.0	51.9	56.8	59.3	63.1	66.5	70.6	75.7	79.6	83.0
88	50.8	52.6	57.6	60.1	63.9	67.4	71.5	76.7	80.6	84.0
89	51.5	53.4	58.4	60.9	64.8	68.2	72.4	77.6	81.6	85.0
90	52.3	54.2	59.2	61.8	65.6	69.1	73.3	78.6	82.5	86.0
91	53.0	54.9	60.0	62.6	66.5	70.0	74.2	79.5	83.5	87.0
92	53.8	55.7	60.8	63.4	67.4	70.9	75.1	80.4	84.4	88.0
93	54.5	56.5	61.6	64.2	68.2	71.8	76.0	81.4	85.5	88.9
94	53.5	57.2	62.4	65.1	69.1	72.6	76.9	82.3	86.4	89.9
95	56.1	58.0	63.2	65.9	69.9	73.5	77.8	83.2	87.3	90.9
96	56.8	58.8	64.1	66.7	70.8	74.4	78.7	84.2	88.3	91.9
97	57.6	59.6	64.9	67.6	71.6	75.3	79.6	85.1	89.2	92.9
98	58.4	60.4	65.7	68.4	72.5	76.2	80.5	86.1	90.2	93.8
99	59.1	61.1	66.5	69.2	73.4	77.0	81.4	87.0	91.2	94.8
100	59.9	61.9	67.3	70.1	74.2	77.9	82.4	87.9	92.1	95.8

n / α	0.50	0.40	0.30	0.20	0.10	0.05	0.025	0.01	0.005	0.001
1	0.455	0.708	1.07	1.64	2.71	3.84	5.02	6.63	7.88	10.8
2	1.39	1.83	2.41	3.22	4.61	5.99	7.38	9.21	10.6	13.8
3	2.37	2.95	3.67	4.64	6.25	7.81	9.35	11.3	12.8	16.3
4	3.36	4.04	4.88	5.99	7.78	9.49	11.1	13.3	14.9	18.5
5	4.35	5.13	6.06	7.29	9.24	11.1	12.8	15.1	16.7	20.5
6	5.35	6.21	7.23	8.56	10.6	12.6	14.4	16.8	18.5	22.5
7	6.35	7.28	8.38	9.80	12.0	14.1	16.0	18.5	20.3	24.3
8	7.34	8.35	9.52	11.0	13.4	15.5	17.5	20.1	22.0	26.1
9	8.34	9.41	10.7	12.2	14.7	16.9	19.0	21.7	23.6	27.9
10	9.34	10.5	11.4	13.4	16.0	18.3	20.5	23.2	25.2	29.6
11	10.3	11.5	12.9	14.6	17.3	19.7	21.9	24.7	26.8	31.3
12	11.3	12.6	14.0	15.8	18.5	21.0	23.3	26.2	28.3	32.9
13	12.3	13.6	15.1	17.0	19.8	22.4	24.7	27.7	29.8	34.5
14	13.3	14.7	16.2	18.2	21.1	23.7	26.1	29.1	31.3	36.1
15	14.3	15.7	17.3	19.3	22.3	25.0	27.5	30.6	32.8	37.7
16	15.3	16.8	18.4	20.5	23.5	26.3	28.8	32.0	34.3	39.3
17	16.3	17.8	19.5	21.6	24.8	27.6	30.2	33.4	35.7	40.8
18	17.3	18.9	20.6	22.8	26.0	28.9	31.5	34.8	37.2	42.3
19	18.3	19.9	21.7	23.9	27.2	30.1	32.9	36.2	38.6	43.8
20	19.3	21.0	22.8	25.0	28.4	31.4	34.2	37.6	40.0	45.3

(*Continued*)

n / α	0.50	0.40	0.30	0.20	0.10	0.05	0.025	0.01	0.005	0.001
21	20.3	22.0	23.9	26.2	29.6	32.7	35.5	38.9	41.4	46.8
22	21.3	23.0	24.9	27.3	30.8	33.9	36.8	40.3	42.8	48.3
23	22.3	24.1	26.0	28.4	32.0	35.2	38.1	41.6	44.2	49.7
24	23.3	25.1	27.1	29.6	33.2	36.4	39.4	43.0	45.6	51.2
25	24.3	26.1	28.2	30.7	34.4	37.7	40.6	44.3	46.9	52.6
26	25.3	27.2	29.2	31.8	35.6	38.9	41.9	45.6	48.3	54.1
27	26.3	28.2	30.3	32.9	36.7	40.1	43.2	47.0	49.6	55.5
28	27.3	29.2	31.4	34.0	37.9	41.3	44.5	48.3	51.0	56.9
29	28.3	30.3	32.5	35.1	39.1	42.6	45.7	49.6	52.3	58.3
30	29.3	31.3	33.5	36.3	40.3	43.8	47.0	50.9	53.7	59.7
31	30.3	32.3	34.6	37.4	41.4	45.0	48.2	52.2	55.0	61.1
32	31.3	33.4	35.7	38.5	42.6	46.2	49.5	53.5	56.3	62.5
33	32.2	34.4	36.7	39.6	43.7	47.4	50.7	54.8	57.6	63.9
34	33.3	35.4	37.8	40.7	44.9	48.6	52.0	56.1	59.0	65.2
35	34.3	36.5	38.9	41.8	46.1	49.8	53.2	57.3	60.3	66.6
36	35.3	37.5	39.9	42.9	47.2	51.0	54.4	58.6	61.6	68.0
37	36.3	38.5	41.0	44.0	48.4	52.2	55.7	59.9	62.9	69.3
38	37.3	39.6	42.0	45.1	49.5	53.4	56.9	61.2	64.2	70.7
39	38.3	40.6	43.1	46.2	50.7	54.6	58.1	62.4	65.5	72.1
40	39.3	41.6	44.2	47.3	51.8	55.8	59.3	63.7	66.8	73.4
41	40.3	42.7	45.2	48.4	52.9	56.9	60.6	65.0	68.1	74.7
42	41.3	43.7	46.3	49.5	54.1	58.1	61.8	66.2	69.3	76.1
43	42.3	44.7	47.3	50.5	55.2	59.3	63.0	67.5	70.6	77.4
44	43.3	45.7	48.4	51.6	56.4	60.5	64.2	68.7	71.9	78.7
45	44.3	46.8	49.5	52.7	57.7	61.7	65.4	70.0	73.2	80.1
46	45.3	47.8	50.5	53.8	58.6	62.8	66.6	71.2	74.4	81.4
47	46.3	48.8	51.6	54.9	59.8	64.0	67.8	72.4	75.7	82.7
48	47.3	49.8	52.6	56.0	60.9	65.2	69.0	73.7	77.0	84.0
49	48.3	50.9	53.7	57.1	62.0	66.3	70.2	74.9	78.2	85.4
50	49.3	51.9	54.7	58.2	63.2	67.5	71.4	76.2	79.5	86.7
51	50.3	52.9	55.8	59.2	64.3	68.7	72.6	77.4	80.7	88.0
52	51.3	53.9	56.8	60.3	65.4	69.8	73.8	78.6	82.0	89.3
53	52.3	55.0	57.9	61.4	66.5	71.0	75.0	79.8	83.3	90.6
54	53.3	56.0	58.9	62.5	67.7	72.2	76.2	81.1	84.5	91.9
55	54.3	57.0	60.0	63.6	68.8	73.3	77.4	82.3	85.7	93.2
56	55.3	58.0	61.0	64.7	69.9	74.5	78.6	83.5	87.0	94.5
57	56.3	59.1	62.1	65.7	71.0	75.6	79.8	84.7	88.2	95.8
58	57.3	60.1	63.1	66.8	72.2	76.8	80.9	86.0	89.5	97.0
59	58.3	61.1	64.1	67.9	73.3	77.9	82.1	87.2	90.7	98.3
60	59.3	62.1	65.2	69.0	74.4	79.1	83.3	88.4	92.0	99.6
61	60.3	63.2	66.3	70.0	75.5	80.2	84.5	89.6	93.2	100.9
62	61.3	64.2	67.3	71.1	76.6	81.4	85.7	90.8	94.4	102.2
63	62.3	65.2	68.4	72.2	77.7	82.5	86.8	92.0	95.6	103.4
64	63.3	66.2	69.4	73.3	78.9	83.7	88.0	93.2	96.9	104.9
65	64.3	67.2	70.5	74.4	80.0	84.8	89.2	94.4	98.1	106.0
66	65.3	68.3	71.5	75.4	81.1	86.0	90.3	95.6	99.3	107.3
67	66.3	69.3	72.6	76.5	82.2	87.1	91.5	96.8	100.6	108.5
68	67.3	70.3	73.6	77.6	83.3	88.3	92.7	98.0	101.8	109.8

n α	0.50	0.40	0.30	0.20	0.10	0.05	0.025	0.01	0.005	0.001
69	68.3	71.3	74.6	78.6	84.4	89.4	93.9	99.2	103.0	111.1
70	69.3	72.4	75.7	79.7	85.5	90.5	95.0	100.4	104.2	112.3
71	70.3	73.4	76.7	80.8	86.6	91.7	96.2	101.6	105.4	113.6
72	71.3	74.4	77.8	81.9	87.7	92.8	97.4	102.8	106.6	114.8
73	72.3	75.4	78.8	82.9	88.8	93.9	98.5	104.0	107.9	116.1
74	73.3	76.4	79.9	84.0	90.0	95.1	99.7	105.2	109.1	117.3
75	74.3	77.5	80.9	85.1	91.1	96.2	100.8	106.4	110.3	118.6
76	75.3	78.5	82.0	86.1	92.2	97.4	102.0	107.6	111.5	119.9
77	76.3	79.5	83.0	87.2	93.3	98.5	103.2	108.8	112.7	121.1
78	77.3	80.5	84.0	88.3	94.4	99.6	104.3	110.0	113.9	122.3
79	78.3	81.5	85.1	89.3	95.5	100.7	105.5	111.1	115.1	123.6
80	79.3	82.6	86.1	90.4	96.6	101.9	106.6	112.3	116.3	124.3
81	80.3	83.6	87.2	91.5	97.7	103.0	107.8	113.5	117.5	126.1
82	81.3	84.6	88.2	92.5	98.8	104.1	108.9	114.7	118.7	127.3
83	82.3	85.6	89.2	93.6	99.9	105.3	110.1	115.9	119.9	128.6
84	83.3	86.6	90.3	94.7	101.0	106.4	111.2	117.1	121.1	129.8
85	84.3	87.7	91.3	95.7	102.1	107.5	112.4	118.2	122.3	131.0
86	85.3	88.7	92.4	96.8	103.2	108.6	113.5	119.4	123.5	132.3
87	86.3	89.7	93.4	97.9	104.3	109.8	114.7	120.6	124.7	133.5
88	87.3	90.7	94.4	98.9	105.4	110.9	115.8	121.8	125.9	134.7
89	88.3	91.7	95.5	100.0	106.5	112.0	117.0	122.9	127.1	136.0
90	89.3	92.8	96.5	101.1	107.6	113.1	118.1	124.1	128.3	137.2
91	90.3	93.8	97.6	102.1	108.7	114.3	119.3	125.3	129.5	138.4
92	91.3	94.8	98.6	103.2	109.8	115.4	120.4	126.5	130.7	139.7
93	92.3	95.8	99.6	104.2	110.9	116.5	121.6	127.6	131.9	140.9
94	93.3	96.8	100.7	105.3	111.9	117.6	122.7	128.8	133.1	142.1
95	94.3	97.9	101.7	106.4	113.0	118.8	123.9	130.0	134.2	143.3
96	95.3	98.9	102.8	107.4	114.1	119.9	125.0	131.1	135.4	144.6
97	96.3	99.9	103.8	108.5	115.2	121.0	126.1	132.3	136.6	145.8
98	97.3	100.9	104.8	109.5	116.3	122.1	127.3	133.5	137.8	147.0
99	98.3	101.9	105.9	110.6	117.4	123.2	128.4	134.6	139.0	148.2
100	99.3	102.9	106.9	111.7	118.5	124.3	129.6	135.8	140.2	149.4

References

ABAQUS v.6.5 Online documentation collection. http://www.intrinsys.com/software/simulia/abaqus-unified-fea (2007).

Abernethy R.B., *The new Weibull handbook*, Robert B. Abernethy, North Palm Beach (1994).

Abramowitz M. and I.A. Stegun, *Handbook of mathematical functions/with formulas, graphs, and mathematical tables*, Dover Publications, Mineola (1972).

Altshuller G.S., *Creativity as an exact science: The theory of the solution of inventive problems*, Gordon and Breach Science Publishing, New York (1984).

Altshuller G.S., *And suddenly the inventor appeared, TRIZ, the theory of inventive problem solving*, Translation from Russian by Lev Shulyak, Technical Innovation Center, Worcester, MA (1996).

Altshuller G.S., *The innovation algorithm, TRIZ, systematic innovation and technical creativity*, Technical Innovation Center, Inc., Worcester 1999.

Anderson T.L., *Fracture mechanics: Fundamentals and applications*, Taylor and Francis, Boca Raton (2005).

Andrews J.D. and T.R. Moss, *Reliability and risk assessment*, Professional Engineering Publishing, London (2002).

Ang A.H.S. and W.H. Tang, *Probability concepts in engineering planning and design, vol. 1, Basic principles*, John Wiley & Sons, Inc., New York (1975).

Ashby M.F., *Material selection in mechanical design*, 3rd ed., Elsevier, Oxford (2005).

Ashby M.F. and D.R.H. Jones, *Engineering materials 1: An introduction to their properties & applications*, Butterworth-Heinemann, Oxford (2000).

Barlow R.E. and F. Proschan *Mathematical theory of reliability*, John Wiley & Sons, Inc., New York (1965).

Barlow R.E. and F. Proschan, *Statistical theory of reliability and life testing*, Rinehart and Winston, Inc., New York (1975).

Barron R.F. and B.R. Barron, *Design for thermal stresses*, Wiley, New York (2012).

Batdorf S.B. and J.G. Crose, A statistical theory for the fracture of brittle structures subjected to non-uniform polyaxial stresses, *Journal of Applied Mechanics*, **41**, 459–464 (1974).

Bazovsky I., *Reliability theory and practice*, Prentice-Hall, Inc., Englewood Cliffs (1961).

Beasley M., *Reliability for engineers: An introduction*, Macmillan Education Ltd, London (1991).

Bedford T. and R. Cooke, *Probabilistic risk analysis, foundations and methods*, Cambridge University Press, Cambridge (2001).

Bellman R., *Dynamic programming*, Princeton University Press, Princeton (1957).

Reliability and Risk Models: Setting Reliability Requirements, Second Edition. Michael Todinov.

Bergman B., On the variability of the fracture stress of brittle materials, *Journal of Materials Science Letters*, **4**, 1143–1146 (1985).

Bernoulli D., Exposition of a new theory on the measurement of risk, *Papers of the Imperial Academy of Sciences in Petersburg*, **5**, 175–192 (1738). (Translated from Latin and republished in *Econometrica* 22 (1), 23–36 (1954).)

Bernoulli J., *Wahrscheinlichkeitsrechnung (Ars conjectandi, 1713). Ostwalds Klassiker der exakten Wissenschaften*, W. Engelmann, Leipzig (1899).

Bessis J., *Risk management in banking*, 2nd ed., John Wiley & Sons, Ltd, Chichester (2002).

Billinton R. and R.N. Allan, *Reliability evaluation of engineering systems*, 2nd ed., Plenum Press, New York (1992).

Bird G.C. and D. Saynor, The effect of peening shot size on the performance of carbon-steel springs, *Journal of Mechanical Working Technology*, **10** (2), 175–185 (1984).

Blischke W.R. and D.N. Murthy, *Reliability: Modelling, prediction, and optimisation*, John Wiley & Sons, Inc., New York (2000).

Booker J.D., M. Raines and K.G. Swift, *Designing capable and reliable products*, Butterworth-Heinemann, Oxford (2001).

Box G.E.P. and M.E. Muller, A note on the generation of random normal deviates, *Annals of Mathematical Statistics*, **28**, 610–611 (1958).

Budinski K.G., *Engineering materials: Properties and selection*, 5th ed., Prentice-Hall, Inc., Englewood Cliffs (1996).

Budynas R.G., *Advanced strength and applied stress analysis*, 2nd ed., McGraw-Hill, New York (1999).

Bury K.V., *Statistical models in applied science*, John Wiley & Sons, Inc., New York (1975).

Carter A.D.S., *Mechanical reliability*, Macmillan Education Ltd, London (1986).

Carter A.D.S., *Mechanical reliability and design*, Macmillan Press Ltd, London (1997).

Chapman C. and S. Ward, *Project risk management*, 2nd ed., John Wiley & Sons, Ltd, Chichester (2003).

Chatfield C., *Problem solving: A statistician's guide*, Chapman & Hall, London (1998).

Christian J.W., *The theory of transformations in metals and alloys*, Pergamon Press, Oxford (1965).

Colbourn C.J., *The combinatorics of network reliability*, Oxford University Press, New York (1987).

Collins J.A., *Mechanical design of machine elements and machines*, John Wiley & Sons, Inc., New York (2003).

Cormen T.H., T.C.E. Leiserson, R.L. Rivest and C. Stein, *Introduction to algorithms*, 2nd ed., MIT Press and McGraw-Hill, Cambridge, MA (2001).

Cullity B.D., *Elements of X-ray diffraction*, 2nd ed., Addison-Wesley, Reading (1978).

Dasgupta A. and M. Pecht, Material failure mechanisms and damage models, *IEEE Transactions on Reliability*, **40** (5), 531–536 (1991).

Dasgupta S., C. Papadimitriou and U. Vazirani, *Algorithms*, McGraw Hill, Boston (2008).

DeGroot M., *Probability and statistics*, Addison-Wesley, Reading (1989).

Denardo E.V., *The science of decision making*, John Wiley & Sons, Inc., New York (2002).

Dhillon B.S. and C. Singh, *Engineering reliability: New techniques and applications*, John Wiley & Sons, Inc., New York (1981).

Dianqing L., Z. Shengkun and T. Wenyong, Risk based inspection planning for ship structures subjected to corrosion deterioration, *Proceedings of the probabilistic safety assessment and management conference PSAM7-ESREL'04*, Springer, Berlin, vol. **6**, pp. 3336–3342 (2004).

Dodson B., *Weibull analysis*, ASQC Quality Press, Milwaukee (1994).

Dowling N.E., *Mechanical behaviour of materials*, Prentice Hall, Upper Saddle River (1999).

Draper N.R. and H. Smith, *Applied regression analysis*, 2nd ed., John Wiley & Sons, Inc., New York (1981).

Drenick R.F., The failure low of complex equipment, *Journal of the Society for Industrial and Applied Mathematics*, **8**, 680–690 (1960).

Ebeling C.E., *An introduction to reliability and maintainability engineering*, McGraw-Hill, New York (1997).

Ellis R., D.Gulick, *Calculus: one and several variables*, Harcourt Brace Jovanovich, New York (1991).

Epstein R.A., *The theory of gambling and statistical logic*, Elsevier, Amsterdam (2009).

Erhard G., *Designing with plastics*, Hanser, Munich (2006).

Everitt B.S. and D.J. Hand, *Finite mixture distributions*, Chapman and Hall, London (1981).

Ewalds H.L. and R.J.H. Wanhill, *Fracture mechanics*, 1st ed., Edward Arnold, London (1984).

Ford L.R. and D.R. Fulkerson, Maximal flow through a network, *Canadian Journal of Mathematics*, **8** (5), 399–404 (1956).

Fowlkes W.Y. and C.M. Creveling, *Engineering methods for robust product design: Using Taguchi methods in technology and product development*, Addison-Wesley, Reading 1995.

French M., *Conceptual design for engineers*, 3rd ed., Springer-Verlag London Ltd, London (1999).

Freudenthal A.M., Safety and the probability of structural failure, *American Society of Civil Engineers Transactions*, paper No. 2843, 1337–1397 (1954).

Freudenthal A.M., Statistical approach to brittle fracture, in *Fracture* vol. **II**, Ed. H. Liebowitz, Academic Press, New York, pp. 591–619 (1968).

Gildersleeve M.J., Relationship between decarburisation and fatigue strength of through hardened and carburising steels, *Materials Science and Technology*, **7**, 307–310 (1991).

Glitzmann P. and V. Klee, Polytopes: On some complexity of some basic problems in computational convexity II. Volume and mixed volumes, in *Polytopes: Abstract, convex and computational*, Ed. T. Bisztriczky, P. McMullen, R. Schneider and A.W. Weiss, Kluwer, Dordrecht (1994).

Gnedenko B.V., *The theory of probability*, Chelsea Publishing Company, New York (1962).

Grosh D.L., *A primer of reliability theory*, John Wiley & Sons, Inc., New York (1989).

Gross D. and C.M. Harris, *Fundamentals of queuing theory*, 2nd ed., John Wiley & Sons, Inc., New York (1985).

Gumbel E.J., *Statistics of extremes*, Columbia University Press, New York (1958).

Hador J. and W.R. Russel, Rules for ordering uncertain prospects, *American Economic Review*, **59**, 25–34 (1969).

Hansch G. and H. Leuy, The efficiency of choices involving risk, *Review of Economic Studies*, **36**, 335–346 (1969).

Harry M.J. and J.R. Lawson, *Six sigma producibility analysis and process characterisation*, Addison-Wesley, Reading (1992).

Haugen E.B., *Probabilistic mechanical design*, John Wiley & Sons, Inc., New York (1980).

Heitmann W.E., T.G. Oakwood and G. Krauss, Continuous heat treatment of automotive suspension spring steels, in *Fundamentals and applications of microalloying forging steels: Proceeding of a symposium sponsored by the Ferrous Metallurgy Committee of TMS*, Ed. C.J. Van Tyne, G. Krauss and D.K. Matlock, Minerals, Metals & Materials Society, Warrendale (1996).

Henley E.J. and H. Kumamoto, *Reliability engineering and risk assessment*, Prentice-Hall, Inc., Englewood Cliffs (1981).

Hertzberg R.W., *Deformation and fracture mechanics of engineering materials*, 4th ed., John Wiley & Sons, Inc., New York (1996).

Hillson D., Extending the risk process to manage opportunities, *International Journal of Project Management*, **20**, 235–240 (2002).

Hobbs G.K., *Accelerated reliability engineering, HALT and HASS*, John Wiley & Sons, Ltd., Chichester (2000).

Horowitz E. and S. Sahni, *Computer algorithms*, Computer Science Press, Rockville (1997).

Hoyland A. and M. Rausand, *System reliability theory*, John Wiley & Sons, Inc., New York (1994).

International Electrotechnical Commission (IEC), International vocabulary, Chapter 191: Dependability and quality of service, IEC 50 (191), (1991).

Ishikawa T., A. Hattori, H. Kawano, T. Nagao and A. Kobayashi, Development of reduction technique of thermal stress Induced in CFRP bonded steel plates, Asia-Pacific conference on FRP in structures, APFIS, Sapporo, Japan, Paper No. F2B04 (2012).

Jais C., B. Werner and D. Das, Reliability predictions - continued reliance on a misleading approach, *Proceedings of the Reliability and Maintainability Symposium (RAMS)*, IEEE, Orlando, FL, pp. 1–6 (2013).

Juran J.M. and F.M. Gryna, *Juran's quality control handbook*, 4th ed., McGraw-Hill, New York 1988.

Kececioglu D. and J.A. Jacks, The Arrhenius, Eyring, inverse power law and combination models in accelerated life testing, *Reliability Engineering*, **8**, 1–6 (1984).

Knuth D.E., *The art of computer programming, vol. 2, seminumerical algorithms*, 3rd ed., Addison-Wesley, Reading (1997).

Kuo W., V.R. Prasad, F.A. Tillman and C.L. Hwang, *Optimal reliability design*, Cambridge University Press, New York (2001).

L'Ecuyer P., Efficient and portable random number generators, *Communications of the ACM*, **31**, 742–749, 1988.

Lehmer D.H., Mathematical methods in large-scale computing units, in *Proceedings of the 2nd annual symposium on large-scale digital computing machinery*, Harvard University Press, Cambridge, MA, pp. 141–145 (1951).

Lewis E.E., *Introduction to reliability engineering*, John Wiley & Sons, Inc., New York 1996.

Mattson E., *Basic corrosion technology for scientists and engineers*, Wiley (Halsted Press), New York (1989).

McMahon C.J. and M. Cohen, Initiation of cleavage in polycrystalline iron, *Acta Metallurgica*, **13** (6), 591–604 (1965).

Meeker W.Q. and L.A. Escobar, *Statistical methods for reliability data*, John Wiley & Sons, Inc., New York (1998).

Melchers R.E., *Structural reliability analysis and prediction*, 2nd ed., John Wiley & Sons, Ltd, Chichester (1999).

Metcalfe A.V., *Statistics in engineering: A practical approach*, Chapman & Hall, London (1994).

MIL-HDBK-217F, *Reliability prediction of electronic equipment*, US Department of Defence, Washington, DC (1991).

MIL-STD-1629A, *US Department of Defence procedure for performing a failure mode and effects analysis*, US Department of Defence, Washington, DC (1977).

Miller K.J., Materials science perspective of metal fatigue resistance, *Materials Science and Technology*, **9**, 453–462 (1993).

Miller I. and M. Miller, *John E. Freund's mathematical statistics*, 6th ed., Prentice Hall International, Inc., Upper Saddle River (1999).

Miner M.A., Cumulative damage in fatigue, *Journal of Applied Mechanics*, **12**, 159–164 (1945).

Montgomery D.C., G.C. Runger and N.F. Hubele, *Engineering statistics*, 2nd ed., John Wiley & Sons, Inc., New York (2001).

Moore P., *The business of risk*, Cambridge University Press, Cambridge (1983).

Nelson W., *Accelerated testing, statistical models, test plans and data analysis*, Wiley, New York (2004).

Nielsen T.D. and J.Y. Jaffray, Dynamic decision making without expected utility: An operational approach, *European Journal of Operational Research*, **169**, 226–246 (2006).

Niku-Lari A., Shot-peening, in the first international conference on shot peening, Paris, 14–17 September, pp. 1–27, Pergamon Press, Oxford (1981).

O'Connor P.D.T, *Practical reliability engineering*, 4th ed., John Wiley & Sons, Ltd., New York (2003).

Ohring M., *Engineering materials science*, Academic Press, Inc., San Diego (1995).

OREDA, *Offshore reliability data*, DNV Technica, Hovik (1992).

Orlov P.I., *Osnovi construirovania*, vol. **1**, Mashinostroenie (in Russian), Moskva (1988).

Pahl G., W. Beitz, J. Feldhusen and K.H. Grote, *Engineering design*, Springer, Berlin (2007).

Paris P.C. and F. Erdogan, A critical analysis of crack propagation laws, *Journal of Basic Engineering*, **85**, 528–534 (1963).

Paris P.C., M.P. Gomez and W.P. Anderson, A rational analytic theory of fatigue, *The Trend in Engineering*, **13**, 9–14 (1961).

Park S.K. and K.W. Miller, Random number generators: Good ones are hard to find, *Communications of the ACM*, **31** (10), 1192–1201 (1988).

Parzen E., *Modern theory of probability and its applications*, John Wiley & Sons, Inc., New York (1960).

Peng L.C. and T.L. Peng, Thermal insulation and pipe stress, *Hydrocarbon Processing*, **77** 5, 111–113 (1998).

Peterson R.E., *Stress concentration factors*, Wiley, New York (1974).

Pflug G., Some remarks on value-at-risk and conditional-value-at-risk, in *Probabilistic constrained optimization: Methodology and applications*, Ed. R. Uryasev, Kluwer Academic Publishers, Dordrecht (2000).

Pickford J. (Ed.), *Mastering risk, vol.1: Concepts*, Pearson Education Ltd, Harlow (2001).

Pierre D.A., *Optimisation theory with applications*, Courier Dover Publications, Mineola (1986).

Pigman J.G., K.R. Agent, J.G. Mayes and C.V. Zeger, Optimal highway safety improvement investments by dynamic programming, Kentucky Transportation Center Research Report, Paper 902, p. 412 (1974).

Porter A., *Accelerated testing and validation*, Newnes, Oxford (2004).

Ramakumar R., *Engineering reliability, fundamentals and applications*, Prentice Hall, Englewood Cliffs (1993).

ReliaSoft, Accelerated life testing on-line reference, ReliaSoft's eTextbook for accelerated life testing data analysis. http://reliawiki.com/index.php/Accelerated_Life_Testing_Data_Analysis_Reference (2007).

Reniers G.L. and K. Sorensen, An approach for optimal allocation of safety resources: Using the knapsack problem to take aggregated cost-efficient preventive measures, *Risk Analysis*, **33** (11), 2056–2067 (2013).

Richardson J.W. and J.L. Outlaw, Ranking risky alternatives: innovations in subjective utility analysis, Risk Analysis VI, *Proceedings of the 6th international conference on computer simulation risk analysis and hazard mitigation*, 5–7 May 2008, Greece (Cephalonia), pp. 213–224 (2008).

Richter A., M.L. Brandeau and D.K. Owens, An analysis of optimal resource allocation for prevention of infection with human immunodeficiency virus (HIV) in injection drug users and non-users, *Medical Decision Making*, **19** (2), 167–179 (1999).

Roberts F.S., *Measurement theory with applications to decision-making, utility and the social sciences*, Addison-Wesley Publishing Company, Reading (1979).

Rockafellar R.T. and S. Uryasev, Conditional value-at-risk for general loss distributions, *Journal of Banking and Finance*, **26**, 1443–1471 (2002).

Rosenfield A.R., *George R. Irwin Symposium on cleavage fracture*, Ed. Kwai S. Chan, The Minerals, Metals and Materials Society, Materials Park, pp. 229–236 (1997).

Ross S.M., *Simulation*, 2nd ed., Harcourt Academic Press, San Diego (1997).

Ross S., Adding risks: Samuelson's fallacy of large numbers revisited. *Journal of Financial and Quantitative Analysis*, **34** (3), 323–339 (1999).

Ross S.M., *Introduction to probability models*, 7th ed., A Harcourt Science and Technology Company, San Diego (2000).

Rubinstein R.Y., *Simulation and the Monte-Carlo method*, John Wiley & Sons, Inc., New York (1981).

Samuel A. and J. Weir, *Introduction to engineering design: Modelling, synthesis and problem solving strategies*, Elsevier, London 1999.

Samuelson P.A., Risk and uncertainty: A fallacy of large numbers, *Scientia*, **98**, 108–113 (1963).

Schneider J.J. and S. Kirkpatrick, *Stochastic optimization*, Springer, Berlin (2006).

Smith D.J., *Reliability engineering*, Putman Publishing, El Segundo, CA (1972).

Sobol I.M., *A primer for the Monte-Carlo method*, CRC Press, Boca Raton (1994).

Sommerville D.M.Y., *An introduction to the geometry of N dimensions*, Dover, New York (1958).

Starmer C., Developments in non-expected utility theory: The hunt for a descriptive theory of choice under risk, *Journal of Economic Literature*, **38**, 332–382 (2000).

Sundararajan C. (Raj), *Guide to reliability engineering: Data analysis, applications, implementations and management*, Van Nostrand Reinhold, New York (1991).

Sutton I.S., *Process reliability and risk management*, Van Nostrand Reinhold, New York (1992).

Thompson W.A., *Point process models with applications to safety and reliability*, Chapman & Hall, London (1988).

Thompson G., *Improving maintainability and reliability through design*, Professional Engineering Publishing Ltd, London (1999).

Ting J.C. and F.V. Lawrence, Modelling the long-life fatigue behaviour of a cast aluminium alloy, *Fatigue and Fracture of Engineering Materials and Structures*, **16** (6), 631–647 (1993).

Tobin J., Liquidity preference as behaviour towards risk, *Review of Economic Studies*, **25**, 65–86 (1958).

Todinov M.T., Influence of some parameters on the residual stresses from quenching, *Modelling and Simulation in Materials Science and Engineering*, **7**, 25–41 (1999a).

Todinov M.T., Maximum principal tensile stress and fatigue crack origin for compression springs, *International Journal of Mechanical Sciences*, **41**, 357–370 (1999b).

Todinov M.T., Probability of fracture initiated by defects, *Materials Science & Engineering A*, **A276**, 39–47 (2000a).

Todinov M.T., Residual stresses at the surface of automotive suspension springs, *Journal of Materials Science*, **35**, 3313–3320 (2000b).

Todinov M.T., Necessary and sufficient condition for additivity in the sense of the Palmgren-Miner rule, *Computational Materials Science*, **21**, 101–110 (2001a).

Todinov M.T., An efficient method for estimating from sparse data the parameters of the impact energy variation in the ductile-to-brittle transition region, *International Journal of Fracture*, **111**, 131–150 (2001b).

Todinov M.T., Probability distribution of fatigue life controlled by defects, *Computers & Structures*, **79**, 313–318 (2001c).

Todinov M.T., Distribution mixtures from sampling of inhomogeneous microstructures: Variance and probability bounds of the properties, *Nuclear Engineering and Design*, **214**, 195–204 (2002).

Todinov M.T., Modelling consequences from failure and material properties by distribution mixtures, *Nuclear Engineering and Design*, **224**, 233–244 (2003).

Todinov M.T., Reliability governed by the relative locations of random variables in a finite interval, *IEEE Transactions on Reliability*, **53** (2), 226–237 (2004a).

Todinov M.T., Setting reliability requirements based on minimum failure-free operating periods, *Quality and Reliability Engineering International*, **20**, 273–287 (2004b).

Todinov M.T., A new reliability measure based on specified minimum distances before the locations of random variables in a finite interval, *Reliability Engineering and System Safety*, **86**, 95–103 (2004c).

Todinov M.T., Reliability analysis and setting reliability requirements based on the cost of failure, *International Journal of Reliability, Quality and Safety Engineering*, **11** (3), 273–299 (2004d).

Todinov M.T., Uncertainty and risk associated with the Charpy impact energy of multi-run welds, *Nuclear Engineering and Design*, **231**, 27–38 (2004e).

Todinov M.T., Limiting the probability of failure for components containing flaws, *Computational Materials Science*, **32**, 156–166 (2005).

Todinov M.T., Equations and a fast algorithm for determining the probability of failure initiated by flaws, *International Journal of Solids and Structures*, **43**, 5182–5195 (2006a).

Todinov M.T., Reliability analysis based on the losses from failures, *Risk Analysis*, **26** (2), 311–335 (2006b).

Todinov M.T., Reliability analysis of complex systems based on the losses from failures, *International Journal of Reliability, Quality and Safety Engineering*, **13** (2), 1–22 (2006c).

Todinov M.T., *Risk-based reliability analysis and generic principles for risk reduction*, Elsevier, Amsterdam (2007).

Todinov M.T., A comparative method for improving the reliability of components, *Nuclear Engineering and Design*, **239**, 214–220 (2009a).

Todinov M.T., Robust design using upper bound variance theorem, *International Journal of Performability Engineering*, **5** (4), 446–462 (2009b).

Todinov M.T. Is Weibull distribution the correct model for predicting probability of failure initiated by non-interacting flaws, *International Journal of Solids and Structures*, **46**, 887–901 (2009c).

Todinov M.T., The cumulative stress hazard density as an alternative of the Weibull model, *International Journal of Solids and Structures*, **47**, 3286–3296 (2010).

Todinov M.T., Virtual accelerated life testing of complex systems, in *Intelligent decision systems in large-scale distributed environment*, Ed. P. Bouvry, H. González-Vélez and J. Kolodziej, Springer, Berlin, pp. 293–314 (2011).

Todinov M.T., *Flow networks*, Elsevier, Amsterdam (2013a).

Todinov M.T., Parasitic flow loops in networks, *International Journal of Operations Research*, **10** (3), 109–122 (2013b).

Todinov M.T., New models for optimal reduction of technical risks, *Engineering Optimization*, **45** (6), 719–743 (2013c).

Todinov M.T., Dominated parasitic flow loops in networks, *International Journal of Operations Research*, **11** (1), 1–17 (2014a).

Todinov M.T., Optimal allocation of limited resources among discrete risk-reduction options, *Artificial Intelligence Research*, **3** (4), 15–27 (2014b).

Todinov M.T and S. Same, A fracture condition incorporating the most unfavourable orientation of the crack, *International Journal of Mechanics and Materials in Design*, DOI: 10.1007/s10999-014-9258-x (2014).

Todinov M.T., Reducing risk through segmentation, permutations, time and space exposure, inverse states and separation, *International Journal of Risk and Contingency Management*, **4** (3), pp.1–21, (2015).

Todinov M.T. and E. Weli, Optimal risk reduction in the railway industry, by using dynamic programming, *WASET, International Journal of Mechanical, Aerospace, Industrial and Mechatronics Engineering*, **79** (7), 415–419 (2013).

Todinov M.T., M. Novovic, P. Bowen and J.F. Knott, Modelling the impact energy in the ductile/brittle transition region of C-Mn multi-run welds, *Materials Science & Engineering A*, **A287**, 116–124 (2000).

Trivedi K.S., *Probability and statistics with reliability, queuing and computer science applications*, 2nd ed., John Wiley & Sons, Ltd, Chichester (2002).

Trustrum K. and A. De S. Jayatilaka, Applicability of Weibull analysis for brittle materials, *Journal of Materials Science*, **18**, 2765–2770 (1983).

Tuckwell H.C., *Elementary applications of probability theory*, Chapman & Hall, London (1988).

Uicker J.J. Jr, G.R. Pennock and J.E. Shigley, *Theory of machines and mechanisms*, 3rd ed., Oxford University Press, New York (2003).

Vinogradov O., *Introduction to mechanical reliability: A designer's approach*, Hemisphere Publishing Corporation, New York (1991).

Von Neumann J. and O. Morgenstern, *Theory of games and economic behaviour*, Princeton University Press, Princeton (1944).

Vose D., *Risk analysis, a quantitative guide*, 2nd ed., John Wiley & Sons Ltd., New York (2000).

Wald A., *Statistical decision functions*, John Wiley & Sons, Inc., New York (1950).

Weibull W., A statistical distribution of wide applicability, *Journal of Applied Mechanics*, **18**, 293–297 (1951).

Weli E. and M. Todinov, A new approach to risk reduction in the railway industry, in *Infrastructure risk and resilience: Transportation, 2013*, The Institution of Engineering and Technology, London, pp. 47–52 (2013).

Wirth N., *Algorithms+data structures = programs*, Prentice-Hall, Englewood Cliffs (1976).

Index

Reliability and Risk Models: Setting Reliability Requirements, Second Edition. Michael Todinov.